IET ENERGY ENGINEERING 169

Small Wind and Hydrokinetic Turbines

Other volumes in this series:

Volume 1	**Power Circuit Breaker Theory and Design** C.H. Flurscheim (Editor)
Volume 4	**Industrial Microwave Heating** A.C. Metaxas and R.J. Meredith
Volume 7	**Insulators for High Voltages** J.S.T. Looms
Volume 8	**Variable Frequency AC Motor Drive Systems** D. Finney
Volume 10	**SF_6 Switchgear** H.M. Ryan and G.R. Jones
Volume 11	**Conduction and Induction Heating** E.J. Davies
Volume 13	**Statistical Techniques for High Voltage Engineering** W. Hauschild and W. Mosch
Volume 14	**Uninterruptible Power Supplies** J. Platts and J.D. St Aubyn (Editors)
Volume 15	**Digital Protection for Power Systems** A.T. Johns and S.K. Salman
Volume 16	**Electricity Economics and Planning** T.W. Berrie
Volume 18	**Vacuum Switchgear** A. Greenwood
Volume 19	**Electrical Safety: A guide to causes and prevention of hazards** J. Maxwell Adams
Volume 21	**Electricity Distribution Network Design, 2nd Edition** E. Lakervi and E. J. Holmes
Volume 22	**Artificial Intelligence Techniques in Power Systems** K. Warwick, A.O. Ekwue and R. Aggarwal (Editors)
Volume 24	**Power System Commissioning and Maintenance Practice** K. Harker
Volume 25	**Engineers' Handbook of Industrial Microwave Heating** R.J. Meredith
Volume 26	**Small Electric Motors** H. Moczala *et al.*
Volume 27	**AC–DC Power System Analysis** J. Arrillaga and B.C. Smith
Volume 29	**High Voltage Direct Current Transmission, 2nd Edition** J. Arrillaga
Volume 30	**Flexible AC Transmission Systems (FACTS)** Y.-H. Song (Editor)
Volume 31	**Embedded Generation** N. Jenkins *et al.*
Volume 32	**High Voltage Engineering and Testing, 2nd Edition** H.M. Ryan (Editor)
Volume 33	**Overvoltage Protection of Low-Voltage Systems, Revised Edition** P. Hasse
Volume 36	**Voltage Quality in Electrical Power Systems** J. Schlabbach *et al.*
Volume 37	**Electrical Steels for Rotating Machines** P. Beckley
Volume 38	**The Electric Car: Development and future of battery, hybrid and fuel-cell cars** M. Westbrook
Volume 39	**Power Systems Electromagnetic Transients Simulation** J. Arrillaga and N. Watson
Volume 40	**Advances in High Voltage Engineering** M. Haddad and D. Warne
Volume 41	**Electrical Operation of Electrostatic Precipitators** K. Parker
Volume 43	**Thermal Power Plant Simulation and Control** D. Flynn
Volume 44	**Economic Evaluation of Projects in the Electricity Supply Industry** H. Khatib
Volume 45	**Propulsion Systems for Hybrid Vehicles** J. Miller
Volume 46	**Distribution Switchgear** S. Stewart
Volume 47	**Protection of Electricity Distribution Networks, 2nd Edition** J. Gers and E. Holmes
Volume 48	**Wood Pole Overhead Lines** B. Wareing
Volume 49	**Electric Fuses, 3rd Edition** A. Wright and G. Newbery
Volume 50	**Wind Power Integration: Connection and system operational aspects** B. Fox *et al.*
Volume 51	**Short Circuit Currents** J. Schlabbach
Volume 52	**Nuclear Power** J. Wood
Volume 53	**Condition Assessment of High Voltage Insulation in Power System Equipment** R.E. James and Q. Su
Volume 55	**Local Energy: Distributed generation of heat and power** J. Wood
Volume 56	**Condition Monitoring of Rotating Electrical Machines** P. Tavner, L. Ran, J. Penman and H. Sedding
Volume 57	**The Control Techniques Drives and Controls Handbook, 2nd Edition** B. Drury
Volume 58	**Lightning Protection** V. Cooray (Editor)
Volume 59	**Ultracapacitor Applications** J.M. Miller
Volume 62	**Lightning Electromagnetics** V. Cooray

Volume 63 **Energy Storage for Power Systems, 2nd Edition** A. Ter-Gazarian
Volume 65 **Protection of Electricity Distribution Networks, 3rd Edition** J. Gers
Volume 66 **High Voltage Engineering Testing, 3rd Edition** H. Ryan (Editor)
Volume 67 **Multicore Simulation of Power System Transients** F.M. Uriate
Volume 68 **Distribution System Analysis and Automation** J. Gers
Volume 69 **The Lightening Flash, 2nd Edition** V. Cooray (Editor)
Volume 70 **Economic Evaluation of Projects in the Electricity Supply Industry, 3rd Edition** H. Khatib
Volume 72 **Control Circuits in Power Electronics: Practical issues in design and implementation** M. Castilla (Editor)
Volume 73 **Wide Area Monitoring, Protection and Control Systems: The enabler for Smarter Grids** A. Vaccaro and A. Zobaa (Editors)
Volume 74 **Power Electronic Converters and Systems: Frontiers and applications** A.M. Trzynadlowski (Editor)
Volume 75 **Power Distribution Automation** B. Das (Editor)
Volume 76 **Power System Stability: Modelling, analysis and control** A.A. Sallam and Om P. Malik
Volume 78 **Numerical Analysis of Power System Transients and Dynamics** A. Ametani (Editor)
Volume 79 **Vehicle-to-Grid: Linking electric vehicles to the smart grid** J. Lu and J. Hossain (Editors)
Volume 81 **Cyber-Physical-Social Systems and Constructs in Electric Power Engineering** S. Suryanarayanan, R. Roche and T.M. Hansen (Editors)
Volume 82 **Periodic Control of Power Electronic Converters** F. Blaabjerg, K. Zhou, D. Wang and Y. Yang
Volume 86 **Advances in Power System Modelling, Control and Stability Analysis** F. Milano (Editor)
Volume 87 **Cogeneration: Technologies, optimisation and implementation** C.A. Frangopoulos (Editor)
Volume 88 **Smarter Energy: From smart metering to the smart grid** H. Sun, N. Hatziargyriou, H.V. Poor, L. Carpanini and M.A. Sánchez Fornié (Editors)
Volume 89 **Hydrogen Production, Separation and Purification for Energy** A. Basile, F. Dalena, J. Tong and T.N. Veziroğlu (Editors)
Volume 90 **Clean Energy Microgrids** S. Obara and J. Morel (Editors)
Volume 91 **Fuzzy Logic Control in Energy Systems with Design Applications in MATLAB®/Simulink®** İ.H. Altaş
Volume 92 **Power Quality in Future Electrical Power Systems** A.F. Zobaa and S.H.E.A. Aleem (Editors)
Volume 93 **Cogeneration and District Energy Systems: Modelling, analysis and optimization** M.A. Rosen and S. Koohi-Fayegh
Volume 94 **Introduction to the Smart Grid: Concepts, technologies and evolution** S. K. Salman
Volume 95 **Communication, Control and Security Challenges for the Smart Grid** S.M. Muyeen and S. Rahman (Editors)
Volume 96 **Industrial Power Systems with Distributed and Embedded Generation** R. Belu
Volume 97 **Synchronized Phasor Measurements for Smart Grids** M.J.B. Reddy and D.K. Mohanta (Editors)
Volume 98 **Large Scale Grid Integration of Renewable Energy Sources** A. Moreno-Munoz (Editor)
Volume 100 **Modeling and Dynamic Behaviour of Hydropower Plants** N. Kishor and J. Fraile-Ardanuy (Editors)
Volume 101 **Methane and Hydrogen for Energy Storage** R. Carriveau and D.S-K. Ting
Volume 104 **Power Transformer Condition Monitoring and Diagnosis** A. Abu-Siada (Editor)
Volume 106 **Surface Passivation of Industrial Crystalline Silicon Solar Cells** J. John (Editor)
Volume 107 **Bifacial Photovoltaics: Technology, applications and economics** J. Libal and R. Kopecek (Editors)

Volume 108 **Fault Diagnosis of Induction Motors** J. Faiz, V. Ghorbanian and G. Joksimović
Volume 110 **High Voltage Power Network Construction** K. Harker
Volume 111 **Energy Storage at Different Voltage Levels: Technology, integration, and market aspects** A.F. Zobaa, P.F. Ribeiro, S.H.A. Aleem and S.N. Afifi (Editors)
Volume 112 **Wireless Power Transfer: Theory, technology and application** N. Shinohara
Volume 114 **Lightning-Induced Effects in Electrical and Telecommunication Systems** Y. Baba and V.A. Rakov
Volume 115 **DC Distribution Systems and Microgrids** T. Dragičević, F. Blaabjerg and P. Wheeler
Volume 116 **Modelling and Simulation of HVDC Transmission** M. Han (Editor)
Volume 117 **Structural Control and Fault Detection of Wind Turbine Systems** H. R. Karimi
Volume 119 **Thermal Power Plant Control and Instrumentation: The control of boilers and HRSGs, 2nd Edition** D. Lindsley, J. Grist and D. Parker
Volume 120 **Fault Diagnosis for Robust Inverter Power Drives** A. Ginart (Editor)
Volume 121 **Monitoring and Control using Synchrophasors in Power Systems with Renewables** I. Kamwa and C. Lu (Editors)
Volume 123 **Power Systems Electromagnetic Transients Simulation, 2nd Edition** N. Watson and J. Arrillaga
Volume 124 **Power Market Transformation** B. Murray
Volume 125 **Wind Energy Modeling and Simulation, Volume 1: Atmosphere and plant** P. Veers (Editor)
Volume 126 **Diagnosis and Fault Tolerance of Electrical Machines, Power Electronics and Drives** A.J.M. Cardoso
Volume 128 **Characterization of Wide Bandgap Power Semiconductor Devices** F. Wang, Z. Zhang and E.A. Jones
Volume 129 **Renewable Energy from the Oceans: From wave, tidal and gradient systems to offshore wind and solar** D. Coiro and T. Sant (Editors)
Volume 130 **Wind and Solar Based Energy Systems for Communities** R. Carriveau and D. S-K. Ting (Editors)
Volume 131 **Metaheuristic Optimization in Power Engineering** J. Radosavljević
Volume 132 **Power Line Communication Systems for Smart Grids** I.R.S. Casella and A. Anpalagan
Volume 139 **Variability, Scalability and Stability of Microgrids** S.M. Muyeen, S.M. Islam and F. Blaabjerg (Editors)
Volume 145 **Condition Monitoring of Rotating Electrical Machines** P. Tavner, L. Ran, C. Crabtree
Volume 146 **Energy Storage for Power Systems, 3rd Edition** A.G. Ter-Gazarian
Volume 147 **Distribution Systems Analysis and Automation, 2nd Edition** J. Gers
Volume 152 **Power Electronic Devices: Applications, failure mechanisms and reliability** F. Iannuzzo (Editor)
Volume 153 **Signal Processing for Fault Detection and Diagnosis in Electric Machines and Systems** Mohamed Benbouzid (Editor)
Volume 155 **Energy Generation and Efficiency Technologies for Green Residential Buildings** D. Ting and R. Carriveau (Editors)
Volume 157 **Electrical Steels, 2 Volumes** A. Moses, K. Jenkins, Philip Anderson and H. Stanbury
Volume 158 **Advanced Dielectric Materials for Electrostatic Capacitors** Q. Li (Editor)
Volume 159 **Transforming the Grid Towards Fully Renewable Energy** O. Probst, S. Castellanos and R. Palacios (Editors)
Volume 160 **Microgrids for Rural Areas: Research and case studies** R.K. Chauhan, K. Chauhan and S.N. Singh (Editors)
Volume 166 **Advanced Characterization of Thin Film Solar Cells** N. Haegel and M. Al-Jassim (Editors)
Volume 167 **Power Grids with Renewable Energy: Storage, integration and digitalization** A.S. Sallam and Om P. Malik
Volume 172 **Lighting Interaction with Power Systems, 2 Volumes** A. Piantini (Editor)
Volume 905 **Power System Protection, 4 Volumes**

Small Wind and Hydrokinetic Turbines

Edited by
Philip Clausen, Jonathan Whale and David Wood

The Institution of Engineering and Technology

Published by The Institution of Engineering and Technology, London, United Kingdom

The Institution of Engineering and Technology is registered as a Charity in England & Wales (no. 211014) and Scotland (no. SC038698).

First published 2021

The Institution of Engineering and Technology
Michael Faraday House
Six Hills Way, Stevenage
Herts, SG1 2AY, United Kingdom

www.theiet.org

British Library Cataloguing in Publication Data
A catalogue record for this product is available from the British Library

ISBN 978-1-83953-071-5 (Hardback)
ISBN 978-1-83953-072-2 (PDF)

Typeset in India by MPS Limited

Contents

About the editors

Philip Clausen is currently the acting head of the School of Engineering at The University of Newcastle, Australia. His research expertise includes small wind turbine dynamics and fatigue testing, and issues related to small wind turbine blades.

Jonathan Whale is a Senior Lecturer in Engineering and Energy at Murdoch University, Australia. Previously, he worked for the Australian Cooperative Research Centre for Renewable Energy on commercialisation of wind turbines. He has been a member of International Energy Agency Task groups on small and distributed wind turbines and has contributed to international design standards for wind turbines. His research focuses on the impact of inflow turbulence on the performance of wind turbines.

David Wood is the Schulich chair in renewable energy at the University of Calgary, Canada. His research activities include renewable energy, primarily wind turbine aerodynamics, and renewable energy integration. He is the author of a book on small wind turbines and has co-authored numerous journal papers and book chapters on various aspects of their analysis, design, and manufacture. He is a Director of the Wind Energy Institute of Canada.

Introduction

Wind turbines extract kinetic energy from the wind. Hydroturbines can either extract energy from water flow or use the potential energy due to a difference in elevation or “head”. This book considers kinetic-energy-extracting wind and hydro turbines although two later chapters will show that the boundary with head-driven hydroturbines is somewhat artificial. Putting aside questions of the time period over which the velocity, V, is averaged, the power available in any fluid flow is proportional to $\rho V^3 A$, where ρ is the density, and A is usually taken as the swept area of the blades, which is one measure of the size of the turbine. The density of water is around 800 times that of air, whereas typical water velocities may be only a fifth or a tenth of the wind speed at the windy site. It follows that a wind turbine and hydrokinetic turbine of similar size will, on average, produce similar orders of magnitude of power. The use of “Small” in the title of this book indicates that we consider sub-megawatt turbines rather than wind turbines designed for large onshore and offshore wind farms. Most turbines described in the following chapters have power outputs measured in kilowatts. The restriction on size allows us to describe and exploit the commonality of turbines for the two fluids.

The cubic dependence of available power on V implies the importance of siting a turbine to maximise V. This is the topic of Chapter 1, for air, and 2, for water. The problem for small wind turbines can be the cost of measuring the wind speed and the complexity of the often highly turbulent flow, which is difficult to compute accurately by numerical simulation. It also means that publicly available wind data, usually in the form of wind atlases, are of limited value as they give wind speeds at heights much greater than those applicable for small wind turbines and extrapolation or interpolation is error-prone. The techniques available are discussed in detail in Chapter 1. For hydrokinetic turbines, the most common public data is river volume flow rate or “discharge” with units of length^3/sec. Converting discharge to V is difficult to do accurately, as explained in Chapter 2. Resource assessment for both fluids is unsatisfactory and is complicated by the strongly nonlinear dependence of power on velocity. This has another consequence: the period over which the velocity is measured at the installation site must be comparable to that used in determining the relationship between power and speed by the manufacturer or independent test agency. One benefit of commonality could well be to encourage the extension of standardised averaging periods and test procedures for wind turbines to hydrokinetic turbines with appropriate modification.

To set the context for the remainder of the book, Chapter 3 surveys the available hydrokinetic turbines. Examples of small wind turbines are described in later chapters.

Both water and air at typical atmospheric conditions are Newtonian fluids and so obey the Navier–Stokes equations and the laws of aerodynamics and hydrodynamics. The principles of operation of wind and hydrokinetic turbines are, therefore, largely the same, and this similarity is exploited in the next chapters that discuss the modelling of horizontal- and vertical-axis turbines. Chapter 4 describes the computational modelling of vertical-axis wind turbines, and Chapter 5 their testing in wind tunnels. Chapter 6 describes the most common and oldest form of horizontal-axis wind turbines, the water pumping windmill whose "modern" form developed towards the end of the 19th Century. Surprisingly, windmill aerodynamics has not previously studied in detail. The major difference between air and water is that cavitation can occur near the blades in the latter if the local pressure falls below the vapour pressure. This is most common on hydrokinetic turbines located near the water surface as any increase in depth increases the static pressure. Cavitation is to be avoided as it causes blade erosion and unsteady loads; a considerable portion of Chapter 7 on horizontal-axis turbine modelling explains how to do this. That chapter introduces blade element-momentum theory (BEMT), which is often called the "workhorse" of turbine analysis. BEMT is considered briefly in Chapter 4 for vertical-axis turbines in the context of more sophisticated computational fluid dynamics analysis.

Chapter 8 describes the relatively new field of energy harvesting whereby a (usually) small amount of power can be produced for a critical application when other power sources cannot be used.

To paraphrase Albert Einstein: theory without practice is lame and practice without theory is blind. The next section of the book describes field testing and the experience of turbine manufacturers and installers. We believe that these experiences have not been well documented in the literature and invite the reader to ponder the relationship between theory and practice that is provided by these chapters. They cover basic field testing to determine wind turbine performance in Chapter 9, and sophisticated assessment of the turbine loads under the heading of "aeroelastic" analysis in Chapter 10. This analysis is commonplace for large turbines but in its infancy for small wind and hydrokinetic turbines. Chapters 11 and 12 were written by commercial developers of small hydrokinetic turbines. The following two chapters give a similar perspective for wind turbines, describing Arctic applications in Chapter 13 and a commercially available diffuser-augmented wind turbine in Chapter 14. Chapter 15 describes the complexities and challenges of testing tidal hydrokinetic turbines in the enormous tides of the Bay of Fundy, Canada. This chapter also provides the photograph on the book cover.

One of the major differences between large and small turbines is that the basic technology of the former – three upwind blades rotating on a horizontal axis – is fixed whereas there is a rich diversity of small turbine designs. This book provides examples of energy-harvesting devices, vertical-axis and diffuser-augmented turbines to display that diversity and to give the reader an appreciation of the challenges of moving from theory to practice. One particularly important challenge for turbines of all sizes is the need to move away from unsustainable materials for turbine blades, primarily petroleum resin and fibreglass. The development of natural materials as described in Chapter 16 is a step in this direction, made easier by the simple fact that small experimental blades are easier to manufacture than large ones.

Chapter 1
Wind resource assessment for small wind turbines

Takaaki Kono[1] and Jonathan Whale[2]

Wind resource assessment (WRA) is the process of estimating a wind turbine's future energy production and is a key factor in the successful installation, operation, and performance of a small wind turbine (SWT). The definition of SWTs varies from country to country. For example, the US Department of Energy and Japan's Nippon Kaiji Kyokai (ClassNK) define SWTs as having a power rating less than or equal to 100 kW [1] and 20 kW [2], respectively. The International Electrotechnical Commission (IEC) SWT design standard IEC61400-2 considers parameters other than power rating and defines SWTs as having a rotor swept area smaller than or equal to 200 m^2, generating electricity at a voltage below 1,000-V AC or 1,500-V DC for both on-grid and off-grid applications [3]. The level of detail and analysis in WRA typically increases with turbine size and cost. On the upper end of the SWT range, a simplified WRA might be followed by a more detailed analysis and feasibility study that includes on-site data collection to reduce risk.

This chapter starts with an overview of the general wind project development life cycle (a cycle not limited to SWTs). This is followed by a detailed description of the steps in the general WRA process. Finally, the specific features of WRA for SWTs are described.

1.1 General life cycle of wind energy project

Figure 1.1 shows the seven phases in the life cycle of a general wind energy project [4].

The first phase is "prefeasibility analysis and prospecting." The purpose of prefeasibility analysis is to identify areas that have the potential for feasible wind projects. It utilizes publicly available wind resource maps along with considerations such as tariffs, land availability, licensing process, transmission, logistics, cost of project, and others. Prospecting involves site visits to the regions that satisfy the prefeasibility requirements in order to collect additional data regarding terrain,

[1]Faculty of Mechanical Engineering, Kanazawa University, Japan
[2]Discipline of Engineering and Energy, College of Science, Health, Engineering and Education, Murdoch University, Australia

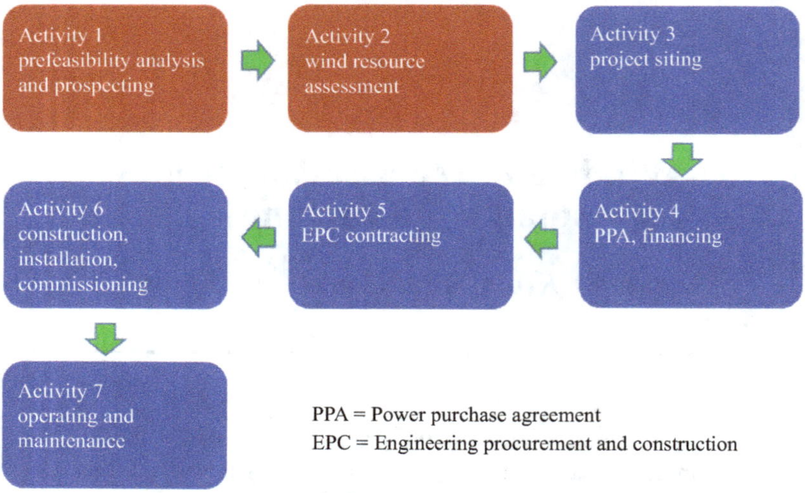

PPA = Power purchase agreement
EPC = Engineering procurement and construction

Figure 1.1 Activities in the life cycle of a wind energy project

vegetation, landownership, and other factors. The outcome of prospecting is a shortlist of areas with the greatest potential and locations for wind measurement in these areas.

The second phase is "WRA." WRA is the quantification of wind resources to compute parameters such as average wind speed, average wind energy density, and the average annual energy production (AEP) of a proposed wind power plant. A preliminary WRA is started in the previous phase. Most of this phase's activities include wind measurement and AEP estimation based on wind flow modeling. More details of WRA are explained in the next section.

The third phase is "project siting." Project siting includes all the activities that precede construction, such as environmental impact assessment, obstruction to aviation airspace analysis, interference with radar and telecommunications analysis, site planning and engineering, visual impact assessment, logistics planning, grid interconnection, and permitting and licensing. These activities may be conducted in parallel with the WRA phase.

The fourth phase is "power purchase agreement (PPA)." This phase mainly focuses on the financial aspects of a wind project. The first step is PPA negotiation with a purchasing utility, followed by agreements with equity and debt financiers. This phase takes place after the WRA and siting phases.

The fifth phase is "engineering procurement and construction (EPC) contracting." This phase begins with the selection of an EPC contractor and the selection of turbines for the project. In this phase, detailed project plans are made along with detailed engineering of the foundation, roads, collection system, substation, grid interconnection, logistics, etc.

The sixth phase is "construction, installation, and commissioning." In this phase, the wind power plant site is prepared, roads are constructed, foundations of turbines are built, towers are erected, the nacelle and rotor are raised, and the wind turbines are commissioned.

The seventh phase is "operation and maintenance." After the wind turbines are commissioned, ongoing operation and maintenance activities are conducted. A supervisory control and data acquisition system is often used to monitor the operation of the wind farm and to alert users of maintenance issues of turbines.

1.2 General wind resource assessment process

The five steps of the WRA process [5] are illustrated in Figure 1.2. WRA starts with a preliminary assessment or prospecting. In this step, based on publicly available wind resource maps and wind data, alternate sites are evaluated for adequate wind speed. If the site is acceptable, then an on-site wind measurement campaign is carried out. After wind data is collected for a sufficient period, a detailed WRA process begins. If the wind measurement location and the turbine location are not the same, spatial extrapolation is carried out to estimate the wind conditions at the turbine location using the measured wind data. The next step is to extrapolate the measured wind data along the temporal dimension and create a long-term wind dataset that covers the time-period covered by a long-term reference wind dataset. The last step is to compute the AEP. The details of each step of a general WRA are laid out in the following subsections.

At the time of writing, progress is being made on a specific international WRA standard. The scope of the IEC61400-15 standard is to define a framework for assessment and reporting of the wind resource, energy yield, and site suitability input conditions for onshore and offshore wind power plants. The standard is likely to be split into two parts with IEC61400-15-1 covering Step 1 in Figure 1.2 and IEC61400-15-2 covering Steps 2–5 in Figure 1.2, including uncertainty calculations [6].

1.2.1 Preliminary assessment (Step 1 of WRA)

Most projects start without on-site measurement of wind data. During the prospecting phase of a wind project, a preliminary WRA is conducted. In the prospecting phase, a developer starts with a wind resource map such as the one shown in Figure 1.3. The Japanese wind resource map [7] in Figure 1.3 provides annual average wind speed, shape, and scale parameters for the Weibull probability density function, which is used to model the wind speed probability distribution, and wind roses at 30, 50, and 70 m above the ground. This map was developed by using the Local Area Wind Energy

Figure 1.2 Five steps of WRA process

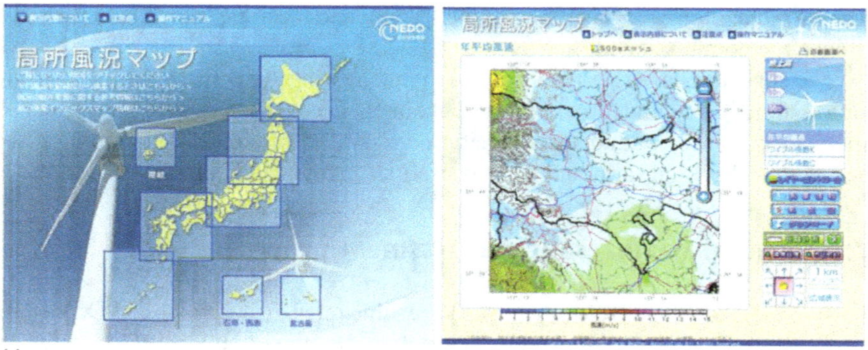

(a) (b)

Figure 1.3 Japanese wind resource map: (a) home page and (b) 30 m above ground based on 500-m mesh

Prediction System [8], in which a three-stage nesting method was utilized: numerical models in the field of mesoscale meteorology were used for the first domain (5 km×5 km horizontal mesh resolution) to the third domain (500 m×500 m horizontal mesh resolution).

A global wind resource map is provided for free by the World Bank [9]. In addition, onshore wind resource maps have been provided free of charge by several countries, including Australia [10], Austria [11], Canada [12], Ireland [13], Japan [7], Republic of Korea [14], United States of America [15], etc. These free maps can provide a general indication of good or poor wind resources, however, may have significantly low site-specific accuracy, since they do not provide high enough resolution nor include information on complex terrain, ground cover, and other local aspects. Therefore, these maps should be used as a tool to identify regions with good wind resources, and not for the serious viability of a wind project. Generally, paid maps or services can often provide higher resolution and more flexibility with zooming, orientation, and additional features.

After identifying areas that are rich in wind resources using the wind resource maps, basic due diligence is performed to narrow down the choices to the most promising areas that are worthy of further consideration. This due diligence may include performing a basic desktop analysis, which uses the following data:

- *Proximity to power transmission lines and road network*. This can be done by superimposing a map of transmission and road networks on the wind resource map. Areas that are close to infrastructure are preferred.
- *Terrain features and environmental and/or cultural value of regions under consideration*. A wind-rich area may be eliminated if it is located in high mountain ranges, a national forest, or environmentally or culturally sensitive zones.
- *Availability of land (size and type of ownership)*. If there is an area suggested from the basic desktop analysis, a site visit is carried out to perform site

assessments. Three main goals of the visits are (i) to confirm the assumptions and data used in the desktop analysis, (ii) to obtain additional information not available in a map or geographic information system format, and (iii) to select places to install wind monitoring systems. The following aspects are examined:

- *Flagging of local vegetation for the strength and direction of prevailing wind.* The Griggs-Putnam index for conifers defines eight classes of tree deformation regarding the ranges of average prevailing wind speed [16,17]. The Barsch index applies to hardwoods [18].
- *Conversations with local people regarding wind resource.* There is a possibility to know approximately the strong wind spots in the local area by collating information obtained from local people [19].
- *Terrain features affecting wind characteristics.* Elevated areas not only experience higher wind speeds due to their increased height in the wind profile but also, given the size and shape of the landform, may cause local acceleration of the wind speed as shown in Figure 1.4 [20]. Zones of strong turbulence formed in the leeward of a ridge as shown in Figure 1.4 [20] and near a cliff should be avoided when siting anemometers and wind turbines.
- *Obstacles affecting wind characteristics.* In the wake of obstacles, such as buildings and trees, wind speed tends to be low and turbulence intensity tends to be high. To avoid the negative effects of the obstacles, it is recommended to install a turbine tower at least two times higher than the highest obstacle or 20 times the highest obstacle horizontally in the dominant wind direction [17,21]. It is important to estimate the future height of trees.
- *Soil type and condition.* Although the site assessment does not normally include a soil test, it is important to check at least some information about obvious soil issues that could affect the design and construction of the foundation, such as intermittent water, sand, unstable slopes, rocks, expansive clay, and frost depth [22].
- *Availability and quality of roads.* From the viewpoint of delivering wind turbines, towers, and construction equipment, availability and quality of roads must be evaluated [22].

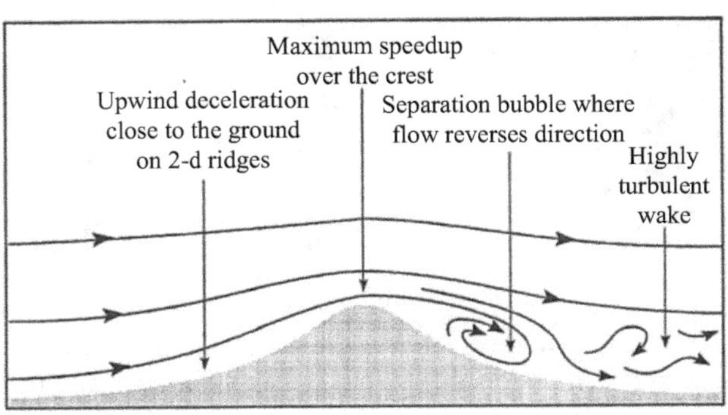

Figure 1.4 Wind characteristics over a hill [20]

1.2.2 On-site wind measurement (Step 2 of WRA)

For the general wind project, in order to avoid seasonal biases, on-site wind measurement for a year or more is required in the project area. There are two types of on-site wind measurements: in situ and remote sensing. In situ measurements are conducted with meteorological towers. There are two options for remote sensing: sonic detection and ranging, which is based on sound waves and light detection and ranging, which is based on light waves. Both types of measurement provide wind speed, wind direction, temperature, barometric pressure, and a few other atmospheric conditions. Generally, remote sensing is not used for SWT projects. Therefore, this chapter describes only wind measurement based on meteorological towers.

Three types of anemometers (e.g., Figure 1.5) are used for the measurement of horizontal wind speed [23]. Of these, the cup anemometer is the most commonly used because of its low cost and relatively good accuracy. In particular settings, however, both propeller and sonic anemometers are used.

- *Cup anemometer*. This instrument is made up of three or four cups connected to a vertical shaft. The wind causes the cup assembly to turn because the drag coefficient of the cup's concave side is greater than that of its convex side. The cup rotation is almost linearly proportional to the wind speed over a specified range. A transducer in the anemometer converts this rotational movement into an electrical signal, which is sent via a wire to a data logger. The data logger measures the frequency or magnitude of the signal and applies a predetermined multiplier and offset to convert the signal to a wind speed. The advantages of cup anemometers are that they are reasonably low cost, measure wind from any direction and are typically accurate to within ±2%, after calibration in wind tunnels. One disadvantage with this type of anemometer is that it may

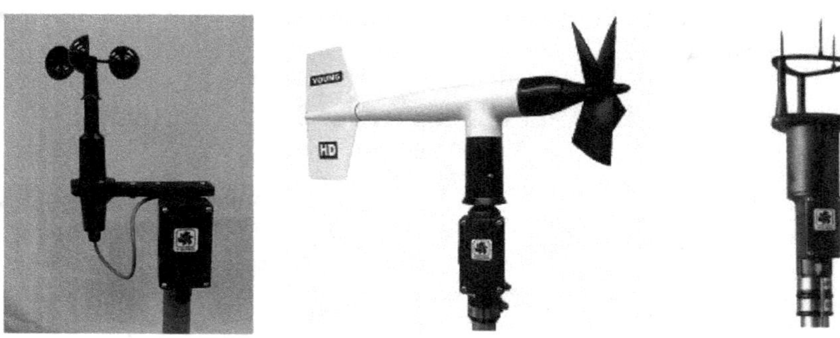

(a) (b) (c)

Figure 1.5 Commonly used anemometers for horizontal wind speed measurement: (a) cup anemometer (CYG-3102), (b) propeller anemometer with wind vane (CYG-5108), (c) sonic anemometer (CYG-86000). Courtesy of Climatec Inc.

record higher speeds during sudden drops in wind speed due to the inertia of the anemometer. The inertia of the cup anemometer influences the response time of the instrument and is quantified by the "distance constant," or the distance the air travels past the anemometer during the time taken for the instrument to reach 63% of equilibrium speed after a step change in wind speed. Another disadvantage is that the values of horizontal wind speed produced by the instrument are also slightly influenced by the vertical components of wind flow.

- *Propeller or prop–vane anemometer*. This comprises a propeller mounted on a horizontal shaft that is oriented into the wind through the use of a tail vane. The propeller anemometer also produces an electrical signal proportional to wind speed. This type of anemometer can record slightly lower speeds than cup anemometers under turbulent conditions. This so-called underspeeding is caused by its tendency to oscillate around the central direction or to lag behind sudden wind direction shifts, resulting in the propeller not always pointing directly into the wind. Since propeller anemometers are usually built from lighter materials than cup anemometers they tend to have, lower distance constants are more responsive and have lower starting thresholds. The downside of this is that the propeller anemometer is not as robust as the cup anemometer and is more likely to suffer from structural fatigue.
- *Sonic anemometer*. This instrument, which has no moving parts, measures the wind speed of ultrasound transmitted between fixed points using pairs of transducers. Some sonic anemometers measure wind in two dimensions, whereas others measure in three. Since they have no rotational inertia, sonic anemometers are more responsive to rapid changes of wind speed and wind direction than cup or propeller anemometers. Sonics are particularly useful in measuring 3D flows, where they can sample data at high rates in order to capture turbulence behavior. However, they are more expensive than other types of anemometers and consume more power.

A wind vane (e.g., Figure 1.6) is usually used to measure wind direction. In the most familiar type, a horizontal tail connected to a vertical shaft rotates to align with the wind. It is recommended that wind vanes be installed on at least two monitoring levels, to determine the wind direction with adequate redundancy. Ideally, they should not be mounted on the same booms or even at the same heights as the anemometers, since the sensor wake or "shadow" could interfere with obtaining accurate speed readings. Mounting the wind vanes 1 or 2 m below the anemometers is customary. In addition, a resolution greater than or equal to 1° is recommended.

Air temperature is normally measured 2–3 m above the ground level or near the hub height, or both levels. In most locations, the difference between the average air temperature near the ground level and that at the hub height is generally within 1°C. Air temperature is used to estimate air density, which influences the calculation of the power production of a wind turbine. An ambient air temperature sensor is usually composed of three parts: the transducer, a temperature probe, and a radiation shield

Figure 1.6 Wind vane (CYG-3302). Courtesy of Climatec Inc.

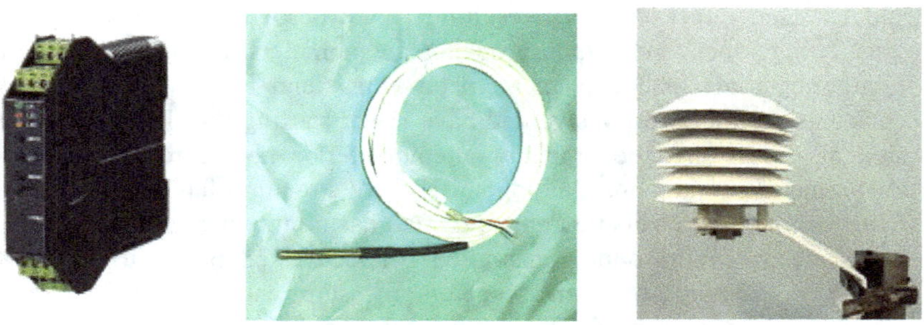

(a) (b) (c)

Figure 1.7 Air temperature sensor: (a) transducer (C-PT-CVM5), (b) temperature probe (C-HPT), (c) radiation shield (CYG-41303). Courtesy of Climatec Inc.

(e.g., Figure 1.7). The NREL Wind Resource Assessment Handbook recommends mounting the temperature sensor at least one tower diameter out from the tower to minimize tower heating effects and to mount the sensor on the side of the mast facing the prevailing wind direction to ensure good ventilation of the sensor [24].

Knowing the barometric pressure along with the air temperature can improve the accuracy of air density estimation, since normal variations in pressure at the same temperature can affect air density by about 1%. Most barometric pressure sensors (e.g., Figure 1.8) utilize a piezoelectric transducer that sends a DC voltage to a data logger and may require an external power source. It is necessary to expose the transducer to the ambient outside air pressure and must not be installed in an airtight enclosure. With a pressure port, dynamic pressure errors due to wind can be minimized.

With regard to towers for wind resource monitoring, there are two basic structural types: tabular and lattice. For both types, there are three versions: tilt-up, telescoping, and fixed. Moreover, towers can be either self-supporting or stabilized by guy wires anchored in the ground. For most sites, tilt-up tubular towers are preferred for WRA

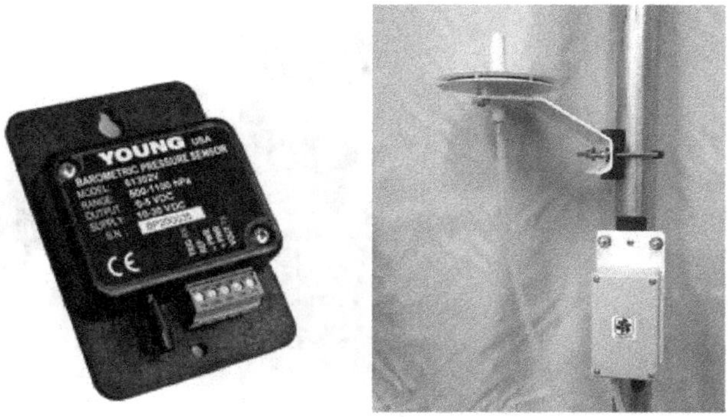

(a) (b)

Figure 1.8 Barometric pressure sensor in a case and pressure port: (a) barometric pressure sensor (CYG-61302), (b) pressure port (CYG61002) and case (CYG-61360). Courtesy of Climatec Inc.

since they are relatively easy to raise and lower, require minimal ground preparation, and are relatively inexpensive.

Anemometers and wind vanes are most often installed on booms that extend horizontally from the side of the tower. Near the boom's end, a short vertical mast or pillar is attached to hold the sensor above the boom. Except when the sensors are directly downwind of the tower, the boom and pillar should be long enough so that the influence of both the tower shadow and the boom shadow on the speed and direction readings is kept small. According to the NREL Wind Resource Assessment Handbook [24], the wind sensors should be at least six tower diameters out from the tower for tubular towers and at least three tower widths out from tower for lattice towers. For triangular lattice towers the tower width is measured as the length of one face.

Anemometers can also be installed at the top of the tower as shown in Figure 1.9. This has the advantage of avoiding the problem of tower wake, although care must still be taken to use a long-enough vertical boom to minimize the impact of the tower top on the observed wind speed. The NREL Wind Resource Assessment Handbook recommends a vertical boom length on top of the tower of at least 0.3 m [24]. Lighting protection equipment, such as a lightning rod, must be installed on the tower, in such a way that the interference with the anemometers is avoided and connected to the common ground.

For all parameters except wind direction, the average value in each 10-min interval is recorded. This averaging period of 10-min is recommended by international wind standards and is chosen since it falls within the "spectral gap" of the frequency spectrum of kinetic energy of the horizontal wind speed. By selecting an averaging period in this gap then short-term fluctuations are included in the wind speed standard deviation's statistics and longer term fluctuations appear as slow changes in the average wind speed values. Regarding the wind direction, the average is defined as a vector resultant value, which is the direction

Figure 1.9 Example of top mounted anemometers

with resolved components given by the means of the northerly and easterly speeds. Averages are used in evaluating wind speed variability, as well as wind speed and direction frequency distributions. The standard deviation should be determined for both wind speed and wind direction and is defined as the population standard deviation (σ) for all 1-s samples within each 10-min interval [21]. The standard deviations of wind speed and wind direction are indicators of turbulence. The maximum and minimum values measured during each interval should be recorded for all parameters. This is particularly important for the maximum 3-s gust, which can affect whether a turbine design is considered suitable for the site.

1.2.3 Spatial extrapolation of measured data (Step 3 of WRA)

For a single wind turbine installation, the wind measurement location is usually the same as the turbine location. If the two locations are not the same, the results from the measurements must be extrapolated to the planned wind turbine locations. Usually for spatial extrapolation, computer models are used, which are primed with wind speed data measured on-site. For mildly changing terrain, thermal stability, and terrain, in which roughness around the measured sites are similar to roughness around the turbine sites, linear wind flow models are utilized for the spatial extrapolation and AEP computation. If assumptions for the linear wind flow models are not valid, computational fluid dynamics (CFDs) modeling is performed.

The linear flow models are based on a linearization of the Navier–Stokes equations, which was originally introduced by Jackson and Hunt [25]. They were designed to be used reliably in neutral atmospheric conditions over terrain with adequately gentle slopes to ensure fully attached flow conditions. The governing equations are as follows:

$$\frac{\partial \widetilde{u}_i}{\partial x_i} = 0, \quad \text{for } i = 1, 2, 3 \tag{1.1}$$

$$U_j \frac{\partial \widetilde{u}_i}{\partial x_j} = \frac{\partial \widetilde{p}}{\partial x_i} + K_h \frac{\partial}{\partial x_j}\left(\frac{\partial \widetilde{u}_i}{\partial x_j}\right) + K_v \frac{\partial^2 \widetilde{u}_i}{\partial x_3^2}, \quad \text{for } i = 1, 2, 3 \text{ and } j = 1, 2, \tag{1.2}$$

where U_j ($j = 1, 2$) are the horizontal velocity components of the unperturbed flow, $\tilde{u}_i$ ($i = 1, 2, 3$) are the velocity perturbation components, and $\tilde{p}$ is the pressure perturbation. K_h and K_v are the effective kinematic viscosities in the horizontal and vertical directions, respectively. Linear models perform reasonably well where wind is not significantly influenced by steep slopes, flow separation, thermally driven flows, and other dynamic and nonlinear atmospheric boundary layer phenomena [26].

Among the linear flow models, the Wind Atlas Analysis and Application Program (WAsP) [27], a Jackson–Hunt model developed by the Risø National Laboratory of Denmark, has been widely used [26]. The "WAsP method" proceeds in two stages. First, the observed wind at an anemometer on a meteorological tower is used to derive the background wind field, which represents the site-independent wind resource that exists in the absence of terrain. Second, the process is subsequently reversed, utilizing the background wind as an input to predict the wind profile at other locations. In addition to implementing the basic Jackson–Hunt approach, WAsP has several modules that address various needs in wind flow modeling including the ability to incorporate the effects of surface roughness changes and obstacles.

The significant difference between CFD and linear flow models is that CFD models solve a more complete form of the Navier–Stokes equations. Therefore, CFD models require considerably more computational resources than linear flow models. Compared to the linear flow models, the CFD models can predict flow detachment and reattachments in the separation zone in most cases. However, if the Reynolds-averaged Navier–Stokes-type turbulence models are used, the accuracy of the results in this region is questionable. To obtain reasonably accurate results in this region, large-eddy simulation type turbulence models, which are much more computationally expensive, should be used. Typical CFD models for atmospheric flow simulation applications follow the single wind direction approach representing a sector from the discretized wind rose. Flow simulations for each sector, considering the effects of orography and roughness, result in speed-up factors [26].

1.2.4 Temporal extrapolation of measured data (Step 4 of WRA)

Due to the interannual variability of wind speed, various studies have indicated that wind data measured over just a few years are insufficient to reflect the average conditions present over the lifetime of a wind project (typically 20 years) [28]. A process known as measure–correlate–predict (MCP) is typically utilized to relate and adjust on-site measurements to a long-term reference [29]. The MCP has the following steps [5]:

1. Obtain on-site wind speed measurement for a year or more and long-term reference data for 20 or more years. It is recommended to obtain as many long-term datasets as possible.
2. Correlate site measurement data with long-term reference datasets for a concurrent time period. If the correlation is acceptable, these reference datasets are selected for the next step.
3. Based on the correlation factors derived from step 2, estimate the wind speed at the on-site measurement location for the historical period, which covers the duration of reference time series. This is the hindcasting step.
4. Convert the hindcast into a forecast, if necessary, or use the hindcast for the AEP computations.

In most practical applications, the correlate and predict steps are conducted iteratively with several sources of long-term reference datasets.

In most cases, it is very difficult to obtain reliable, consistent long-term measurement data. In such a case, a feasible alternative is reanalysis data. The most commonly used reanalysis data is NCEP/NCAR data from NCEP (National Center for Environmental Prediction) and NCAR (National Center for Atmospheric Research) [30]. These institutes continuously conduct a global reanalysis of weather data. For the initialization of a global numerical weather prediction model, data from synoptic weather stations, radiosondes, pilot balloons, aircrafts, ships, buoys, satellites, etc. are collected, controlled, gridded and prepared. By running past-periods on a global scale, the data is reanalyzed in a globally consistent manner, which enables it to be utilized in certain situations as long-term reference data. The period of available reanalysis data can be much longer than the period of available on-site measurement data.

1.2.5 Annual energy production computation (Step 5 of WRA)

Using the extrapolated wind speed at the wind turbine location and hub height, along with the turbine's power curve (e.g., Figure 1.10), we can compute the gross AEP of a specific wind turbine, AEP_{gross}, by

$$AEP_{gross}\ (\text{kWh}) = \frac{\Sigma_{i=1}^{i_{\max}} P(v_i)}{i_{\max}} \times \frac{\Delta_t}{1\ (\text{h})} \times 8,760\ (\text{h}), \tag{1.3}$$

where $P()$ is the power output of the chosen turbine, v_i is the time series of the predicted wind speed with a time spacing of Δ_t, and $i_{\max}$ is the number of data points [5].

To compute the net AEP of a specific wind turbine, AEP_{net}, it is necessary to consider various factors that could reduce the actual energy produced:

$$AEP_{net} = AEP_{gross} \times R_{available} \times C_{correction}. \tag{1.4}$$

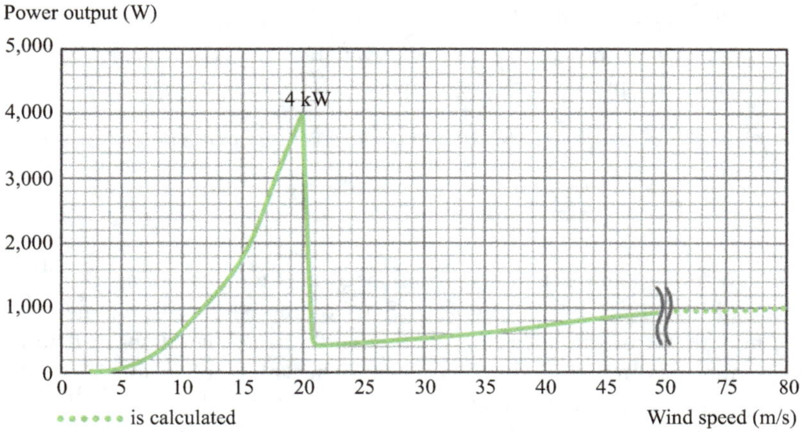

Figure 1.10 Power curve for Zephyr Airdolphin GTO. Courtesy of Zephyr Inc

where $R_{available}$ is the available ratio, which is the ratio of the total time that the wind turbine can operate, excluding the downtime for faults and services to the total time that the wind turbine can operate without considering the downtime. $C_{correction}$ is the coefficient for power output correction that takes into account the reduction of power output due to blade soiling, control error/hysteresis and yaw error losses caused by strong turbulence, icing losses, grid outages, wire losses, inverter losses, and other factors. Generally, wire and inverter losses for SWTs are considered in the certified power curve. Control and yaw error losses are also considered in measured power curves; however, the consideration of these losses is not sufficient for a highly turbulent site. The other losses depend highly on the wind turbine or local environment [22].

1.3 Specific features of WRA for SWTs

When on-site wind measurements are undertaken for SWTs, shorter averaging periods than the 10-min international standard for wind turbines are often used, due to the lower moment of inertia of SWT rotors. The international wind turbine power performance standard, IEC61400-12-1 [31] suggests a 1-min averaging period for SWT monitoring to capture the changes in power output in response to changes in wind conditions.

SWT installations do not have the budgets of large wind farm projects, and it is common to find cases in which on-site wind measurement is not performed due to the cost and time needed to collect a year of wind resource data. In particular, it is rare to conduct on-site wind measurements for rooftop SWTs. This section describes WRA without on-site wind measurement and WRA over the roof of a building.

1.3.1 WRA without on-site wind measurement

Even if on-site wind measurement is not available, the AEP of an SWT can be roughly estimated using wind resource maps as follows:

1. Compute the wind speed at the height of the SWT using the power law equation:

$$V = V_{ref} \times \left(\frac{H_{hub}}{H_{ref}}\right)^{\alpha}, \tag{1.5}$$

where V is the annual average wind speed at the height of an SWT, V_{ref} is the wind speed at the location of the SWT on the wind resource map, H_{hub} is the hub height of the SWT, H_{ref} is the height of the wind speed data of the map, and α is the wind shear exponent. The values of α are 0.10–0.14 for flat grassland and coastal areas, 0.17–0.25 for fields and paddies, and 0.25–0.50 for city areas [19]. The more detailed values of α are summarized in reports [21,22].
2. If the shape and scale parameters for the Weibull probability density function (k and c, respectively) are available from the wind resource map, estimate the occurrence frequency of wind speed u_j using

$$f(u_j) = \frac{k}{c}\left(\frac{u_j}{c}\right)^{k-1} \exp\left(-\left(\frac{u_j}{c}\right)^{k}\right)\Delta_u \tag{1.6}$$

from $j = 1$ to $j_{\max}$. Here, u_1 is usually 0.5 or 1 m/s, $u_{j\max}$ is the cutout wind speed of the SWT, and Δ_u is the unit width of wind speed bins, which is usually 0.5 or 1 m/s [23]. If the Weibull parameters are not available from the wind resource map, estimate the values of $f(u_j)$ using the Rayleigh distributions as follows:

$$f(u_j) = \frac{\pi}{2}\frac{u_j}{V^2}\exp\left(-\frac{\pi}{4}\left(\frac{u_j}{V}\right)^{2}\right)\Delta_u. \tag{1.7}$$

The Rayleigh distribution is derived by substituting $k = 2$ into (1.6) and is often utilized due to its simplicity, as it requires only the average wind speed V to roughly estimate the occurrence frequency of wind speed.
3. Compute the gross AEP of the SWT as follows:

$$AEP_{gross}(\text{kWh}) = \Sigma_{j=1}^{j_{\max}}\left(P(u_j) \times f(u_j) \times 8{,}760(\text{h})\right). \tag{1.8}$$

If a wind rose at the SWT's location is available from the wind resource map, the net AEP of the SWT can be computed using the following equation:

$$AEP_{net}(\text{kWh}) = \Sigma_{l=1}^{l_{\max}}\left(AEP_{gross} \times R_{available} \times C_{correction}(\theta_l) \times \frac{\Delta_\theta}{360^\circ}\right), \tag{1.9}$$

where $l_{\max}$ is the total number of wind direction sectors, and $C_{correction}$, which is the coefficient for power output correction, is a function of wind direction θ_l, and Δ_θ is the unit angle of wind direction sectors. If there are obstacles that generate significant turbulence at the hub height of the SWT, in sector θ_l and upwind of the SWT, $C_{correction}$ (θ_l) should be reduced from 100%. Turbulence can reduce the AEP of a horizontal axis wind turbine by approximately 15%–25% [22]. According to the measurement campaigns by Dundalk Institute of Technology and Ashikaga Institute of Technology in a report from International Energy Agency Wind Task 27 [32], turbulence significantly reduces the power output of the horizontal axis wind turbines at high wind speed ranges, although turbulence increases the power output of the SWTs at low wind speed ranges. With regard to vertical axis wind turbines (VAWTs), a recent study based on wind tunnel experiments reported that with an increase in inflow turbulence intensity, the power coefficients of H-type Darrieus VAWTs were increased [33]; therefore, it might not be necessary to reduce the value of $C_{correction}\,(\theta_l)$ for VAWTs. Since existing research on the effects of turbulence intensity on the performance of VAWTs are extremely limited, further studies are expected.

One alternative option to the use of a relevant wind resource map would be to use wind measurements at a nearby reference site to the candidate site such as at a local airport or weather stations. However, the wind measurements at these reference sites are not intended for WRA; and sensors are not always at appropriate heights or at locations that are free of obstructions. In addition, these sites are usually far from the site of interest, with considerably different orography, tree cover and monitoring heights, making these data of questionable usefulness [22].

In addition, there are some simplified methods that can be used to estimate AEP. These methods often make assumptions regarding the turbine power curve, wind frequency distribution, and ratio of furling to average wind speed. One example is use of the equation:

$$AEP\,(\text{kWh}) = 8.8 C_o A V^3, \tag{1.10}$$

where A is the rotor swept area, V is the average wind speed at the site, and C_o is an average power coefficient, determined from four points on the power curve between cut-in wind speed and rated wind speed [34]. This technique can be subject to errors since there are many assumptions made. Caution must be exercised if the turbine characteristics differ significantly from the assumed characteristics in this method.

Finally, in some cases the SWT manufacturer has published plots of average energy per day versus mean wind speed at the site. In this case, all that is needed is to adapt the manufacturer's data (which is often produced at sea level conditions) to the conditions of the site you intend to install the turbine at. However, the customer must be aware that the manufacturer's data can be misleading, and the results are entirely dependent on the assumptions that the manufacturer has made regarding the wind regime at the site.

1.3.2 WRA over the roof of a building

Wind conditions over the roof of a building can drastically change both horizontally and vertically in a short distance due to the occurrence of separated airflow from the edges of the roof [35]. In addition, wind conditions over the roof of a building are significantly affected by the roof shape, horizontal shape of the building, wind direction, surrounding buildings, and other factors [35]. Therefore, to estimate the AEP of an SWT on the roof of a building at a reasonable level of accuracy, it is necessary to conduct an on-site wind measurement by setting an anemometer at or near the candidate position of the SWT rotor. Regarding good practice in conducting rooftop wind monitoring campaigns, Tabrizi *et al.* [36] reported that conservative estimates of turbulence in the complex environment of a warehouse rooftop in an industrial park could be achieved using a sampling rate of 10 Hz and an averaging period of 10 min.

Although it is not yet common to conduct CFD simulations for the purpose of estimating the AEP of a rooftop SWT, CFD simulations have been utilized in many studies on wind conditions over the roof of a building (e.g., [35,37,38]). Knowledge of the characteristics of wind conditions over the roof of a simple shape building can be helpful for selecting the positions of SWT rotors. In the following, CFD results for wind conditions over the roof of a high-rise cuboid building (height: 80 m, width: 40 m) [38] are described.

Figure 1.11 shows the horizontal distributions of the mean wind velocity and the standard deviation of horizontal wind velocity at 4, 8, and 12 m above the roof surface. The prevailing wind direction is from left to right. At the windward corners of the roof, wind conditions are generally favorable (i.e., high wind speed and low turbulence intensity), even at relatively low heights. This is because the separation of the airflow at the windward corners of the roof is small.

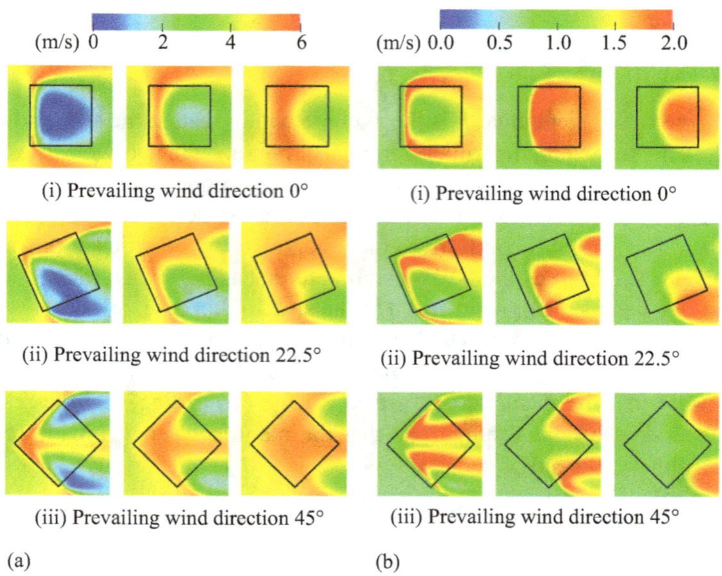

Figure 1.11 Horizontal distributions of (a) the mean horizontal wind velocity and (b) the standard deviation of the horizontal wind velocity at 4 m (left), 8 m (middle), and 12 m (right) above the roof surface

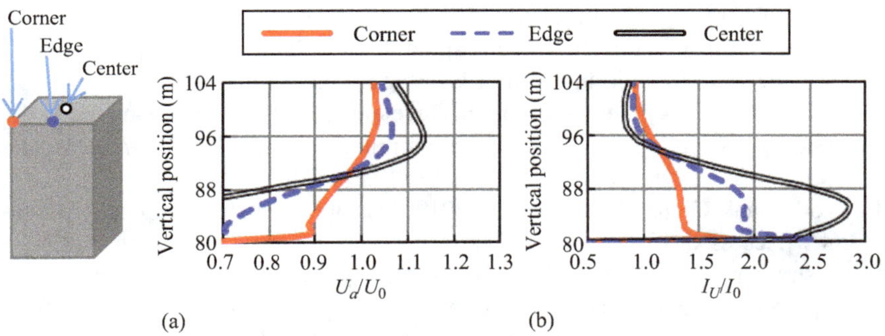

Figure 1.12 Vertical profiles of (a) horizontal mean wind velocity ratios and (b) turbulence intensity ratios at representative locations on the roof

Figure 1.12 shows the vertical profiles of horizontal velocity ratios and turbulence intensity ratios at representative locations on the roof under the condition of no prevailing wind (i.e., the frequency occurrence is the same for all wind directions). Here, U_a and I_U are the horizontal mean wind velocity and turbulence intensity, respectively; U_0 and I_0 are the horizontal wind velocity and turbulence intensity under the condition where there is no building. Near the roof surface, the wind conditions are relatively favorable at the corner of the roof. In contrast, at higher heights, the center of the roof is the most optimal location for installing SWTs.

When installing rooftop SWTs, it is necessary to consider the building-scale structural impacts that may require extensive engineering, building reinforcement, noise management, and system safety analysis [22], as well as wind conditions.

References

[1] U.S. Department of Energy (DOE). Distributed Wind Market Report 2018. U.S. DOE, 2018.

[2] Nippon Kaiji Kyokai (ClassNK). Type Certification for Wind Turbines. Available from: https://www.classnk.or.jp/hp/en/authentication/renewableenergy/windturbine.html [Accessed: 15-February-2021].

[3] International Electrotechnical Commission (IEC) 61400-2: 2013. Wind Turbines-Part 2: Design Requirements for Small Wind Turbines. International Electrotechnical Commission (IEC): Geneva, Switzerland, 2013.

[4] Asian Development Bank. Guidelines for Wind Resource Assessment: Best Practices for Countries Initiating Wind Development. Asian Development Bank, 2014.

[5] Jain P. *Wind Energy Engineering, Second Edition*. McGraw-Hill Education: New York, USA, 2016.

[6] Lee J and Fields J. An overview of Wind Energy Production Prediction Bias, Losses and Uncertainties. Wind Energy Science, 2020. Available from https://doi.org/10.5194/wes-2020-85 [Accessed: 25-February-2021].

[7] New Energy and Industrial Technology Development Organization (NEDO). Local Wind Resource Map, 2006. Available from: https://appraw1.infoc.nedo.go.jp/nedo/index.html [Accessed: 22-August-2021].

[8] Murakami S, Mochida A and Kato S. Development of local area wind prediction system for selecting suitable site for windmill. *Journal of Wind Engineering and Industrial Aerodynamics*, 91(12–15). pp. 1759–1776, 2003.

[9] World Bank. Global Wind Atlas 3.0, 2019. Available from: https://globalwindatlas.info/ [Accessed: 15-Feburary-2021].

[10] Australian Government. Maps of Average Velocity, 2011. Available from: http://bom.gov.au/climate/averages/maps.shtml [Accessed: 15-Feburary-2021].

[11] Energiewerkstatt. Wind Map of Austria, 2011. Available from: https://windatlas.at/ [Accessed: 15-Feburary-2021].

[12] Government of Canada. Environment and Climate Change Canada's Wind Energy Atlas, 2004. Available form: http://windatlas.ca/index-en.php [Accessed: 15-Feburary-2021].

[13] Government of Ireland. SEAI's Wind Atlas, 2013. Available from: https://seai.ie/technologies/seai-maps/wind-atlas-map/ [Accessed: 15-Feburary-2021].

[14] Korea Institute of Energy Research. Renewable Energy Resource Maps, 2015. Available from: https://kier-solar.org/user/map/map_gallery.do [Accessed: 15-Feburary-2021].

[15] National Renewable Energy Laboratory (NREL). WINDExchange Wind Energy Maps and Data, 2012. Available from: https://windexchange.energy.gov/maps-data [Accessed: 15-Feburary-2021].

[16] Wegley HL, Ramsdell JV, Orgill MM and Drake RL. Siting Handbook for Small Wind Energy Conversion Systems, 1980. Available from: https://osti.gov/biblio/5490541 [Accessed: 15-Feburary-2021].

[17] U.S. Department of Energy (DOE). Small Wind Guidebook, 2019. Available from: https://windexchange.energy.gov/small-wind-guidebook [Accessed: 15-Feburary-2021].

[18] Hiester TR and Pennell WT. Meteorological Aspects of Siting Large Wind Turbines. United States, 1981. Available from: https://osti.gov/biblio/6657537-meteorological-aspects-siting-large-wind-turbines [Accessed: 15-Feburary-2021].

[19] Japan Small Wind Association. Small Wind Turbine Guide Introductions, 2012. [In Japanese].

[20] Finnigan J, Ayotte K, Harman I, *et al.* Boundary-layer flow over complex topography. Boundary-Layer Meteorology, 177. pp. 247–313, 2020.

[21] IEA Wind. Expert Group Report on Recommended Practices Micro-Siting Small Wind Turbines for Highly Turbulent Sites, 2018. Available from: https://higherlogicdownload.s3.amazonaws.com/IEAWIND/c90e3186-96d9-46cd-b06d-3ce3684613c5/UploadedImages/Reports/IEA_Wind_RP_19_Micro_Siting_Small_Wind_Turbines_in_Highly_Turbulent_Sites.pdf [Accessed: 15-Feburary-2021].

[22] Olsen T and Preus R. Small Wind Site Assessment Guidelines. NREL/TP-5000-63696. National Renewable Energy Laboratory (NREL), 2015.

Available from: https://nrel.gov/docs/fy15osti/63696.pdf [Accessed: 15-Feburary-2021].

[23] Brower MC. Wind Resource Assessment: A Practical Guide to Developing a Wind Project. Wiley: New Jersey, USA, 2012.

[24] National Renewable Energy Laboratory (NREL). Wind Resource Assessment Handbook – Fundamentals for Conducting a Successful Monitoring Program. Prepared by AWS Scientific, Inc For National Renewable Energy Laboratory, 1997. Available from: https://nrel.gov/docs/legosti/fy97/22223.pdf [Accessed: 24-February-2021].

[25] Jackson PS and Hunt JCR. Turbulent wind flow over a low hill. Quarterly Journal of the Royal Meteorological Society, 101. pp. 929–955, 1975.

[26] IEC TR 61400-12-4: 2020. Wind Energy Generation Systems – Part 12-4: Numerical Site Calibration for Power Performance Testing of Wind Turbines. IEC: Geneva, Switzerland, 2020.

[27] Mortensen NG, Heathfield DN, Myllerup L, Landberg L and Rathmann O. Getting started with WAsP 9. Technical Report Risø-I-2571(EN), Risø National Laboratory, 2007. Available from: https://core.ac.uk/download/pdf/13802297.pdf [Accessed: 15-Feburary-2021].

[28] Troen I. A high resolution spectral model for flow in complex terrain. *Proc. 9th Symposium on Turbulence and Diffusion*, Roskilde, Denmark, 1990.

[29] Carta JA, Velázquez S and Cabrera P. A review of measure-correlate-predict (MCP) methods used to estimate long-term wind characteristics at a target site. *Renewable and Sustainable Energy Reviews*, 27. pp. 362–400, 2013.

[30] Langreder W. Chapter 2 Wind Resource and Site Assessment, Wind Power Generation and Wind Turbine Design (Edited by Wei Tong), WIT Press: Southampton, UK, 2010.

[31] IEC TR 61400-12-1: 2017. Wind Energy Generation Systems – Part 12-1: Power Performance Measurements of Electricity Producing Wind Turbines. IEC: Geneva, Switzerland, 2017.

[32] IEA Wind. Compendium of IEA Wind TCP Task 27 Case Studies, 2018. Available from: https://community.ieawind.org/viewdocument/compendium-of-iea-wind-tcp-task-27-1 [Accessed: 1-March-2021].

[33] Molina AC, De Troyer T, Massai T, Vergaerde A, Runacres MC and Bartoli G. Effect of turbulence on the performance of VAWTs: An experimental study in two different wind tunnels. *Journal of Wind Engineering and Industrial Aerodynamics*, 193, 2019.

[34] Berrill T. *Introduction to Renewable Energy Technologies – Resource Book*. Renewable Energy Centre, Brisbane Institute of TAFE, Queensland, Australia, 1997.

[35] Toja-Silva F, Kono T, Peralta C, Lopez-Garcia O and Chen J. A review of computational fluid dynamics (CFD) simulations of the wind flow around buildings for urban wind energy exploitation. *Journal of Wind Engineering and Industrial Aerodynamics*, 180. pp. 66–87, 2018.

[36] Tabrizi A, Whale J, Lyons T and Urmee T. Rooftop wind monitoring campaigns for small wind turbine applications: Effect of sampling rate and averaging period. *Renewable Energy*, 77. pp. 320–330, 2015.
[37] Tabrizi A, Whale J, Lyons T and Urmee T. Performance and safety of rooftop wind turbines: Use of CFD to gain insight into inflow conditions. *Renewable Energy*, 67. pp. 242–251, 2014.
[38] Kono T, Kogaki T and Kiwata T. Numerical investigation of wind conditions for roof-mounted wind turbines: Effects of wind direction and horizontal aspect ratio of a high-rise cuboid building. *Energies*, 9(11). p. 907, 2016.

Chapter 2
Resource assessment for hydrokinetic turbines

Muluken Temesgen Tigabu[1], *David Wood*[2] *and Bimrew Tamrat Admasu*[3]

The performance of a wind or hydrokinetic turbine (HKT) depends on its "power curve"—the output power versus wind or water speed—and the magnitude of the resource. This chapter considers the assessment of the resource for HKTs and raises some interesting issues of the interaction with the turbine power curve. HKTs are a comparably new energy conversion technology, and there are few resource assessment tools available especially in comparison to wind turbines. The main aim of this chapter is to discuss the techniques of resource estimation for rivers and human-made water structures such as irrigation canals. The chapter presents the main features of existing models used for HKT resource prediction through hydraulic stimulation and then hydrological models for validation. The main outcome of resource estimation is the water velocity which can be combined with an HKT power curve, or used to provide the power density (PD), the available power per square meter of turbine rotor area. The two chief difficulties of resource assessment are described: the first is that most hydrological data is provided as volume flow rates in m^3s that can be related to the velocity only if the river cross-section is available, which is not the case in general. Second, the velocity varies with time and the time period used to determine the average velocity is crucial to the resource assessment.

2.1 Introduction

Renewable energy is often intermittent and the dependence of power output on resource level is often highly nonlinear. These features make it difficult to estimate the performance of renewable energy technologies at a particular site in terms of, say, the annual energy production, AEP, measured in kWh/year. It is then instructive to look at the resource assessment capability of freely available renewable energy

[1]BahirDar Energy Center, BahirDar Institute of Technology, BahirDar University, BahirDar, Ethiopia
[2]Department of Mechanical and Manufacturing Engineering, University of Calgary, Calgary, Alberta, Canada
[3]Faculty of Mechanical and Industrial Engineering, BahirDar Institute of Technology, BahirDar University, Ethiopia

project tools such as the system advisor model, SAM.* SAM has resource data for photovoltaic and distributed wind projects but nothing for HKTs in rivers or canals.

The main aim of this chapter is to describe the current state of resource prediction for HKTs. This usually means estimating the PD with units of kW/m^2 where the area (m^2) relates to the swept area of the turbine blades. We concentrate on the PD as it is a simple, but not necessarily a complete, measure of the resource. We show later that PD depends on the velocity and discuss the issues related to the time over which the velocity is averaged. We describe the main issues of HKT resource assessment, highlight useful and mostly free software, present predictions and measurements, and show the vast resource potential of HKTs. In general, HKTs harness the kinetic energy of rivers, human-made water structures, tides, and ocean currents. This chapter considers only the first two. The term "hydrokinetic" is often used inconsistently to describe a range of energy converters; for example, Khan *et al.* [1] used it for river energy conversion, whereas it means ultralow-head hydro turbine in Zhou and Deng [2], and zero head hydro turbine in Suwa and Masud [3]. In this chapter, the term will be used only for HKTs as defined earlier.

Rivers have long been used as a source of drinking water, means of transportation, and for irrigation. These uses are complementary to the installation of HKTs. It is noteworthy that many rivers throughout the world contain abundant potential to generate electricity by using HKTs. One of the advantages of such technologies is for remote regions in off-grid areas where the operating cost of fuel-based power generation is high. The analysis in this chapter assumes that only one HKT will be installed at the site(s) being investigated and that the power extracted is too small to influence the downstream resource. Multiple turbines can interfere with each other, e.g., Okulov *et al.* [4], and reduce the resource especially when the rotor swept area is large compared to the cross-section of the river or canal.

2.2 The context of hydrokinetic turbines

HKTs can be deployed in many locations in rivers and human-made water structures. The basic parameters for the deployment of HKTs are the local velocity, and the local channel width and depth. The last two determine the position of maximum velocity that maximizes the power and also will lead to blockage effects if the flow area is not large compared to the swept area of the blades. The general characteristics of the HKT resource also depend on the shape of a river channel, and this shape depends on how much water has been flowing in it for how long. There are two important aspects of the flow characteristics: (1) the "discharge" or volume flow rate of water, (Q) in (m^3/s), that is often the only data available for many rivers. Q is approximately constant over the year for some rivers and their tributaries but other rivers show strong seasonal variation in Q; (2) the velocity (V) in (m/s). In practice, V is usually averaged over some time period, T, that we denote as $\overline{V}_T$ if T is known,

*https://sam.nrel.gov/

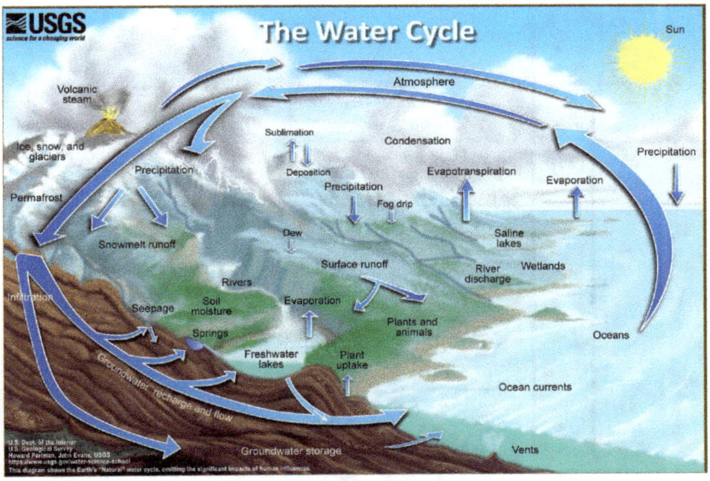

Figure 2.1 The water cycle

and often over the river cross-section. This further averaging is denoted as $\langle \overline{V}_T \rangle$. The relationship between Q, strictly Q_T, and $\langle \overline{V}_T \rangle$ comes through the cross-sectional area, A, of the river that clearly varies along the length of the river and can vary daily or seasonally. The average velocity at any cross-section is just $\langle \overline{V}_T \rangle = Q_T/A$. All these characteristics are dependent on the nature of the water cycle as shown in Figure 2.1 taken from.[†] The water cycle is the movement of water on, in, and above the Earth, which has been operating for billions of years.

The flow of water in a river or canal is an open-channel flow where the free surface is subjected to atmospheric pressure. Free surface effects are associated with the Froude number, Fr, that is the ratio of inertial and gravitational forces and defined by

$$Fr = \frac{\langle \overline{V} \rangle}{\sqrt{gd}} \tag{2.1}$$

where g is gravity, and d is the depth of the river. Fr is used to characterize the flow as critical $Fr = 1$, subcritical $Fr < 1$, and supercritical $Fr > 1$. For HKT installation, it is usual to avoid $Fr \geq 1$ as gravity waves may be produced by a change in the flow depth that will increase the structural and hydrodynamic load on the HKTs. The velocity at a given cross-section at a given instant of time varies due to the pressure increase with depth, the shear stress, particularly at the bottom and sides of the river or channel, and the no-slip boundary condition at the same locations. A generic variation is illustrated in Figure 2.2 and detailed measurements of the variation are given by Neary *et al.* [5]. The average velocity has three components,

[†]https://www.usgs.gov

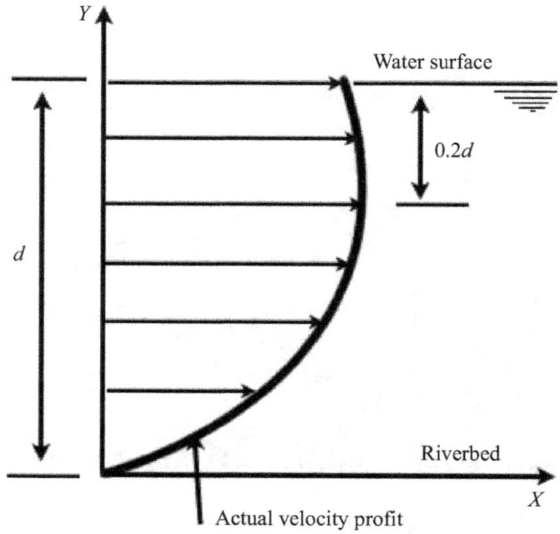

Figure 2.2 Velocity distribution

i.e., $\overline{V_x}$, $\overline{V_y}$, and $\overline{V_z}$, where x and y are shown in Figure 2.2, and z is normal to the page. The cross-stream velocities, $\overline{V_y}$ and $\overline{V_z}$, are usually small and can be neglected leaving the velocity in the flow direction, $\overline{V_x}$, as the important one for resource assessment. HKT deployment usually requires finding the location of the maximum velocity. From Figure 2.2, this point is located $0.2d$ from the free surface.

2.3 Power density estimation for HKTs

Simple non-dimensionalization of turbine output power, P, suggests that it can be expressed as

$$P = \frac{1}{2} C_P \rho A_R V^3 \tag{2.2}$$

where C_P is the turbine power coefficient that typically lies between 0.2 and 0.4. V is the velocity, ρ is the water density, and A_R is the swept area of the rotor. Defining $PD = P/A_R$ shows that the resource is measured by V but in a strongly nonlinear fashion. A further complication is that, in practice, both P and V are averaged in time to determine the turbine power curve, $\overline{P}$ versus $\overline{V}$. Some effects of averaging can be illustrated by considering wind turbines for which power curve determination and resource assessment are more advanced and standardized than for HKTs. By the International Electrotechnical Commission Standard EC 61400-12-1:2017, the power curve of a large wind turbine is determined by averaging the power and speed over 10-min periods and then "binning" the measurements in small increments of average speed, and averaging within the bins to get a single curve. This curve, like a power curve for an HKT, and as implied by (2.2), is approximately cubic in wind speed for speeds less than around 10 m/s. For resource assessment, wind speeds measured over the same time can be binned to provide the probability distribution of speed. Software like SAM combines the power curve and probability distribution to determine the AEP or the capacity factor, CF, defined as the average output power divided by the turbine's rated (maximum) power. The CF and AEP are directly related. For most wind turbines, it turns out that the AEP is almost linear in the average wind speed because the highly nonlinear power curve is modulated by the wind speed probability distribution. In other words, a standard power curve cannot be used with the long-time average wind speed to determine the AEP or CF. It follows that resource assessment for wind or water power must provide velocities measured over times comparable to those used to determine the power curve. To put this another way: from the previous discussion of the AEP, increasing the averaging time turns the cubic relationship between $\overline{P}$ and $\overline{V}$ into an approximately linear one. Thus, for example, estimating $\overline{V}$ from monthly discharges is unlikely to accurately determine the resource. Combining the effects of the averaging time with the spatial variability of V means that accurate estimates of PD are difficult to achieve.

Determining the site performance of HKTs is more challenging than for wind turbines due to the lack of short-period-averaged water velocity measurements made over a long time. Using an assumed probability distribution function (pdf) of wind speed is common practice and can also be adopted for HKTs. Pdfs such as the Type 1 extreme and log–normal were used for HKT assessment by Punys *et al.* [6] and Weibull and Rayleigh pdfs are very often used for wind resource assessment. For example, the AEP can be determined from the Weibull probability distribution, $f(V)$, which is given by

$$f(V) = \frac{k}{c}\left(\frac{V}{c}\right)^{k-1} \exp\left[-\left(\frac{V}{c}\right)^{k}\right] \tag{2.3}$$

where k is the "shape parameter" whose value is often about 2, and c is the "scale parameter" that is directly related to the long-term $\overline{V}$. The power produced in a given period (T) such as for a year (AEP) can be calculated as

$$AEP = T\int_0^\infty P(V)f(V)dV \tag{2.4}$$

Lalander *et al.* [7] used log–normal and Weibull distributions to estimate the power available from tidal currents and rivers. For "regulated" rivers, in which the flow is controlled by a reservoir, the log–normal distribution was accurate for many site. For unregulated rivers, the distribution was, in general, difficult to match to any theoretical pdf. They suggested that a simple measure of suitability was the ratio of long-term $\overline{V}$ to the rated speed of the turbine.

2.4 Resource prediction models

Resource prediction models can be applied at a specific site or a wider scale of river networks. They should be cost-effective and fast to implement. The preeminent requirement for the installation of HKTs is to find enough resourceful sites after deciding what constitutes such a site [1]. It is also important that the hydrological data and the analysis software are freely available to allow detailed resource assessment in developing countries with limited resources.

PD determination is based on two main methods that are (1) hydrological and (2) hydraulics, Punys *et al.* [6]. The first requires the direct measurement of the river flow. However, as mentioned earlier, hydrological data is mainly in the form of Q but the important parameter is the velocity profile of the river. Due to this, and the issues raised previously about the spatial and temporal averaging of the velocity, the hydrological approach does not lead to a full understanding of the river flow field and is of limited use in estimating the PD.

To estimate the PD, modeling the velocity field is mandatory; however, its characteristics are very complex due to the spatial and temporal variations. There are different types of models available to predict these variations, as summarized in Figure 2.3.

2.4.1 Analytical methods

These are best suited for predictable water flow in simple geometries such as irrigation canals and other human-made waterways. To determine the velocity, analytical methods can be employed, one of such analytical model is called the Chezy formula defined as

$$\langle \overline{V} \rangle = C\sqrt{R_l S} \tag{2.5}$$

where C is Chezy coefficient ($C = \sqrt{8g/n}$) and n is the roughness coefficient, R_l is the hydraulic radius, and S is the slope of the canal bed. Clearly, $\langle \overline{V} \rangle$ can only be a long-time average velocity because all terms on the right are independent of time. R_l can be computed as

$$R_l = \frac{A}{W_P} \tag{2.6}$$

where A is cross-sectional area and W_P is wetted perimeter. In Figure 2.4, $A = WD + XD$ and $W_P = W + 2L$, where $L = \sqrt{X^2 + D^2}$.

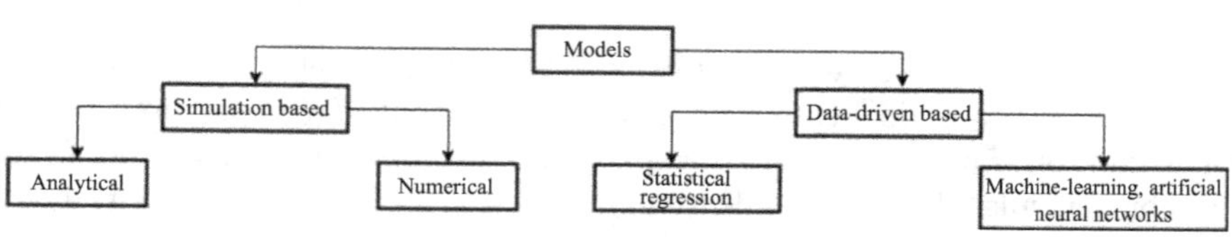

Figure 2.3 Velocity prediction models [8]

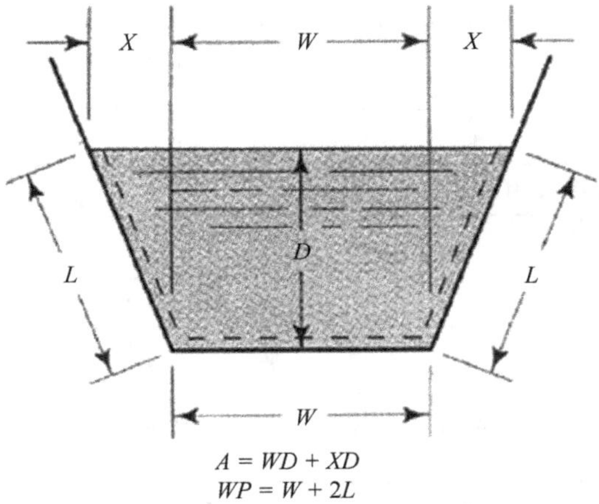

Figure 2.4 Typical cross-section of a common irrigation canal

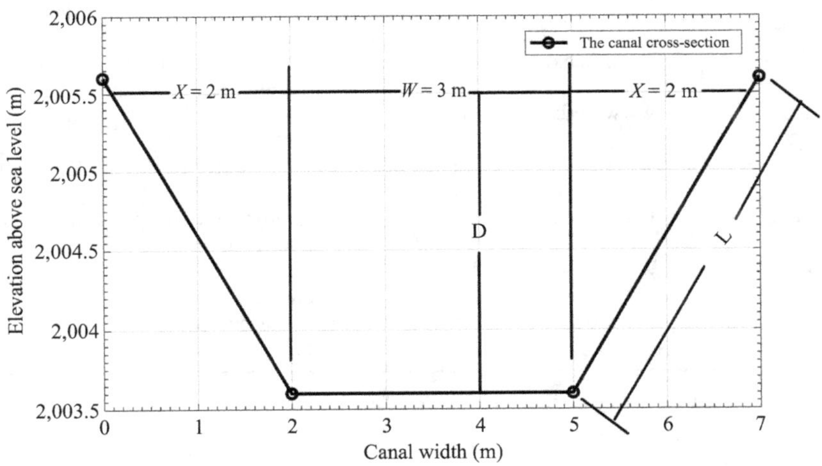

Figure 2.5 Typical cross-section of Koga irrigation canal. Figure adapted with permission from [9], Elsevier

To demonstrate the applicability of the analytical method, a comparison is made by calculating $\langle\overline{V}\rangle$ using (2.5) with the measured $\langle\overline{V}_T\rangle$, averaged over 2-min, by [9] for an irrigation canal whose cross-section is shown in Figure 2.5 which also contains the required parameters for the analytical model. Table 2.1 compares the analytical results with measurements at three different depths in the canal. The analytical method estimates $\langle\overline{V}\rangle$ with an average relative error accuracy of 25%, which is high considering the simple, known geometry of the canal. The implication is that the analytical approach should be used only for preliminary resource assessment.

Table 2.1 Comparison of analytical method with measured $\langle \overline{V}_T \rangle$ in the canal of figure with $C = 65.9\ m^{1/2}/s$ and $S = 1/1{,}000$

Station (m)	Measured			Analytical			
	D (m)	**W (m)**	**$\langle \bar{V} \rangle$ (m/s)**	**A**	**W_P (m)**	**R_l (m)**	**$\langle V \rangle$ (m/s)**
0	2	3	1.4	10	8.66	1.15	2.24
3,253	1.7	3	1.7	8.5	8.24	1.03	2.11
9,800	1.5	3	1.9	7.5	8.0	0.93	2.02

Table 2.2 Software for resource assessment of HKTs

Name	Availability	Functionality
CCHE2D/3D	www.ncche.olemiss.edu/	Simulation of river flows with depth-average and turbulence model
Delft3D	www.oss.deltares.nl	River flow analysis of water level, $\overline{V}$, and coastal hydrodynamics modeling
HEC-RAS	www.hec.usace.army.mil	1D and 2D river analysis system coupled with GIS for $\overline{V}$
HYDROKAL	www.hydrocalc2000.com	Open-channel flow, riverine cross-section geometry, $\overline{V}$, F_r
H2D2	www.gre-ehn.ete.inrs.ca/H2D2	Shallow-water solver for $\overline{V}$, Q, and water levels
MIKE21	www.dhi.se	Hydrodynamic modeling of coastal or marine areas
MWSWAT	www.swat.tamu.edu	Plug-in for GIS software for waste and soil assessment
TELEMAC-2D/3D	www.opentelemac.org	Free-surface two-dimensional flows, turbulence simulation

Note: All websites were last accessed December 29, 2020.

2.4.2 Numerical models

Analytical simulation studies are often based on 1D flow simulation where the velocity is taken to be a function only of the distance along the river. Theoretically, more accurate numerical models can be used, such as CCHE2D or CCHE3D, to predict the flow field. These are numerical models to solve the continuity equation and the 2D or 3D momentum equations, respectively, see Table 2.2. Toniolo [10] used CCHE2D to assess the resource potential of the Kvichak River, Alaska is shown in Figure 2.6, where zero velocities at the flow edges and maximum velocities at the midsection of the river were simulated.

There are other freely available different numerical and geographic information system (GIS)-based prediction models available for resource estimation, Table 2.2.

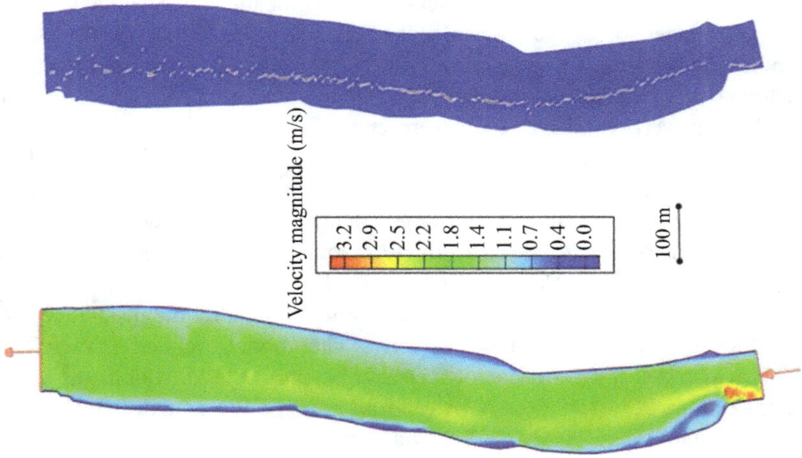

Figure 2.6 (a) Velocity distribution and (b) the lines showed the location of maximum velocity. Figure adapted with permission from [10], Elsevier

Resource assessment can be done at the global, regional, and site-specific levels with different levels of complexity, type, and size. At the regional level, Korkovelos *et al.* [11] used open-source GIS dataset from the ArcGIS software to evaluate the resource for small-scale hydro-power over a 712,615-km river network in 44 countries in sub-Saharan Africa. The main GIS dataset used were Digital Elevation Map (DEM), Global River Network, Global Streamflow Characteristics Dataset, and Administrative Boundaries. Using the geospatial tool, a total of 15,599 potential sites were identified across the subcontinent with a total potential of 25,221 MW. Despite possible inaccuracy and uncertainty, and long computation times to process the large amount of data, geospatial tools provide useful estimations of the resource at the country or regional level. Despite these regional or larger area studies, especially for data-scarce regions using global data, e.g., Magaju *et al.* [12], HKT resource assessment needs a site-specific study with a determination of the velocity profile to find the location of maximum resource. This must be done over a representative time especially when there is a strong seasonality in the water flow. As a result, one of the keys for conducting the resource assessment is the detailed knowledge of the temporal and spatial variations of the velocity. However, detailed information about the flow field is difficult to find, because it is dependent on the discharge, cross-section, slope, and roughness of the river system. Hence, finding an accurate and detailed database containing such information is often a major challenge, dos Santos *et al.* [13].

Many previous numerical studies of rivers and canals have been performed to understand the open-channel hydraulics, sedimentation, environmental, and morphological impacts. Hence, the available numerical models are often used for purposes that are not directly related to HKT resource assessment, Spasojevic and Holly [14]. Efforts have been made to use open-channel hydraulic numerical models for resource prediction of HKTs, e.g., Fouz *et al.* [15] who used a numerical model using Delft3D, see Table 2.2, for HKT resource assessment in estuarine areas in Spain and Portugal. The Delft3D model solves the Navier–Stokes equations for shallow water using the Boussinesq assumptions. To validate the model, measurements of the velocity were made at one site for 22 days using an acoustic Doppler current profiler (ADCP). Once the simulations were validated, they were applied for the entire estuarine area to determine the resource potential for each season. Intra-annual variation can have a significant effect on the average PD, which can be ten times higher in winter than in summer and autumn, [15], implying a large change in, say, the monthly *CF*. Other similar models also were used to estimate resource potential for HKTs such as CCHE2D by Toniolo [10]. Based on the CCHE2D model, Duvoy and Toniolo [16] developed the first tool for river resource assessment to predict the instantaneous PD and the location of the maximum velocity. The module, called HYDROKAL, Table 2.2, was tested for the Tanana River near Nenana, Alaska which was found to have a maximum PD of 13,500 W/m^2 and an average PD of 6,500 W/m^2. A 2D numerical model, MIKE21, Table 2.2, was used by Lalander and Leijon [17] to compare the simulated velocity with measured data to map the HKT resource. In order to do that, the current velocity was sampled with an ADCP in a Swedish river over 1 month. The numerical models give a good correlation between measured hourly averaged ADCP data with a correlation coefficient, R^2, of 0.94. A TELEMAC-2D numerical model—see Table 2.2—was used by Lalander [18] to estimate the velocity distribution in a river measured using an ADCP. However, the measurements and simulations showed significant differences. The model used to estimate the velocity was shown to give a large error due to the fact that the velocity in the middle of the river was much higher than the cross-sectional average velocity. It was recommended that measurements be made of the cross-sectional velocity variations at several sections.

HKTs can be installed the downstream of conventional hydro-power plants to take advantage of the remaining kinetic energy, da Silva Holanda *et al.* [19]. The assessment was made using H2D2 listed in Table 2.2. To validate the numerical model, field surveys were conducted to measure the velocities by ADCP in 2012 and 2013. The model reasonably simulated the observed velocities with a minimum error of 0.83%, an average error of 8.77%, and a maximum error of 22%. HKT resource potential of a lowland river in Lithuania was assessed by Punys *et al.* [6] using a hydrological and hydraulic model to predict the PD. Stream gauging stations records (river stage, flow, and velocity area) were used. The one-dimensional hydraulic software HEC-RAS 4.1 (see Table 2.2) was used with the GIS extension GeoRAS to perform river calculations. A high-resolution DEM of the river was employed. This study concluded that to assess the PD, a 1D hydraulic model is sufficient. Compared to hydrological method, hydraulic modeling is more detailed and is useful for assessing the PD at cross-sections spaced along the entire length of the river. This requires, however, channel bathymetry in digital form or sufficient measured cross sections. Ladokun *et al.* [20] made a regional scale assessment of the PD potentials of some rivers in the lower Niger river basin. The methodology employed was hydrological model and a spatial tool MWSWAT, an open-source interface to SWAT using the GIS system, Table 2.2. From the hydrological model, dos Santos *et al.* [13] estimated the PD using CFD in the form of ANSYS CFX, and bathymetric measurements from a multifrequency echo sounder. ADCP measurements to find the maximum velocity points showed the simulations to be accurate for the Jamari River in the Amazon river basin where the HKT potential estimated as 109.5 kW.

Despite the advanced techniques just described, most numerical models determine the mean velocity averaged over the cross-sectional area that is always less than the maximum velocity. Hence, using simulated results may significantly underestimate the PD at any river cross-section.

Tigabu *et al.* [9] performed a measurement campaign at Bahir Dar Abay River, in Ethiopia, where 2-min average water speeds were measured. The aim was to use an averaging time for speed that is comparable to that used in determining the power curve of HKTs. Reference [9] determined the relationship between $\overline{V}_T$ for $T = 2$ min to those for $T = 2$, 10, 30 min, 1, 2, and 5 h. They found that as T increased, the variation becomes more significant; for example, for $T = 5$ h, the difference reaches 17%.

2.4.3 Statistical methods

Assuming a relationship between discharge and flow, $\langle\overline{V}\rangle = f(\overline{Q})$, statistical methodology can be employed to estimate the functional dependence usually from limited measurements of $\langle\overline{V}\rangle$. This correlation then can be used to estimate the temporal variation of the HKTs potential for time periods for which Q has been measured but $\langle\overline{V}\rangle$ has not been measured. Using regression analysis, [19] estimated that $\langle\overline{V}\rangle = 0.0749\overline{Q}^{0.3137}$ but did not specify T. Punys *et al.* [6] found the relationships to be $\langle\overline{V}\rangle = 0.1125\overline{Q}^{0.382}$ for $T = 1$ year. Using curve fitting [9] determined the empirical relation as $\langle\overline{V}\rangle = 0.02\overline{Q}^{0.75}$ for different rivers for $T = 1$ day. A similar analysis was made for irrigation canals by [21] but they did not specify T.

2.4.4 Machine learning and artificial neural networks

The next frontier in the resource assessment of HKTs is the application of artificial intelligence (AI) and machine learning to the PD using time series data. AI has been used in hydrology for river flow forecasting, flood pattern forecasting, and water resources management. The techniques used in hydrology include artificial neural networks (ANNs), adaptive neural-based fuzzy inference system techniques, genetic programming models, and support vector machines. Using the coefficient of correlation (R^2) and other evaluation measures, Wang *et al.* [22] assessed the accuracy of AI methods in forecasting monthly Q. The training data of monthly Q from January 1953 to December 2004 were used, and validation was done using data from January 2000 to December 2004. The result illustrated that the AI methods are powerful tools to model the discharge time series and can give good prediction. For example, the monthly Q prediction using ANNs showed $R^2 = 0.92$ during training and 0.932 during validation against the observed Q.

Among AI techniques, ANNs have become successful river discharge prediction methods [23]. ANNs can predict long-term river Q, [24, 25], and daily and monthly river discharge, Cheng *et al.* [26], and hourly Q, Yaseen *et al.* [27]. To the authors' knowledge, however, the use of AI for HKT resource assessment has not been attempted. It is hoped that this deficiency is soon rectified.

2.4.5 Velocity variation in time

In rivers, V, Q, and the cross-section can vary on timescales from inter-annual down to minutes. The inter-annual or seasonal variations have already been shown to have significant effects on the estimation of PD. The estimation of the long-term average water velocity normally takes more than 20 years of data. Seasonal variations usually depend on changes in rainfall. Mostly the maximum velocity occurs in the rainy seasons; Table 2.3 illustrates the variation of monthly averaged water velocity for a river in a region of Ethiopia with a pronounced rainy season. Clearly, the variation in $\overline{V}$ cannot be represented by a single month of data.

V can also vary on shorter timescales. Averaging times less than that used to determine the HKT power curve are unimportant. In an unstandardized industry, however, the time periods used by different manufacturers or third parties to determine turbine power curves are likely to vary. This variation is unhelpful unless there is a range of averaging times that do not alter the power curve as well as $\overline{V}$. If so, then

$$\frac{\overline{P}(T_2)}{A_R} \approx \frac{\rho}{2} \int_0^{T_2} \overline{V}^3_{T_1}(t)dt \tag{2.7}$$

where T_2 is the time interval for determining PD that must be greater than that used for determining the power curve, T_1. Figure 2.7 illustrates the time variation of $\overline{V}$ determined by averaging over 2 min for a total period of 90 min by [9]. Since (2.7) is in terms of the cube of a velocity that changes with t, then PD will depend on the mean square of the fluctuations in $\overline{V}$ averaged over T_1. In other words, the PD is affected by the turbulence in the river on timescales greater than that used to determine the power curve.

Turbulence in any fluid flow is usually measured by the turbulence level. The instantaneous velocity (V) is defined as the sum of a mean $\overline{V}$ and velocity fluctuation, v_i:

$$V = \overline{V} + v \tag{2.8}$$

The turbulence level is defined as

$$I_u = \frac{\sigma}{\overline{V}} = \frac{\sqrt{\overline{v^2}}}{\overline{V}} \tag{2.9}$$

where σ is the standard deviation of the velocity fluctuations. The importance of turbulence for PD estimation is difficult to assess. Making the assumption that C_P

Table 2.3 Variation of monthly average water velocity $\langle\overline{V_m}\rangle$ [9]

Month	Jan	Feb	Mar	Apr	May	Jun	Jul	Aug	Sep	Oct	Nov	Dec
$\langle\overline{V_m}\rangle$ (m/s)	1.04	0.92	0.85	0.84	1.09	1.61	1.97	2.12	1.92	1.45	1.17	0.99

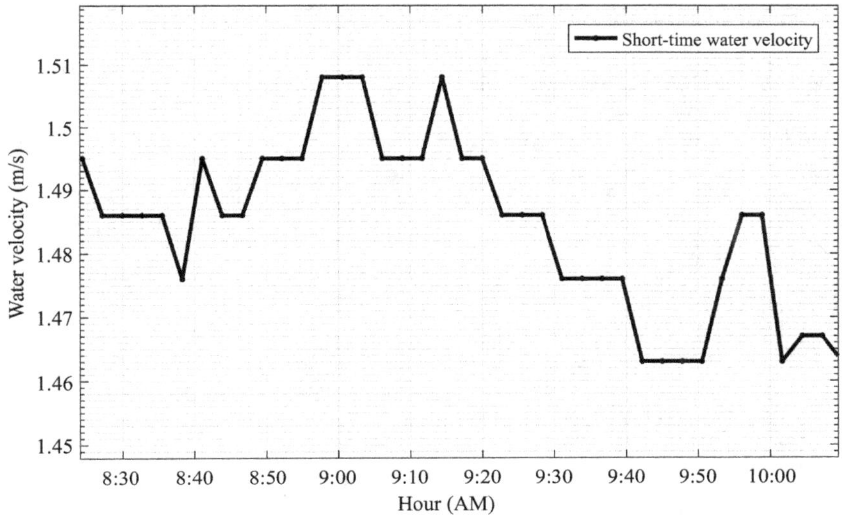

Figure 2.7 Short-term variation of water velocity. Figure adapted with permission from [9], Elsevier

remains constant and enforcing $T_2 \gg T_1$, the PD determined by averaging over the same time, $\overline{PD}_{T_2}$, is derived in [5] as

$$\overline{PD} = \frac{1}{2}\rho\left(1 + 3I_u^2\right)\overline{V}^3 \tag{2.10}$$

where the subscript T_2 is omitted from both sides. The *increase* in PD from the term $3I_u^2$ can be substantial and there is evidence from small wind turbine field tests that it can occur in practice. An increase in PD is reasonable on the grounds that it represents extra kinetic energy in the river, but the success in extracting it depends on the turbine design, if the experience with wind turbines is a guide.

The determination of v is fundamental to I_u, and $\overline{PD}$. However, there is a difficulty in obtaining meaningful variability of v even if advanced measurement devices such as ADCPs are used. Alternatively, analytical and numerical approaches can be used to estimate, say, $\overline{v^2}$. Such estimates were compared by Nezu [28] with measurement who suggested that analytical and numerical models give useful first-order approximations of say $\overline{v^2}$. For example, [5] gives $\sqrt{\overline{v^2}}/u_* = 2.3\exp(-z/D)$, where the friction velocity $u_* = \sqrt{\tau/\rho}$, where ρ is the shear stress on the bed, D is depth, and z is the height above the bed.

2.4.6 Variation in cross-sectional area

Maximizing energy production form an HKT requires placing it at the location of maximum V. However, $V = 0$ at the solid boundaries and usually reaches a maximum along the cross-section of a midplane. For HKTs whose swept area is much less than

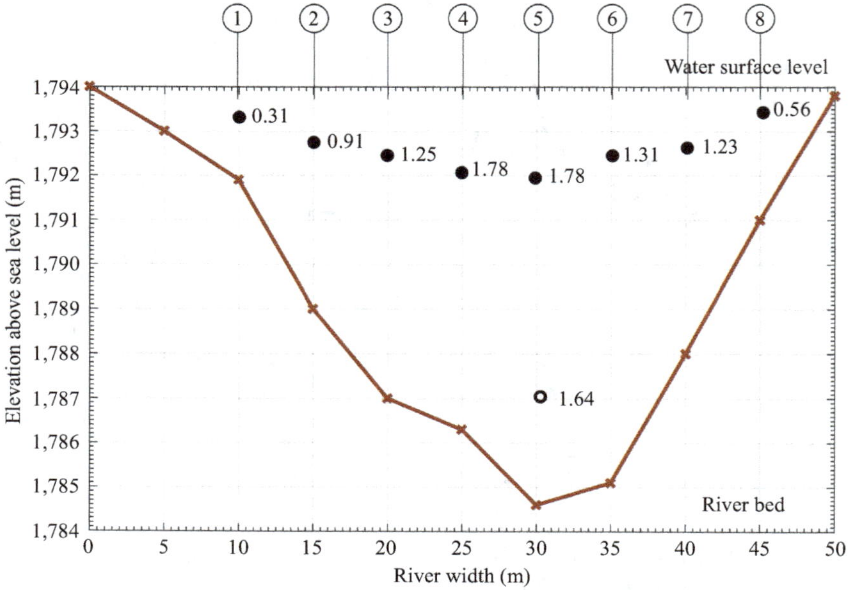

Figure 2.8 Cross-sectional area variation of water velocity (in m/s) [9]

the width or depth of the river cross-section, the spatial variation of $\overline{V}$ must be known to find the possible maximum *PD*. To illustrate this, the measured V of a river by [9] at eight equispaced locations at a typical cross-section is shown in Figure 2.8. The measurements of V were taken at $0.2d$, where d is measured from the upper water surface level to the river bed, and the solid circles indicate the measurement position. Another measurement at $\sim 0.8d$ across the midplane section (open circle) was taken to show the variation of V in the vertical direction.

Clearly, the figure demonstrates significant variation in PD in the cross-sectional area. How that variation changes with time and how it effects HKT performance when the swept area is comparable to the river cross-sectional area are issues that have only begun to be addressed.

2.5 Conclusion

The performance of an HKT at a particular site depends on its nonlinear power curve in combination with the pdf of the water speed. This chapter considers only the simplest case of a single turbine that is small compared to the typical dimension of the river cross-section. In other words, we do not consider large and/or multiple turbines that could interact with each other. Even in the simplest case, there are two main obstacles to accurate prediction of power output: the uncertain and non-standard methodology of power curve determination and the paucity of detailed data on the probability distribution of water speed. The two obstacles interact when the time used to average the water speed becomes large compared to the time used to bin the power curve data.

This chapter surveys the available resource assessment tools, ranging from very simple models that turn the discharge or flow rate, Q, into estimates of the available PD that we showed is only a function of the water speed V. In the absence of detailed information on river cross-sectional geometry, there is no universal relation between Q and V that is unfortunate given that Q may be the only long-term data available for a river. Further, the relation can only hold for some average measure of V whereas what is needed at any location is the maximum velocity. More detailed hydrological and hydraulic models were discussed. These are more accurate in principle than postulating a relationship between V and Q, but there is very little data available on the power output from installed HKTs to make definite statements. Fortunately, many of these complex computational tools are freely available, and this may well lead to improvements in performance prediction. Methods based on machine learning have started to be applied to resource prediction but there is much to be done.

The chapter concludes with some simple statements about the magnitude of turbulence and its significance for the performance of an HKT. It is telling that in writing our review, we did not find any actual performance data for an installed HKT that could be used to validate a resource assessment. This is an area where much work needs doing.

References

[1] Khan MJ, Iqbal MT, and Quaicoe JE. River current energy conversion systems: Progress, prospects and challenges. *Renewable and Sustainable Energy Reviews*. 2008;12(8):2177–2177. Available from: www.sciencedirect.com/science/article/pii/S136403210700069X.

[2] Zhou D and Deng ZD. Ultra-low-head hydroelectric technology: A review. *Renewable and Sustainable Energy Reviews*. 2017;78:23–30. Available from: www.sciencedirect.com/science/article/pii/S1364032117305968.

[3] Suwa Y and Masud IA. 4-6-1 Design of a zero head hydro-turbine for energy harvesting in Nigeria. In: *Proceedings of the Annual Conference of The Japan Institute of Energy*. vol.27. The Japan Institute of Energy; 2018. p. 172–173.

[4] Okulov V, Naumov I, Kabardin I, *et al.* Experiments on line arrays of horizontal-axis hydroturbines. *Renewable Energy*. 2021;163:15–21.

[5] Neary VS, Gunawan B, and Sale DC. Turbulent inflow characteristics for hydrokinetic energy conversion in rivers. *Renewable and Sustainable Energy Reviews*. 2013;26:437–445. Available from: www.sciencedirect.com/science/article/pii/S1364032113003365.

[6] Punys P, Adamonyte I, Kvaraciejus A, *et al.* Riverine hydrokinetic resource assessment. A case study of a lowland river in Lithuania. *Renewable and Sustainable Energy Reviews*. 2015;50:643–652. Available from: www.sciencedirect.com/science/article/pii/S1364032115004256.

[7] Lalander E, Grabbe M, and Leijon M. On the velocity distribution for hydrokinetic energy conversion from tidal currents and rivers. *Journal of*

Renewable and Sustainable Energy. 2013;5(2):023115. Available from: https://doi.org/10.1063/1.4795398.

[8] Fleming S and Gupta H. The physics of river prediction. *Physics Today*. 2020;73(7):46–46.

[9] Tigabu MT, Wood DH, and Admasu BT. Resource assessment for hydrokinetic turbines in Ethiopian rivers and irrigation canals. *Energy for Sustainable Development*. 2020;58:209–224. Available from: www.sciencedirect.com/science/article/pii/S0973082620302866.

[10] Toniolo H. Hydrokinetic assessment of the Kvichak River near Igiugig, Alaska, using a two-dimensional hydrodynamic model. *Energy and Power Engineering*. 2012;4(06):422. Available from: www.SciRP.org/journal/epe.

[11] Korkovelos A, Mentis D, Siyal SH, *et al.* A geospatial assessment of small-scale hydropower potential in sub-Saharan Africa. *Energies*. 2018;11(11). Available from: www.mdpi.com/1996-1073/11/11/3100.

[12] Magaju D, Cattapan A, and Franca M. Identification of run-of-river hydropower investments in data scarce regions using global data. *Energy for Sustainable Development*. 2020;58:30–41. Available from: www.sciencedirect.com/science/article/pii/S0973082620302623.

[13] dos Santos IFS, Camacho RGR, Tiago Filho GL, *et al.* Energy potential and economic analysis of hydrokinetic turbines implementation in rivers: An approach using numerical predictions (CFD) and experimental data. *Renewable Energy*. 2019;143:648–662. Available from: www.sciencedirect.com/science/article/pii/S0960148119306706.

[14] Spasojevic M and Holly F. Two- and three-dimensional numerical simulation of mobile-bed hydrodynamics and sedimentation. *Sedimentation Engineering: Processes, Measurements, Modeling, and Practice*. 2008:683–761.

[15] Fouz DM, Carballo R, Ramos V, *et al.* Hydrokinetic energy exploitation under combined river and tidal flow. *Renewable Energy*. 2019;143:558–568. Available from: www.sciencedirect.com/science/article/pii/S0960148119306949.

[16] Duvoy P and Toniolo H. HYDROKAL: A module for in-stream hydrokinetic resource assessment. *Computers and Geosciences*. 2012;39:171–181. Available from: www.sciencedirect.com/science/article/pii/S009830041100207X.

[17] Lalander E and Leijon M. Numerical modeling of a river site for in-stream energy converters. In: *Proceedings of the 8th European Wave and Tidal Energy Conference, EWTEC09*, Uppsala, Sweden. vol. 710; 2009. p. 826832.

[18] Lalander E. Hydrokinetic Resource Assessment: Measurements and Models. Uppsala University; 2013.

[19] da Silva Holanda P, Blanco CJC, Mesquita ALA, *et al.* Assessment of hydrokinetic energy resources downstream of hydropower plants. *Renewable Energy*. 2017;101:1203–1214. Available from: www.sciencedirect.com/science/article/pii/S0960148116308758.

[20] Ladokun L, Ajao K, and Sule B. Regional scale assessment of the gross hydrokinetic energy potentials of some rivers in Lower Niger River Basin, Nigeria. *Nigerian Journal of Technology*. 2015;34(2):421–428.

[21] Botto A, Claps P, Ganora D, *et al.* Regional-scale assessment of energy potential from hydrokinetic turbines used in irrigation channels. In: *SEEP 2010 Conference Proceedings*, Bari, Italy; 2010.

[22] Wang WC, Chau KW, Cheng CT, *et al.* A comparison of performance of several artificial intelligence methods for forecasting monthly discharge time series. *Journal of Hydrology*. 2009;374(3):294–294. Available from: www.sciencedirect.com/science/article/pii/S0022169409003527.

[23] Kai Feng Z and jing Niu W. Hybrid artificial neural network and cooperation search algorithm for nonlinear river flow time series forecasting in humid and semi-humid regions. *Knowledge-Based Systems*. 2021;211:106580. Available from: www.sciencedirect.com/science/article/pii/S0950705120307097.

[24] Tan QF, Lei XH, Wang X, *et al.* An adaptive middle and long-term runoff forecast model using EEMD-ANN hybrid approach. *Journal of Hydrology*. 2018;567:767–780. Available from: www.sciencedirect.com/science/article/pii/S0022169418300155.

[25] Elsafi SH. Artificial neural networks (ANNs) for flood forecasting at Dongola Station in the River Nile, Sudan. *Alexandria Engineering Journal*. 2014;53 (3):655–655. Available from: www.sciencedirect.com/science/article/pii/S1110016814000660.

[26] Cheng M, Fang F, Kinouchi T, *et al.* Long lead-time daily and monthly streamflow forecasting using machine learning methods. *Journal of Hydrology*. 2020;590:125376. Available from: www.sciencedirect.com/science/article/pii/S0022169420308362.

[27] Yaseen ZM, Naganna SR, Sa'adi Z, *et al.* Hourly river flow forecasting: Application of emotional neural network versus multiple machine learning paradigms. *Water Resources Management*. 2020;34(3):1075–1091. Available from: https://doi.org/10.1007/s11269-020-02484-w.

[28] Nezu I. Turbulence in Open-Channel Flows. IAHR-Monograph; 1993. Available from: https://ci.nii.ac.jp/naid/10003099278/en/.

Chapter 3
Small hydrokinetic turbines

Brian Kirke[1]

3.1 Introduction

Electrical energy can be produced from the kinetic energy of flowing water in tides, rivers and potentially ocean currents. Tidal energy can be harnessed either from the potential energy in tidal rise and fall, or from the kinetic energy in flowing water by means of hydrokinetic turbines (HKTs). The energy in rivers can also be harnessed either from potential energy (conventional hydro) or from the kinetic energy in flowing water (hydrokinetic energy). This chapter deals with small-scale (up to 50-kW rated capacity) HKTs designed to extract energy from rivers, tides and possibly ocean currents with minimal drop in surface elevation. It covers much of the same material as [1], a review of HKTs for moderate-sized rivers, but tidal options are also included in the present work. There are few tidal sites requiring less than 50 kW, but arrays of small turbines have been proposed to generate the large output required for grid-connected sites.

HKTs which capture kinetic energy from flowing water rather than potential energy from a drop in surface level have attracted increasing attention in recent years because they offer the promise of tidal, ocean current and river energy with minimal environmental impact, unlike tidal barrages which can cause environmental problems and conventional hydropower which involves damming a river, which may inundate fertile river valleys, displace communities and deprive downstream users of water, fish and river transport. Much has been written about the design of HKTs and the literature abounds with reviews, for example [2–5]. Kirke and Coiro [2] included tidal rise and fall as well as hydrokinetic schemes. The study of Laws and Epps [3], published in 2016, cites cost reduction as the first barrier to development. It reviews 'the cutting-edge research across all steps of project development'. It lists 163 references and is focused mainly on large-scale development. The study of Niebuhr *et al.* [4] reviews small-scale technology, with particular focus on 'enhancement', i.e., increasing flow velocity. It also lists cost reduction as the first barrier to development. Of those companies that reach the

[1]STEM (Science, Technology, Engineering and Mathematics Unit), University of South Australia, Adelaide, Australia

prototype or demonstration stage, many fail, possibly because they are not affordable for potential users or are too expensive to compete with other options. According to Dr Barbara Sexon of Thropton Energy Services (pers. comm.), 'machines made in Europe and USA or Canada are likely to be too expensive for the people who really need them'. And according to Voytek Klaptocz of Mavi Innovations in Canada, who developed an HKT but decided it was not a financially viable business to be in, 'the greatest challenge to this sector is not technical but financial' (pers. comm.). This chapter reviews existing and proposed technologies, sites and reasons for success or failure.

3.2 The need/potential market

Although small HKTs will never have a big impact on worldwide energy supplies, they have applications for electrical power supply for many small, remote off-grid communities around the world where it is difficult to maintain diesel fuel supplies, where the climate may not favour solar or wind, and where there is not enough elevation for conventional micro-hydro. Examples are shown in Figure 3.1, after [6], and Figure 3.2.

Compared with wind and solar, tidal power has the advantage of predictability. Tidal flows vary with lunar cycles while rivers flow 24/7, although they may dwindle or cease completely in the dry season in some places. Besides rivers and tidal flows, there are also potential niche applications such as canals, outflows from thermal power stations and wastewater treatment plants, and sites downstream of hydro dams where flow is fairly constant and local villages may not benefit from the power generated.

For widespread adoption, they must be affordable, reliable and easily deployed. Few of many products have achieved these aims, with the possible

Figure 3.1 A typical village in Sarawak, located right next to a river [6]

Figure 3.2 A small eco-resort in western Canada with a narrow, fast tidal flow (photo by the author)

exception of the Garman turbine developed by Thropton Energy Services [7] for relatively deep rivers such as the Nile, and small turbines developed by Smart Hydro [8] for higher velocity sites.

3.3 Potential problems with HKTs

Possible problems with small HKTs include

- capital cost and capacity factor;
- low power density due to low velocity at sites where power is needed;
- shallow water where flow is fast;
- rocky stream bed reducing available space for turbine;
- the need to provide for boat passage in some rivers;
- floating debris clogging turbines;
- damage from floating logs;
- unexpectedly high loading from eddies and flash floods;
- difficulty deploying and retrieving turbines.

According to Clayton Bear, President, CEO and Chief Technology Officer of New Energy Corporation, 'The issue we face with the remote market is not with the cost of the hardware, it is more related to costs associated with deployment and more importantly after sales service and support. In order to install such systems in a remote location a whole regional support infrastructure must be put in place'.

Where an installation cannot be justified on purely economic grounds, large expenditure might be justified for any of a number of reasons, as follows:

1. A means of showcasing an organisation's green credentials, for example an eco-resort.
2. A one-off installation in extreme conditions where there is no viable alternative.

Figure 3.3 The ORPC (Ocean Renewable Power company) 22-kW submersible turbine

3. A prototype from which a much lower cost production model might be developed.
4. A prototype with potential for scaling up, like the 50-kW wind turbines that were standard in the 1980s, from which multi-megawatt turbines have evolved.
5. Expensive, one-off units have been demonstrated in a few places where there are strong river or tidal currents and commercial viability may not be a priority, including First Nations communities in Alaska and Canada, and luxury eco-resorts in Canada. However, the question must be asked as to whether there are enough such sites to justify tooling up for a production run which might bring down unit costs.

3.4 Differences between hydrokinetic and wind energy

The basic concept of hydrokinetic energy may appear the same as that of wind turbines, i.e. capturing kinetic energy from a moving fluid, but there are major differences, and these can favour turbine designs radically different from the conventional axial flow turbine, as discussed next.

3.4.1 Limited flow depth

Although depths at the best tidal sites such as the Pentland Firth off the north coast of Scotland may exceed 40 m, few river sites are deep enough and flow fast enough for multi-kW turbines. One exception is the Kvichak River in Alaska where a 22-kW turbine (Figure 3.3) has been installed by ORPC, the Ocean Renewable Power company [9] to supply the community of Igiugig, but most river sites where flow velocities are high enough to be viable have depths of only 1 or 2 m [1,10]. This limits the diameter of axial flow turbine rotors.

3.4.2 Tidal and river flow velocities are lower than wind

Tidal flow velocities are much lower than wind speeds, seldom exceeding about 3–4 m/s, and few rivers flow at more than 2 m/s, while wind turbines are typically designed to develop rated power at wind speeds ranging from 12 to 17 m/s, i.e. typically about four times that of tidal currents. Power is basically the product of force and change of velocity across the turbine. Therefore, the lower the flow velocity, the greater the force necessary to produce a given power output, so for the same power output, the force on an HKT would be typically about four times that on a wind turbine, and for this reason most HKTs are designed to develop rated power in flows of around 3 m/s, although these are almost never encountered in rivers – except perhaps during floods – and off-grid tidal sites are rare.

3.4.3 Ocean current flow velocities are even lower

Although velocities are lower, ocean currents contain enormous amounts of energy. The Gulf Stream off Eastern Florida is among the fastest ocean currents on the planet, but according to [11], the average ocean current speed is about 1.5 m/s (presumably in the most favourable location). Even in the fastest tidal flows such as the Pentland Firth and the Bay of Fundy, tidal HKTs are at present far from commercial viability, so it may be a long time before costs come down to the point where conventional HKTs are viable in ocean currents.

3.4.4 Much higher fluid density

The density of water is about 830 times higher than that of air (wind), so even with much lower velocities, power densities are typically higher.

3.4.5 Cavitation

Unlike wind turbine blades, HKT blade velocities are limited by the danger of cavitation if the cavitation number σ is below 1. $\sigma = (P_T - P_V)/(1/2\rho V^2)$ where P_T is the pressure at the depth of the turbine, P_V is the vapour pressure, ρ is the water density and V is the velocity of the water relative to the blade [12]. For practical purposes, this means that blade tip speed should be kept below about 12 m/s.

3.4.6 Predictability

Unlike wind, tidal flow velocities are generally predictable, but this is not always the case, especially where the bathymetry is irregular and flows are turbulent. For example, a 1-MW Open Hydro-turbine failed in the Bay of Fundy in 2009 due to much higher than expected maximum current velocities [13]: 'all 12 turbine rotor blades were destroyed by tidal flows that were two and a half times stronger than for what the turbine was designed'.

3.4.7 Unidirectional and bidirectional flow

Tidal flow, unlike wind, is bidirectional at most but not all sites. According to [14], writing of the Pentland Firth, 'ebb and flood streams did not fully overlap, and ... the tidal streams were more complicated, turbulent, and variable than existing models suggest'. River flow is always in the same direction, although floods may change the course of some rivers to some extent.

3.4.8 Limited flow field: blockage

Wind turbines operate in what is effectively an infinite flow field, and according to Betz's law, the upper limit of power extraction for a turbine in open flow is 16/27 or 59.3% of the kinetic energy flux in a flow area equal to the swept area of the turbine. A turbine's performance coefficient C_P is the ratio of actual power delivered to the kinetic energy flux. But at some tidal and most river sites, the channel cross-sectional area is limited, so that an array of turbines may block a significant portion of the flow area. In these cases the assumptions on which the Betz (also known as Lanchester–Betz) Limit is based no longer apply and much more power can be extracted by utilising a small drop in level across turbines, i.e. some potential energy is harvested. This fact and its implications are discussed in Section 3.11.

3.5 Turbine types

Most HKTs are axial flow, similar to scaled-down wind turbines, or based on the Darrieus crossflow principle. A few Savonius rotor HKTs have been built, some waterwheels, and some experimental systems using vortex shedding or flapping blades have been demonstrated at laboratory level. Figure 3.4 shows a Schottel axial flow HKT and a New Energy vertical axis Darrieus HKT, with the term 'swept area' illustrated.

3.5.1 Axial flow turbines

The three-blade axial flow turbine on a monopole support structure dominates wind turbine technology, and HKT technology has until recently followed suit. One of

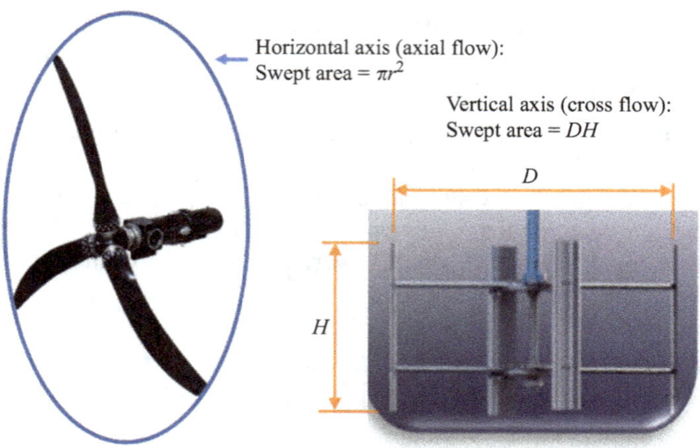

Figure 3.4 A Schottel axial flow HKT [15] and a New Energy vertical axis Darrieus HKT [16]

the earliest HKTs was the Garman inclined axis turbine, shown in Figure 3.5 being manufactured and assembled in Sudan. These were designed initially for pumping water for irrigation or community supply, but also for generating electricity. The long drive shaft allowed the generator to be mounted above water, typically on a boat or raft, with the turbine trailing downstream. The rotor and one bearing were the only components below the water level.

Thropton Energy Services [7] developed a range of turbines for local manufacture with the emphasis on local self-reliance and affordability. Sizes range from 2.2- to 4-m diameter, able to operate in depths of at least 1.75 m/s and a design flow velocity of 0.9 m/s and operating flow range of 0.5–2 m/s. According to their website, they have 20-year experience in this field and have worked in the United Kingdom, Sudan, Somalia, Egypt, Colombia and Peru.

The initial development of these turbines and their use was reported in Peter Garman's book 'Water Current Turbines: a Fieldworkers' Guide' ISBN 0 946688 27 3, available from Practical Action Publishing. Thropton Energy has produced a complete set of engineering drawings and a detailed manual covering assembly, installation and maintenance of the Garman Turbine which can be shared with organisation's wishing to produce the technology, subject to contract.

Figure 3.6 shows another affordability initiative – a low-cost axial flow HKT developed at the University of Malaysia, Sarawak Campus (UNIMAS), using a 0.6-m diameter two-blade fan costing \$20 as a turbine and mass-produced aquaculture aerator floats also costing \$20 each, with the rotor mounted under the floats and a two-stage belt and bicycle chain drive to the generator.

Figure 3.5 A Garman HKT in Sudan (photo by Giles Stacey, supplied courtesy Dr B. Sexon, Thropton Energy Services)

Figure 3.6 A small axial flow turbine developed at the University of Malaysia, Sarawak Campus (UNIMAS) (photo courtesy M. Anyi, UNIMAS)

This turbine produced about 92 W in a 1.26-m/s flow, equivalent to 2.2 kWh per day, with a water-to-wire efficiency of 34%. With battery storage, this would meet the service level of 2 kWh per day and 700-W peak demand deemed sufficient for a typical rural household in Sarawak [17], but where applications need more power, multiple small turbines like this could be deployed side by side, perhaps with low-speed generators, inclined shafts and direct drive like the Garman turbine shown in Figure 3.5 to eliminate the two-stage belt and chain drive. Turbines could be mounted either on a common platform with a common generator, or as autonomous units so that failure of one turbine due to flash floods would not bring the whole system down.

Figure 3.7 shows a Mako HKT. The Mako tidal turbine business is being restructured under a new ownership and branding but will continue technology development in Australia with turbines 2–3-m diameter and priced at A $20,000–70,000 with outputs of 5–15 kW each, designed to be linked together in arrays to maximise electricity production at a site. The company will focus on opportunities in Australia, the United Kingdom/Europe and the USA then other markets (Douglas Hunt, CEO, pers. comm., 13 February 2021).

Figure 3.8 shows a 1-m diameter ducted axial flow Smart Hydro river HKT from Germany. It is claimed [8] that more than 40 of these units have been sold, making it probably the most successful river turbine so far produced. However, it weighs 380 kg, making it difficult to launch, as shown, requires 2.8-m/s current

Figure 3.7 A Mako turbine [18]

Figure 3.8 A 1-m diameter ducted Smart Hydro HKT [8]

velocity to achieve its 5-kW rated power output and costs $16,400, so it would probably have to be subsidised to be affordable in the Third World countries.

Although axial flow HKTs currently (2021) dominate the field, it is by no means clear that this design is inherently better than any other. For optimum performance an axial flow turbine must yaw to follow reversing tidal flow direction, and blades must be tapered and twisted, and blade geometry can be optimised for only one tip-speed ratio (the ratio of blade tip speed to ambient flow velocity), so if the flow velocity decreases the turbine, RPM should ideally decrease in proportion. The blades are cantilevered from the hub, so they are subject to very high-bending moments and the inner part of each blade near the hub is mainly structural, contributing little to power output. The generator is located under water, making it vulnerable to leakage and hard to access for maintenance.

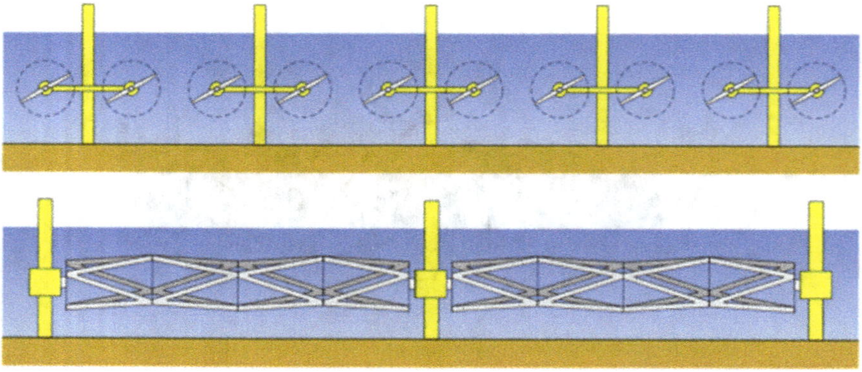

Figure 3.9 Multiple axial flow turbines contrasted with long horizontal axis crossflow turbines [19]

3.5.2 Crossflow turbines

These may be either the vertical axis as shown in Figure 3.4 or the horizontal axis as shown in Figures 3.3 and 3.9. The generator of a vertical axis HKT can easily be placed above water level where it is readily accessible, as shown in Figure 3.4, and being insensitive to flow direction, there is no need for a yaw mechanism to follow reversing tidal flows. Whether mounted with the axis vertical or horizontal, the rectangular swept area allows them to be mounted in close-packed arrays or as long, small diameter turbines. Figure 3.9, after [19], contrasts an array of axial flow turbines with a proposed long horizontal axis crossflow turbine design named 'THAWT' (transverse horizontal axis water turbine).

3.5.2.1 Vertical axis Darrieus turbines

Figure 3.4 shows a Canadian New Energy Corporation [16] vertical axis Darrieus HKT, and Figure 3.10 shows a 25-kW model raised out of the water. This company started developing HKTs in 2003, starting with a 5-kW model and progressing to the 25-kW model shown, with plans for larger sizes up to 250 kW, all based on the same vertical axis configuration. Despite a great deal of testing, demonstration and product refinement, it appears that there have been few actual installations, possibly because of cost barriers.

3.5.2.2 Horizontal axis Darrieus turbines

Although most Darrieus HKTs are mounted with an axis vertical like their wind turbine ancestors, and this enables direct drive to a generator above water level, a horizontal axis orientation has the advantage that a wide swept area can be intercepted in a shallow flow, at least in reversing tides and unidirectional river flows (unlike omnidirectional wind). An example of a horizontal axis Darrieus HKT is shown in Figure 3.11, from Idénergie [20].

The Idénergie turbine is basically a pair of small Darrieus turbines driving a generator directly. It costs about $10,000, can operate in 0.6-m depth and 1–3-m/s

Figure 3.10 A 25-kW New Energy Corp vertical axis Darrieus HKT in Canada, raised out of the water

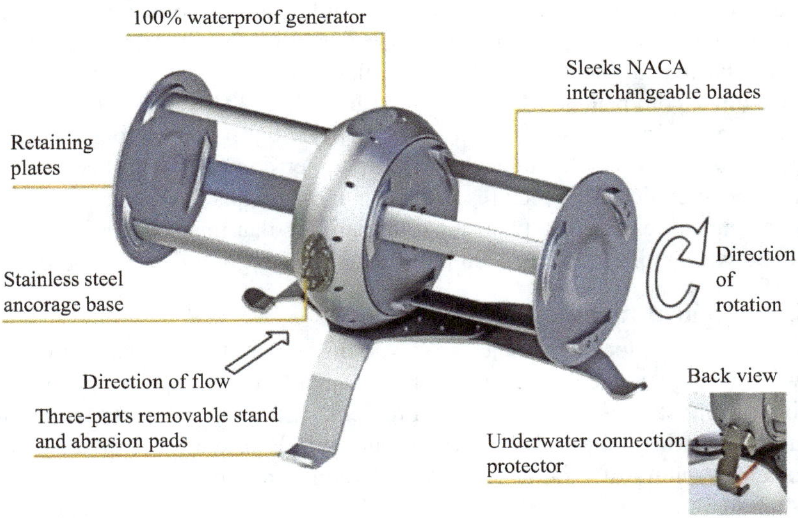

Figure 3.11 The Idénergie turbine

flow, and is claimed to be able to generate 'up to 12 kWh per day' – so rated power is presumably 500 W in 3 m/s flow. According to a 2016 post on their website, they were '*currently actively seeking partners to assist ... with ... commercialization efforts*'. The only installations mentioned were financed by the Build in Canada Innovation Program and at the time of writing (February 2021), there has been no news on their website since 2017, so it has apparently not been a commercial success.

Figure 3.12 The New Energetics Inc. turbine

The 'orthogonal' turbine (Figure 3.12) is a variant on the Darrieus, with a long shaft and several short blades at different azimuth angles, in some ways resembling a discontinuous helical turbine in that torque ripple is reduced in comparison to the standard Darrieus turbines with only three blades. There is a lapsed patent https://patents.google.com/patent/US8007235, and until late 2020, there was a website, New Energetics Inc., complete with turbine prices, but it can no longer be found. The inventor, Professor Lyatkher, now has another company with a website [21] but with no specifications, pictures or prices, and attempts to contact him recently have been unsuccessful. The present author believes that this concept has potential for shallow river sites and has been building and testing low-cost prototypes.

3.5.2.3 Helical turbines

Helical (so-called Gorlov) turbines are a variant on the Darrieus. Normally mounted with an axis vertical, they have the advantage that torque ripple is nearly eliminated, but starting torque is still a problem. The ORPC [9] has built a 25-kW submersible horizontal axis helical HKT, shown in Figure 3.3, supplying the First Nations community of Igiugig in Alaska. Although millions of dollars have been spent on the development of this turbine [22], there is no mention of multiple units being built except for various prototype arrangements, a tidal version and a river version, and it is not clear what the unit cost of a production model would be.

3.5.2.4 Variable pitch Darrieus turbines

Disadvantages of the basic fixed pitch Darrieus turbine include a lack of starting torque, shaking and torque ripple. Helical blades can solve the torque ripple problem but cannot improve starting torque significantly. Variable pitch can improve all of these factors [23]. Numerous variable pitch systems have been proposed and a few prototypes have been demonstrated, but no products have been found on the market.

3.5.2.5 Savonius rotors

These share most of the advantages of Darrieus HKTs, and they have the advantages over Darrieus turbines of simplicity, high-starting torque and possible debris shedding, but they are essentially drag machines with high solidity, low efficiency and in their standard form they turn slowly, requiring a large step-up gear ratio to drive a generator. A standard Savonius rotor and a modified Benesh rotor are shown in Figure 3.13.

A detailed CFD optimisation of a Savonius rotor with twisted blades by Kumar and Saini [24] predicted a maximum efficiency C_P of 0.426, and an optimisation by Rahai [25] produced an impressive performance curve compared with the standard and Benesh rotors, as shown in Figure 3.14.

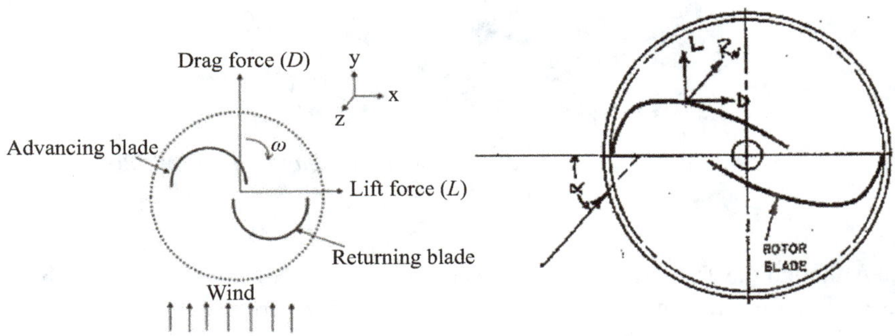

Figure 3.13 Left, standard Savonius rotor and right, modified Benesh rotor

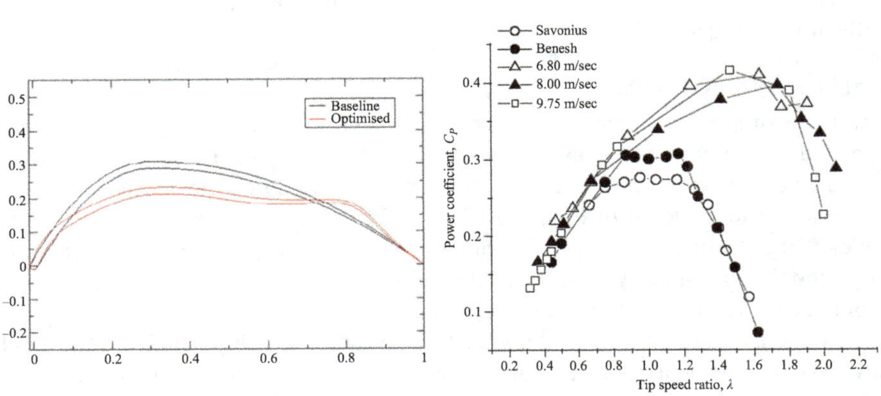

Figure 3.14 Rahai's optimised profile and performance compared with conventional Savonius and Benesh rotors

Figure 3.15 The EcoCinetic vertical axis Savonius rotor [26]

Although the TSR range is still low compared with axial and Darrieus turbines, these C_P figures are comparable with the best axial flow turbines, and if they can be verified independently, then this form of HKT will be well worth developing. However, attempts to contact Dr Rahai have been unsuccessful.

Two companies appear to be actively developing Savonius HKTs. The EcoCinetic turbine shown in Figure 3.15 is a vertical axis Savonius rotor, claimed to deliver 2 kW in a 3-m/s current [26].

The Waterotor [27] is a modified horizontal axis Savonius rotor with a deflector plate, shown in Figure 3.16.

3.5.2.6 Waterwheels

These are simple in principle but large and unwieldy in relation to the power produced, and they turn even more slowly than Savonius rotors, so they need a large step-up gear ratio to drive a generator. Some one-off units have been built, for example the unit shown in Figure 3.17. The Barsha Pump shown in Figure 3.18, developed at Delft University of Technology (Netherlands) by aQysta, which aims to provide sustainable hydro-powered pumping solutions that can have a positive economic, environmental and social impact.

The only evidence of waterwheel commercial development for electrical power is the Greenenergy Hydrocat shown in Figure 3.19. Several YouTube videos were found describing it, for example [30], and a pdf was also found, giving dimensions, prices and output, ranging from a 1.25-m diameter rotor 1.5-m wide, claimed to generate 183 W in 1-m/s current and costing €14,700 to a 40-kW model

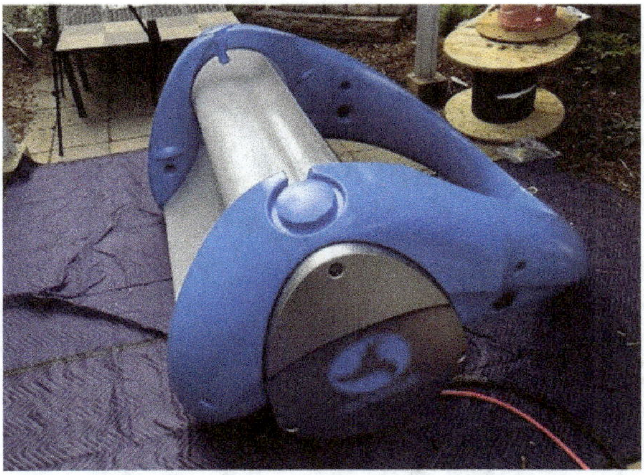

Figure 3.16 The Waterotor, a modified horizontal axis Savonius rotor with a deflector plate [27]

Figure 3.17 A floating waterwheel generator in the Congo [28]

costing €66,400. But the pdf has disappeared and there is no contact point for potential purchasers, so it must be assumed that these devices are not on the market.

3.5.3 Flying wing

A new concept being developed by Minesto [31] is shown in Figure 3.20. A wing 'flies' through the water like an underwater kite at speeds many times that of the current, so that a small, inexpensive turbine mounted on the wing experiences high-relative velocities and can generate correspondingly high power. These may offer better economics in deep, slow flows.

Figure 3.18 The Barsha Pump: a waterwheel driving a coil pump [29]

Figure 3.19 The Greenenergy Hydrocat [30]

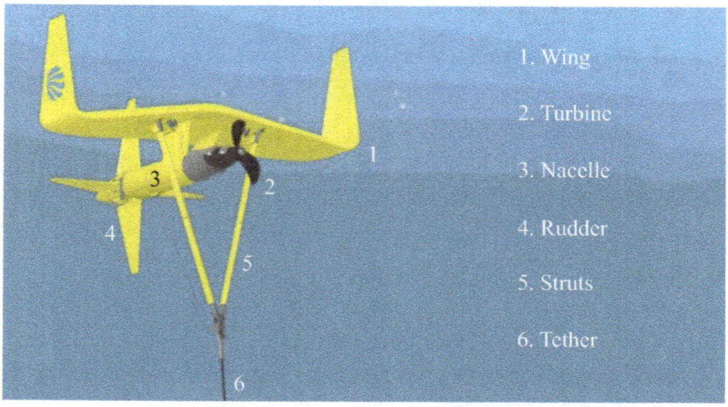

Figure 3.20 The Minesto flying wind turbine [30]

3.6 Ducts and diffusers

A duct around a turbine or a diffuser downstream of a turbine in open flow can increase its power output significantly, both by increasing the pressure drop across the turbine and by increasing the flow through it. This is often confused with what happens in a Venturi in a closed pipe, in which the velocity (and kinetic energy) increases and pressure decreases in the converging upstream section, while the reverse occurs in the diverging downstream section. Energy is simply converted from pressure to velocity and back again, with only a small loss of total energy provided it is designed correctly so there is no flow separation. To extract energy from a Venturi, there must be a pressure drop.

Rawlings [32] has shown a 73% power increase with a symmetrical two-way duct around a vertical axis Darrieus turbine, which if repeated across a channel as shown schematically in Figure 3.21 could serve also as structural columns supporting a tidal bridge.

Gilbert and Foreman [33] reported a power augmentation factor of 4.25 for a laboratory wind turbine with a diffuser, i.e. it produced 4.25 times as much power as the same turbine in open flow. They used slots in the diffuser to draw in high-energy flow from outside the diffuser for boundary layer control and were able to achieve this result with a short, wide-angle diffuser, a much more economical arrangement than the long diffusers studied by a number of other investigators. Tidal Energy P/L [34] applied the same principle to a vertical axis HKT and

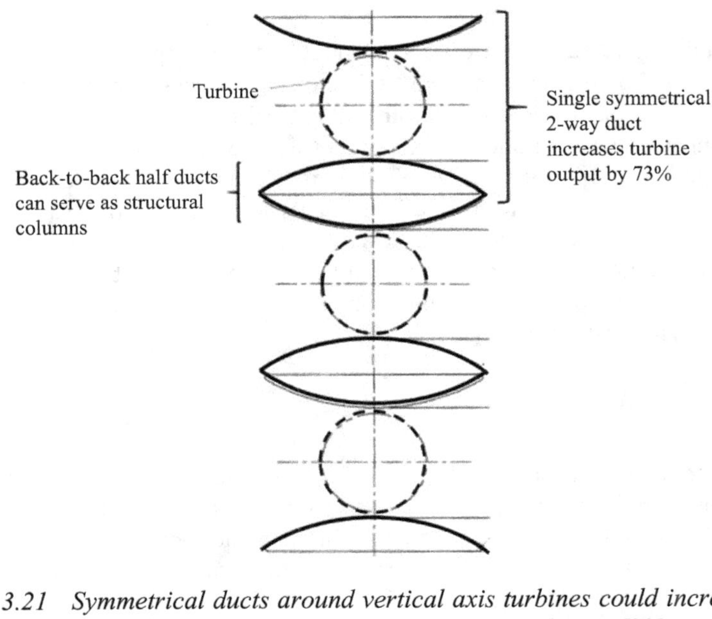

Figure 3.21 Symmetrical ducts around vertical axis turbines could increase power output while serving as structural columns [32]

Figure 3.22 A vertical axis turbine (yellow and green blades) in slatted duct, partially raised for easy viewing

measured approximately 3-fold increase in power with a rectangular slatted diffuser shown in Figure 3.22, from [35]. However, the increased load on the turbine and mounting structure and the cost of the duct or diffuser offset any gain in power output and it remains to be seen whether ducts and diffusers will improve cost-effectiveness.

3.7 Oscillating foils

As shown in Figure 3.23, these offer an alternative to rotating blades. According to [36], 'The maximum blade speed of the hydrofoil is several times lower than the blade tip speed of a conventional turbine at optimal operation, which may reduce harmful interactions with aquatic animals. In addition, the device is structurally more robust because it does not rely on the fast rotation of long blades. Since the hydrofoil can be designed to have a high aspect ratio, it is also advantageous in operation in shallow water channels'. C_P up to about 30% was reported but 'the hydrofoil was actuated by a linear motor for heaving motion and by a rotary stepper motor for pitching motion Passive pitching or heaving motion induced by the free stream was not considered'.

3.8 Vortex shedding

Another oscillating technology, the VIVACE Converter, uses forces exerted on a cylinder by vortex shedding. According to [37], multiple cylinders as shown in Figure 3.24 can work synergistically and achieve higher power-to-volume density than conventional HKTs and high-power conversion efficiency, estimated at 88.6% of the Betz limit (i.e. $C_P = 52\%$) at a flow speed of 1.08 m/s, which is superior to the peak efficiency of an isolated wind or tidal turbine.

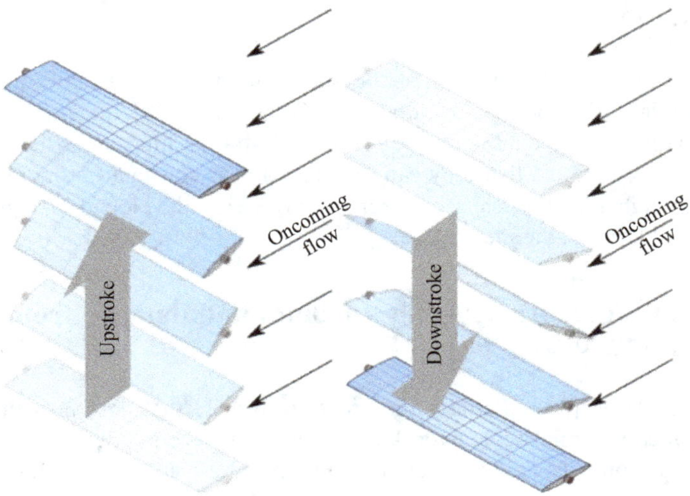

Figure 3.23 Oscillating hydrofoils [36]

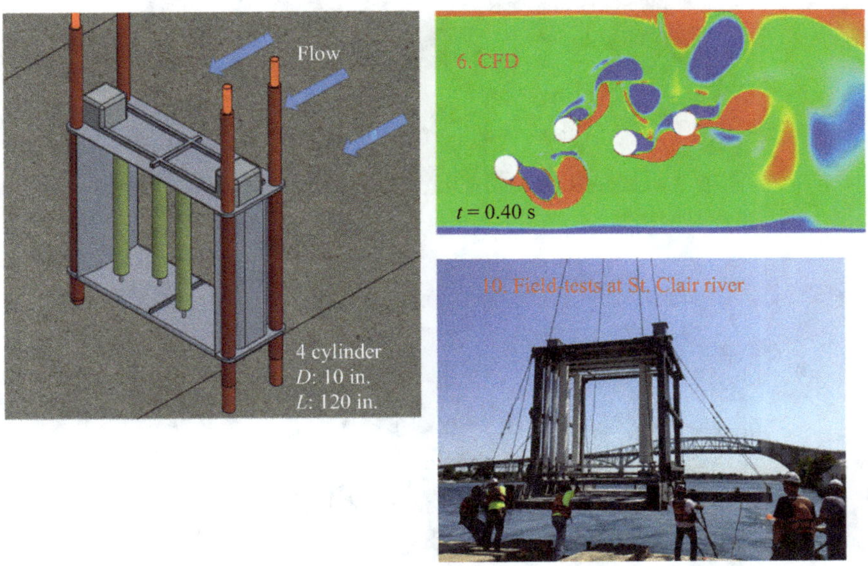

Figure 3.24 Left, 4-kW river model concept; upper right, CFD and lower right, field test [37]

3.9 Tidal sails

Another concept suitable for wide, shallow sites that uses linear rather than rotary motion is tidal sails, being developed by Tidal Sails [38], in which multiple hydrofoils attached to a continuous belt 'sail' in a straight line across the current as shown in Figure 3.25, so that they can maintain a near constant optimum angle of attack. Although the tidal sails concept is only being developed at a megawatt scale, it should work at a small scale.

3.10 Floating debris clogging turbines and damage from floating logs

According to Tyler [39], 'Perhaps the greatest obstacle that confronts the implementation of commercial-scale hydrokinetic devices in rivers is debris'. This can be a problem in some rivers, as shown in Figures 3.26 and 3.27. Mesh screens upstream of the turbine have been proposed, but are likely to simply trap debris,

Figure 3.25 Tidal sails [38]

Figure 3.26 Debris accumulation on a pontoon-mounted HKT in Alaska [39]

Figure 3.27 Weed caught on shaft and blade of a turbine

Figure 3.28 A blade of an axial flow turbine designed to swing back in the plane of rotation and shed debris whenever it is generating negative torque (M. Anyni, pers. comm.)

building up a barrier and reducing flow through the turbine. Smart Hydro [8] uses an array of parallel wires strung at an acute angle to the flow, upstream of the turbine, as shown in Figure 3.8. They also use swept back blades like those shown in Figure 3.6, which are claimed to prevent clogging by debris. One possible solution described by Anyi and Kirke [6] is to allow a blade of an axial flow turbine to swing back in the plane of rotation and shed debris whenever it is generating negative torque, as shown in Figure 3.28.

3.11 Blockage and water surface gradient

Water with a free surface will only flow if there is a gradient, either from tidal rise and fall or from a riverbed gradient. The flow velocity is limited by turbulence and bottom friction: the rougher the bottom the steeper the surface gradient for a given flow velocity, or conversely, the slower the flow for a given gradient. If a turbine or any other obstruction is placed in the flow, the effective resistance to flow is increased, the upstream level must increase, and the average flow velocity must decrease.

If the flow velocity through the turbine or array decreases, there must be a compensating increase in velocity somewhere else to maintain the same volumetric flow. This can be seen in Figure 3.29, a modelled representation of flow through a horizontal axis crossflow turbine after [40]. The flow velocity was reduced immediately downstream of the turbine, indicating some extraction of kinetic energy, but the velocity above and below the turbine and at the surface downstream was considerably greater and there was a small drop in the surface level downstream of the turbine, indicating that the *average* velocity and therefore the total kinetic energy flux had *increased*, so most of the power must have come from *potential* energy.

If the swept area of the turbines is small in relation to the total flow area, i.e. the blockage ratio is small, the increase in velocity around the turbine may appear negligible and the surface level far downstream may appear to be unchanged, leading to a widespread, but erroneous belief that HKTs extract energy from flowing water with no change in elevation. For example, [41] states that 'HKTs are zero head turbines' and [42] states that 'a river current device takes water kinetic energy to generate electricity without changing the water level'. But when the blockage becomes significant, changes in water level can no longer be ignored. As stated previously, the level upstream will rise immediately the upstream of the turbines by an amount depending on the degree of blockage and reducing further upstream, forming a backwater curve as shown in Figure 3.30 (originally from [43] but no longer accessible), showing that an obstruction such as a bank of turbines can affect upstream levels for a considerable distance. If turbines are spaced too closely along a river, the downstream turbines will reduce the power output of upstream turbines, and the rise in level may cause local flooding.

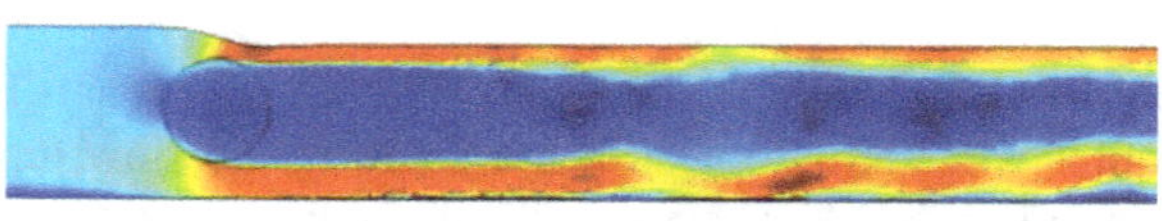

Figure 3.29 Modelled velocity contour plot for a horizontal axis crossflow turbine represented by the circle on the left, blocking 62.5% of the upstream flow area, flow from left to right. High-speed flow regions are shown in red and low-speed regions in dark blue, after [40]

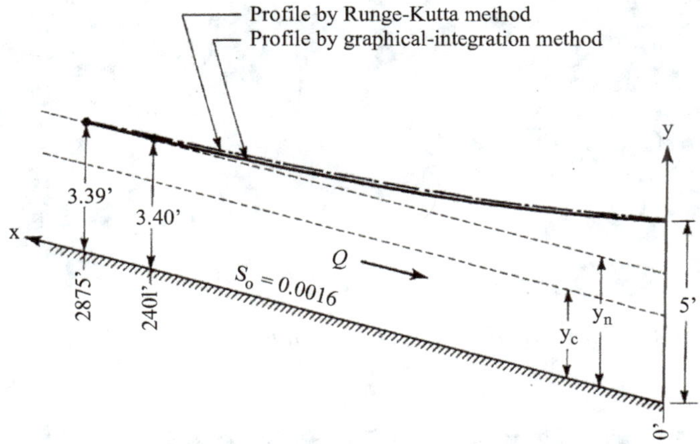

Figure 3.30 Backwater curve, after [43] (no longer on the website)

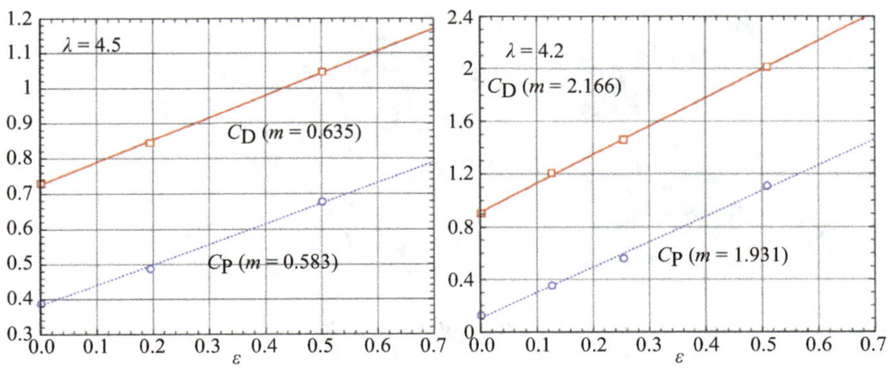

Figure 3.31 The effect of channel blockage on drag coefficient C_D and performance coefficient C_P for axial flow HKTs (left) and crossflow HKTs (right), after [44]

3.11.1 Apparently exceeding the Betz limit

Numerous authors have stated that the efficiency of individual turbines blocking a significant part of the flow can exceed the Betz limit of 59.3% of the kinetic energy in the swept area of the turbine, which is the normally accepted limit for turbines in open flow. For example, [40] measured an apparent C_P up to nearly 1.5 with a model horizontal axis Darrieus turbine in a laboratory flow tank and [44] found a similar effect for crossflow HKTs at high blockage, although the effect was less for axial flow turbines, as shown in Figure 3.31.

Figure 3.32 A small vertical axis HKT operating at high blockage in Nepal [16]

Figure 3.33 The river at Long Anyat village, Sarawak: foreground deep and slow, downstream shallow and fast, where a high blockage HKT could be placed (photo and annotation by the author)

This finding is of considerable practical importance where it is desired to extract maximum power from a moderate-sized river. Figure 3.32, from [16], shows a small vertical axis HKT operating at high blockage in Nepal, and Figure 3.33, from [1], shows how a horizontal axis crossflow turbine could be deployed in a similar way.

The fact that so-called HKTs at high blockage extract potential energy while kinetic energy increases, as explained earlier and illustrated in Figure 3.29, shows that the distinction between very low head hydro-turbines and HKTs can be somewhat blurred. Figure 3.34 shows how the total specific energy is made up of depth y (potential energy) plus velocity head or kinetic energy $V^2/2g$, so that potential energy can be extracted from subcritical flow (Froude number <1) with a drop in elevation accompanied by an increase in kinetic energy.

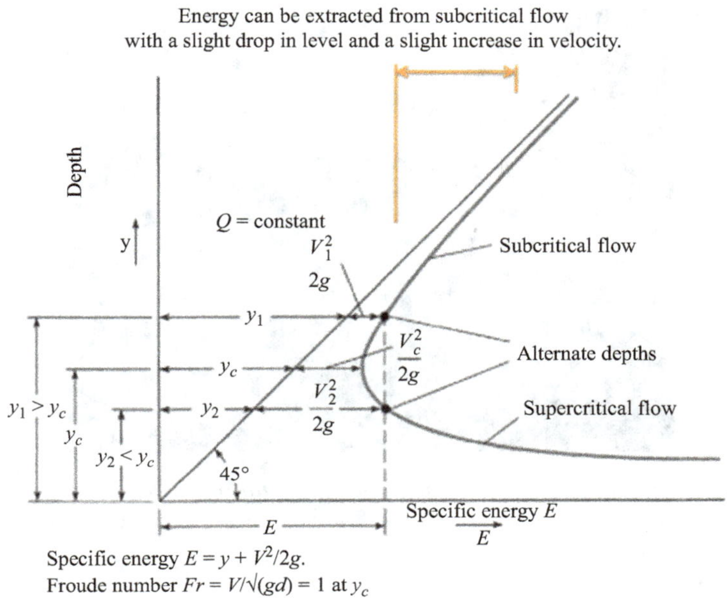

Figure 3.34 Energy diagram showing how the total specific energy (horizontal axis) is made up of depth y (potential energy) plus velocity head or kinetic energy $V^2/2g$. Adapted from [45]

3.12 Tidal and river level variation

There is usually, but not always, a big variation in water level at sites where there is a strong tidal flow, and the water level in many rivers also varies considerably with wet and dry weather, so provision must be made for these changes in level when mooring HKTs. Figure 3.35 shows floating structures and structures built on stilts well above the low water level on the Ucayali River at Pucallpa in Peru, indicating that the level may rise several metres. A bank of turbines may be designed to present significant blockage and raise the water level considerably during periods of low flow, but the blockage may be insignificant in the much larger flow area when the level is high and so may not be a problem.

3.13 Fast-flowing rivers

In contrast to large, relatively slow-flowing rivers such as the Mississippi, the Chili River which flows through Arequipa, Peru, carries a relatively small discharge, about 8 m^3/s except in the wet season when it increases considerably, but it has a steep slope: there is a 770-m drop in elevation through the city alone. Thus, the potential energy dissipated in turbulence and bottom friction in that stretch of the

Figure 3.35 Floating structures and structures built on stilts well above the low water level on the Ucayali River at Pucallpa in Peru (photo by the author)

Figure 3.36 Chili River in Arequipa, Peru (photo by the author)

river alone is $Q\rho gh = 8 \times 1{,}000 \times 9.8\ \text{N} \times 770\ \text{m} = 60\ \text{MW}$. But it is shallow and rocky, as shown in Figure 3.36, like most fast-flowing rivers (since they would otherwise scour), so in its present state, there is not much potential for HKTs. However, it would be possible to place small weir structures with low head turbines at intervals like that in Figure 3.32, which shows a New Energy HKT in Nepal.

3.14 Canals

Canals in relatively flat country are designed with minimum slope to carry the design flow, and any blockage by turbines would be likely to cause upstream

Figure 3.37 A drop structure in a canal in Peru (photo by the author)

flooding. But some canals such as that shown in Figure 3.37 have drop structures to limit flow velocity, and there is potential to capture some energy in these structures.

3.15 Economics

The cost of the turbine itself is only a small part of the complete installation, which includes gearing, generator, mounting structure, anchoring, electrical cabling, inverter. Small complete systems of about 5-kW rated capacity in 3-m/s flows such as those shown in Figures 3.4 and 3.8 typically cost tens of thousands of USD for supply only, making the installed cost prohibitive for all but one-off demonstrations. Hopefully, the cost will come down as experience is gained, as has been the case with wind turbines, but there may be a need for a radical rethink of some design features. The small river HKT shown in Figure 3.5 is an example, where off-the-shelf items have been adapted to serve as the key components of turbine rotor, pontoon and generator.

3.16 Actual progress to date

To date, only a few small HKTs have been demonstrated or proposed in inland waterways, for several reasons:

1. Few natural rivers have both the flow velocity and depth to accommodate significant-sized turbines. Most deployments have been in constructed or excavated channels or canals downstream from existing hydro dams such as that at Duncan Dam, British Columbia, Canada [46].

2. The high cost, as outlined earlier.
3. Turbine designers want to avoid reports of embarrassing failures, so they design for high-velocity flows and loads many times those likely to be encountered in normal operation.
4. Manufacturers want to present their products in a favourable light, so they generally quote rated power in flow velocities around 3 m/s. For example, [8,16] quote rated output at 3 and 3.1 m/s, respectively. But such flow velocities are very rare in natural rivers where the depth is sufficient for significant-sized turbines. This issue has been discussed in [1,10].

3.17 Future prospects

Investors and designers should develop simple, reliable, affordable turbines (Model T Fords rather than Sherman tanks) with overload protection for extreme conditions, in the same way that wind turbines have overload protection. There is a huge and profitable potential market for such turbines to deliver useful power at thousands of sites in many moderate-sized rivers around the world.

References

[1] Kirke, B.K. Hydrokinetic turbines for moderate sized rivers. *Energy for Sustainable Development* 58 (2020), 182–195. https://doi.org/10.1016/j.esd.2020.08.003.

[2] Kirke, B.K. and Coiro, D.P. (2019). *Renewable Energy from the Oceans: From Wave, Tidal and Gradient Systems to Offshore Wind and Solar, Chapter 3, Tidal and current energy.* IET. 10.1049/PBPO129E.

[3] Laws, N.D. and Epps, B.P. Hydrokinetic energy conversion: Technology, research, and outlook. *Renewable and Sustainable Energy Reviews* 57 (2016) 1245–1259.

[4] Niebuhr, C.M., van Dijk, M., Neary, V.S., and Bhagwan, J.N. A review of hydrokinetic turbines and enhancement techniques for canal installations: Technology, applicability and potential. *Renewable and Sustainable Energy Reviews* 113 (2019) 109240.

[5] Manoj Sood, M. and Singal, S.K. Development of hydrokinetic energy technology: A review. *International Journal of Energy Research* 43 (2019) 5552–5571.

[6] Anyi, M. and Kirke, B.K. Tests on a non-clogging hydrokinetic turbine. *Energy for Sustainable Development* 25 (2015) 50–55.

[7] Thropton Energy Services. www.throptonenergy.co.uk/ Accessed 21 August 2021.

[8] Smart Hydro. www.smart-hydro.de/decentralized-rural-electrification-projects-worldwide/ Accessed 7 February 2021.

[9] Ocean Renewable Power Corporation (ORPC). www.orpc.co/ Accessed 7 February 2021.

[10] Kirke, B.K. Hydrokinetic and ultra-low head turbines in rivers: a reality check. *Energy for Sustainable Development* 52 (2019) 1–10.

[11] Sniderman, D. (2012). Energy from Florida's Ocean Currents. www.asme.org/engineering-topics/articles/renewable-energy/energy-from-florida-s-ocean-currents Accessed 18 October 2020.

[12] Bahaj, A.S., Molland, A.F., Chaplin, J.R., and Batten, W.M.J. Power and thrust measurements of marine current turbines under various hydrodynamic flow conditions in a cavitation tunnel and a towing tank. *Renewable Energy* 32, Issue 3 (2007) 407–426. https://doi.org/10.1016/j.renene.2006.01.012.

[13] Failed tidal turbine explained at symposium. www.cbc.ca/news/canada/nova-scotia/nova-scotia-company-aims-remove-failed-tidal-turbine-2021-1.5710913 Accessed 14 March 2018, no longer accessible.

[14] Goddijn-Murphy, L., Woolf, D.K, and Easton, M.C. Current patterns in the inner sound (Pentland Firth) from underway ADCP data. *Journal of Atmospheric and Oceanic Technology* 30 (2013) 96–111. https://journals.ametsoc.org/view/journals/atot/30/1/jtech-d-11-00223_1.xml Accessed 18 October 2020.

[15] Schottel Tidal Generator. www.blackrocktidalpower.com/fileadmin/data_BRTP/pdf/STG-datasheet.pdf Accessed 7 February 2021.

[16] New Energy Corporation. www.newenergycorp.ca/gallery.html Accessed 7 February 2021.

[17] Fauzi Shahab, S.M. "Rural Power Supply Strategy," Townhall National Blue Ocean Strategy, Sarawak, 2017. [Online]. Available: http://pmsarawak.treasury.gov.my/wp-content/uploads/Townhall NBOS/Rural Power Supply Strategy.pdf Accessed 7 February 2021.

[18] Mako. https://oceanenergygroup.org.au/mako-tidal-turbines-a-new-era-in-energy-generation-1877/ Accessed 7 February 2021.

[19] Kepler Energy. http://keplerenergy.co.uk/index.html. No longer accessible.

[20] Idénergie Inc., Montreal, Canada. http://idenergie.ca/en/news/ Accessed 7 February 2021.

[21] British American Turbines. http://baturbines.com/tidal.html Accessed 7 February 2021.

[22] www.orpc.co/our-solutions/turbine-generator-unit Accessed 7 February 2021.

[23] Lazauskas, L and Kirke, B.K. Modeling passive variable pitch cross flow hydrokinetic turbines to maximize performance and smooth operation. *Renewable Energy* 45 (2012) 41–50.

[24] Kumar, A and Saini, R.P. Performance analysis of a single stage modified Savonius hydrokinetic turbine having twisted blades. *Renewable Energy* 113 (2017) 461–478.

[25] Rahai, H.R. (2005). *Development of Optimum Design Configuration and Performance for Vertical Axis Wind Turbine*. California Energy Commission Energy Innovations Small Grant Program, California State University, Long Beach.

[26] Les hydroliennes EcoCinetic. http://hydrolien.fr/hydrolienne-domestique/ Accessed 7 February 2021.
[27] Waterotor. https://waterotor.com/ Accessed 7 February 2021.
[28] Anyi, M. (2013). Water Current Energy for Remote Community: Design and Testing of a Clog-free Horizontal Axis Hydrokinetic Turbine System. PhD Thesis, University of South Australia.
[29] The Barsha Pump. www.aqysta.com/products/barsha-pump/.
[30] Greenenergy Hydrocat. www.youtube.com/watch?v=p4scG3cuL7I Accessed 7 February 2021.
[31] Minesto. https://minesto.com/our-technology Accessed 18 October 2020.
[32] Rawlings, G.W. (2008). Parametric Characterization of an Experimental Vertical Axis Hydro Turbine, BASc, University of British Columbia, 2005.
[33] Gilbert, B.L. and Foreman, K.M. Experiments with a diffuser-augmented model wind turbine. *Journal of Energy Resources Technology, Transactions of the ASME* 105 (1983) 46–53.
[34] Tidal Energy P/L, Palm Beach, Australia. www.tidalenergy.com.au/.
[35] Kirke, B.K. (2003). Developments in Ducted Water Current Turbines. http://citeseerx.ist.psu.edu/viewdoc/download?doi=10.1.1.531.3501&rep=rep1&type=pdf.
[36] Kim, D., Strom, B., Mandre, S., and Breuer, K. Energy harvesting performance and flow structure of an oscillating hydrofoil with finite span. *Journal of Fluids and Structures* 70 (2017) 314–326.
[37] Bernitsas, M.M. A no-blade, no-rotor, current energy converter (of friendly vortices and instabilities). OMAE 2020 39th Int Conf on Ocean, Offshore and Arctic Engineering, Aug 3–7, 2020, Fort Lauderdale, FL.
[38] Tidal Sails. http://tidalsails.com/
[39] Tyler, R.N. (2011). River Debris: Causes, Impacts, and Mitigation Techniques. Alaska Center for Energy and Power, April 13, 2011.
[40] McAdam, R.A., Houlsby, G.T., and Oldfield, M.L.G. Experimental measurements of the hydrodynamic performance and structural loading of the Transverse Horizontal Axis Water Turbine: Part 1. *Renewable Energy* 59 (2013) 105–114.
[41] Vermaak, H.J., Kusakana, K., and Koko, S.P. Status of micro-hydrokinetic river technology in rural applications: A review of literature. *Renewable and Sustainable Energy Reviews* 29 (2014) 625–633.
[42] Devoy, P. and Toniolo, H. HYDROKAL: A module for in-stream hydrokinetic resource assessment. *Computers & Geosciences* 39 (2012) 171–181.
[43] Lin, W. and Gray, D.M. (1971). Calculation of Backwater Curves by the Runge-Kutta Method, pp. 1–20. www.usask.ca/hydrology/ListPubs.php?SortKey=2&AuthorMatch=lin.
[44] Kinsey, T. and Dumas, G. Impact of channel blockage on the performance of axial and cross-flow hydrokinetic turbines. *Renewable Energy* 103 (2017) 239–254.

[45] Roberson, J.A. and Crowe, C.T. (1997). *Engineering Fluid Mechanics*, 6th Ed. John Wiley & Sons, Inc. New York, Chichester, Brisbane, Toronto, Singapore, Weinheim.

[46] Instream Energy Systems World's First Full-Scale VAHT Pilot Project in a Discharge Channel. www.instreamenergy.com/duncan-dam-british-columbia.

Chapter 4

Computational methods for vertical axis wind turbines

Anders Goude[1] *and Victor Mendoza*[1]

There are two major types of vertical axis wind turbines (VAWTs): the Savonius rotor, which is based on the drag force and the Darrieus rotor, which is based on the lift force. The vertical axis turbine has the option to be modeled in two-dimensional (2D), which offer significant speed improvements and can give good approximations of the turbine forces, but cannot handle all interactions within the turbine and are less suitable for wake modeling. The Savonius rotor is commonly modeled with traditional computational fluid dynamics (CFD) software, as flow separation is crucial to its operation. While CFD models with resolved boundary layers also can be applied to the Darrieus, it is computationally demanding and simplified models, where the blades are modeled by external force models, are commonly applied. The most simplified model is the streamtube model, which uses a very simplified stationary model for the flow through the turbine. A more advanced approach is the actuator cylinder model, where the stationary flow is modeled through Euler's equation, giving a more realistic flow field. If time-dependent solutions are desired, there are the options to use either the actuator line model (ALM) or the vortex method, where the ALMs typically are implemented within traditional CFD solvers, while the vortex method often is implemented as a Lagrangian method that uses the vorticity as discretization parameter. The simplified models are primary useful for turbines with a high tip speed ratio, while the deep stall that can occur at a low tip speed ratio often requires traditional CFD solvers that resolve the boundary layers.

4.1 Introduction

The VAWT, also called crossflow turbine, is a turbine where the axis of rotation is perpendicular to the flow direction. The common configuration is a vertical axis of rotation (hence its name) with a horizontal flow direction. This has the advantage that the turbine will work for any horizontal flow and does not need any yaw mechanism to adjust its orientation. Other advantages include that one can place the generator on the ground, which can give easier maintenance and a lower center of gravity, especially

[1]Department of Electrical Engineering, Uppsala University, Sweden

important for far offshore floating turbines deployed at deeper water depths (potential trend of the wind energy industry) [1], and is a reason for the renewed interest in VAWTs. The main disadvantages are that the turbines available today have lower power coefficients than the state-of-the-art horizontal axis wind turbine (HAWT) and that the turbine generates pulsating forces, which is a source for fatigue loads. However, far offshore, the turbines operating conditions are considerably different from onshore, where HAWTs dominate the market due to their technology maturity, and the concept of floating VAWTs could be more suitable presenting a potential reduction in the overall cost of energy compared to the conventional HAWT.

There are two distinct types of VAWTs: the turbines based on the drag force (often called Savonius rotors) and the Darrieus-type turbines, which are based on the lift force, as shown in Figure 4.1. Savonius rotors are characterized by having large blades, of similar size as the cross-sectional area of the turbine. This will give a significant weight of the blades, and the Savonius turbine is mainly used for small scales where weight is not as important (weight scales as R^3 while the cross-sectional area scales as R^2). The Savonius rotors are simple to construct and have robust aerodynamics. The design is therefore suitable when reliability is important. Disadvantages are that they do not have as high-power coefficients as lift-based turbines. The large blades also give a large turbine area, which is negative with respect to storm loads, as the thrust force reduction of shutting down the turbine is significantly smaller than what can be achieved for lift-based turbines. This turbine concept has been widely employed for micro generation in urban environments (where the wind conditions are characterized by high variability, turbulence, and a low mean speed) for powering streets lamps together with solar panels and small battery banks [3]. Recently, novel Savonius rotor concepts with twisted blades have been employed for non-conventional applications, such as energy production in extremely cold conditions, as depicted in Figure 4.2 [4].

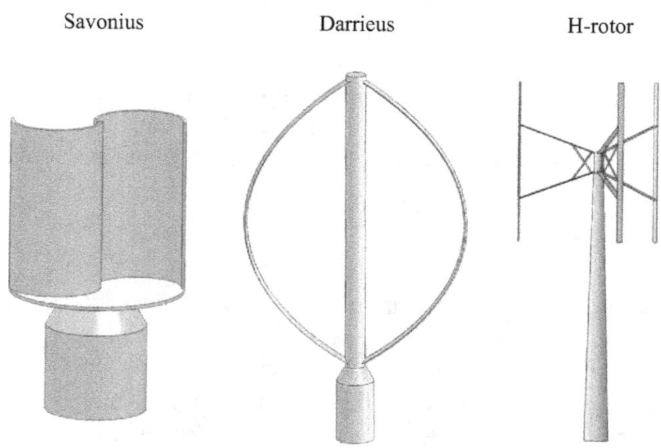

Figure 4.1 Different types of vertical axis turbines, where the Savonius rotor (left) is drag based, while the curved bladed Darrieus rotor (middle) and straight bladed Darrieus rotor, also called H-rotor (right) are lift based. From [2]

Figure 4.2 SAVANT: Savonius turbine with twisted blades for energy production in extremely cold conditions (5–10 W depending on wind availability). From [4]

The Darrieus rotor uses the lift force to generate power and has small airfoil-shaped blades. There are several blade-type designs here. The biggest commercial deployment of these turbines were of the curved bladed design by the company FloWind in North America, mainly in the 1980s [5]. This curved bladed design is to reduce the centrifugal stresses in the blades, by forming the blades according to the shape of a spinning rope. The advantages of this design, except for the reduced centrifugal loads, are that it may be possible to build them without additional struts to attach the blades (although some designs used struts), and the blades do not have any clear blade tip with the associated losses. Disadvantages are that the turbine radius varies with height, which causes the blades to operate in unfavorable conditions close to the ends and reduces the turbine area, and the blades are more complicated to manufacture, and the difficulty to build towers, as the tower only can extend to the lowest point of the turbine due to the blade attachment.

The second blade design is the straight bladed turbine, also called H-rotor, which has straight vertical blades as shown in Figure 4.3. This requires struts to attach the blades, which generate additional turbine components and aerodynamic drag. The advantage is that the cross-sectional area is larger and the blades are easy to manufacture. A negative aspect is that the blade tips can generate losses. A general disadvantage of both the straight bladed and the curved bladed designs is that they generate pulsating thrust force and torque. A way to overcome this is the

Figure 4.3 A 12-kW VAWT, designed and built by the Division of Electricity at Uppsala University. The turbine is equipped with load cells used for force measurements. From [6]

helical turbine design, where the blades are curved around the surface of the cylinder representing the swept area of the turbine, with the disadvantage that they are harder to manufacture.

4.2 VAWT theory

The power of a wind turbine is given by

$$P = \frac{1}{2} C_P \rho A V_\infty^3, \tag{4.1}$$

where C_P is the power coefficient, ρ is the density of air, V_∞ is the free-stream velocity and A is the turbine area. For the VAWT, one typically uses the cross-sectional area ($A = HD$ for the straight bladed turbine where H is the blade height and D is the diameter), as this is the area corresponding to the available energy in the wind. One common parameter used for the VAWT is the tip speed ratio λ, defined as

$$\lambda = \frac{R\Omega}{V_\infty}, \tag{4.2}$$

where R is the turbine radius and Ω is the rotational speed. A major difference between the Savonius rotor and the Darrieus rotor is that Savonius rotors operate at a significantly lower λ than the Darrieus rotor. Another common parameter, mainly used for the Darrieus rotor, is the solidity

$$\sigma = \frac{cN_b}{R}, \tag{4.3}$$

where c is the blade chord and N_b is the number of blades. Note that there are different definitions of the solidity (the difference is which length one should divide with where both radii, diameter or circumference can be used), and one should be careful with comparing solidities from different sources without first checking the definition.

4.2.1 Savonius

The conventional Savonius turbine rotor (with an S-shape blade cross-section) is considered a drag-driven machine. The manufacturing process can be described as sliding a hallow cylinder (pipe) along the central plane into two semicircular parts, which have an offset between them. The acting flow on the Savonius rotor produces higher drag force on the concave surface compared to the convex surface resulting in the rotation, as shown in Figure 4.4.

Aerodynamic studies of the Savonius rotor typically require full-geometry CFD modeling and more details of this turbine will be found in Section 4.8.1.

4.2.2 Darrieus

The Darrieus rotor works by having blades that move in a circle according to Figure 4.5. The text will focus on the common design that the blades are relatively small, compared to the turbine size. This will allow us to treat the blades as airfoils operating in the flow, and airfoil theory can be used to describe the working principles of the turbine. For simplicity, we will start with a 2D description of the turbine and later go into more complicated phenomena.

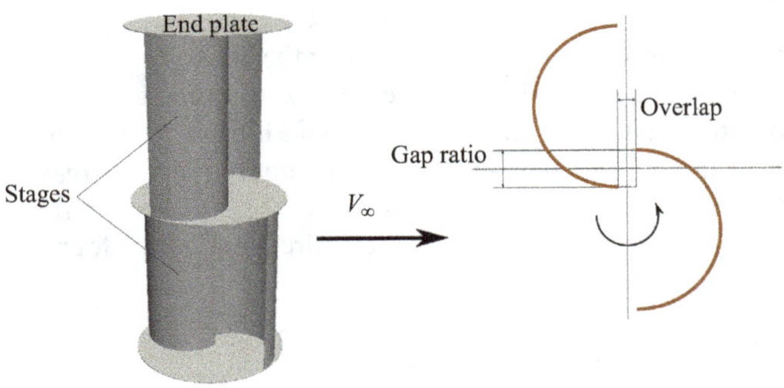

Figure 4.4 Schematic with some main design parameters for a Savonius rotor

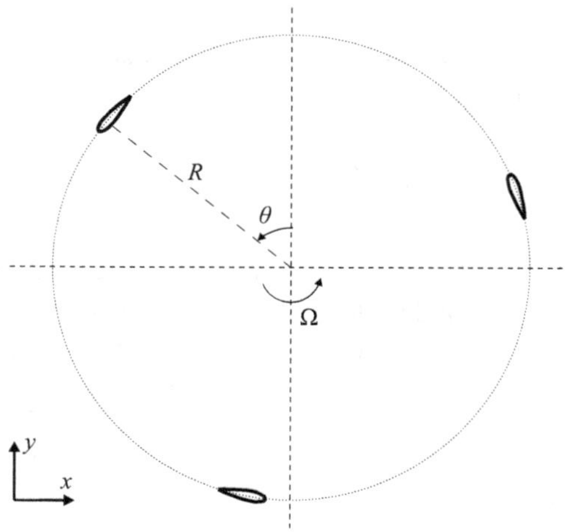

Figure 4.5 2D view of a 3-bladed Darrieus rotor

4.2.2.1 Basic 2D theory

Assume that the asymptotic flow velocity V_∞ is in the positive $\hat{x}$-direction $\left(\vec{V}_\infty = V_\infty \hat{x}\right)$. Due to the turbine wake, we will have a reduced flow velocity at the turbine $\left(\vec{V}_{\text{induced}}\right)$. This velocity will vary, depending on the turbine position. We can now define a local flow velocity as

$$\vec{V}_{local} = V_{local,x}\hat{x} + V_{local,y}\hat{y} = \vec{V}_\infty + \vec{V}_{induced}. \tag{4.4}$$

In the general case, the velocity will both have an $\hat{x}$ and a $\hat{y}$ component. The turbine blade will be moving with the velocity

$$\vec{V}_{blade} = R\Omega\hat{\theta}. \tag{4.5}$$

The relative velocity can be written as the combination of these two

$$\vec{V}_{rel} = \vec{V}_{local} - \vec{V}_{blade}. \tag{4.6}$$

Now, write all velocities in the reference frame of the blade (i.e. rotate the coordinate system with $-\theta^{\circ}$). Note that we have defined $\theta = 0$ according to Figure 4.5, which is different from the traditional definition of cylindrical coordinates to make the upwind pass of the turbine start at $\theta = 0$. The definition of θ may vary among different references. In this work, we will maintain the base vectors from cylindrical coordinates with $\hat{r}$ being positive outward (both directions can be found in literature). This gives

$$\begin{aligned}\vec{V}_{rel} = &\left(R\Omega + V_{local,x}\cos\theta + V_{local,y}\sin\theta\right)\hat{x}' \\ &+ \left(-V_{local,x}\sin\theta + V_{local,y}\cos\theta\right)\hat{y}',\end{aligned} \tag{4.7}$$

which written in polar form will give the angle of attack and the magnitude of the relative wind speed. Assuming that $-90° \leq \varphi \leq 90°$, these will become

$$V_{rel} = \sqrt{\left(R\Omega + V_{local,x}\cos\theta + V_{local,y}\sin\theta\right)^2 + \left(-V_{local,x}\sin\theta + V_{local,y}\cos\theta\right)^2}, \tag{4.8}$$

$$\varphi = \arctan\frac{-V_{local,x}\sin\theta - V_{local,y}\cos\theta}{R\Omega + V_{local,x}\cos\theta + V_{local,y}\sin\theta}. \tag{4.9}$$

Note that for a VAWT operating at $\lambda < 1$, one will reach flow velocities with negative $\hat{x}'$ component, which will make the expression for the relative wind angle invalid, as it is outside the 180° branch of arctan. Also note that the sign in definition of relative wind angle varies between sources. One should therefore always take care when comparing these equations from different sources.

The blade can have a pitch angle θ_p, see Figure 4.6 and the angle of attack is

$$\alpha = \varphi + \theta_p. \tag{4.10}$$

Once again, the reader should be aware of that the positive direction of θ_p varies between references as well.

The blades will have a lift force F_L and a drag force F_D, and the corresponding lift coefficient C_L and drag coefficient C_D associated with them. When analyzing the turbine, it is more convenient to use the normal force F_N and the tangential force F_T, which are obtained by rotating the lift- and drag coefficients using the relative wind angle

$$C_N = C_L\cos\varphi + C_D\sin\varphi, \tag{4.11}$$

$$C_T = C_L\sin\varphi - C_D\cos\varphi, \tag{4.12}$$

and the forces are obtained as

$$F_N = \frac{1}{2}C_N\rho A_{blade}V_{rel}^2, \tag{4.13}$$

$$F_T = \frac{1}{2}C_T\rho A_{blade}V_{rel}^2, \tag{4.14}$$

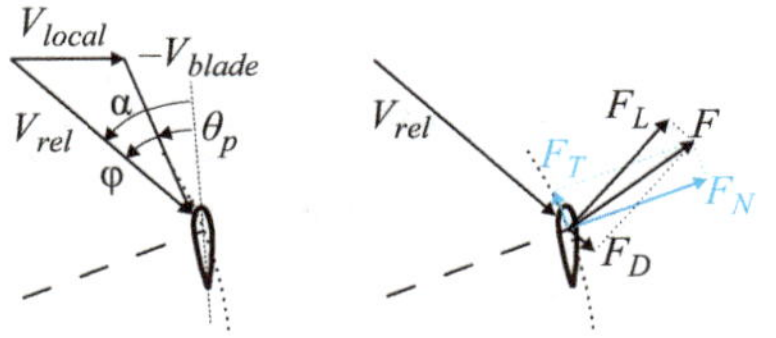

Figure 4.6 View of blade at downwind part of turbine, illustrating the velocity vectors in (4.6) and the angles in (4.10) (left) and the force F and its different compositions (F_L and F_D) or (F_N and F_T) (right)

where $A_{blade} = cH$ is the blade area and c is the chord length (assuming constant chord length). Here, the normal force is positive in the radial direction (outward) and the tangential force is positive in the direction of the blade (the same definition as for cylindrical coordinates). The turbine power can be calculated from the mean tangential force as

$$P_{mean} = RF_{T,mean}\Omega N_b, \tag{4.15}$$

assuming that the radius is constant, where N_b is the number of blades.

4.2.2.2 Flow curvature and blade attachment point

The curved path that the blades are moving in will affect the aerodynamics of the blades, causing them to appear to have a slightly different shape than they would have for a straight path, see Figure 4.7. This was originally studied by Migliore [7]. For small blades, one can approximate the blade as having a small camber, hence giving a slight addition to the lift force (for blades that are large compared to the turbine, there will also be effects of a difference in blade velocity between the upper and lower sides). Flow curvature effects are significant enough that it is recommended to account for them when implementing a simulation model for VAWTs.

Due to the curved blade motion, the attachment point for the blade will also affect the blade characteristics. In connection with the flow curvature, the directions of the normal and tangential forces become unambiguous, as the directions depend on where on the blade they are defined. Here, we will make the definition that they are always in the point where the force is acting, which for a traditional airfoil in straight flow would be the quarter chord position, assuming that the blade does not stall. This choice is made to ensure that the tangential force is the only contribution to the turbine torque, while the normal force should always be in the radial direction. Now, define the relative attachment point x_0 as 0 if the blade would be attached at the leading edge and 1 if the blade would be attached at the trailing edge (i.e. the quarter chord position becomes $x_0 = 0.25$). Assuming that the force is acting on the quarter chord position, the distance to this position is

$$R_{new} = \sqrt{R^2 + c(x_0 - 0.25)^2}, \tag{4.16}$$

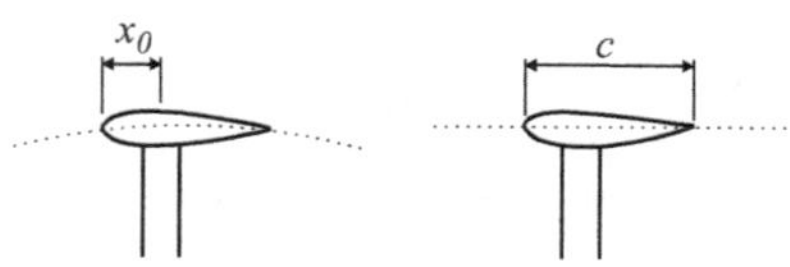

Figure 4.7 Illustration of virtual camber. Left: The blade moving in the circular path of the VAWT. Right: Effective camber if the path would be straight. c is the chord length and x_0 is the normalized attachment point

the angle is

$$\theta_{p,correction} = \arctan\left(\frac{c}{R}(x_0 - 0.25)\right), \tag{4.17}$$

and we can consider a shift in attachment point the same as applying a constant pitch angle to the blade (with a small correction to the radius).

It is more difficult to correct for the virtual camber effect by analytical corrections. The challenge is that the new airfoil virtually has a different shape than the original airfoil. Hence, there will be additional uncertainties if one uses studies on the original airfoil to predict performance in curved flow. This has been studied in [8]. Additionally, the flow speed can vary over the blade due to variations in the velocity field, which makes angle of attack estimates hard as there is not a unique value to use. For simplified models, it is still useful to have estimates to this curvature effect. One correction presented by Sharpe [9] suggests to correct the normal force by

$$C_{N,correction} = -\frac{a_s \Omega c}{4V_{rel}}, \tag{4.18}$$

where a_s is the slope of the lift force curve:

$$C_L \approx a_s \alpha + C_{L0}. \tag{4.19}$$

This expression is designed for blades with attachment point $x_0 = 0.5$, and translated to blades attached at the quarter chord position, the corresponding approximation would be (assuming $V_{\mathrm{rel}} \approx r\Omega$)

$$C_{N,correction} = -\frac{a_s \Omega c}{2V_{rel}}. \tag{4.20}$$

The expression is only suitable when the blade does not stall. Another suggestion is to implement it through a corrected angle of attack, similar to the attachment point correction, which becomes (assuming quarter chord attachment point)

$$\alpha_{correction} = -\frac{\Omega c}{2V_{rel}}. \tag{4.21}$$

The corrections given in this section should be added to the original values (derivation can be found in [2]). Once again, the readers are urged to be careful with the sign of the expressions as it varies between sources.

4.2.2.3 Dynamic stall

Vertical axis turbines that rotate at a low λ will experience dynamic stall (unless active pitching is used to prevent it). This can be seen from (4.9), where the relative wind angle becomes larger with a decreased rotational speed. As it varies between positive and negative values, stall cannot be prevented with a fixed pitch angle for sufficiently low λ. Exactly at which λ that the turbine will stall depends on many factors, such as Reynolds number, choice of airfoil, the turbine solidity, but an approximation is that the turbine can start reaching sufficiently high angles of attack for stall below $\lambda \approx 4$.

The angle of attack is constantly changing for the VAWT blade, as can be seen from (4.9). If a blade is placed in a flow with a given pitch angle, there is a time delay before the circulation around the blade reaches its stationary value. For a blade that is changing its angle of attack, this will be represented by a lag in the circulation, and therefore a lag in the forces of the blades. This effect applies even when the angles of attack are below the stall angle. When the angle of attack becomes large enough to enter the stall regime, the dynamic effect becomes much more prominent. Now, the blade can briefly experience a higher lift force than in the static case. This is due to the leading edge vortex that starts to form when the blade enters the stall regime. During the formation of the leading edge vortex, the blade can experience the higher lift force until the vortex has detached from the blade. In a similar way, when the blade pitches down below the static stall angle, it will take time for the flow to reattach, causing a lower lift force until the reattachment is complete. This gives a hysteresis behavior, which is illustrated in Figure 4.8.

Modeling of dynamic stall is a major challenge when simulating VAWTs [11,12]. Traditionally, models for dynamic stall are calibrated for sinusoidal changes in angle of attack. However, the angle of attack variations experienced from the VAWT are not sinusoidal. Equation (4.9) is asymmetric, even for a constant wind speed. When the wake is included, additional variations to the angle of attack are introduced, making it even harder to have an analytic model that can predict dynamic stall for all cases. Simulation models can also experience small higher frequency oscillations in the wind speed (e.g. due to turbulence and model uncertainties). Using models that only depend on the rate of change of the angle of attack (such as the Gormont model [13] that is common for VAWT modeling) can therefore experience issues, while models, such as Leishman–Beddoes models [14,15], which use lag functions and hence have a history, are less sensitive. Implementations of the Gormont model for VAWTs can be found in

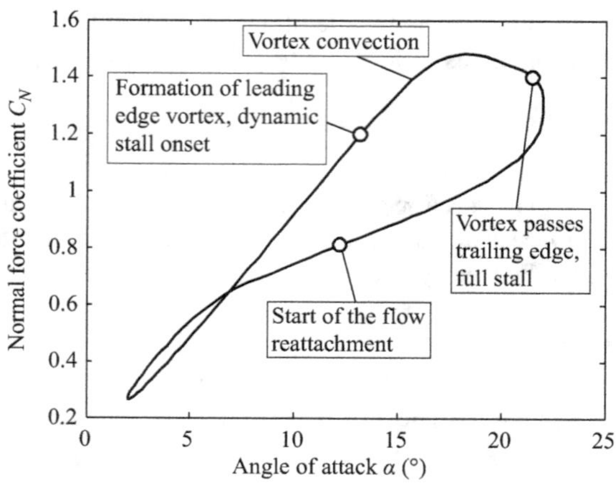

Figure 4.8 Illustration of dynamic stall. The figure shows how the normal force coefficient can vary when a bland increases its angle of attack into the stall region and leaves it again. From [10]

[16], a Leishman–Beddoes-type model in [17], and comparisons of the two can be found in [18,19]. Dynamic stall models do also have a limit in how deep stall that can be modeled, making it harder to model VAWTs at low λ.

One additional issue with dynamic stall models is that the curved blade path gives different behavior upwind and downwind. Upwind, the leading edge vortex will be on the inside of the blade, and the blade will curve toward this vortex. This can cause the vortex to remain attached for a longer time, compared to straight flow. Downwind, the opposite applies, and the blade curves away from the leading edge vortex, causing it to detach earlier. Hence, one can expect dynamic stall effects to be more prominent upwind and less prominent downwind. Dynamic stall in VAWTs is also characterized by the release of the leading edge vortex upwind, which then collides with the blade downwind. This is illustrated in Figure 4.9. Studies have shown that dynamic stall effects were hardly seen downwind [21] due to the early detachment of the leading edge vortex and that modifying the dynamic stall model to reduce stall effects downwind improves the accuracy [19].

4.2.2.4 3D blade shapes

The theory in Section 4.2.2.1 assumes that the blades are straight and vertical. This is not always the case, for example, for a curved bladed turbine. It is commonly assumed that the flow in the blade span direction does not affect the performance of the blade, which allows for 2D theories to apply. Assume that $\hat{t}$ is the base vector in the tangential direction (of the rotation), $\hat{s}$ is the base vector in the span direction, and $\hat{n}'$ is the base vector with direction $\hat{t} \times \hat{s}$ (here we define it with positive direction outward to make it match the 2D theory, see Figure 4.10). This is not the same normal direction as before, as it will have a vertical component (unless the span vector is vertical). If we define

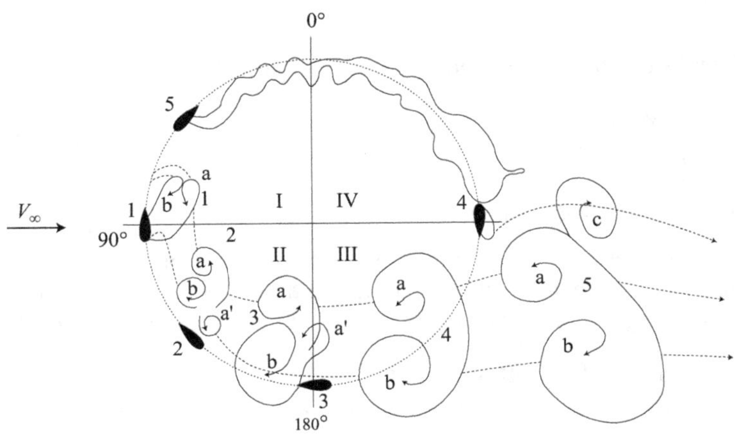

Figure 4.9 The propagation of the leading edge vortex generated by dynamic stall. From [19,20]

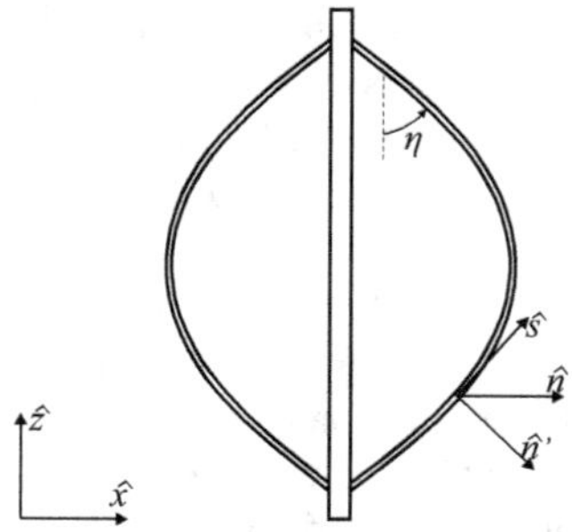

Figure 4.10 A curved bladed Darrieus rotor, showing the definition of the blade angle η and the different basis vectors $\hat{n}'$, $\hat{n}$, and $\hat{s}$

$$\vec{V}_{local,2D} = \left(\vec{V}_{local} \cdot \hat{t}\right) + \left(\vec{V}_{local} \cdot \hat{n}'\right)\hat{n}', \tag{4.22}$$

we can use the same equations as for the 2D theory (with $\hat{t} = -\hat{x}'$, $\hat{n}' = \hat{y}'$) to calculate the relative wind direction and speed. Using these expressions with the assumption that the blade has slope η according to Figure 4.10, we can obtain the corrected expressions as

$$V_{rel,x'} = \left(R\Omega + V_{local,x}\cos\theta + V_{local,y}\sin\theta\right)^2, \tag{4.23}$$

$$V_{rel,y'} = \left(\left(-V_{local,x}\sin\theta + V_{local,y}\cos\theta\right)\cos\eta + V_{local,z}\sin\eta\right)^2, \tag{4.24}$$

$$V_{rel} = \sqrt{V_{rel,x'}^2 + V_{rel,y'}^2}, \tag{4.25}$$

$$\varphi = \arctan\left(\frac{V_{rel,y'}}{V_{rel,x'}}\right), \tag{4.26}$$

and the results can be transformed back to a cylindrical coordinate

$$C_N = C_L\cos\varphi\cos\eta + C_D\sin\varphi\cos\eta, \tag{4.27}$$

$$C_T = C_L\sin\varphi - C_D\cos\varphi, \tag{4.28}$$

$$C_z = C_L\cos\varphi\sin\eta + C_D\sin\varphi\sin\eta, \tag{4.29}$$

where C_z is the vertical force component. Here, we have omitted the case of the helical turbine where there is a slope in the tangential direction as well, and the reader is referred to using (4.22) for the general case where the blade can have any direction.

4.2.2.5 Wake and tip effects

According to the continuity equation, combined with momentum conservation, we know that the local flow velocity $\vec{V}_{local}$ at the turbine is decreased, compared to the asymptotic speed, and thereby, the flow will expand. Compared to the horizontal axis turbine, the VAWT has a much more complicated wake. One reason is that the blades are sweeping a cylindrical shape (instead of a flat shape) and crosses the flow two times, hence interacting with their own wake, making the downwind sweep more complicated. Another reason is that the wake is not symmetric in the same way as for the HAWT (except for approximate mirror symmetry around the midplane if wind shear is ignored). For 2D flow, the wake is slightly asymmetric as the blade is moving with the flow on one side, while on the other side, the blade is moving against the flow, giving asymmetric relative wind angles, see (4.9). The shape of the wake mainly becomes complicated due to three-dimensional (3D) effects, as the wake expands differently in the horizontal and vertical directions. The wake expansion will depend on the turbine aspect ratio (*H*/*D*), where a high aspect ratio mainly will have expansion horizontally, while the vertical component of the expansion becomes more significant for low aspect ratio turbines. To further add to this, depending on where most energy is absorbed (upwind or downwind), the wake will expand differently. A large energy absorption upwind causes the wake to expand more horizontally and can even contract vertically, while a large absorption downwind (combined with a small absorption) can cause the opposite, i.e. a large vertical expansion and possibly a horizontal contraction [22]. This is illustrated in Figure 4.11, where the turbine has been simulated for three different pitch angles, which causes a shift in the power absorption between upwind and downwind. This shift also changes the lateral force on the turbine, which causes a change in the wake deflection as seen in the figure by comparing the two pitched cases.

One should also consider the blade tips for 3D models. Here, there is difference between the straight- and curved bladed turbines, as the curved bladed design will have its blade tips at a very small radius, which therefore can be neglected, while

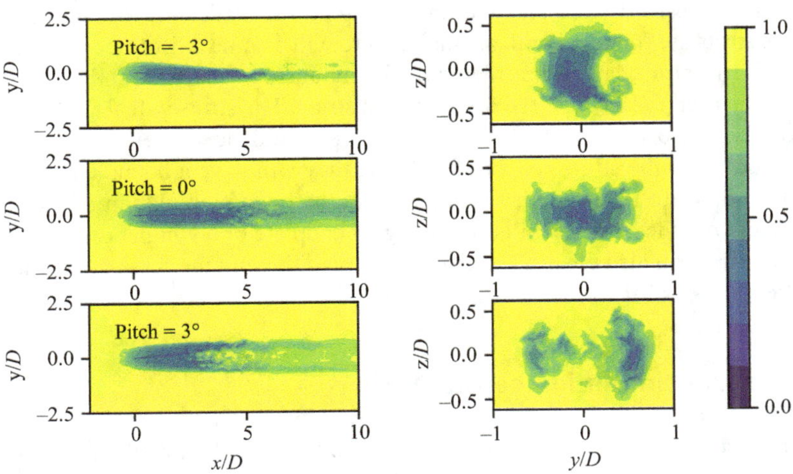

Figure 4.11 Actuator line model simulation of a 5-bladed turbine ($H = 3.5$ m, $D = 6$ m, $c = 0.18$ m) at $\lambda = 4$. The figures show both the horizontal cross-sections for the center of the turbine (left) and vertical cross-sections at $x/D = 4$ (right). Details about the turbine can be found in [23]

the straight bladed turbine will have blade tips at a full radius [24]. This will generate tip vortices, as the blade forces approach zero at the tip. These tip vortices will be different, compared to traditional finite wing theory (which assumes infinite length of the tip vortices), as the length of the tip vortex will be limited by the turbine diameter. The blade can also interact with the tip vortices from the other blades and its own vortices released upwind during the downwind pass.

4.2.2.6 Struts

Struts are needed to attach the blades for straight bladed (and helical) turbines and are optionally used for curved bladed turbines. The easiest case to model is the streamlined horizontal struts with horizontal flow. These should have relatively low drag and no lift force [25]. Hence, a simplified model can be given by setting the lift force to zero, the relative flow velocity to

$$V_{rel} = |r\Omega + V_{local,x}\cos\theta + V_{local,y}\sin\theta|, \tag{4.30}$$

where r is a given position along the strut. One can also use that $C_T = C_D$ and integrate the losses over the whole strut. The torque from a single strut at a given position θ along the revolution would then be given by

$$T = \int_0^R r\frac{1}{2}C_D\rho c(r)V_{rel}(r)^2 dr. \tag{4.31}$$

There can also be additional losses due to the strut attachment point, where both can experience lost lift force and additional drag on the blade.

If the struts are mounted with an angle, there will also be a lift force generated. In principle, the struts can be considered blades mounted at an angle, and (4.25) and (4.26) can be used to calculate the forces on the struts. The struts will also affect the shape of the wake due to the lift force, which should be modeled in accurate models. Further, if the struts generate lift force, they will have a bound circulation. As the tip of the struts is attached to the blades, the bound circulation of the strut will interact with the blade (similar to having an endplate or a winglet on the strut). Hence, the strut can also change the circulation (and thereby the force according to Kutta–Joukowski's lift formula) of the blades.

4.3 General simulation considerations

There are several factors to consider when choosing a simulation model to use. One aspect is which simulation model does the best job at simulating the turbine, with respect to how long the simulations take to do the job. The first question to ask is what the purpose of the modeling is? Some options can be to design a turbine for good performance, to calculate the forces for the mechanical design, aeroelastic modeling, modeling of the wake for farm studies, etc. As an example for turbine design, it is common to initially use a fast simulation model (such as choosing an approximate chord length for your turbine), while the more accurate models can be used later to fine-tune the turbine design.

One other aspect is which models that are available to use, and if new models are needed, how complicated they are to implement. There are significant differences in the work needed to implement different models, and one should be careful which functionality that is actually necessary for the case of interest.

One of the major challenges with turbine modeling in general is that there are many different scales to consider. At the smallest scale, we have the flow in the boundary layers on the turbine blades. Challenges here are to properly model blade forces, and especially the flow separation related to dynamic stall will play an important role in the turbine performance. On a larger scale, we have the turbine as a whole. Here, we have the interaction between the blades and the turbine's own wake. We also have the tip vortices and the interaction between the struts and the blades. On an even larger level, we have the turbine wake and the interaction between turbines. One can also include the interaction with the terrain here. Resolving all these different scales at the same time using traditional CFD is not practical, and if the larger scales are of interest, it is common to introduce some approximations for the smaller scales.

4.3.1 Model approximations

There are many options for simplifying the simulations to achieve models that can run faster. The main aim here is to go through some of the options that are available, which should be decided upon when designing the model.

4.3.1.1 2D or 3D

By making the assumption that the flow mainly occurs in the horizontal plane, and that the turbine also only moves in this plane, one can obtain significant information about the VAWT from 2D modeling. This is especially useful for turbines with a high aspect ratio, where the 3D effects due to tip vortices are small. The main advantage with 2D is that the simulations are much faster to run (and some models can be notably easier to implement in 2D as well). Simulations with a vortex code in [26] showed that the predictions of the total force on a blade is relatively close between 2D and 3D models, indicating that 2D models can be used for estimating blade forces. They can also be used to initially design turbine parameters, such as chord length. Another common application can be models that need to resolve the smallest scales (e.g. modeling of dynamic stall for the VAWT), where 3D simulations become very expensive.

The negative aspects of 2D modeling is that tip effects and wake effects are hard to model. Tip effects can to some degree be handled by analytical correction models for known blade shapes, see Section 4.2.2.5, but more complicated geometries like blades with end-plates or winglets need 3D models. When it comes to the struts, the question is how much lift force they generate. For horizontal struts, the losses can be corrected for with models such as (4.31), but if the struts start generating lift force, 3D models are required.

The turbine wake will be more stable in 2D (i.e. wake decay will occur later), and the effects of variations between horizontal and vertical expansion as mentioned in

Section 4.2.2.5 cannot be modeled. Hence, for wake modeling (and thereby farm- and terrain modeling), it is recommended to use 3D.

4.3.1.2 Blade modeling

There are two main options for the blade forces. The first is to directly solve the entire flow around the blade with CFD. This has the advantage that it works with any blade shapes, while the disadvantage is that it will require very small time-steps (related to highly refined mesh cells regions through the Courant number). More details can be found in Section 4.8.

The second approach is to replace the blades with a model for blade forces. This is one of the most common approximations to use for turbine modeling, as it removes the smallest scale (mostly present in boundary layer regions) and therefore significantly increases the model speed. The blade force model commonly extracts the flow velocity at the blade positions ($\vec{V}_{local}$) from the flow model and uses the equations in Section 4.2.2 to calculate the relative wind speed and angle of attack, which are used as input for the models. This will give a lift and a drag coefficient that can be used to calculate the force. This approximation is also commonly used for modeling of horizontal axis turbines, but it is more complicated for the VAWT due to the complex aerodynamics. The blade force model is commonly 2D and for 3D simulations, and the 2D model is applied for each blade segment. The disadvantages for this model are that one is limited to blades for which one of these models exists. The easiest way is to use a direct lookup table. While table data is available for many airfoils for low angles of attack, it is rare to find table data for the entire 360° needed for low λ. One popular choice for 4-digit symmetrical NACA profiles is [27]. It is however recommended to include a dynamic stall model, see Section 4.2.2.3.

4.3.1.3 Stationary or time dependent

One possible way to reduce the computational time is to approximate the solution with a steady-state model for the flow velocity (the turbine will still be rotating). The idea is to distribute the force over the swept surface and the flow field is modeled using the time-averaged forces. Stationary models are typically combined with a blade force model that quickly can evaluate blade forces during one revolution. One can still use time-dependent blade force models (such as dynamic stall models), while they operate on a fixed velocity field. The disadvantage is that not all phenomena will be captured without considering the time-dependent flow field. As an example, the VAWT will generate distinct tip vortices, while a stationary flow field will have to use an average flow for these.

4.4 Streamtube models

This is one of the simplest and fastest models that are being used for VAWTs. It is based on a stationary analytical flow model and uses a force model to determine the blade forces. The principle is the same as the blade element momentum models commonly used for HAWTs, with corrections to handle that the turbine blades cross their own wake. The main idea is to assume that the flow can be confined in a streamtube. From the traditional Betz theory, by naming the inlet flow speed V_∞ and the outlet V_e of the streamtube, both at atmospheric pressure, we have the speed of the turbine disk as

$$V_{disk} = \frac{V_\infty + V_e}{2}. \tag{4.32}$$

Now, assume that the area of streamtube at the turbine disk is A_{disk}, the force can be obtained as

$$F_{disk} = \rho A_{disk} V_{disk} (V_\infty - V_e). \tag{4.33}$$

The force can also be obtained through the equations in Section 4.2.2. (by using $V_{local} = V_{disk}$), by calculating the thrust force

$$F_{disk} = -F_N \sin\theta + F_T \cos\theta. \tag{4.34}$$

These equations can be solved, which is the basis of the streamtube model (note that both F_N and F_T are functions of V_{disk}).

The single streamtube model was originally developed by Templin [28]. The model was later improved by Lapin [29], who introduced the double-streamtube model. The idea here is that one should use one streamtube for the upwind part, and one for the downwind part, assuming that the pressure in between is the atmospheric pressure, which allows for the outlet of the first streamtube to be the inlet of the second. This will allow for different flow speeds upwind and downwind. Another improvement was the multiple streamtube model by Strickland [30], who instead wanted to account for flow velocity variation between the center and the sides of the turbine. These two were later combined into the double multiple streamtube model that both use multiple streamtubes, and one set of streamtubes upwind and another downwind. As the double multiple streamtube outperforms the earlier versions, and still is relatively simple to implement and runs fast, only that model will be covered here. Two different double multiple streamtube models were created in the early 1980s: one by Paraschivoiu [31], which assumes that the flow velocity only has an $\hat{x}$ component throughout the entire turbine. One by Read and Sharpe [32], who instead assume that the streamlines expand linearly through the turbine, and the expansion can be calculated by fulfilling the continuity equation with the assumption that the streamlines are straight in the middle of the turbine, and that all flow expansions only occur in the horizontal plane. The model by Read and Sharpe will therefore have a velocity in the $\hat{y}$ direction (but $V_{local,z} = 0$) in (4.25) and (4.26). As the flow expands both in the vertical and the horizontal directions, and this expansion depends on the aspect ratio and the proportion between how much energy is absorbed upwind and downwind, which of these two models that will work best will vary from case to case.

A brief implementation of the method by Paraschivoiu will be given here. Start by dividing the turbine into an array of streamtubes according to Figure 4.12.

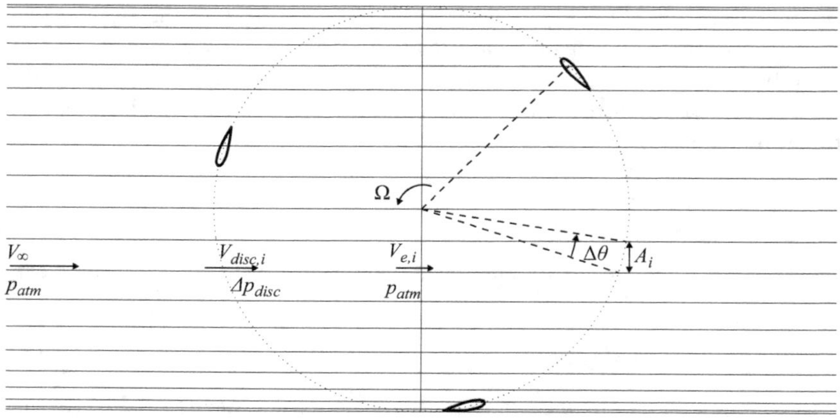

Figure 4.12 The discretization used in the streamtube model. Here, the velocities and pressures are only shown for the upwind part

Vertically, an even grid size of Δz is common to use. In the horizontal plane, one can either chose even grid size in the $\hat{y}$ direction, or one can use an even step size in $\Delta\theta$ along the swept surface. The later will be used here, as it gives a better resolution of the turbine forces close to the sides. Now, for the upwind part, the procedure is as follows:

1. Define an interference factor u_i as $V_{disk,i} = u_i V_\theta$ for each streamtube, with size $A_i = \Delta z \times \Delta\theta|\sin\theta_i|$ where θ_i is the position of the center of the streamtube at the turbine disk. Initialize u_i to a suitable starting guess (1 can be used).
2. With $V_{local,x} = V_{disk,i}$, calculate flow relative flow velocity and angle with (4.25) and (4.26).
3. Apply all desired correction models, such as corrections due to flow curvature, and then calculate angle of attack according to (4.10).
4. With the angle of attack calculated, use the force model and calculate lift and drag coefficients, and convert them to normal and tangential forces with (4.27)–(4.29) and (4.11)–(4.14) with $A_{blade} = c\Delta z/\cos\eta$ according to step 1. Calculate the thrust force according to (4.34). This is done for every streamtube.
5. The new interference factor is now obtained from

$$1 - \frac{1}{u_i} = \frac{1}{2V_{disk,i}^2 \rho R_i |\sin\theta_i| \Delta z \Delta\theta} = F_{disk,i}, \tag{4.35}$$

   iterated from step 2 until convergence.
6. Calculate the velocity at the end of the streamtube as $V_{e,i} = (2u_i - 1)V_\infty$. Now, the same procedure can be done again for the downwind side, but with $V'_{disk,i} = u'_i V_{e,i}$ instead for step 1 (note that all variables, including this u'_i will have new values for the downwind part).

The previous iterative procedure may not always converge (especially around the stall angle and depending on the force model used). If convergence issues occur, one solution is to apply a low-pass filter to the changes in u_i. In the model of Paraschivoiu with straight streamtubes, it is possible to solve the upwind part first and then solve the downwind part after, i.e. the model assumes that nothing that happens downwind can affect the solution of the upwind half. If corrections to the flow direction due to flow expansion are to be included as in [32], the upwind and downwind regions have to be solved combined as there will be feedback from the downwind region to the upwind region. Note that one should check the validity of the model in terms of velocities in the streamtubes. The streamtube approximation is only valid for interference factors above 0.5. This is extra important to check for the upwind part, as an interference factor below 0.5 will cause a negative flow velocity as starting velocity for the downwind part.

One of the weaknesses with the streamtube model is that it does not have any real flow description (i.e. the flow is only known at the turbine disk and in the transition region between the upwind and downwind regions). Hence, many additional models are required to get good results. The curvature corrections are already mentioned in Section 4.2.2.2. One additional correction model is the 3D tip corrections, where Sharpe used screw propeller corrections [9] while Paraschivoiu also used corrections from finite wing theory. See [24] for details about these corrections. Wind shear can be included in the model by adjusting the inlet velocity of the streamtubes to match a given velocity profile. Struts are a bit complicated to include if they are mounted at an angle and generate lift force, as some streamtubes now will be intersected by blades or struts 4 times instead of 2. The coupling effects between struts and blades can also not be modeled by the streamtube model.

The advantages with the streamtube model are that it is very fast to run, and also relatively easy to implement in its basic form (but can become complicated with all the extra correction models required). It can handle 3D turbine shapes, such as curved bladed turbines. The streamtube model is mainly useful for quick studies of blade forces and powers and is useful as an initial design tool to quickly iteration through many designs. It is limited to simple geometries and cannot handle interactions between turbines. The physics behind it can also be questioned, such as that the implementation shown here with straight streamtubes violates the continuity equation or that the downwind part does not affect the upwind part.

4.5 Actuator cylinder models

The actuator cylinder is a stationary model for the VAWT. The benefits, compared to the double multiple streamtube model, is that this model has a more realistic model for the flow. The model was originally developed by Madsen [33]. It has similarities with the traditional actuator disk model used for HAWTs, where the turbine is approximated as a flat disk with a pressure drop. As the VAWT swept area is cylindrical shaped, the turbine will be represented by a cylindrical surface, for which the force can be distributed over. The actuator cylinder model is based on the stationary Euler's equations

$$\rho\left(\vec{V}\cdot\nabla\right)\vec{V} = -\nabla p + \vec{f}, \tag{4.36}$$

combined with the continuity equation. The actuator cylinder is represented by the external force $\vec{f}$, which is the mean force over the swept area of the turbine, and can be obtained in a similar way as the normal and tangential forces are calculated for the streamtube model, with the difference that the flow speed is obtained from a solution to (4.36). Similar to the streamtube model, if the flow is determined from the solution to Euler's equation, but will also depend on the forces from the blade force model used, this model will have to be iterated to find a solution. The details on how to implement the solution are described in [34] and are omitted here.

The suggested approach is to split the velocity field into a linear component and a nonlinear component, where the nonlinear component is solved iteratively. A full solution is reported to take 30–60 s in 2D according to [34] for a single wind speed there is the possibility to find good approximations to the nonlinear part as described in [34], which significantly improves the speed, this makes this model a suitable choice if many simulations need to be performed.

One limitation of the steady-state model is that it will not resolve individual tip vortices, which will introduce an error for 3D flows. A steady-state model will also not capture the leading edge vortices due to dynamic stall, which can have an important effect on the downwind half of the turbine. The use of a blade force model to calculate the forces will have the same limitations as it has for the streamtube model, in respect to that it only works for known blade profiles. While flow expansion effects are captured by the use of (4.36), the effects of flow curvature will need to be corrected for manually. There will also be limitations to complex geometries, as interactions such as those between struts that generate lift and the blades will have to be modeled externally.

4.6 Actuator line models

The ALM is a time-dependent model where the blade is approximated as a force distribution along the blade span. The method is commonly used for HAWTs for wake modeling [35]. The method uses a blade force model, similar to the previously mentioned models. The idea is to replace the blade with an equivalent force, which, compared to traditional CFD, removes the need to model the flow in the boundary layer. The model is therefore significantly faster than CFD simulations that resolve the blades, but the time-dependent flow field solution will be significantly slower than streamtube models or the actuator cylinder model.

As a time-dependent model, the turbine will have to be simulated for many revolutions until a wake has developed behind the turbine. The required number of revolutions varies depending on several factors, of which the thrust force is a major contributor. A large thrust force will generate a more significant wake and will hence require more revolutions for convergence. One should also be aware that 2D simulations will in general require more revolutions for convergence than 3D simulations. Another important factor to consider for all CFD-based models is the size of the simulation domain. The domain needs to be sufficiently long for a proper wake to develop. The domain also has to be wide enough to avoid artificially increasing the power due to blockage. It was shown by Garret and Cummins [36] that the peak power in a channel is $16/27\,(1-\varepsilon)^{-2}$ where ε is the blockage ratio, defined as the turbine cross-sectional area/simulation domain cross-sectional area, showing that the blockage ratio needs to be small to avoid numerical errors.

The implementation for the ALM for the VAWT is very similar to how it is implemented for the HAWT and will therefore only briefly be described here. For details, please see e.g. [35,37]. Start by dividing the blades into segments in the spanwise direction. For each segment at each time-step, the flow speed can be sampled and equations in Section 4.2.2.4 can be used to calculate relative wind speed, angle of attack that are needed for the blade force model. With the force known, it can be applied to the Navier–Stokes equations as an external force (similar to (4.36)), see [35,38] for details. It is common to use a fixed mesh for actuator line simulations. In this case, the mesh cell where the blade segment currently is located needs to be identified. To avoid numerical instabilities, one can distribute the force over the mesh cells near the location of the blade segment using a Gaussian distribution centered at the force center of the blade (typically the quarter chord position).

One important factor to consider is that a force will generate a circulation according to Kutta–Joukowski's lift theorem. This circulation will not be centered exactly at the blade position due to numerical errors, combined with the fact that the blade is moving through the mesh. Hence, this circulation can interfere with the velocity sampling and methods to reduce this error are described in [39]. This also puts a limit on the time-step and it is generally recommended that the blade should move a shorter distance than the mesh size at each time-step.

The flow modeling in the ALM has issues resolving the flow close to the blade tip (the lift force will not automatically approach zero at the tip). Hence, one generally needs to apply a 3D correction model to force the lift to zero at the blade tip. Methods for handling this can be found in [40–42]. This gives the limitation that the ALM only works well for blade shapes that have a good 3D tip correction model.

A challenge with the ALM is how to handle struts for the VAWT, especially when they generate lift force. A simple approach is to assume that the struts are independent blades and solve them separately. A problem occurs at the attachment between the blades and the struts, as the circulation from the struts will interact with the circulation of the blade. The authors do not know of any correction model for VAWTs that properly handles this interaction effect. Hence, there will be a modeling error at the attachment when the circulation of the strut propagate into the blade.

One limitation of the ALM is that it cannot capture the leading edge vortices generated from dynamic stall (see Figure 4.9). The source of this limitation is that the blade is modeled as force distribution, which generates equivalent circulation around it. Even if the dynamic stall model properly can calculate the lift and drag forces due to dynamic stall, the leading edge vortex will not be introduced into the flow. This means that the interaction between the leading edge vortex and the blade

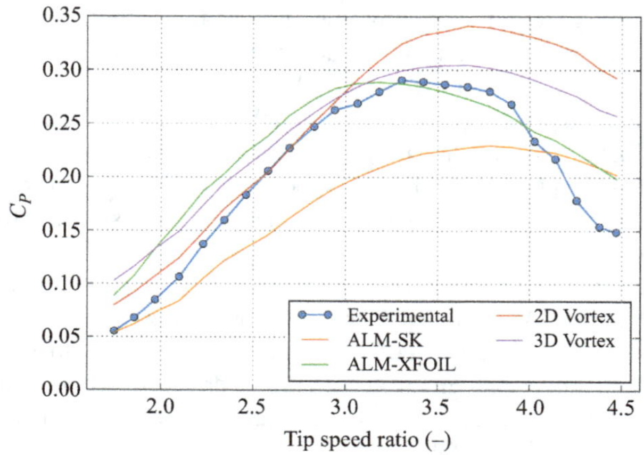

Figure 4.13 C_p for a VAWT as a function of λ. The ALM considers force coefficients from the report of Sheldahl and Klimas (SK) [27] and XFOIL program [43]. From [26]

at the downwind part will be neglected by the ALM (this is neglected by all previous models as well).

It is necessary to highlight that the results obtained from the ALM are extremely sensitive to the chosen coefficients for the force modeling, as shown in Figure 4.13, where the same turbine has been simulated using both the data set by Sheldahl and Klimas [27] and using XFOIL [43], showing large differences. The figure also shows the same turbine simulated with the vortex method (see Section 4.7) with the Sheldahl and Klimas data set, which illustrates that small changes in the evaluated angle of attack give large differences in turbine performance. The 2D method in this figure does not include the struts that are mounted with an angle and hence generate lift force and thereby torque. This is beneficial for low λ, but negative at high λ, and explains why the 2D model predicts a lower power at low λ.

ALMs are typically implemented as an addition to existing CFD software. This has the advantage that all the capabilities of the CFD software can be used within this model. This is a significant strength of the model. First, it makes it easier to implement it as there are available CFD software that can be used, and second, the capabilities that can be used from the existing software make this model extra useful to model wind turbines in complex wind fields. This makes the model popular for wake modeling and allows for modeling of farms of VAWTs. The disadvantage is that the simulations can be time-consuming, at least if the wake needs to be resolved properly.

4.7 Vortex model

The vortex model is an alternative to solving the traditional Navier–Stokes equations, which is mainly seen for time-dependent simulations. The vorticity is defined as $\vec{\omega} = \nabla \times \vec{V}$ and one can easily verify that $\nabla \cdot \vec{\omega} = \nabla \cdot \left(\nabla \times \vec{V}\right) = 0$ (vector identity), i.e. the vorticity field is divergence free. The vorticity equation is based on taking the curl of Navier–Stokes equations, which gives

$$\frac{\partial \vec{\omega}}{\partial t} + \left(\vec{V} \cdot V\right)\vec{\omega} = \left(\vec{\omega} \cdot \nabla\right)\vec{V} + \nu \nabla^2 \vec{\omega}. \tag{4.37}$$

The left hand side is the vortex convection, which states that the vorticity will be propagated with the flow velocity. $\left(\vec{\omega} \cdot \nabla\right)\vec{V}$ is the vortex stretching, which ensures that the vorticity field stays divergence free. $\nu \nabla^2 \vec{\omega}$ is the viscous diffusion. The advantages of using the vorticity as discretization parameter is that it divergence free, which means that it is a conserved quantity and vorticity can only be created at the boundaries of the system. The vortex method makes it easier to keep the vorticity conserved (compared to the traditional velocity as discretization variable which can have artificial dissipation of vorticity). This has the potential to give more stable simulations, which should allow for fewer discretization elements and larger time-steps. A second advantage is if one uses a homogeneous inflow with zero vorticity. Here, only the wake will contain vorticity, and hence, only the wake needs to contain discretization elements, which significantly can reduce the number of discretization elements. This benefit does not apply if wind shear and atmospheric turbulence are included. For more information about vortex methods in general, see [44].

Vortex methods are commonly implemented in a Lagrangian way (optionally with remeshing onto a grid). In a purely Lagrangian implementation, the vortices are released from the turbine blades and propagated with the flow velocity. Here, the vortex particle will be represented by its position $\vec{x}_i$ and its circulation $\vec{\Gamma}_i$. The velocity can be reconstructed using Biot Savart's law

$$\vec{V}(\vec{r}) = \sum_{i=1}^{N} \frac{1}{4\pi} \frac{\vec{\Gamma}_i \times (\vec{r} - \vec{r}_i)}{|\vec{r} - \vec{r}_i|^3} g\left(\frac{|\vec{r} - \vec{r}_i|}{\delta}\right), \tag{4.38}$$

where the function g is a kernel function with cutoff parameter δ that is used to remove the singularity as $\vec{r} - \vec{r}_i$. The other approach is to utilize the stream function ($\vec{V} = \nabla \times \vec{\Psi} \Rightarrow \nabla^2 \vec{\Psi} = -\vec{\omega}$) and solve Poisson's equation, which for a regular grid can be solved using fast Poisson solvers that rely on the fast Fourier transform (FFT) [45,46].

If Biot Savart's law is used, the simulation domain will be infinite. Hence, the model is best suited for the case where wind shear is neglected (as that would mean vorticity everywhere that in an infinite domain is challenging to handle). A direct solution of (4.38) would require $O(N^2)$ operations, where N in the number of vortices. The method can be accelerated by using, e.g. the fast multipole method (FMM) [47], which can achieve approximately $O(N)$ performance. Still, a characteristic of the method is that it runs fast in the beginning where the number of vortices is fewer and slows down at later stages. The approach with Poisson's equation using fast Poisson solvers is instead based on a periodic domain.

The asymptotic run-time is that of the FFT $O(N \log N)$, although the constant term in front is much smaller that for the FMM. The disadvantage is the regular grid that increases the number of vortices required.

Vortex methods are significantly easier to implement in 2D as the stretching term vanishes in 2D (the vorticity vector is orthogonal to the velocity vector). The divergence-free condition is easily satisfied in 2D, as it is naturally satisfied by conserving the vorticity of the system. The natural implementation in 2D is to let the vortices be particles. In the 3D case, there are two popular implementations: the vortex particle method and the vortex filament method [44]. The vortex particle method is based on point particles, while the filament method typically is implemented as a vortex lattice behind the blades. In the filament method, one connects the nodes of the lattice by filaments (see Figure 4.14) and allows the nodes to move with the flow velocity. The advantage of the filament method is that the vorticity field will stay divergence free, and the stretching is handled automatically when the nodes change their relative positions. Hence, a basic implementation of the filament method is easier to implement and has been implemented for VAWTs in, e.g., [48,49]. The disadvantage is that there is no natural way of including diffusion into the method (except for core spreading that only works to some degree until remeshing is required, which is problematic in the filament approximation). Some distance downwind, where the wake starts to destabilize, the node points will drift very far away from each other, creating too long filaments. While one can add extra node points along the filaments (which means more computational elements), the method will have issues with the turbulent wake due to the lack of diffusion. Hence, the method can be used in the near wake and for blade force calculations but is questionable for wake studies and turbine

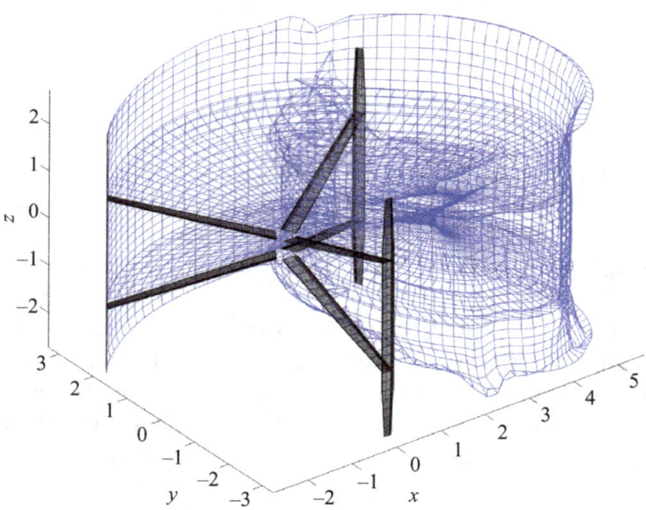

Figure 4.14 Example of a vortex filament lattice generated behind one of the blades and its support arms. From [26]

to turbine interaction. A disadvantage of filaments is also that they are harder to accelerate with methods such as the FMM, as long filaments will have to be divided into smaller parts for the algorithm to work; hence, the speedups are not as good for filaments as for point vortices.

The other approach is vortex particles, which has been implemented for VAWTs in [50]. This has the difficulty that it does not automatically satisfy the divergence-free property. Hence, vortex particle methods in 3D will require both vortex stretching calculations, a method for divergence relaxation to remove artificial divergence and it is common to also use some remeshing model to redistribute the particles onto a more regular lattice. These extra steps make the model more complicated to implement. The advantage is that diffusion can be performed, which allows one to avoid the problem of the filament method with too long filaments.

One interesting aspect of vortex methods is the possibilities for blade modeling. Vortex methods are typically used with some simplification for the blade force modeling. Here, the simplest option is to assume a lifting line approximation where the blade is a single vortex line (point in 2D), which can be implemented as a line of filaments with its strength representing the bound circulation. Here, one can sample the flow velocity at the center of the line. The benefit of the vortex method is that one knows exactly where this line is, and hence, one can ensure that the bound circulation does not interfere with the velocity sampling. With this approximation, flow curvature effects need to be added after, and the solution can experience issues close to the blade tips. A more advanced option, which can be recommended, is to include a panel method [51] to model the blade. The panel method is used to solve the Laplace equation and hence ensure that the no-penetration boundary condition is satisfied for the blade. The simplest panel method is to model the blade as a flat plate along the mean chord line (can be implemented as a lattice of filaments with the no-penetration boundary condition fulfilled in its center), or more advanced models can also model the blade surface. These methods will handle flow curvature effects and 3D tip effects directly, and the method can therefore be used for complicated blade geometries. With the panel method, it is common to solve the Kutta-condition to calculate the blade circulation. This will give the circulation for attached flow. Here, one can either use this value direction, but this will significantly overestimate the turbine performance, or one can apply a blade force model, which can include stall effects, see Section 4.2.2.3. Here, one needs to convert the circulation back into an angle of attack, which can be done, either by assuming a linear relationship, or one can create a lookup table by running the code for a fixed wing with known angles of attack in a constant straight flow. A third interesting potential for blade force modeling is the double-wake method that combines the potential flow model with an integral boundary layer method. Here, the boundary layer equations are used to calculate the flow in the attached part of the boundary layer, and where the separation point is estimated, vorticity is released to force flow separation. This has the potential to give a model that works without the limitations of the blade force models, which only works for known airfoils where data is available. It can also introduce the leading edge vortex into the flow, which otherwise is missing from the other simplified models. This has been implemented for VAWTs in [52] in 2D.

The advantages with the vortex method are that it has the potential to run faster than corresponding CFD models based on the finite element/volume method, which offer more flexible blade modeling than, for example, the ALM. A disadvantage is that it is complicated to generalize the model to complex terrains and atmospheric boundary layers.

4.8 CFD with resolved boundary layers

CFD modeling with fully resolved body-fitted grid is one of the most reliable approaches for studying the complex and unsteady aerodynamics of VAWTs. The main advantage of this approach, compared to the models described in the previous sections, is the capability of solving the boundary layer of the body geometries without any approximation (or simplification), and therefore being able to properly reproduce detailed phenomena around and within the rotor region with a high level of accuracy.

This method has a considerable computational cost since it locally solves the governing Navier–Stokes equations in highly refined mesh regions near to the blade boundary layers as shown in Figure 4.15. The implementation of this model is unfeasible, in terms of calculation time, for studying large-scale domains such as wind farms.

The approach requires a moving mesh to reproduce the motion of the rotor. This can be considered a disadvantage compared to the simplified models in terms of the complexity for setting up cases. The conventional configuration consists of a cylindrical moving mesh inside a fixed mesh, with the implementation of different mesh refinement regions (usually the finer mesh cells are in the zones near the rotor), as illustrated in Figure 4.16. This can be compared to the ALM, which also employs a CFD flow solver, but where the elements are moving through a fixed mesh.

Among the main numerical aspects that have direct influence into the accuracy of the results, it can be mentioned the mesh size and structure (i.e. hexagonal, tetrahedral), time discretization, and rotational velocity (tip speed ratio), as well as the turbulence model.

In the stage of preliminary investigation of the turbine performance, there is the need for simulations with relatively low computational cost. The first approach is the 2D simulation; however, the results should be carefully interpreted due to the lack of

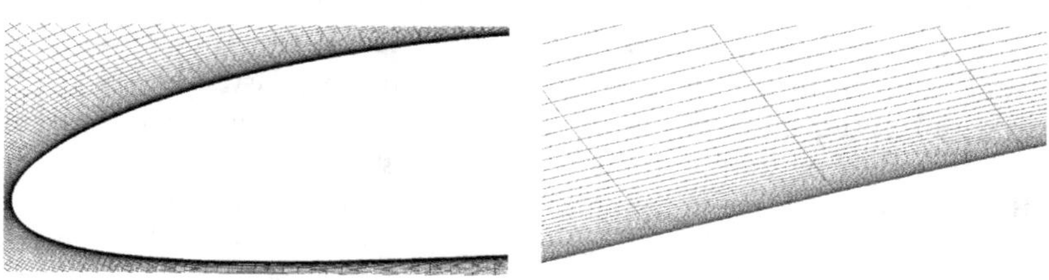

Figure 4.15 Typical hexagonal mesh configuration for an NACA airfoil CFD simulation: (left) overall view and (right) boundary layer detail

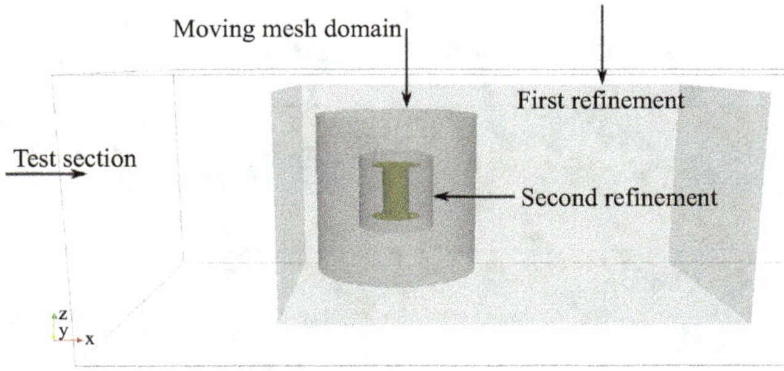

Figure 4.16 Conventional configuration of a chamber domain for a VAWT CFD simulation, showing the fixed outer domain and the rotating inner domain as well as examples on mesh refinement regions. From [4]

the 3D effects. In particular, the performance of the turbines is sensitive to the losses and other complex phenomena that are discussed in the following sections.

4.8.1 Savonius

While Savonius rotors have a relatively simple design, their flow field structure is characterized by a complex aerodynamics. Solving the Navier–Stokes equations by using CFD simulations with resolved boundary layers is the required method to analyze the resulting flow within and around the rotor. The reliability of numerical modeling for this kind of turbine is directly related to an accurate prediction of the flow separation point on the outer surface (and therefore a proper estimation of the resulting torque). Other models do not have the capability to do this. Moreover, the complexity of the flow pattern increases for nonconventional geometries, such as rotors with twisted blades, in which the released vorticity is more pronounced. In Figures 4.17 and 4.18, the normalized streamwise velocity and the turbulent kinetic energy for both straight and twisted bladed Savonius turbines operating at optimal λ are revealed, respectively. This allows to identify the overall structure of the resulting wake, as well as the influence of the generated vorticity on it. The resulting flow is asymmetric with the main flow direction. Vortical structures are generated in blade tips and end plates, which are dissipated along the main flow direction. A slower wake recovery process is observed for the turbines with twisted blades, as well as the incoming flow deflection in the direction parallel to the rotational axis (large change of direction). To visualize the vortex formation, isolines of the Q-criterion are employed, which are defined as the flow regions with a positive second invariant of the velocity [53,54]. It is observed that larger turbulence levels produce a faster wake recovery process due to the improvement of the mixing process and momentum transfer.

Among the main design parameters that can be properly reproduced only by using 3D CFD modeling, it is highlighted as follows:

- **The aspect ratio,** which is defined as the ratio of the turbine height and its diameter, playing an essential role for maximizing the rotor performance.

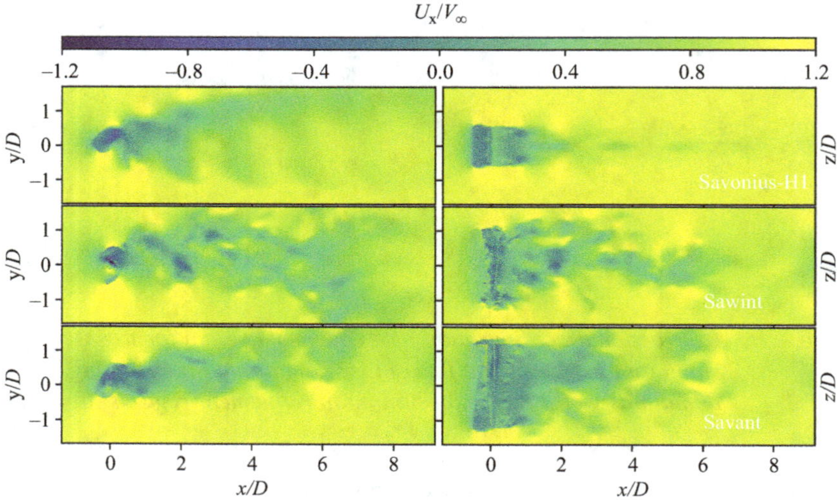

Figure 4.17 *Instantaneous normalized streamwise velocity in the horizontal and vertical middle plane, at optimal λ for both straight (SAVONIUS H1) and twisted (SAWINT and SAVANT) bladed Savonius rotors. From [4]*

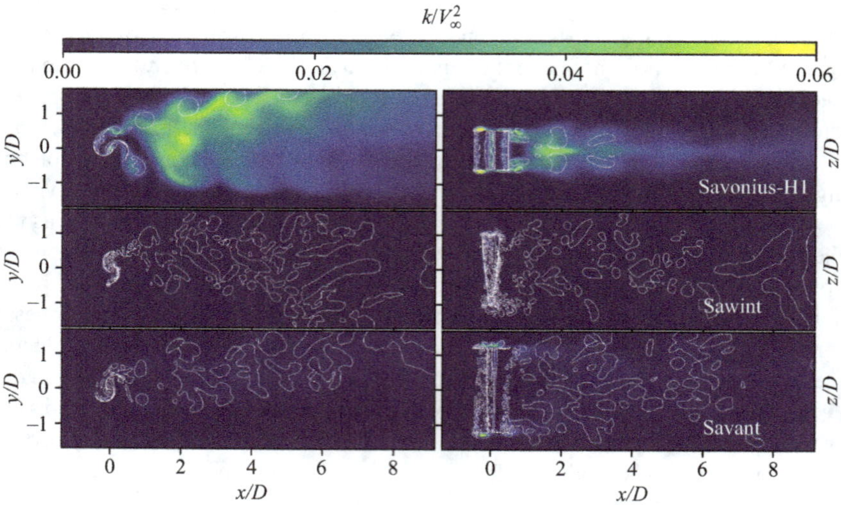

Figure 4.18 *Normalized time-averaged turbulent kinetic energy (TKE) in the horizontal and vertical middle plane, at optimal λ for both straight (SAVONIUS H1) and twisted (SAWINT and SAVANT) bladed Savonius rotors. The isolines represent a Q-value of 10. From [4]*

- **Top and bottom end plates,** which help to prevent flow leakage from the concave surface of the blades toward the main external flow. This keeps satisfactory levels of pressure difference between concave and convex sides of the blades along the height of the rotor.
- **Multistaging,** which is an alternative to reduce the large torque fluctuations with the rotor angle by mounting single-stage rotors one above the other with an equal phase shift.

Figure 4.4 depicts some of these main parameters. Other design parameters such as the gap and overlap ratios, number of blades, blade profile, Reynolds number, tip speed ratio can still be analyzed by using 2D CFD simulations, with the assumption that the aspect ratio of the studied rotor is large enough to neglect 3D effects.

4.8.2 Darrieus

The Darrieus rotor is more time-consuming to simulate than the Savonius rotor with CFD when the blade boundary layers are to be resolved due to the small size of the blades compared to the turbine diameter. The main challenge is that one needs a very fine mesh at the blade surface to refine the boundary layers, which gives small time-steps. At the same time, the turbine has to run for many revolutions to properly build up a wake, which will make the total number of time-steps large. It was shown in [55] using 2D simulations that a very fine mesh is required for mesh independence. The domain size requirements and number of revolutions needed should be similar to the ALM (see Section 4.6), as the physics behind the limitations are the same. This has been studied in, e.g. [56], suggesting 20–30 revolutions and at least 10 turbine diameter length from the turbine to both inlet and outlet for convergence. The reader should be aware that these values will vary depending on the turbine design, tip speed ratio, and the desired level of convergence. If the main aim is to calculate blade forces/turbine performance, one technique to increase the speed of the simulations is to run the initial simulations with a lower mesh resolution to generate the wake, or possibly using a simplified model such as the ALM to generate a wake, which can be used as initial condition for the CFD simulation. The required accuracy of the wake decreases further downwind from the turbine; hence, one can use less accuracy for the initial parts of the simulation. The wake is mainly dependent on the lift force (for high enough λ that stall is not important), so it is important to have a good description of this force during low accuracy initial simulations for wake generation.

Simulations of VAWTs have been performed with both LES [57,58] and RANS [59,60]. For more details about the choice of method, see [61]. An overview of different turbulence models to choose from for RANS can be found in [62], with the recommendation to use a k–ω to SST-based model for best results. CFD simulations are still commonly performed in 2D due to the required computational resources if a larger number of simulations are needed, while 3D models can be used if only a few simulations are needed. An example of a case where 3D is needed is shown in Figure 4.19, where the difference in vorticity is shown with respect to if struts are included or not.

Figure 4.19 CFD modeling of vortical structure for a Darrieus by using an isosurface $Q = 6$: with (left) and without (right) considering struts. This shows the added complexity of the vorticity generated from the turbine when struts that generate lift are included. Courtesy of Aya Aihara, Uppsala University

The main advantage of CFD simulations with resolved boundary layers is that they work for all geometries. Also, simulation software is available, both commercial and open source options, that can run these turbines. This method works in many cases, where the simplified models show poor accuracy, such as high-solidity turbines operating at low λ and where separated flow is important for performance prediction. It is also suitable for complex geometries that are complicated to handle with simplified models and for design of custom airfoils for VAWTs. The main disadvantage is the computational time required to achieve sufficient accuracy, and it is recommended to have access to a computational cluster if this method is chosen.

4.9 Model validations

A key aspect for all simulation models is to validate the results. Common options here are either to use power coefficient data, force measurements or wake measurements for validation. The most suitable method will depend on the intended application and the availability of experimental data. The first option to use power coefficients has the advantage that this has been measured for many vertical axis turbines and is important if the aim is to design a turbine with high efficiency. The disadvantage is that it only gives one value for each simulation, while the power coefficient is a combination of many different aerodynamic phenomena. Hence, a good agreement in power coefficient does not ensure accurate simulations. The value of the power coefficient will depend on the losses of the system. Especially for simplified models, it is common to neglect minor losses, such as losses from blade–strut attachments, and in 2D, many losses are neglected. Unless all details of the turbine are included, one can therefore expect that the simulations should predict a higher power coefficient than the experiments.

Validations with force measurements of blade forces have the advantage that a blade is a more isolated system that the entire turbine, and one can therefore perform much more detailed validations. One will also have force values for each azimuth angle. If possible, we recommend to perform this validation. The disadvantage is the limited availability of data. Some examples of data that can be used for Darrieus turbines are the open field measurements of a curved bladed rotor by Akins [63] and the normal force measurements on the turbine in Figure 4.3 by Dyachuk *et al.* [64], and also the wind tunnel measurements of McLaren *et al.* [65] and Peng *et al.* [66].

The third option is to validate the velocity field, which is important for wake studies and has the advantage that each simulation allows for detailed validation studies. These measurements are generally performed in wind tunnels (or towing tanks) to allow for controlled measurements and some examples of measurements of Darrieus turbines are [67–70]. The disadvantage is that the data mainly is for low Reynolds numbers, which can affect the stall characteristics of the blade.

In some particular situations for the simplified models, when there is no data available for the specific studied case, one option is to validate using results of advanced models (such as CFD simulations with resolved boundary layers).

As a final remark, validations using full-scale data are complicated, due to both large computational costs and uncertainties related to the measurements as there are many parameters that contribute to the total measurement error and the lack of available data. It can therefore be a good idea to validate the simulation model for a single pitching blade first. While this will not be fully representative, it is a good check that the models produce realistic forces for known profiles. There are plenty of measurements of blades in wind tunnels with more controlled circumstances than for blades in VAWT motion.

4.10 Discussion and conclusions

All simulation models mentioned here have advantages and disadvantages. The aspects that will determine which model to use is which output that is needed from the models, how many simulations that have to be performed and the available computational resources vs the actual time available (and which models that are available). For the Savonius rotor, we recommend to fully resolve the flow with CFD simulations and the main choice is 2D vs 3D simulations, where the significantly lower computational time of 2D simulations can be used for initial designs, while 3D simulations can be used to finalize the design.

The Darrieus rotor has many more options as there are several simplified models available. The most general model uses CFD with the blade boundary layers resolved. It has the advantage that it works for all geometric shapes of the blades and struts, and it does not need any additional correction models nor any experimental blade data, and existing software is readily available. The disadvantage is that it is very time-consuming, especially in 3D when sufficient mesh resolution is used to get accurate forces. A main challenge for the simplified models is the need for a blade force model, and unless there already is a calibrated model for the airfoil in question, one will have to be created, either through blade measurements, or one could use CFD models (or generate coefficients with a simplified tool such as XFOIL). These models are often only accurate for limited angles of attack, and the performance decreases in deep stall. Due to the complicated blade motion, it is harder to get a well-calibrated blade force model for the VAWT, compared to the HAWT, and the challenge to model the blade forces increases with decreasing λ due to deeper dynamic stall. Hence, the limitation of this blade force model will limit the accuracy of the simplified models. However, the blade force models are still sensitive to the estimated angles of attack from the simplified models (see Figure 4.13), and hence it is important to have an accurate flow model to properly simulate the turbine.

Assuming that the simplified models can be used, the most suitable model to use will depend on the application. The streamtube model is fastest and is good when many simulations are needed. It can therefore be a good model to start with when designing a turbine, as it will give an approximate design that is acceptable, and it will give good approximations to the forces experienced on the turbine blades. The actuator cylinder model can also be suited for cases where a larger number of simulations are needed. ALMs and vortex models have similar levels of approximations and are useful when refining a turbine design and for studying 3D effects. The vortex model has benefits in terms of how to model the turbine and handle different turbine geometries, while the ALM has benefits when modeling complex winds and atmospheric boundary layers. CFD models with resolved boundary layers are useful for studying complex behaviors that the simplified models cannot handle and to study airfoil design where the blade force models do not work.

References

[1] Eriksson S, Bernhoff H, and Leijon M. Evaluation of different turbine concepts for wind power. *Renewable and Sustainable Energy Reviews*. 2008; 12(5): 1419–1434.

[2] Goude A. Fluid Mechanics of Vertical Axis Turbines: Simulations and Model Development [dissertation]. Uppsala University, Electricity; 2012.

[3] Jang H, Paek I, Kim S, *et al.* Performance prediction and validation of a small-capacity twisted Savonius wind turbine. *Energies*. 2019;12(9):1721.

[4] Mendoza V, Katsidoniotaki E, and Bernhoff H. Numerical study of a novel concept for manufacturing Savonius turbines with twisted blades. *Energies*. 2020;13(8):1874.

[5] Ashwill TD, Sutherland HJ, and Berg DE. *A Retrospective of VAWT Technology*. SandiaLabs., Albuquerque, NM (USA); 2012.

[6] Rossander M, Dyachuk E, Apelfröjd S, *et al.* Evaluation of a blade force measurement system for a vertical axis wind turbine using load cells. *Energies*. 2015;8(6):5973–5996.

[7] Migliore PG, Wolfe WP, and Fanucci JB. Flow curvature effects on Darrieus turbine blade aerodynamics. *Journal of Energy*. 1980;4(2):49–55. Available from: https://doi.org/10.2514/3.62459.

[8] Rainbird JM, Bianchini A, Balduzzi F, *et al.* On the influence of virtual camber effect on airfoil polars for use in simulations of Darrieus wind turbines. *Energy*

Conversion and Management. 2015;106:373–384. Available from: www.sciencedirect.com/science/article/pii/S0196890415008985.

[9] Sharpe D. Refinements and developments of the multiple streamtube theory for the aerodynamic performance of vertical axis wind turbines. In: *Proc. Sixth BWEA Wind Energy Conference, Reading*; 1984. p. 148–159.

[10] Dyachuk E. Aerodynamics of Vertical Axis Wind Turbines: Development of Simulation Tools and Experiments [dissertation]. Uppsala University, Electricity; 2015.

[11] Laneville A and Vittecoq P. Dynamic stall: The case of the vertical axis wind turbine. *Journal of Solar Energy Engineering*. 1986;108(2):140–145. Available from: https://doi.org/10.1115/1.3268081.

[12] Paraschivoiu I and Allet A. Aerodynamic analysis of the Darrieus wind turbines including dynamic-stall effects. *Journal of Propulsion and Power*. 1988;4(5):472–477. Available from: https://doi.org/10.2514/3.23090.

[13] Gormont RE. A Mathematical Model of Unsteady Aerodynamics and Radial Flow for Application to Helicopter Rotors. USAAVLABS TR 72–67. NTIS; 1973.

[14] Leishman J, Beddoes T, Westland Helicopter Ltd. A generalised model for airfoil unsteady aerodynamic behaviour and dynamic stall using the indicial method. In: *Proceedings From the 42nd Annual Forum of the American Helicopter Society*; 1986. p. 243–266.

[15] Leishman J and Beddoes T. A semi-empirical model for dynamic stall. *Journal of The American Helicopter Society*. 1989;34:3–17.

[16] Masson C, Leclerc C, and Paraschivoiu I. Appropriate dynamic-stall models for performance predictions of VAWTs with NLF blades. *International Journal of Rotating Machinery*. 1998;4:129–139.

[17] Major SR and Paraschivoiu I. Indicial method calculating dynamic stall on a vertical axis wind turbine. *Journal of Propulsion and Power*. 1992;8(4): 909–911. Available from: https://doi.org/10.2514/3.23569.

[18] Wang K, Hansen MOL, and Moan T. Model improvements for evaluating the effect of tower tilting on the aerodynamics of a vertical axis wind turbine. *Wind Energy*. 2015;18(1):91–110. Available from: https://onlinelibrary.wiley.com/doi/abs/10.1002/we.1685.

[19] Dyachuk E and Goude A. Simulating dynamic stall effects for vertical axis wind turbines applying a double multiple streamtube model. *Energies*. 2015;8(2): 1353–1372.

[20] Brochier G, Fraunie P, Beguier C, *et al.* Water channel experiments of dynamic stall on Darrieus wind turbine blades. *Journal of Propulsion and Power*. 1986;2(5):445–449. Available from: https://doi.org/10.2514/3.22927.

[21] Ferreira CS, Van Kuik G, Van Bussel G, *et al.* Visualization by PIV of dynamic stall on a vertical axis wind turbine. *Experiments in Fluids*. 2009;46(1):97–108.

[22] Tavernier DD, Ferreira C, Paulsen U, *et al.* The 3D effects of a vertical-axis wind turbine: Rotor and wake induction. *Journal of Physics: Conference Series*. 2020;1618:052040. Available from: https://doi.org/10.1088\%2F1742-6596\%2F1618\%2F5\%2F052040.

[23] Lundin S, Forslund J, Goude A, *et al.* Experimental demonstration of performance of a vertical axis marine current turbine in a river. *Journal of Renewable and Sustainable Energy*. 2016;8(6):064501. Available from: https://doi.org/10.1063/1.4971817.

[24] Paraschivoiu I. *Wind Turbine Design: With Emphasis on Darrieus Concept*. Polytechnic International Press; 2002.

[25] Moran WA. Giromill Wind Tunnel Test and Analysis. Volume II. Technical Discussion. Final Report, USA; 1977.

[26] Mendoza V and Goude A. Validation of actuator line and vortex models using normal forces measurements of a straight-bladed vertical axis wind turbine. *Energies*. 2020;13(3):511.

[27] Sheldahl RE and Klimas PC. Aerodynamic Characteristics of Seven Symmetrical Airfoil Sections Through 180-degree Angle of Attack for Use in Aerodynamic Analysis of Vertical Axis Wind Turbines; 1981.

[28] Templin R. Aerodynamic Theory for the NRC Vertical Axis Wind Turbine, National Research Council, Canada. LTR-LA-160; 1974.

[29] Lapin E. *Theoretical Performance of Vertical Axis Wind Turbines*. In: American Society of Mechanical Engineers (ASME), Winter Annual Meeting, Houston, Texas; 1975.

[30] Strickland JH. *Darrieus Turbine: A Performance Prediction Model Using Multiple Streamtubes*. SandiaLabs., Albuquerque, NM (USA); 1975.

[31] Paraschivoiu I and Delclaux F. Double multiple streamtube model with recent improvements (for predicting aerodynamic loads and performance of Darrieus vertical axis wind turbines). *Journal of Energy*. 1983;7(3):250–255.

[32] Read S and Sharpe DJ. *An Extended Multiple Streamtube Theory for Vertical Axis Wind Turbines*. WIEN; 1980. p. 65–72.

[33] Madsen H. The Actuator Cylinder – A Flow Model for Vertical Axis Wind Turbines [dissertation]. Institute of Industrial Constructions and Energy Technology, Aalborg University Centre; 1982.

[34] Madsen H, Larsen T, Vita L, *et al.* Implementation of the Actuator Cylinder flow model in the HAWC2 code for aeroelastic simulations on Vertical Axis Wind Turbines. In: 51st AIAA Aerospace Sciences Meeting including the New Horizons Forum and Aerospace Exposition, Texas, USA; 2013.

[35] Sϕrensen JN and Shen WZ. Numerical modeling of wind turbine wakes. *Journal of Fluids Engineering*. 2002;124(2):393–399. Available from: https://doi.org/10.1115/1.1471361.

[36] Garrett C and Cummins P. The efficiency of a turbine in a tidal channel. *Journal of Fluid Mechanics*. 2007;588:243–251.

[37] Bachant P, Goude A, and Wosnik M. Actuator line modeling of vertical-axis turbines. arXiv preprint arXiv:160501449. 2016.

[38] Jha PK, Churchfield MJ, Moriarty PJ, *et al.* Guidelines for volume force distributions within actuator line modeling of wind turbines on large-eddy simulation-type grids. *Journal of Solar Energy Engineering*. 2014;136(3):031003. Available from: https://doi.org/10.1115/1.4026252.

[39] Meyer Forsting AR, Pirrung GR, and Ramos-García N. A vortex-based tip/smearing correction for the actuator line. *Wind Energy Science*. 2019;4(2): 369–383.

[40] Shen WZ, Mikkelsen R, Sørensen JN, *et al.* Tip loss corrections for wind turbine computations. *Wind Energy: An International Journal for Progress and Applications in Wind Power Conversion Technology*. 2005;8(4):457–475.

[41] Glauert H. Airplane propellers. In: *Aerodynamic Theory*. Springer, Berlin; 1935. p. 169–360.

[42] Dağ KO and Sørensen JN. A new tip correction for actuator line computations. *Wind Energy*. 2020;23(2):148–160. Available from: https://onlinelibrary.wiley.com/doi/abs/10.1002/we.2419.

[43] Drela M. XFOIL: An analysis and design system for low Reynolds number airfoils. In: *Low Reynolds Number Aerodynamics*, Proceedings of the Conference Notre Dame, Indiana, USA; 1989. p. 1–12.

[44] Cottet GH and Koumoutsakos P. *Vortex Methods – Theory and Practice*. Cambridge University Press; 2000.

[45] Chatelain P, Bricteux L, Backaert S, *et al.* Vortex particle-mesh methods with immersed lifting lines applied to the large eddy simulation of wind turbine wakes. In: *Proc. Wake Conf.*; 2011. p. 120–125.

[46] Chatelain P and Koumoutsakos P. A Fourier-based elliptic solver for vortical flows with periodic and unbounded directions. *Journal of Computational Physics*. 2010;229(7):2425–2431. Available from: www.sciencedirect.com/science/article/pii/S0021999110000021.

[47] Greengard L and Rokhlin V. A fast algorithm for particle simulations. *Journal of Computational Physics*. 1987;73(2):325–348. Available from: www.sciencedirect.com/science/article/pii/0021999187901409.

[48] Murray J and Barone M. The development of CACTUS, a wind and marine turbine performance simulation code. In: 49th AIAA Aerospace Sciences Meeting including the New Horizons Forum and Aerospace Exposition, Orlando, Florida; 2011. Available from: https://arc.aiaa.org/doi/abs/10.2514/6.2011-147.

[49] Tescione G, Simão Ferreira CJ, and van Bussel GJW. Analysis of a free vortex wake model for the study of the rotor and near wake flow of a vertical axis wind turbine. *Renewable Energy*. 2016;87:552–563. Available from: www.sciencedirect.com/science/article/pii/S0960148115303554.

[50] Chatelain P, Duponcheel M, Caprace DG, *et al.* Vortex particle-mesh simulations of vertical axis wind turbine flows: From the blade aerodynamics to the very far wake. *Journal of Physics: Conference Series*. 2016;753:032007. Available from: https://doi.org/10.1088/1742-6596/753/3/032007.

[51] Katz J and Plotkin A. *Low-Speed Aerodynamics*. 2nd ed. Cambridge Aerospace Series. Cambridge University Press, Cambridge, UK; 2001.

[52] Zanon A, Giannattasio P, and Simão Ferreira CJ. A vortex panel model for the simulation of the wake flow past a vertical axis wind turbine in dynamic stall. *Wind Energy*. 2013;16(5):661–680. Available from: https://onlinelibrary.wiley.com/doi/abs/10.1002/we.1515.

[53] Chakraborty P, Balachandar S, and Adrian RJ. On the relationships between local vortex identification schemes. *Journal of Fluid Mechanics*. 2005; 535:189.

[54] Launder BE and Spalding DB. The numerical computation of turbulent flows. In: *Numerical Prediction of Flow, Heat Transfer, Turbulence and Combustion*. Pergamon press, New York, Oxford, Toronto, Sydney, Paris, Frankfurt; 1983. p. 96–116.

[55] Almohammadi KM, Ingham DB, Ma L, *et al.* Computational fluid dynamics (CFD) mesh independency techniques for a straight blade vertical axis wind turbine. *Energy*. 2013;58:483–493. Available from: www.sciencedirect.com/science/article/pii/S0360544213005100.

[56] Rezaeiha A, Kalkman I, and Blocken B. CFD simulation of a vertical axis wind turbine operating at a moderate tip speed ratio: Guidelines for minimum domain size and azimuthal increment. *Renewable Energy*. 2017;107:373–385. Available from: www.sciencedirect.com/science/article/pii/S0960148117300848.

[57] Li C, Zhu S, Xu Y-l, *et al.* 2.5D large eddy simulation of vertical axis wind turbine in consideration of high angle of attack flow. *Renewable Energy*. 2013;51:317–330. Available from: www.sciencedirect.com/science/ article/pii/S0960148112005770.

[58] Peng HY and Lam HF. Turbulence effects on the wake characteristics and aerodynamic performance of a straight-bladed vertical axis wind turbine by wind tunnel tests and large eddy simulations. *Energy*. 2016;109:557–568. Available from: www.sciencedirect.com/science/article/pii/S0360544216305138.

[59] Alaimo A, Esposito A, Messineo A, *et al.* 3D CFD analysis of a vertical axis wind turbine. *Energies*. 2015;8(4):3013–3033.

[60] Orlandi A, Collu M, Zanforlin S, *et al.* 3D URANS analysis of a vertical axis wind turbine in skewed flows. *Journal of Wind Engineering and Industrial Aerodynamics*. 2015;147:77–84. Available from: www.sciencedirect.com/science/article/pii/S0167610515002263.

[61] Ghasemian M, Ashrafi ZN, and Sedaghat A. A review on computational fluid dynamic simulation techniques for Darrieus vertical axis wind turbines. *Energy Conversion and Management*. 2017;149:87–100. Available from: www.sciencedirect.com/science/article/pii/S0196890417306519.

[62] Rezaeiha A, Montazeri H, and Blocken B. On the accuracy of turbulence models for CFD simulations of vertical axis wind turbines. *Energy*. 2019;180:838–857. Available from: www.sciencedirect.com/science/article/pii/S0360544219309168.

[63] Akins RE. Measurements of Surface Pressures on an Operating Vertical-Axis Wind Turbine. NASA STI/Recon Technical Report N; 1989.

[64] Dyachuk E, Rossander M, Goude A, *et al.* Measurements of the aerodynamic normal forces on a 12-kW straight-bladed vertical axis wind turbine. *Energies*. 2015;8(8):8482–8496. Available from: www.mdpi.com/1996-1073/8/8/8482.

[65] McLaren K, Tullis S, and Ziada S. Measurement of high solidity vertical axis wind turbine aerodynamic loads under high vibration response conditions.

Journal of Fluids and Structures. 2012;32:12–26. The 7th International Symposium on Fluid-Structure Interactions, Flow-Sound Interactions, and Flow-Induced Vibrations & Noise. Available from: www.sciencedirect.com/science/article/pii/S0889974612000254.

[66] Peng YX, Xu YL, Zhan S, *et al.* High-solidity straight-bladed vertical axis wind turbine: Aerodynamic force measurements. *Journal of Wind Engineering and Industrial Aerodynamics*. 2019;184:34–48. Available from: www.sciencedirect.com/science/article/pii/S0167610518302617.

[67] Tescione G, Ragni D, He C, *et al.* Near wake flow analysis of a vertical axis wind turbine by stereoscopic particle image velocimetry. *Renewable Energy*. 2014;70:47–61. Special issue on aerodynamics of offshore wind energy systems and wakes. Available from: www.sciencedirect.com/science/article/pii/S0960148114001335.

[68] Rolin VFC and Porté-Agel F. Experimental investigation of vertical-axis wind-turbine wakes in boundary layer flow. *Renewable Energy*. 2018;118:1–13. Available from: www.sciencedirect.com/science/ article/pii/S0960148117310753.

[69] Bachant P and Wosnik M. Characterising the near-wake of a cross-flow turbine. *Journal of Turbulence*. 2015;16(4):392–410. Available from: https://doi.org/10.1080/14685248.2014.1001852.

[70] Peng HY, Lam HF, and Lee CF. Investigation into the wake aerodynamics of a five-straight-bladed vertical axis wind turbine by wind tunnel tests. *Journal of Wind Engineering and Industrial Aerodynamics*. 2016;155:23–35. Available from: www.sciencedirect.com/science/article/pii/S0167610515300325.

Chapter 5
VAWT wind tunnel experiments

Lorenzo Battisti[1]

Experimental testing is an essential part of the wind turbine design path, as it provides early information about aerodynamics, mechanics, performance, and turbine control. According to the size and the budget or the certification purposes, the testing environment can be either natural through field testing, or controlled in dedicated facilities called wind tunnels. The testing procedure provides time series of measured data at a given sampling rate that needs proper data post-processing for a rational use. Following the stochastic nature of wind, the data treatment always needs statistical tools and appropriate uncertainty analysis. The advantage of infield measurements is to perform full-scale tests, from which the real turbine behavior can be simulated and evaluated. The site has to be certified according to Annex A of IEC 61400-12-1, 2005. A few sites have the characteristics suited to host wind turbine test campaigns, and depending on the windiness of the location from 1 to 2 years can be necessary to complete the measurements. On the other hand, wind tunnel tests can only be performed on scaled prototypes, with some dimensional limitations compared to real size turbines. The controlled environment conditions allow repeatable tests and to separate some effects that in open field tests are often superimposed and difficult to disaggregate. In particular, the calibration and validation of numerical codes can be made through controlled experiments. From the point of view of the market-oriented wind turbine development, wind tunnel tests can represent the very early stage of a more wide experimental campaign, followed by infield tests.

5.1 Wind tunnel facility

Wind tunnel facilities consist of some subsystems: a channel to direct the air circulation, a room hosting the turbine where the measurements are taken, also named test chamber, and a power unit to generate the airflow at a generic velocity with opportune devices to adjust the air conditions (basically the turbulence intensity, and the temperature) within the test chamber.

They can be broadly classified as open and closed wind tunnels depending on the fact that the air is renewed continuously or recirculated within a closed circuit

[1]DICAM, University of Trento, Italy

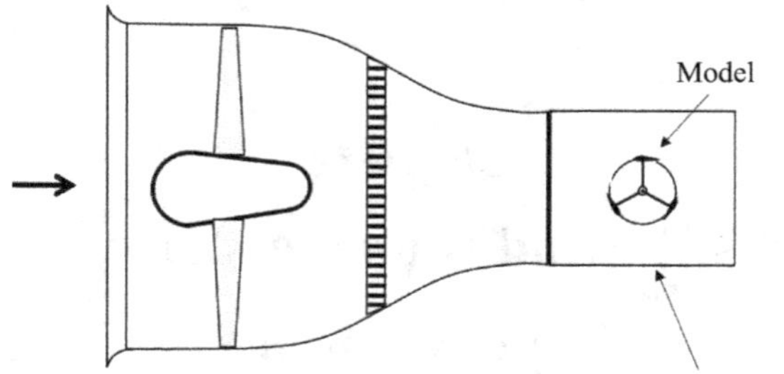

Figure 5.1 Sketch of open-flow circuit wind tunnel

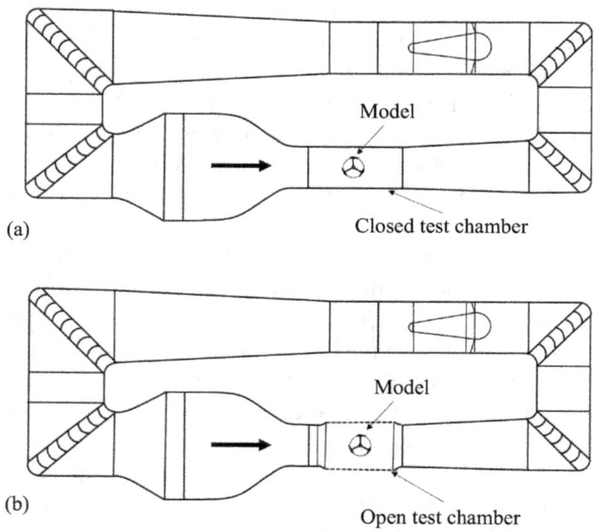

Figure 5.2 Sketch of closed circuit wind tunnels: (a) closed test chamber and (b) open test chamber

respectively. The first solution, shown in Figure 5.1, is simple, less costly, and space-saving, but the feeding air conditions can vary according to local environment climate; therefore, long tests can be affected by a shift from the initial conditions. Closed wind tunnels allow greater control of the test conditions in time, but the thermal power caused by wall-flow friction, and by the compressors or ventilators, need to be extracted from the flow to avoid a progressive variation of temperature, pressure, and humidity of the air. Closed circuits can in turn employ an open or a closed test chamber, as shown in Figure 5.2.

A closed test chamber is obtained by instrumenting a rectilinear portion of the circuit (regardless the facility is open or closed), which realizes a wall-bounded portion of space where the model is positioned, providing stable and controllable

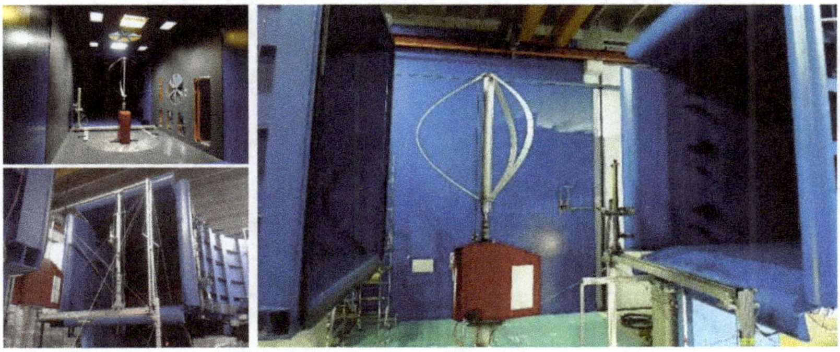

Figure 5.3 Pictures of the test chamber arrangement of the facility of the Polytechnic of Milano Bovisa [1]

flow conditions. Figure 5.3 depicts the test chamber arrangement of the facility of the Polytechnic of Milano Bovisa [1] where most of the data and procedures provided in this chapter have been generated.

An open test chamber is obtained by interrupting the circuit continuity and by testing the turbine prototype in this section (in open-flow wind tunnels, there is also the option to put the model at the piping exit, we speak in this case of open jet). Regardless of the layout employed, the presence of the prototype determines an alteration of the streamtube caused by the prototype–air interaction, which differs from what would occur in open field operation. This interaction has to be taken into consideration as it modifies the pressure, the velocity, and the turbulence intensity fields. As a consequence, the performances and the thrust of the turbine differ from similar open field conditions and some corrective actions on the data should be necessary.

5.1.1 Test rig and measurement setup

Both wind turbine prototype and wind tunnel are instrumented to acquire basic data from the tests [2]. The results and the procedure here presented derive from about 15 years of experimental test carried out on several VAWT architectures, some of them are shown in Figure 5.4.

The sketch of an instrumented turbine prototype is shown in Figure 5.5.

The sensors fitted on the turbine structure are typically strain gauges to measure deformations and forces (longitudinal and transversal thrust), accelerometers, a torque-meter, and an encoder to measure the torque, the rotational speed, and the rotor azimuthal positions, respectively. Occasionally pressure and velocity measurements are also carried out on the turbine blades. Fitting sensors on the rotating part and data delivery to a data acquisition system (DAS) is a stepping difficulty of the measurement campaign needing a slip ring or wireless technology to transfer the data from the rotating to the stationary frame. When the model is small also an aerodynamic balance can be used to measure the 2D or 3D forces acting on the turbine rotor.

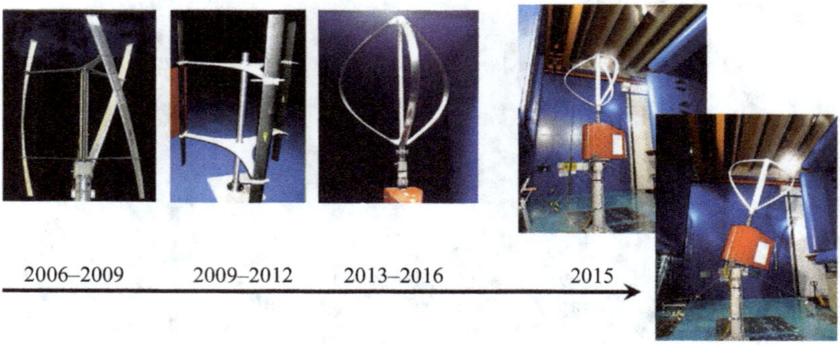

Figure 5.4 VAWT architectures tested in the facility of the Polytechnic of Milano Bovisa [1]

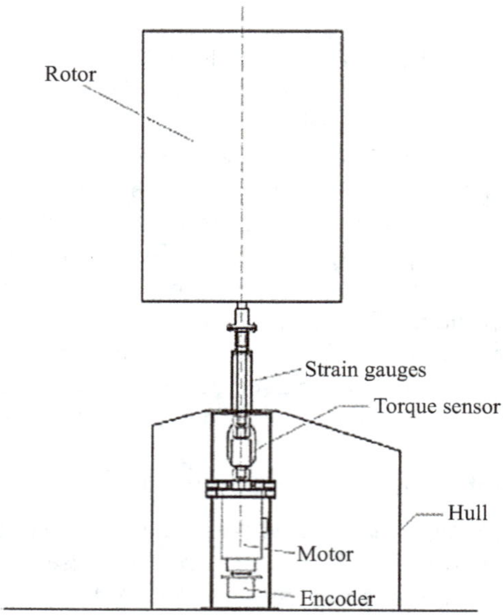

Figure 5.5 Sketch of the turbine test bench

The basic turbine experimental setup consists of the following:

- A torque-meter to measure the mechanical torque transferred to the shaft inserted directly on the power train through elastic joints (alternatively cross-mounted strain gauges can be glued on the shaft).
- An absolute encoder to capture the position of the rotor and the rotational speed of the turbine, positioned at the head of the motor—electric generator through an elastic joint.
- At least two accelerometers to measure the vibrations, installed at the top position of the bearings.

- Some strain gauges to measure the wind-wise and side-to-side deformations of the structure, and therefore the thrust to which the rotor is subjected, through a corresponding number of Wheatstone bridges glued on the shaft.

A suitably shaped hull is built around the test bench to minimize the effect of the test bench blockage and to reduce the wake dimension in the test chamber.

A set of sensors is dedicated to monitoring the wind tunnel operating conditions (velocity, pressure, and temperature) at the inlet of the test chamber and should be synchronized with the turbine DAS. A differential pressure transducer is typically used to measure the flow velocity in the tunnel during measurements in stationary conditions.

Another set of sensors is dedicated to measure the velocity and the pressure at different planes upstream and downstream of the turbine to reconstruct the wake extension. The measurement technique can measure the pressure and/or the velocity of a single spatial point in the flow field. Both aerodynamic pressure probes and hot wire anemometers are used. Directional pneumatic five-hole probes are applied to detect the total and static pressure and the 3D velocity vector. To measure the time-resolved velocity field, two single-sensors, hot wire anemometers are suggested. The sensor wires, mounted normal to the probe stem, have typically a diameter of 5 μm and are operated in constant temperature mode. Uncertainty in the velocity measurements should result at least about 2% after calibration. Hot wire data provides, for each point of the measurement grid, the time averaged and the time-periodic fluctuation of the flow velocity, as a function of the rotor angular position. The time-periodic component is obtained by ensemble averaging, using the encoder mounted on the rotor as a key-phasor. The turbulence intensity can also to be determined, by extracting the unresolved unsteady component from the hot wire data [3].

The whole flow field can be obtained either by displacing the sensor in the space to reconstruct the flow field from the interpolation of discrete measures or can be obtained by advanced techniques as 2D particle image velocimetry [4, 5] and LDA measured on smaller flow volumes. The latter technologies need usually to inseminate the flow with particles carried by the stream, essential for the visualization and data capture. In this way, the streamtube can be reconstructed on different planes past the turbine, and other parameters as the local induced turbulence, the vortex paths, and the integral mass flow, computed.

5.1.2 Wind tunnel tests

Wind turbine tests in wind tunnels are aimed to

- determine the rotor performances;
- analyze the structural behavior;
- analyze the flow field;
- benchmark the theory and the numerical techniques (CFD, lifting line, BEM).

Wind tunnels should not be used to check the correctness of some safety procedure, as turbine-braking capability, or shutdown procedure, because of the inherent risk of these operations that have the potential risk to damage the tunnel structure; therefore, they are preferably performed in the open field. Wind tunnels are typically designed to generate stationary flow conditions that can be varied stepwise with relatively long relaxation times. Because of this feature, there is a limited capability of the wind tunnels to create dynamically variable wind streams. Therefore, some tests, as state transition (i.e. from idling to start-up) or gusts events (to tune the turbine control), cannot be simulated and should be performed in open field conditions.

5.1.3 Performances analysis

The performance analysis concerns the reconstruction of basic functional curves of the turbine as the torque curve (torque vs rotational speed), the power curve (power vs wind velocity), and the power coefficient curve (power coefficient vs tip speed ratio—TSR). It has to be stated that these tests lead to a "quasi-static" wind turbine functional curve, obtained by setting a stepwise variation both of wind and rotational speed parameters. This approach, whilst seemingly inadequate to produce the real dynamic behavior of the turbine, has the advantage to allow analyzing separately some turbine features in controlled and repeatable test conditions.

5.1.3.1 Frequency analysis

Once main aerodynamics forces have been recorded, it is possible to analyze the signal in the frequency domain by using the numeric Fourier transformation. Figure 5.6 (left) shows the spectrum of aerodynamic torque of unfiltered signals (red line) from tests on a 3-bladed turbine. Main unsteady aerodynamic forces at 3P at 20 Hz, where 3P denotes three times a single-blade passing frequency 1P, and their 6P and 9P harmonics at 40 and 60 Hz, respectively, (green lines) are shown. Blue lines indicate mechanical interference at 1P and its harmonics, up to the drive train (DT) resonant frequency. Such a picture indicates that many spurious frequencies could influence the signal and must be cut out to determine real rotor unsteady behavior. Other frequencies of unknown sources have been cut out to improve signal clearness. Figure 5.6 (right) shows an example of the effect of the filter on the phased signal shape.

To verify the reliability of the results, the measurement repeatability has to be checked under the same testing conditions and using the same calibrated gauge. This operation relies on the conversion of data from the time domain to the phased domain. Each recorded sample is described both by a timestamp and by an angular position that is tracked by the absolute encoder. The azimuthal rotor span ($0°-359°$) has been divided into 200 angular bins, then the average value and standard deviation of data falling into each bin are computed; the number of samples in each bin provides sufficient statistics of the phenomena since each bin encloses from 600 to 1,000, according to acquisition length. Figure 5.7 represents an example of the results of the shaft average torque and the standard deviation corresponding to the azimuth position of a single blade taken as reference. This average torque corresponds to the measure of a virtual torque sensor on the shaft phased with a reference blade position.

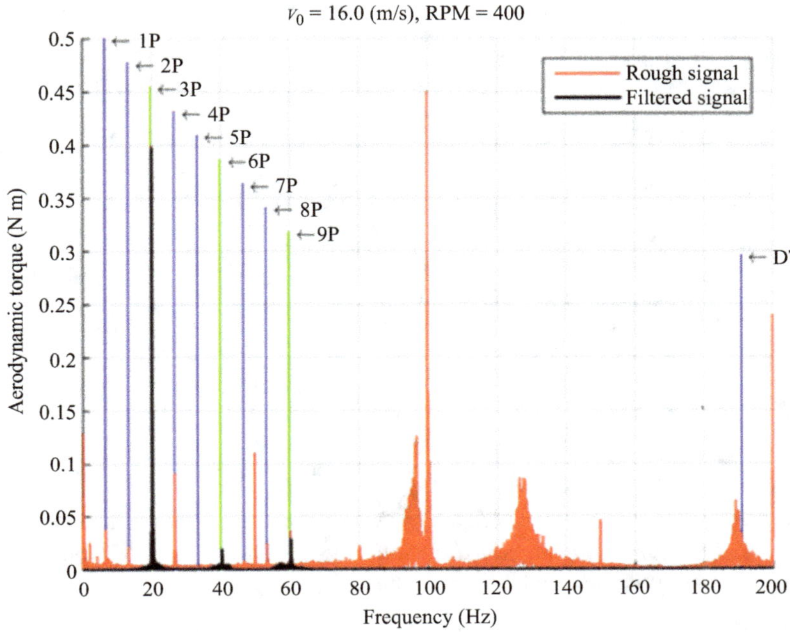

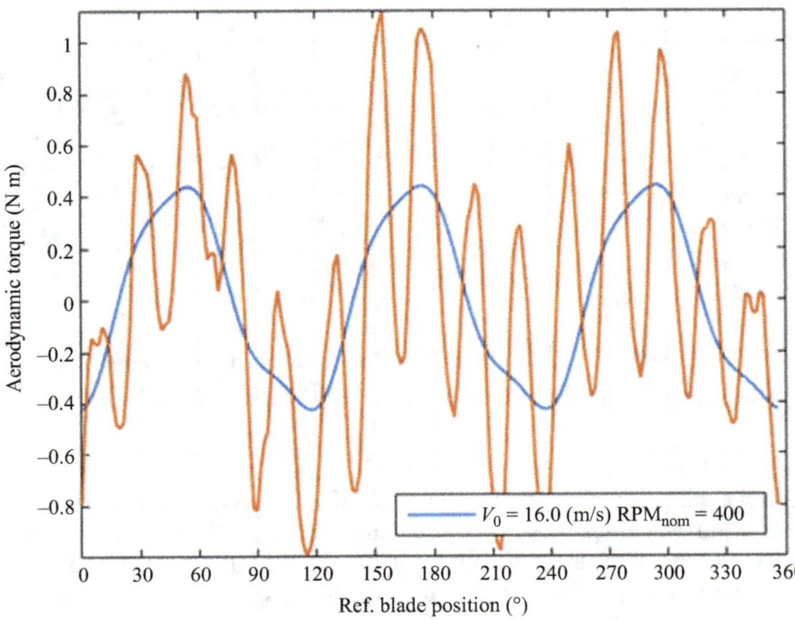

Figure 5.6 Aerodynamic torque in the frequency domain, record taken at $V_0 = 16$ m/s, 400 rpm (left) and comparison between filtered and unfiltered aerodynamic torque phased data (right)

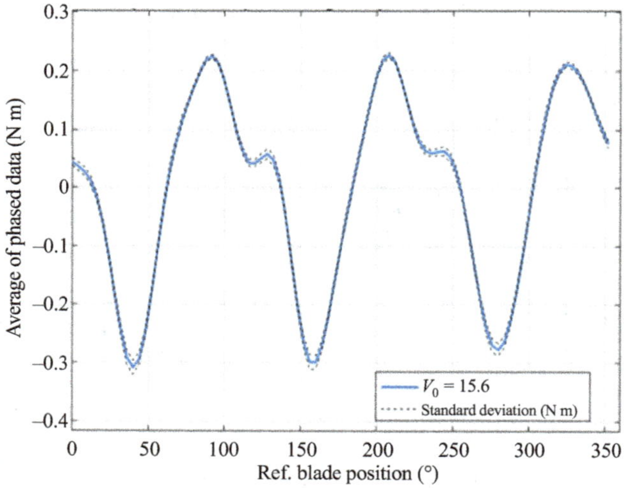

Figure 5.7 Shaft torque (three blades φ-Darrieus averaged data) and the corresponding standard deviation vs a single reference blade azimuthal position, $V_0 = 15.6$ m/s

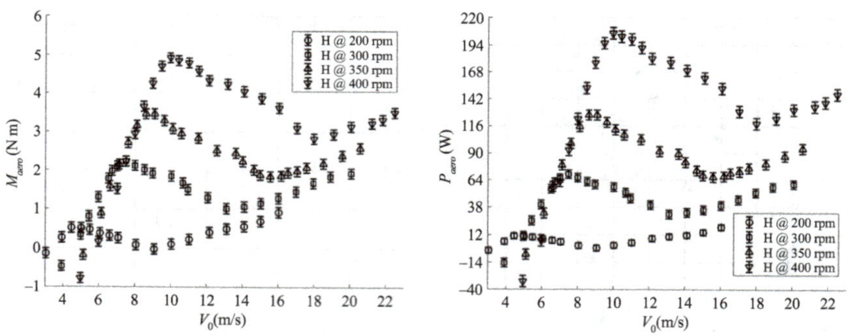

Figure 5.8 Example of torque curves (left) and power curves (right) obtained at different rotational speeds for a number of wind velocities

5.1.3.2 Averaged quantities

The average torque generated as a function of the averaged rotational speed at different wind velocities is shown in Figure 5.8 (left). This data provides the information useful to evaluate the right matching of the turbine rotor to the gearbox (if present) and to parametrize the torque-rotational speed ($M \cdot \omega$) choice of the electrical generator. From the torque curve, the power curve can be readily deduced, being

$$P = M \cdot \omega \tag{5.1}$$

As the power is computed, the shaft power coefficient of the turbine having swept area A_D can be inferred by

$$C_P = \frac{P}{\frac{1}{2}\rho A_D V_0^3} \tag{5.2}$$

Figure 5.8 (right) shows the power as a function of the averaged rotational speed at different wind velocities.

Elaboration of the measurements shown in Figure 5.8 gives the resulting nondimensional representation of C_p–TSR curve, shown in Figure 5.9. The uncertainty band is also shown, computed according to the ISO/IEC Guide 98-3:2008 [6] standard, calculating the type A uncertainty from the average data of each single measurement set and the type B uncertainty from the sensor data-sheets. The uncertainties are composed and the formulation of the uncertainty is carried out using the expanded uncertainty with a confidence interval of 95%. Note that the curves do not collapse in a single one because the Reynolds number varies according to the rotational speed with the related parasitic losses.

5.1.4 Structural analysis

The thrust measurements in the wind-wise direction show a generally monotonous trend that increases with increasing wind speed, while in the transverse direction the measurements show a characteristic wind speed switch point at which the thrust value reverses the sign, becoming slightly negative, as shown in Figure 5.10. Limited to the case in which the rotor rotated at 400 rpm, it is possible to observe that once the conditions for which the turbine operates in stall conditions have been reached (in this case for wind speeds greater than 11 m/s and for TSR values less

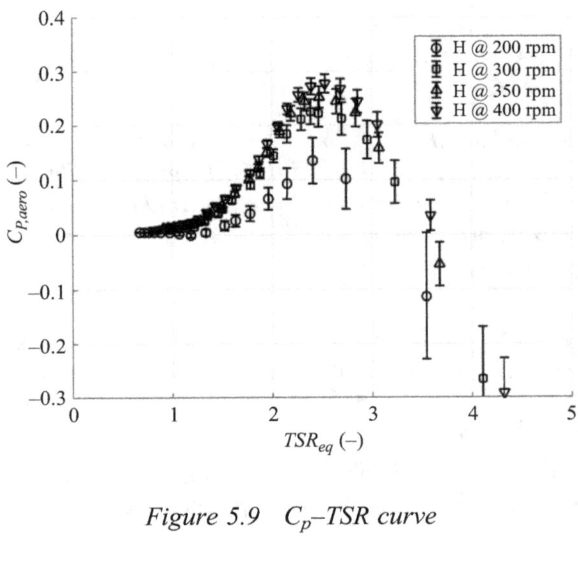

Figure 5.9 C_p*–TSR curve*

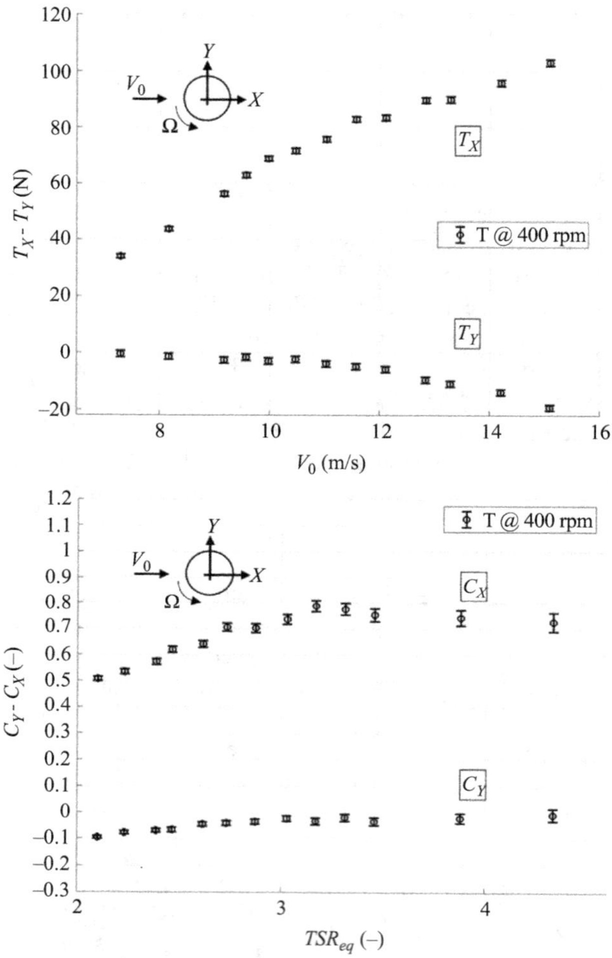

Figure 5.10 Longitudinal thrust T_X and transverse thrust T_Y to the wind direction as a function of wind speed (left); longitudinal thrust coefficient C_X and transverse coefficient C_Y to the wind direction as a function of the equatorial TSR ratio (right). The measurements were made in an open chamber at a rotational speed of 400 rpm

than 3, as confirmed by the power trends of Figure 5.10), there is a more or less marked variation in the slope of the thrust curve for both directions considered.

5.1.5 Flow field analysis

VAWTs present an inherent aerodynamic unsteadiness as a result of the periodic variation of the angle of attack on the blade profiles. Depending on the TSR, the kind of profile adopted, and the Reynolds number, the variation of angle of attack on the blades during each revolution can be sufficiently large to exceed the stall limit of the profile, triggering boundary layer separation for a significant portion of the revolution; these conditions promote an even higher unsteadiness in the turbine wake. As consequence, the vortices shed in the downstream region of the stream-tube undergo periodic rotor phase-locked fluctuations.

Pressure and velocity measurement allow the reconstruction of the flow field, the wake expansion, and the unsteadiness caused by the rotor operational state. Two basic levels of information can be drawn from measurements: time-mean analysis of VAWT wake, and phase-resolved velocity magnitude and turbulence intensity.

The former provides the distributions of flow velocity magnitude downstream of the turbine tracing the wake, which strongly changes its shape and extension as the TSR varies.

The latter allows to distinguish the time-dependent structures, appearing as vertical gradients, from the time-averaged ones, characterized by horizontal gradients. If two or more downstream measurement planes are considered, the effect of the mixing process on the rotor viscous structures can be analyzed. The following figures report the different aerodynamic layout of the wake induced by two different VAWT architectures: the *H*-Darrieus and the ϕ-Darrieus. More information can be found in [7].

Figure 5.11 [7] shows the result of the turbine wake reconstruction at midspan (MS) in the near mean section of an *H*-Darrieus turbine (see Figure 5.13 for details), considering three TSRs at the equatorial section (TSR_{eq}), including high load ($TSR_{eq} = 3.31$), low load ($TSR_{eq} = 1.78$), and very low load ($TSR_{eq} = 1.51$) test conditions. It can be clearly noted that even though the wake appears almost symmetric with respect to the turbine axis ($Y/TSR_{eq} = 0$), the velocity deficit is slightly larger and more uniform on the right side of the wake ($Y/TSR_{eq} > 0$). The asymmetry with respect to the turbine axis is a typical feature of VAWT wakes and follows the working principle of the Darrieus VAWT. Since the tested turbines rotate in counter-clockwise verse, the left side of the wake ($Y/TSR_{eq} < 0$) is generated by the blade profiles moving windward while the right side of the wake is generated by the blade profiles moving leeward.

Figure 5.12 reports for a ϕ-Darrieus the distribution of velocity, turbulence, and deterministic unsteadiness measured at MS for three TSR_{eq} conditions, including high load ($TSR_{eq} = 4.04$), low load ($TSR_{eq} = 2.40$), and very low load ($TSR_{eq} = 1.80$) configurations. As the flat and curved MSs coincide at MS, the equatorial wake measurements are available at a single distance from the rotor. The symmetric and lowly tapered shape of the ϕ-Darrieus curve around the equatorial zone inhibits the onset of 3D effects in rotor aerodynamics at MS. As a result, the equatorial wake profiles for the ϕ-Darrieus turbine strictly resemble the ones observed for the H-shaped turbine (provided that the TSR_{eq} is augmented by approximatively 0.7). The highly tapered shape of the ϕ-Darrieus configuration away from MS induces a dramatic spanwise evolution of the wake profiles.

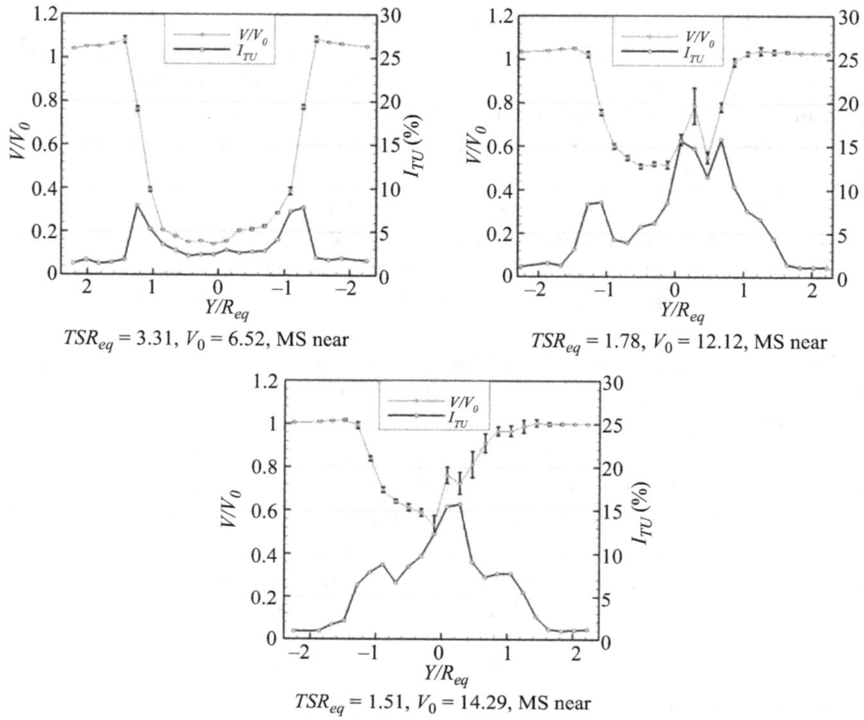

Figure 5.11 Velocity and streamwise turbulence intensity profiles at midspan (MS) of the turbine wake measured in the near mean section for four different TSR_{eq} conditions. The intensity of periodic unsteadiness is also reported as an error bar around the velocity profile. R_{eq} refers to the (constant) radius of the rotor and the transversal coordinate Y is origin for the rotor axis of rotation

Finally, Figure 5.13 shows a comparison of the different shape of the wake and periodic unsteadiness of the tip section of the *H*-Darrieus and ϕ-Darrieus turbine at high TSR_{eq}. The highly tapered shape of the ϕ-Darrieus configuration away from MS induces a dramatic spanwise evolution of the wake profiles. It is apparent that the flow detachment is activated in a small region of the ϕ-Darrieus turbine even at high-load condition, inducing unsteady flows in local regions of the wake. With concerns to unsteady effects, they are evident at high load also for the H-shaped turbine in the tip region, but they are connected to the periodic variation of a large-scale tip vortex. For the ϕ-Darrieus turbine, instead, the region affected by wide oscillations involves a small amount of the flow rate passing through the rotor, reducing the impact of the flow unsteadiness on the overall rotor aerodynamics.

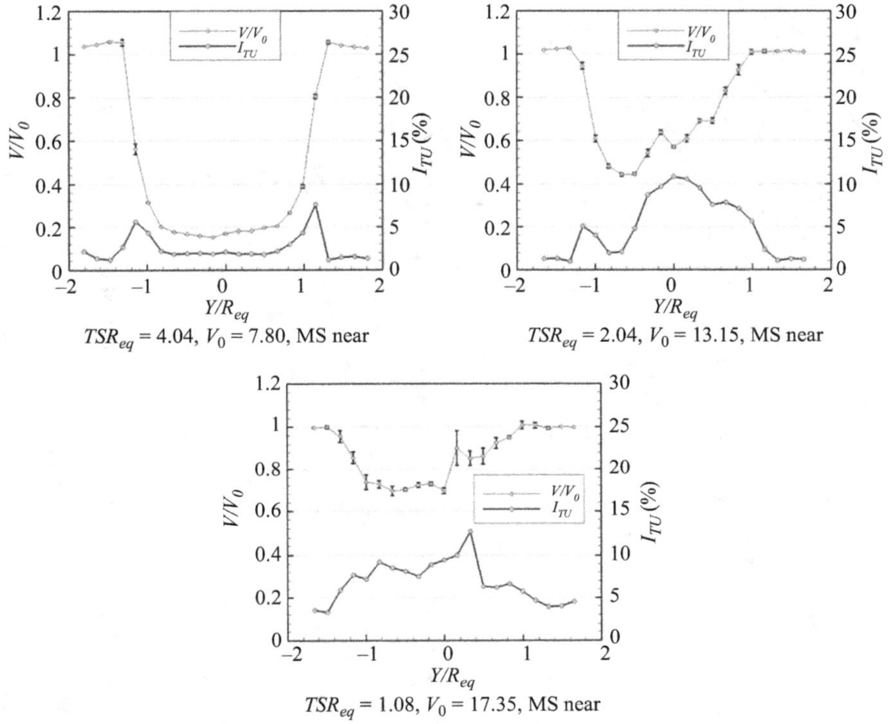

Figure 5.12 Velocity and streamwise turbulence intensity profiles in the equatorial section of the ϕ-Darrieus turbine wake for four different TSR_{eq} conditions. Intensity of periodic unsteadiness is also reported as an error bar around the velocity profile. The transversal coordinate Y is taken null on the rotor axis of rotation

5.1.6 Benchmarking

Experimental measurements are also used to improve the capability prediction and to provide a reliable benchmark for theoretical (actuator disk) and numerical models (CFD, lifting line, BEM) of the turbine, providing some conditions are satisfied. At first, all test conditions need to be declared. For instance, the contribution of bearing losses, thrust at zero revolutions, etc. on the measured power performance of the tested rotor is important information obtained from the experimental data. Also, the location of the test section in the test chamber should be provided for the correct definition of the external simulation domain, as required in the case of CFD code validation. Blockage analysis is also of paramount importance. Whilst this issue could be considered irrelevant when comparing measured data to CFD simulations (thanks to their capability of reproducing the whole measurement environment, implementing also the tunnel surfaces as boundary conditions), it would affect the comparison with other

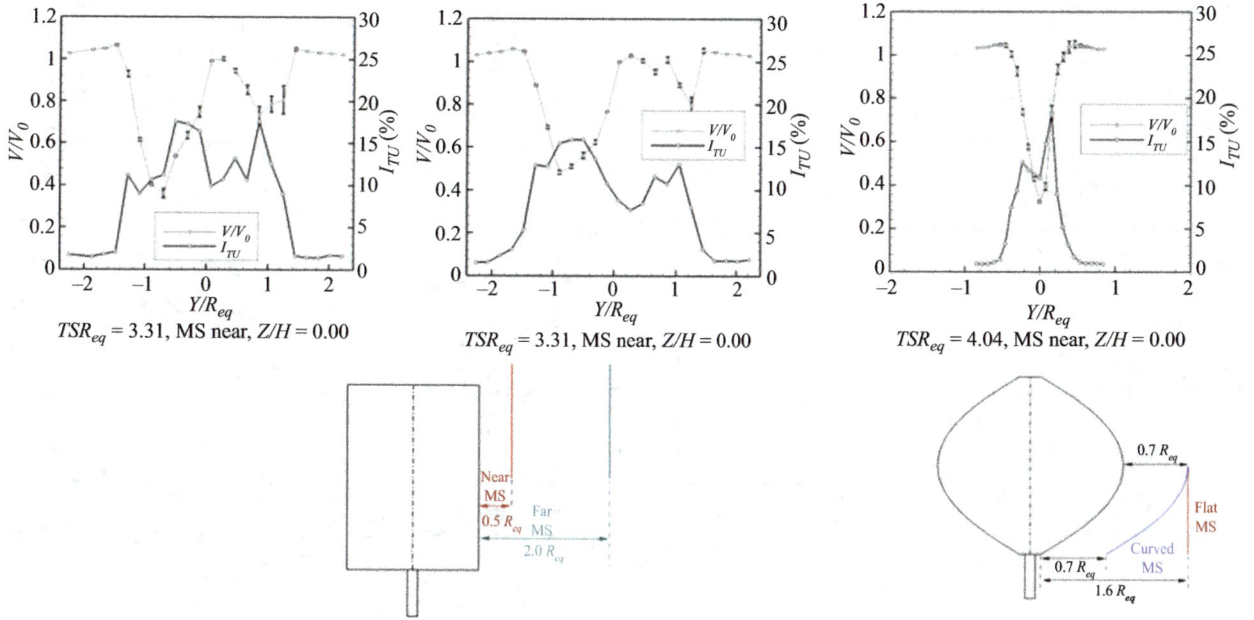

Figure 5.13 *Comparison of the different shape of the wake and periodic unsteadiness of the tip section of the H-Darrieus and ϕ-Darrieus turbine at high TSR_{eq}*

generally accepted simulation tools such as BEM models, as well as extrapolation to real operating conditions. Additionally, a too high blockage ratio, a too low blade Reynolds number and a general lack of information about some important aspects of the flow field (such as the definition of the turbulence intensity of the incoming wind), and rotor geometry (such as the geometry of the spokes and their position with respect to the blade profiles, and the shaft diameter) can lead to a partial unusability of the measurement campaign under the purpose of benchmarking the theory.

5.2 Scaling limits in wind tunnels

Due to the limited dimensions of the test chambers of a wind tunnel, only very small full-scale turbines can be tested. If scaled models are hosted, such limits determine in turn the maximum geometrical downsizing of real turbines. Probably the most challenging issue is the unsatisfactory reproduction of the natural turbulence level, as typical wind tunnels are designed for near-zero turbulence level (less than 0.5%), while the typical 8%–20% infield turbulence level is difficult to reproduced with artificial roughness elements. When wind turbine downsizing is performed, some scaling rules cannot be fully complied with, and some check is needed with the help of similarity rules.

5.2.1 Similarity

The theory of similarity allows to classify machines into types according to their functional and geometric characteristics. Furthermore, the results of this theory allow predicting the behavior of the same machine in different operating conditions or of a prototype machine of different dimensions but geometrically similar to a model machine.

The similarity theory, whose derivation is omitted here, is developed on the basis of the Buckingham's theorem [8], (or Π theorem), which identifies the physical quantities that regulate the outflow through the turbine and that must be satisfied in every point of the flow field. For the specific case of the wind turbine, by setting the subscript *p* to designate the prototype machine and *m* subscript for the model machine, the following similarity function is obtained with the most significant parameters:

$$f\left\{\left(\frac{l_{m,1}}{l_{p,1}},\frac{l_{m,2}}{l_{p,2}},\frac{l_{m,3}}{l_{p,3}}\right),\frac{U}{V_0},\frac{N_pc}{\pi R},\frac{\Delta p^0}{\rho W^2/2},\frac{\rho Wc}{\mu},\frac{W}{W_{sound}},\frac{\sigma_u}{\overline{\overline{V}}}\right\} \tag{5.3}$$

The first three terms are aspect ratios, typically the ratio between the diameter and the height, the diameter and the chord and the diameter, and the transversal dimensions of the wake, followed by the tip speed ratio (*TSR*), the rotor solidity (σ), the Euler number (*Eu*), the Reynolds number referred to the chord (Re_c), the Mach number (*Ma*), and the intensity of the turbulence (*TI*) as a ratio of the standard deviation and the average velocity.

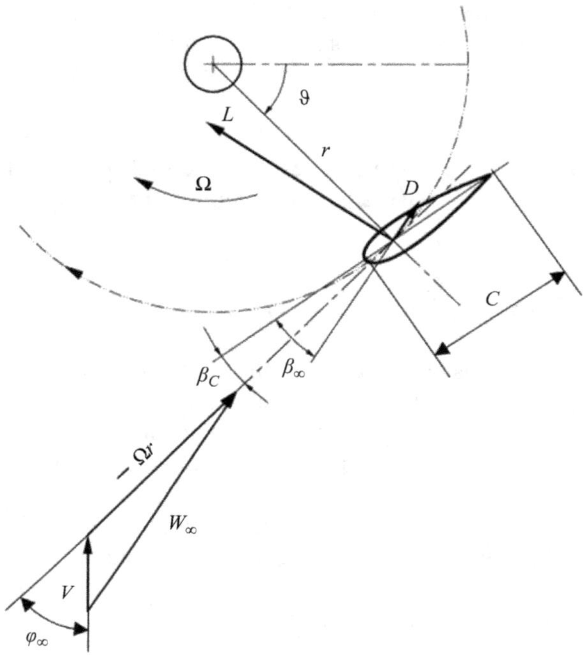

Figure 5.14 Velocity triangles of a VAWT blade

Following the results of the dimensional analysis, it emerges that if a scaled prototype of a wind turbine is created, the values assumed by the *TSR*, σ, Re_c, *Ma*, *TI*, and the aspect ratios should be replicated to ensure that the prototype performs similarly to the real model during the tests.

The similarity of the velocity triangles (see Figure 5.14) dictates that

$$\frac{V_{0,p}}{V_{0,m}} = \frac{V_{D,p}}{V_{D,m}} = \frac{W_{D,p}}{W_{D,m}} = \frac{U_{D,p}}{U_{D,m}} \tag{5.4}$$

From (5.4) we obtain

$$\frac{U_{D,p}}{V_{0,p}} = \frac{U_{D,m}}{V_{0,m}} = TSR \tag{5.5}$$

from which we deduce that the TSR in homologous positions must be the same. Furthermore, for both machines, it holds

$$\frac{U_{D,p}}{V_{D,p}} = \cot_{\varphi p} = \cot_{\varphi m} = \frac{U_{D,m}}{V_{0,m}} = TSR \tag{5.6}$$

therefore, the velocity triangles between prototype and model determines the equality of the flow angles φ_∞ and this equivalence must be valid for homologous profiles, i.e., profiles for which

$$\frac{r_p}{R_p} = \frac{r_m}{R_m} \tag{5.7}$$

where r_p and r_m represent the local blade radial positions, respectively, for the prototype machine and the model machine. Indeed,

$$\frac{V_{D,p}}{V_{D,m}} = \frac{U_{D,p}}{U_{D,m}} = \frac{(\Omega R)_{D,p}}{(\Omega R)_{D,m}} \tag{5.8}$$

If the pitch angles β_c are equal, the angles of attack β_∞ are also equal, since by construction:

$$\beta_\infty = \phi_\infty - \beta_c \tag{5.9}$$

If one recalls the normalized equations of the coefficient of axial force C_X, of torque C_M and the power coefficient C_P produced by the blade element of width dS, we obtain

$$dC_X = \frac{dF_X}{\frac{1}{2}\rho V_0^2 dS} = \frac{\sigma}{\sin^2\varphi_\infty}(C_D \sin\varphi_\infty + C_L \cos\varphi_\infty) \tag{5.10}$$

$$dC_M = \frac{dM}{\frac{1}{2}\rho V_0^2 r dS} = \frac{\sigma}{\sin^2\varphi_\infty}(C_L \sin\varphi_\infty - C_D \cos\varphi_\infty) \tag{5.11}$$

$$dC_P = \frac{dP}{\frac{1}{2}\rho V_0^3 dS} = \frac{\Omega_m r_m}{V_0}\frac{1}{\alpha_m}\frac{\sigma}{\sin^2\varphi_\infty}(C_L \sin\varphi_\infty - C_D \cos\varphi_\infty) \tag{5.12}$$

if the same airfoils are used, the values of φ_∞ are the same, as well as those of C_L and C_D since the incidence values are also the same. It can therefore be written that

$$\frac{dF_{X,p}}{\frac{1}{2}\rho_p V_{0,p}^2 dS_p} = \frac{dF_{X,m}}{\frac{1}{2}\rho_m V_{0,m}^2 dS_m} \tag{5.13}$$

or

$$dF_{X,p} = dF_{X,m}\frac{\rho_p V_{D,p}^2 dS_p}{\rho_m V_{D,m}^2 dS_m} = dF_{X,m}\frac{\rho_p V_{D,p}^2 D_p^2}{\rho_m V_{D,m}^2 D_m^2} \tag{5.14}$$

$$dM_p = dM_m \frac{\rho_p V_{D,p}^2 r_p dS_p}{\rho_m V_{D,m}^2 r_m dS_m} = dM_m \frac{\rho_p V_{D,p}^2 D_p^3}{\rho_m V_{D,m}^2 D_m^3} \tag{5.15}$$

$$dP_p = dP_m \frac{\rho_p V_{D,p}^3 D_p^3}{\rho_m V_{D,m}^3 D_m^3} \tag{5.16}$$

The total quantities are obtained by integrating the previous equations:

$$F_{X,p} = \sum dF_{X,p} = \frac{\rho_p V_{D,p}^2 D_p^2}{\rho_m V_{D,m}^2 D_m^2} \sum dF_{X,m} = \frac{\rho_p V_{D,p}^2 D_p^2}{\rho_m V_{D,m}^2 D_m^2} F_{X,m} \tag{5.17}$$

$$M_p = \sum dM_p = \frac{\rho_p V_{D,p}^2 D_p^3}{\rho_m V_{D,m}^2 D_m^3} \sum dM_m = \frac{\rho_p V_{D,p}^2 D_p^3}{\rho_m V_{D,m}^2 D_m^3} M_m \tag{5.18}$$

$$P_p = \sum dP_p = \frac{\rho_p V_{D,p}^3 D_p^3}{\rho_m V_{D,m}^3 D_m^3} P_m \tag{5.19}$$

With these relationships, once the operating characteristics of the model are known, it is, therefore, possible to correctly set up tests on scaled machines in wind tunnels and establish the functional characteristics of the prototype. These characteristics are, in general, reported in terms of thrust coefficient C_X, torque coefficient C_M, and power C_P as a function of the tip speed ratio *TSR*. These parameters are expressed as

$$C_X = \frac{F_X}{\frac{1}{2}\rho A_D V_0^2} \tag{5.20}$$

$$C_M = \frac{M}{\frac{1}{2}\rho A_D V_0^2 R} \tag{5.21}$$

From the previous text, the limitations of the wind tunnel tests become evident. Once the turbine has been scaled according to geometric similarity, the wakes must also be dimensionally similar. This implies that the blockage in the tunnel must be less than about 5%–10%; otherwise, the thrust coefficient will be modified and the dynamic similarity is lost. Therefore, is the section of the test chamber of the tunnel, rather than the speed that can be achieved within it, which imposes the maximum size of the turbine when tests are carried out on a 1:1 scale. As to the possibility of carrying out tests on scaled turbines, we start from the expression of the Reynolds number based on the chord:

$$Re_c = \frac{\rho W_c}{\mu} \tag{5.22}$$

Table 5.1 shows the comparison between dimensions and functional parameters of a real 3-bladed machine (m) and a scaled prototype (p). Assuming to use a blockage factor of 10%, and to have a test chamber sizing 2 m×2 m (providing $A_T = 4$ m^2), and considering that the real machine has a diameter of $D_m = 3$ m, we obtain that the maximum diameter of the prototype will be equal to $D_p = 0.71$ m. The equality of chord-based Reynolds numbers, considering that we want to operate at the same solidity value (and of the aspect ratio) to have the same operating curve of the real machine, imposes that the prototype chord will

Table 5.1 Effect of geometric scaling in the wind tunnel test design of a prototype machine.

$N_{b,m}$	σ_m	TSR_m	D_m	c_m	$V_{0,m}$	W_m	Ω_m	N_m	$Re_{c,m}$
(–)	(–)	(–)	(m)	(m)	(m/s)	(m/s)	(1/s)	(rpm)	(–)
3	0.127	7	3	0.2	7	49.5	32.7	623.9	673,716
$N_{b,p}$	σ_p	TSR_p	D_p	c_p	$V_{0,p}$	W_p	Ω_p	N_p	$Re_{c,p}$
(–)	(–)	(–)	(m)	(m)	(m/s)	(m/s)	(1/s)	(rpm)	(–)
3	0.127	7	0.71	0.05	29.4	208.1	577.3	11,025	673,716

be about a quarter of the original chord. The machine must therefore rotate at over 11,000 rpm compared to the 624 rpm of the real machine, resulting in a relative speed of over 200 m/s. This leads to an incorrect Mach number (which turns out to be >0.6). Furthermore, this rotation speed is excessive from a structural point of view.

As a last consideration, it must be said that the correct reproduction of the intensity of the environmental turbulence remains extremely complex. Indeed, the need to carry out tests at much higher wind speeds implies correspondingly increasing the value of the standard deviation. The insertion of turbulence promoters in the tunnel is extremely complex and costly and does not, however, effectively reproduce the scale of the turbulence. For this reason, this parameter is often waived and the so-called zero-turbulence power curve is determined, characterized by usually very low levels (<1%) of the wind tunnel. It should then be corrected to take into account the real effect of turbulence in the open field.

5.2.2 *Blockage effect*

Operations of a rotor within a confined environment cause an alteration of the stream tube past the model, and therefore the flow field around the turbine can differ substantially from the one in the open stream (unconfined flow). The effect of such alterations is practically linked to the blockage ratio *B* that is the ratio between the area swept by the turbine rotor and the flow area of the test chamber. The study of the confined rotor is of practical importance for the proper transferability to the field of the measurements made on prototypes. One of the challenging problems is the determination of the actual blockage. Although the blockage *B* based on the geometrical ratio is of immediate evaluation, the parameter seems to be not sufficient to fully consider the variable blockage induced by the dimension of the wake that changes according to the thrust. In particular, the rotating wake of a VAWT is not symmetrical and exhibits a downstream deflection, and the usual blockage correlations [9] developed for static prototypes could be not so effective.

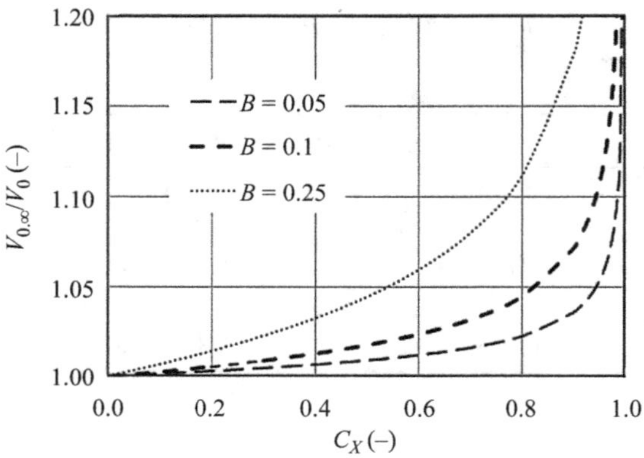

Figure 5.15 Trend of the ratio of the equivalent unconfined speed $V_{0,\infty}$ and tunnel speed V_0 as the thrust coefficient C_X varies according to Glauert

Historically, one of the early but still used correction models is due to Glauert [10] based on the one-dimensional actuator disk theory. The application of the momentum equation to a confined flow with geometric blockage ratio B allows concluding that the thrust can be estimated as the sum of a coefficient due to the performance in an unconfined environment plus a corrective factor that depends on the value of the blockage.

An equivalent unconfined speed $V_{0,\infty}$ can be determined, which corresponds to the speed that gives the same thrust (and therefore power) obtained inside the wind tunnel. The following equation is thus obtained:

$$\frac{V_{0,\infty}}{V_0} = 1 + B\frac{\alpha(1-\alpha)}{2\alpha - 1} + O(B^2) \tag{5.23}$$

The notation $O(B^2)$ indicates higher order terms of B, and $\alpha = V_D/V_0$, where V^D is the velocity at the rotor plane.

According to the momentum conservation principle, the thrust coefficient is

$$C_X = 4\alpha(1-\alpha)$$

(5.23) becomes

$$\frac{V_{0,\infty}}{V_0} = 1 + B\frac{C_X}{4\sqrt{1 - C_X}} + O(B^2) \tag{5.24}$$

As appears from this formulation, when the measured thrust coefficient approaches or exceeds unity, this approach becomes unusable, as shown in Figure 5.15.

By the help of (5.20), the corrected thrust and power coefficients can be obtained:

$$C_{X,\infty} = C_X \cdot \frac{V_0^2}{V_{0,\infty}^2} \tag{5.25}$$

$$C_{P,\infty} = C_P \cdot \frac{V_0^3}{V_{0,\infty}^3} \tag{5.26}$$

Mikkelsen and Sørensen [11] have developed a similar approach to Glauert starting from the same equations, but without linearizing them. With reference to the quantities indicated in Figure 5.16, the result of the analysis states that the actual volumetric efficiency in the tunnel α is given by

$$\alpha = \frac{V_D}{V_0} = \frac{\sigma'}{2\sigma' - 1}\left[\frac{1 - B\sigma'^2}{\frac{B\sigma'(3\sigma - 2)}{1 - 2\sigma'} + 1}\right] = \alpha_\infty F(\sigma', B) \tag{5.27}$$

where $\sigma' = A^3/A^D$. Since in the experiments, the following conditions hold:

$$B \equiv \frac{A_D}{A_T} \leq 1$$
$$\sigma' \equiv \frac{A_3}{A_D} \in \left(1, \frac{1}{B}\right)$$

the function $F(\sigma', B)$ is always <1, and therefore $\alpha < \alpha_\infty$ as shown in Figure 5.17. In the case of the confined turbine, the Froude condition is no longer respected, and

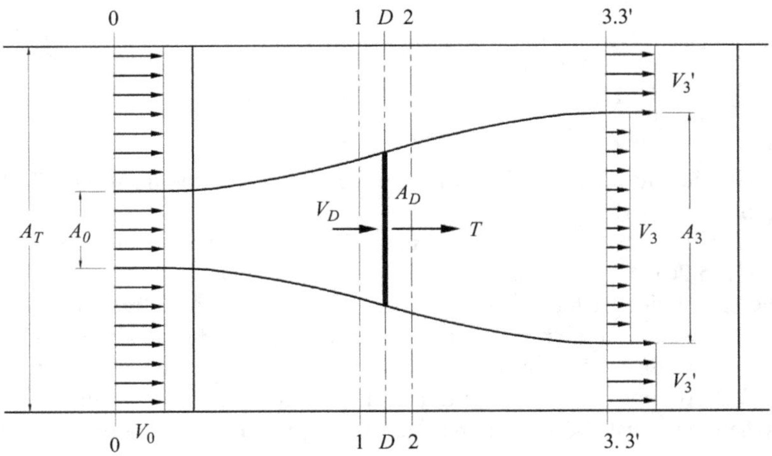

Figure 5.16 Scheme of a ducted rotor

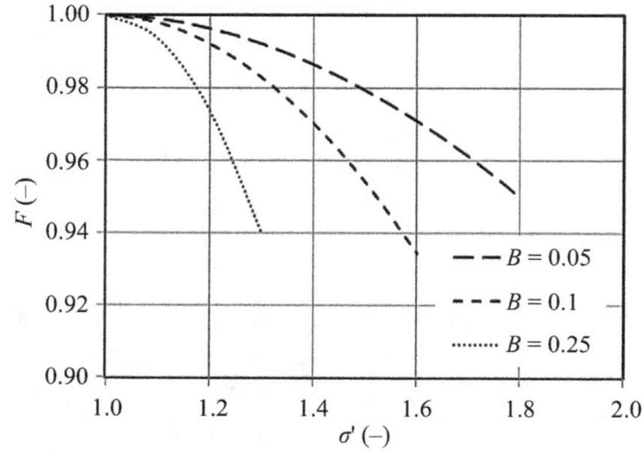

Figure 5.17 F(σ′, B) function as vs σ′ at different blockage values B

if $V_0 = V_{0,\infty}$, the wind velocity at the rotor in the test chamber is lower, while C^X is higher.

The range of validity of the solution is limited by the achievement of the unitary thrust coefficient. The lateral extension of the wake is limited to values less than about twice the area of the disk, beyond which the condition $C^X = 1$ is exceeded.

According to the same approach introduced by Glauert, one can define an equivalent velocity of the undisturbed free vein $V_{0,\infty}$ that produces the same thrust T for a disk velocity that corresponds to that obtainable from (5.27). It follows that

$$C_X = 4\alpha\left(\frac{V_{0,\infty}}{V_0} - \alpha\right) \tag{5.28}$$

therefore,

$$\frac{V_{0,\infty}}{V_0} = \alpha + \frac{1}{4}\frac{C_X}{\alpha} \tag{5.29}$$

and we can therefore deduce the correction coefficients of the thrust coefficient as shown in Figure 5.18.

The estimation of the corrected thrust and power coefficient can be done using (5.25) and (5.26).

The calculation of the equivalent speed in an undisturbed free stream is then carried out based on three parameters that can be measured during the wind tunnel tests, as expansion coefficient of the wake, the geometric blockage, and the measured thrust coefficient. According to this model, the singularity problem does not occur when the thrust coefficient approaches unity, resulting in the main advantage of this approach, even if it requires an additional parameter to be measured, which is the expansion coefficient of the wake, or the pressure, and speed outside the wake.

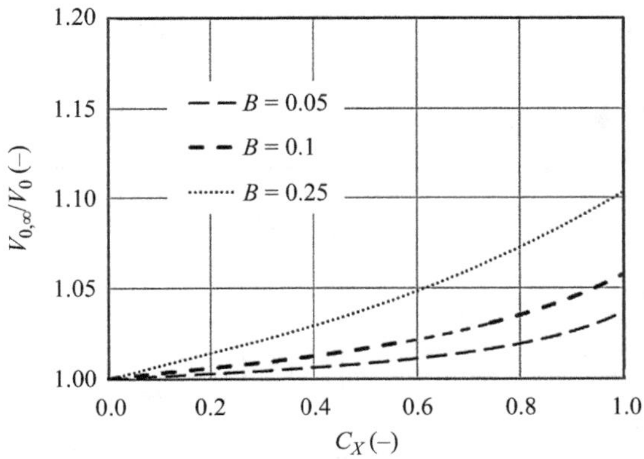

Figure 5.18 Equivalent velocity in undisturbed free vein $V_{0,\infty}$ normalized to V_0 as a function of the thrust coefficient C_X for different ratios of $B = A_D/A_T$

5.2.3 Blockage correction

The optimal condition for an accurate prediction of the blockage effect is to compare the flow condition either in a closed and an open test chamber. In this case, the correction to eliminate the blockage effect from the measurements taken in the closed chamber is obtained from measurements of the axial thrust in the open chamber. The calculation of the blockage is made starting from the thrust measurements along the wind direction (x-axis) normalized to the density value, to consider only the fluid dynamic effects in the analysis. The normalized thrust data T^X/ρ measured in both configurations (open and closed chamber) are then interpolated with a high degree polynomial to best reproduce the trends as illustrated in Figure 5.19. Here the tests conducted at 400 and 600 rpm are shown for comparison [7].

The next step involves determining the equivalent speed in an open chamber in relationship to the measurements made in a closed chamber. From a practical point of view, it is a matter of calculating the normalized thrust T_X/ρ for discrete velocity values in a closed chamber using the corresponding interpolating polynomial relationship. This normalized thrust value is therefore assumed also in the open chamber and the corresponding wind speed is obtained through the respective interpolating polynomial. This speed is considered the equivalent speed in the open field that is assuming the phenomenon of blockage in the open chamber as negligible. The blockage factor is consequently calculated according to the expression:

$$B = \frac{V_{(o.c.)}}{V_{(c.c.)}} - 1$$

As shown in Figure 5.20, the blockage effect increases as the test speed decreases and ranges from about 5% (at 20 m/s) to about 17% (at 4 m/s). At 3 m/s

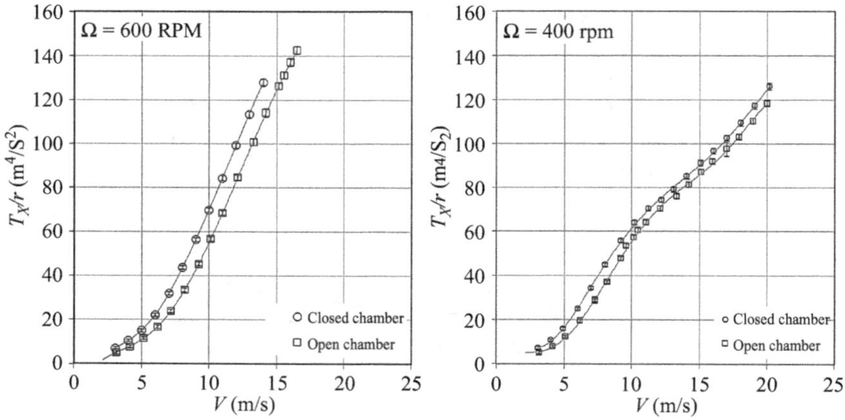

Figure 5.19 Normalized longitudinal thrust T^X/ρ measurements in closed and open chamber as a function of speed and respective interpolating polynomial for tests conducted at 400 and 600 rpm

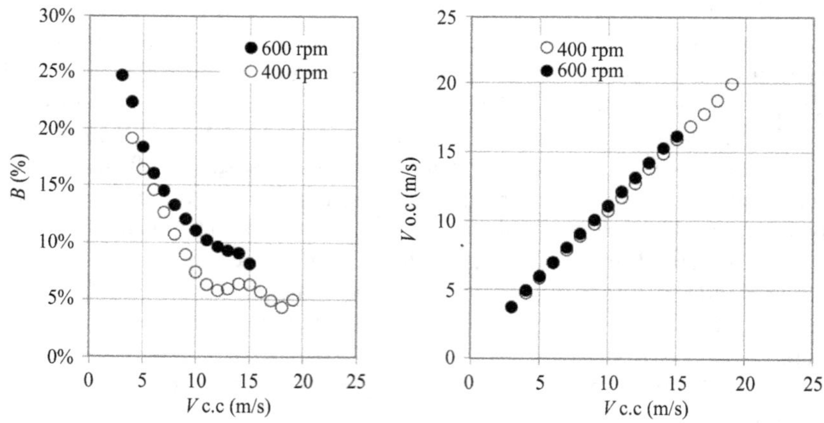

Figure 5.20 Values of the percentage blockage vs wind speed in the closed chamber (left) at 400 and 600 rpm and wind velocity correction (right) vs tunnel wind speed

the blockage phenomenon is even more important requiring further evaluation and analysis of the measured data. In general, observing the diagram on the left of Figure 5.17 and for the case at 400 rpm, it can be stated that the overall blockage trend is caused by the rotor wake: at high wind speeds (low TSR), the wake is characterized by weak speed gradients and a little expansive wake, so blockage is relatively low. As wind speeds decrease (increase in the TSR ratio), the exchange of work is more intense, causing a greater drop in speed in the wake and therefore a greater expansion of the latter, leading to a more marked blockage.

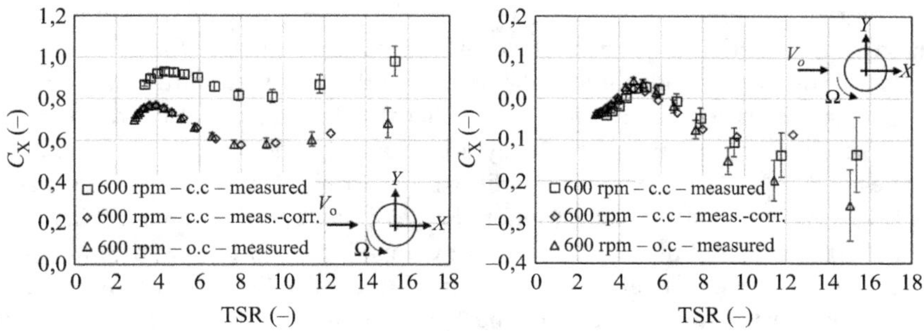

Figure 5.21 Coefficient of longitudinal thrust C_X (left) and transverse C_Y (right) vs wind speed as a function of the equatorial TSR ratio for the rotor operated at 600 rpm (c.c. = closed chamber and o.c. = open chamber)

The blockage factor as previously calculated is then used to correct the wind speed V_0,

$$V_{0,\infty} = V_0(1 + B)$$

and the associated parameters,

$$TSR_{0,\infty} = \frac{TSR_0}{(1 + B)}$$

$$C_{X,0,\infty} = \frac{C_{X,0}}{(1 + B)^2}$$

Figure 5.21 compares the measures of the thrust coefficients in the two directions with the rotor operated at 600 rpm, with and without blockage correction. The shift of the curves of the closed chamber at a lower wind speed is evident. Note that the blockage correlation obtained for C_X is not usable for C_Y. The blockage phenomenon is not very important at low TSR for two reasons. The first is due to a lower thrust coefficient and therefore a lower expansion of the flow that is well contained by the walls of the closed chamber. The second is due to the shape of the C_X curve which, before stalling, shows an increasing trend as the TSR increases that stabilizes during the stall of the rotor.

5.2.4 Conclusion

The execution of tests in a controlled environment is fundamental to deepen the knowledge of the global aerodynamic behavior of the rotors of vertical axis turbines and for the rapid and correct development of projects through the creation of prototypes. Alongside the obvious advantages of carrying out tests in a controlled environment, there are some aspects that must be taken into due consideration in order to obtain valid results. The first one concerns the experimental opportunity to realize a reduced model blockage, since also the use of correction theories determines the introduction of errors in the evaluation of the actual performances. There are also limits to the application of the similarity that emerged in the discussion of the chapter. The second one relates to the fact that the functional curves obtained are of the quasi-static type, that is, they are obtained under conditions of uniform and steady flow. Some transient conditions occurring in the field (i.e., start-up, shutdown, power control, etc.) cannot therefore be duly simulated. Finally, it should be remembered that for an effective comparison of the experimental results with numerical or theoretical methods, the performance of the aerodynamic profiles used in the construction of the rotor must be experimentally validated, since this input is proved to determine the greatest source of disagreement between numerical prediction and experimental results.

References

[1] Wind Tunnel Facility of Polytechnic of Milano, Italy. Available from: http://windtunnel.polimi.it [Accessed: 19-August-2021].

[2] Battisti L., Benini E., Brighenti A., Persico G., and Paradiso B. A review based on evaluation experiences with ten-years activity in VAWT experimental wind tunnel testing. *Green Energy and Technology* – ISSN:1865-3529, 2018.

[3] Dossena V., Persico G., Paradiso B., *et al.* An experimental study of the aerodynamics and performance of a vertical axis wind turbine in a confined and unconfined environment. *Journal of Energy Resources Technology*. 2015, 137:051207-1–051207-12.

[4] Ferreira C.S., Van Kuik G., Van Bussel G., and Scarano F. Visualization by PIV of dynamic stall on a vertical axis wind turbine. *Experiments in Fluids*. 2009, 46:97–108.

[5] Naumov I.V., Mikkelsen R.F., Okulov V.L., and Sørensen J.N. PIV and LDA measurements of the wake behind a wind turbine model. *Journal of Physics: Conference Series*. 2014, 524:012168.

[6] ISO/IEC Guide 98-3:2008. *Uncertainty of measurement Guide to the expression of uncertainty in measurement* (GUM:1995).

[7] Battisti L., Persico G., Dossena V., Brighenti A., and Benini E. Experimental benchmark data for H-shaped and Troposkien VAWT architectures. *Journal of Renewable Energy*. 2018; 125:425–444.

[8] Buckingham E. On physically similar systems; illustrations of the use of dimensional equations. *Physical Review*. 1914, 4(4):345–376.

[9] Maskell E.C. A Theory of the blockage effects on bluff bodies and stalled wings in an enclosed wind tunnel. ARC/R&M-3400, 1963.

[10] Glauert H. Wind tunnel interference on wings, bodies and airscrews. ARC/R & M 1566, 1933.

[11] Mikkelsen R. and Sørensen J.N. Modelling of wind tunnel blockage. *Proc. Global Windpower Conference and Exhibition*, Paris, France; 2002.

Chapter 6
The aerodynamics of water pumping windmills

Itoje H. John[1] and David Wood[1]

Most small wind and hydrokinetic turbines produce electricity but an important alternative is to pump water. Historically, water pumping and grinding of grains have been the main duties of wind turbines, to the extent that the terms "water pumper" and "windmill" were developed for them. In this chapter, we discuss water pumping windmills which are multibladed, low-speed machines directly coupled to a mechanical pump. Probably more of these windmills, examples of which are shown in Figures 6.1 and 6.2, have been built than any other wind turbine design. Water pumpers come in many different forms,* but their basic aerodynamics is still not well known. Although they are still built in many countries, water pumping windmills are particularly appropriate to the developing world, and we will discuss them in that context. The resource assessment, mechanical design of the pump and connection to the rotor, and performance matching are complex issues that we will not cover in this review of the aerodynamics. Interested readers are referred to Meel and Smulders [1], Veldkamp [2], and Kentfield [3].

While there are conflicting speculations about the historical origin of windmills [4, 5], wind has been harnessed as an energy source for centuries. Since the early thirteenth century, wind has been used to pump water in dewatering polders in the Netherlands [6]. Small wind pumps fabricated from wood have been used in France, Portugal, and Spain to pump seawater to produce salt. Thereafter, the American wind pump made of steel with a multibladed, fan-like rotor became the most popular water pumping technology, Figures 6.1 and 6.2. The American windmill has undergone modifications to reduce its weight and cost and improve performance, and several variations have emerged. They are robust and long-living; anecdotal evidence provided to the authors tells of windmills lasting 50 years.

One main attraction of turbines over photovoltaics (PVs) for developing countries is that the former can offer more opportunity for local manufacture. However, developing nations only play a small role in the manufacture of small wind turbines [10]. Sadly, the location of these manufacturing companies in favor of developed countries has created a barrier to the adoption of this technology.

[1]Department of Mechanical & Manufacturing Engineering, University of Calgary, Canada

*This variety is displayed in windmill museums such as www.facebook.com/etzikommuseum and www.facebook.com/The-Windmill-Shed-1438362409755991/.

Figure 6.1 Poldaw 5-m diameter, 18-bladed water pumping windmill [7]

Figure 6.2 Kijito 18-ft diameter, 24-bladed water pumping windmill [8]

Notwithstanding, some mainly windmill manufacturers, such as Kijito in Kenya [8], Jober in Colombia [11], Aureka in India [12], and Poldaw in the UK [7], focus on the developing world.

Water is crucial for life. On a global scale, water use has increased by about 1% per annum since the 1980s, propelled by a mix of population growth,

Table 6.1 Comparison of windmills and PV water pumping.

Windmills	PV water pumping
Long service lives and are easy to dismantle and recycle	Retired PVs are electronic waste and are potentially toxic to the environment because they comprise rare earth metals. Their recyclability is still a subject of debate and is not established in developed or developing countries
Simple to build. The one shown in Figure 6.2 was built on a farm outside of Nairobi, Kenya. Only some of the pump components are imported. Scavenged materials can be used to build, maintain, and repair windmills	Can only be manufactured by a few companies in the world. In other words, PV system investment is exclusively for the privileged class that can afford it
Building and installing in remote locations can create small industries and jobs for locals	Mostly imported to the detriment of the local populace

Adapted from [9].

socioeconomic change, and changing consumption patterns. The need for global water is anticipated to continue to grow at a comparable pace until 2050, accounting for a rise of 20%–30% above the current level, primarily due to increasing demand in the industrial and domestic sectors [13]. According to UNESCO [13], three of every ten people have no access to clean drinking water. Nearly 50% of people who drink water from unsafe sources live in sub-Saharan Africa. Consequently, developing nations with a rapidly growing population have a serious challenge in meeting their water demand. Water pumping using wind turbines or windmills has become less important in modern times partly because of the rise in PV pumpers. Nonetheless, using wind to provide water is experiencing a revival [14]. The inspiring story of William Kamkwamba, who built windmills from scrap materials sourced from his village in Malawi, is a testament to their ease of manufacture and adaptability. His ingenious work was documented in the book *The Boy Who Harnessed the Wind* [15] and was later adapted into a movie† in 2019. Table 6.1 presents a comparison of windmills and PV water pumpers.

6.1 Windmill blades

Water pumping windmills characteristically start under load, and this requires high torque, which is mostly realized by increasing blade number, N or solidity, $\sigma = Nc/(2\pi r)$, where c is the chord and r is the radius, which is in the range of 0.5–0.8 for the classical American windmill [3, 16], of which those in Figures 6.1 and 6.2 are examples. By comparison, a large wind turbine has $\sigma < 0.15$. High solidity complicates the rotor aerodynamic behavior due to the mutual interaction between

†www.netflix.com/ca/title/80200047

adjacent blades and low Reynolds number, *Re*, operation. Even though these windmills have been around for several centuries, their aerodynamic performance is not fully understood, particularly the influence of high solidity.

Windmill blades are generally rolled from thin metal sheet with spars consisting of circular tubes running along most of the blade length. The spars are used to hold the blades and connect them to the rotor hub, as illustrated in Figure 6.3. The parameters describing a circular arc airfoil (CAA) are outlined in Figure 6.5. Quality experimental lift and drag data at appropriate *Re* are required for an accurate design and performance prediction of windmills. However, there is a dearth of aerodynamic data for CAAs at suitable thickness and camber, especially at low *Re*. While computational methods such as computational fluid dynamics (CFD) have been used to predict airfoil performance at high *Re*, it cannot be trusted for $Re < 500{,}000$ [17]. This is because available CFD codes are not sufficiently robust to predict the airfoil performance behavior of low *Re* flow due to the complexities associated with transition from laminar to turbulent flow in the blade boundary layer [18]. Under these circumstances, wind tunnel testing is the most reliable method to obtain the airfoil polar data.

6.2 Windmills compared to wind turbines

Compared to several other technologies for water pumping, reviewed by Gopal *et al.* [20], windmills remain inexpensive and easy to manufacture in developing nations. We will not consider other technologies further, apart from a brief comparison of windmills to conventional wind turbines for water pumping.

Figure 6.3 Cambered circular arc blades from Figure 6.2

An electrical wind pumper produces usually AC power using a generator, which then pumps water directly by connecting to either an AC or DC motor. The choice between a mechanical pump and an electrical wind pump for a given location depends on factors reflected in the advantages and disadvantages of a mechanical wind pump compared to an electrical pump [6, 19]:

- Windmills use many blades made cheaply from curved steel plate. In contrast, electrical wind turbines use typically three fiber-reinforced blades that are relatively expensive.
- Windmills must be located directly over the borehole; the best wind speed location is usually not the best water resource location. Electrical wind turbines can be sited at the best wind location, and power generated from the turbine at any nearby location can easily be transmitted to the pumping site.
- Wind turbines require a generator and associated power electronics. For water pumping, they use centrifugal pumps, which simplifies matching the turbine with the pump; the match is achieved by electronically varying the load. In contrast, windmills use reciprocating or positive displacement pumps; matching the wind turbine to the pump is realized mechanically and can be problematic.
- Surplus electrical power generated by a turbine can be used for lighting and other purposes, and stored in a battery.
- Windmills can start at about 2.5–3.5 m/s and are very suitable for low wind speed regions. In contrast, an electrical wind pump requires a mean speed of 4–5 m/s. They, however, become competitive with their counterpart mechanical wind turbine at a higher mean speed of 5–6 m/s.

Starting at low wind speed is critical for water pumping. Pinilla *et al.* [21] modeled the overall efficiency of windmills and found that the maximum occurred at a wind speed approximately half the cut-in wind speed. This seemingly contradictory statement does make sense: Wood [18] showed that the cut-in wind speed of a small turbine as the wind starts to blow can be twice the cut-out wind speed at which the rotor stops as the wind decreases.

System layouts of both pumping systems are presented in Figure 6.4, showing the main components and the significant differences between each type. While the electric pump performs better relative to the mechanical pump, it is not very suited for low wind speed locations. Further discussion on water pumping windmills will be limited to the direct drive (gearless), mechanical wind pump, the focus of this chapter.

6.3 Windmill aerodynamics

The overall conversion efficiency of water pumping windmills are relatively poor, typically between 7%–27% [19, 21], partly because the basic aerodynamic principles are not well understood and their low rotational speed, as will be explained below. According to Dixon [22], poor matching of pump to rotor and cyclic variation of torque required by the pump are other factors that affect the efficiency of windpump as an energy converter but are outside the scope of the current treatise as earlier

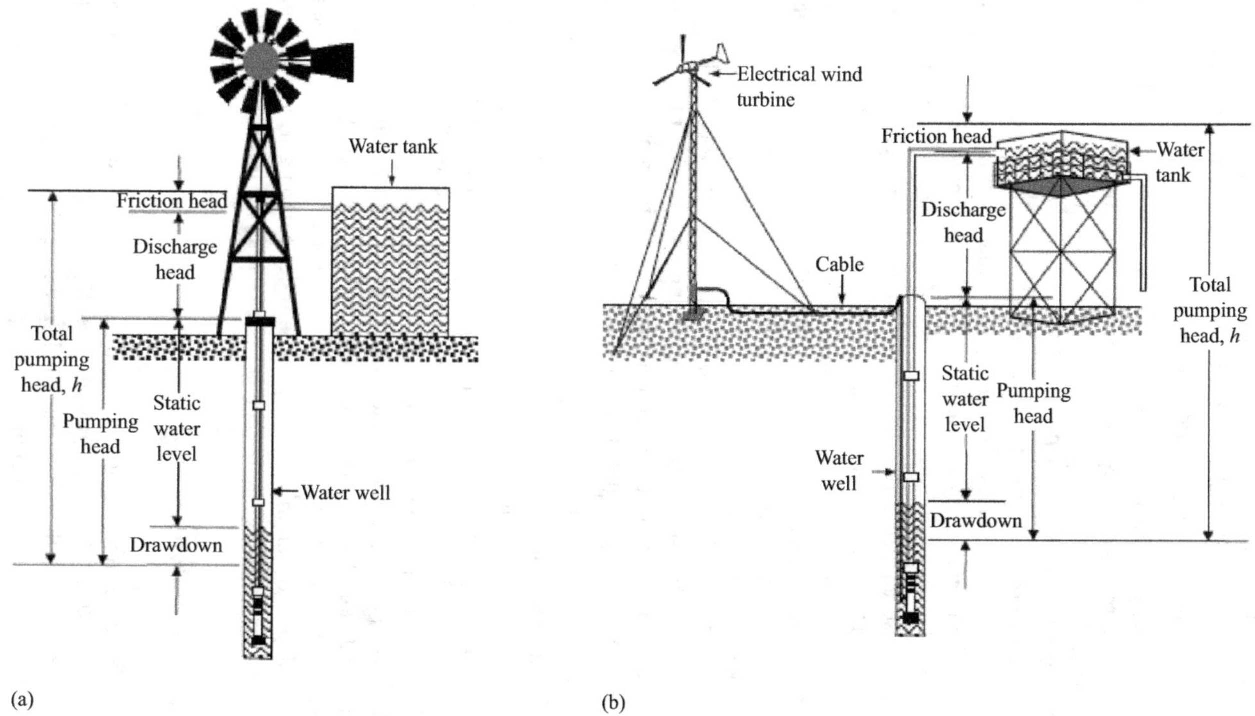

Figure 6.4 Schematic of typical wind pumps: (a) mechanical wind pump and (b) electrical wind pump. Reproduced from [19]

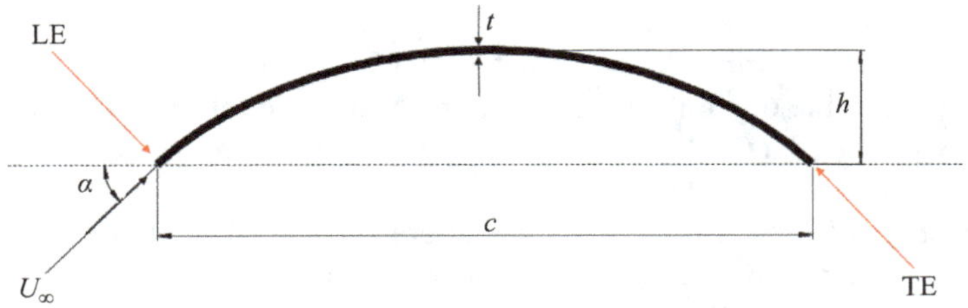

Figure 6.5 Circular-arc airfoil geometrical profile. c, chord; t, thickness; h, camber; LE, leading edge; TE, trailing edge; α, angle of attack; and U_∞, freestream wind speed

Table 6.2 Lift and drag data on circular arc airfoils and flat plates

Ref.	*c* (mm)	*Re*	*t/c* (%)	*h/c* (%)	*AR*	*α* range
[23]	400	6×10^4–3.5×10^5	2	10	5	−10°–90°
[24]	304	2.23×10^5	0.67	0–14	2	−20°–90°
[24]	30	$<1.14 \times 10^4$–1.5×10^4	1	3–9	6	−6°–20°
[26]	30	$<1 \times 10^4$	5	2–12	2	−30°–30°
[27]	100	6.2×10^4	1	4	3	0°–10°
[28]	30	1×10^3–1×10^4	1	3–22	3.3	−5°–25°
[29]	610	1.16×10^5–4.15×10^5	4	21–22	4	−5°–25°

stated. Improving the maximum conversion efficiency also requires a better understanding of high solidity effects and starting performance. We consider windmill aerodynamics in the context of blade element momentum theory as described in Chapter 7, which means that a good starting point is to consider the lift and drag of the airfoils comprising the blades. Designing the blades of SWTs usually requires a compromise between power production and starting torque [18]. Because curved blades are made from thinly rolled metal sheet, the appropriate airfoil section is the CAA whose geometry is defined in Figure 6.5 and is typically used as the airfoil sections with spars (supporting rods) to hold the blades, Figure 6.3.

Unfortunately, there is little data available for the lift, *L*, and drag, *D*, of CAAs in terms of the chord, *c*, camber, *h*, and thickness, *t*, especially at low *Re* and *α* beyond the stall; the latter is necessary to analyze starting. Other essential features that affect the blade's performance are the surface smoothness of the blade and location of the spar on either the pressure or suction sides of the blade, as depicted in Figure 6.3.

CAAs have a distinguished history in aerodynamics going back at least as far as Kutta who used a CAA to derive the famous "Kutta condition" that the flow leaves smoothly from the trailing edge of a non-stalled airfoil, see Figure 8.1 in [30]. The earliest modern studies on CAAs were done by Wallis [31] in 1946 and Ikui *et al.* [32] in 1972. Data of some recent CAA studies are summarized in Table 6.2 where *AR* denotes aspect ratio, the span of the model airfoil divided by *c*.

AR can influence the results as an airfoil, by definition has infinite *AR* but, of course, it is never possible to have that in a wind tunnel. We note that, of the experiments in Table 6.2, only Tezuka *et al.* [27] provide information other than *L* and *D*; they produced a CAA from hypodermic tubing and were able to measure the surface pressure on both sides.

Okamoto *et al.* [25], Mueller and DeLaurier [31], and Okamoto and Azuma [26] showed aerodynamically that cambered thin CAA plates offer better performance than flat plates at $10^3 < Re < 10^4$. An increase in *h/c* increases the maximum lift coefficient, $C^{l,\max}$, lift curve slope, and the corresponding angle of attack, α, due to the likely delay in stall. For example, as shown in Figure 6.6, increasing the camber from 3% to 9% results in an over 55% rise in the magnitude of $C^{l,\max}$, an indication of the significance of camber in windmill blade design. At the same *Re* range, Sunada *et al.* [32] found that thin airfoils with around 5% camber have a high lift–drag ratio, *L/D*, that increases as the maximum camber location moves toward the trailing edge. While [25, 26, 34] measured a limited range of α, Pandey *et al.* [24] considered α between $-20°$ and $90°$ at a single *Re* of 2.23×10^5 and a wide range of camber. Based on their results, 8% camber gives the highest *L/D* that maximizes C^P for 8- and 12-bladed windmill rotors, although the reported theoretical C^P values appear to be optimistic. Similarly, [29, 34] found that for $h/c > 10\%$, flow separation occurs at low α, causing performance deterioration for investigation at a limited range of α.

Several researchers have studied the effects of thickness on CAA, e.g., [25, 33, 34]. As *t/c* decreases, the airfoil performance improves, especially for CAAs with $t/c < 6\%$;

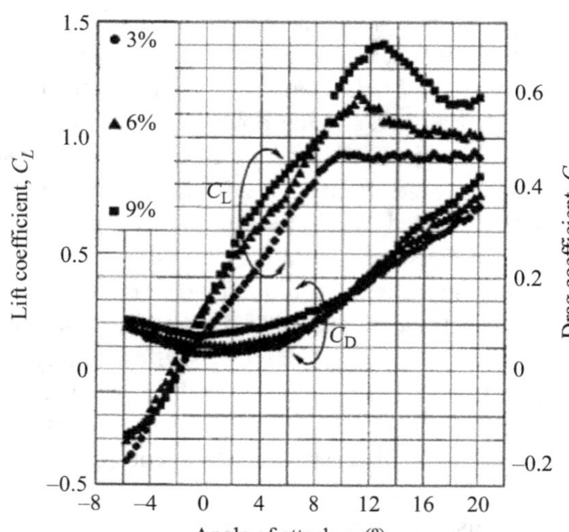

Figure 6.6 Effect of curvature on the aerodynamic characteristics of a rectangular aluminum foil model airfoil with t = 0.3 mm and c = 30 mm. Each symbol refers to a different camber, as shown in the panel on the figure. Reproduced from [25]

that is, the thinner the airfoil, the better its performance. Relatively thick airfoils ($t/c >$ 6%), in the $30{,}000 \leq Re \leq 70{,}000$ regime, can have significant hysteresis effects resulting from laminar separation with transition to turbulent flow that deteriorates their performance [35].

Bruining [23] provided the most extensive data on CAAs at low *Re* covering a wide range of α making the data suitable for BEMT analysis of power production and starting. The measurements were carried out with and without spars at five different positions around the blade profile in a closed jet wind tunnel with a turbulence intensity of 0.02%. Because windmill blades typically use spars, determining the optimal spar position that maximizes *L/D* is important. As shown in Table 6.3, the spar position on the low-pressure side of the blade leads to $L/D \leq 5$, while that on the high-pressure side as in Figure 6.2 produces *L/D* as high as 20. The spars either increase or decrease the C_l and C_d of a clean CAA (i.e. without a spar) depending on the *Re*. Figures 6.7 and 6.8, respectively, compares Bruining CAA's lift and drag performance with and without spars at two *Re*.

Given their slow rotation and high σ, windmill aerodynamic performance differs significantly from conventional wind turbines due to the blade-to-blade interaction, low λ operation, and, consequently, low *Re* [3]. Although studies over the last four decades, e.g., Bragg and Schmidt [36], Nathan [37], Cleijne *et al.* [38], Manwell *et al.* [39], Rijs and Smulders [40], Rijs *et al.* [41], focused on design analysis, field and laboratory testing, and construction of water pumping windmills, they overlooked the impact of solidity on their rotor performance. This factor is also neglected by agencies such as the defunct Consultancy Services Wind Energy Developing Countries (CWD), the former Steering Committee Wind Energy Developing Countries (SWD) of the Netherlands [42], and Intermediate Technology Development Group (ITDG), which is now Practical Action of the United Kingdom [16] established to undertake research and development activities aimed at providing low-cost windpumps in developing regions. CWD's wind turbines for driving water pumps used low solidity and high tip speed ratios. Hence,

Table 6.3 Result of CAA performance at $Re = 10^5$

Spar position	**Conditions for $(C_l/C_d)_{max}$**			**Conditions for stall**		$C_{d,min}$
	α	C_l/C_d	C_l	α_{stall}	$C_{l,max}$	
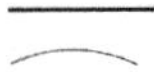	6	23	1.3	12	1.47	0.06
	6	19	1.1	16	1.5	0.06
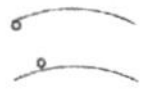	17	3.4	1.12	21	1.28	0.16
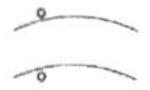	5	18	1.07	12.5	1.3	0.06
	16	4.7	1.26	16	1.27	0.16
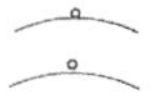	15	4.8	1.28	15	1.28	0.17
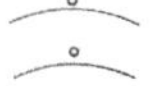	14	4.7	1.2	17	1.24	0.17
	5	18	1.16	12	1.35	0.07
	17.5	1.2	1.2	11	1.4	0.07

Note: The first column shows when a spar was used and its position with respect to blade chord. Reproduced from [22].

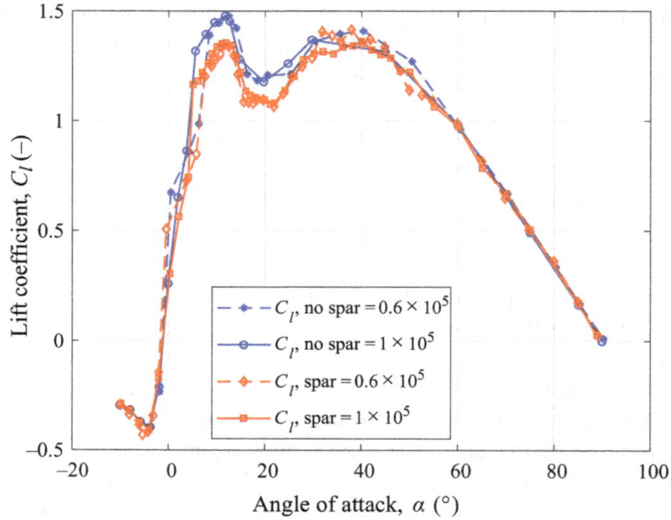

Figure 6.7 Lift coefficients [23]

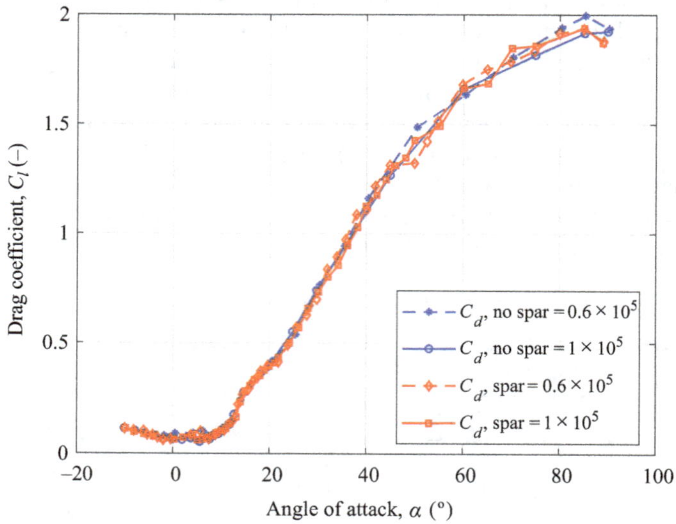

Figure 6.8 Drag coefficients [23]

much of their research efforts were geared towards alleviating the need for high torque at start-up and preventing excessive pump rod accelerations for piston pump types using shorter stroke, cable in tension as opposed to heavy rod [38]. Although these configurations are designed for low wind speed regions where overall water pumping efficiency should be maximized, they are relatively complex. As a result, they are not easy to maintain and repair by the rural populace who are not technically trained.

The effect of variable N or equivalently σ with regards to windmill performance was studied by Wegereef [43] who experimented on a scaled model of an ITDG windmill. He measured the performance of $N = 6$, 12, and 24 bladed rotors with identical blades in a wind tunnel. While the work showed significant increments in starting torque with increased blade root (twist) angles, there was no report on thrust measurements and validation of the torque and power results, rendering the analysis incomplete and necessitating a thorough performance assessment of the model. Sarkis and Pinilla [44] studied a scale model of a commercial fast running low solidity rotor and improved its performance by redesigning its rotor, resulting in a better wind to water conversion efficiency. Intriguingly, not much has been reported on this critical aerodynamic interaction between the rotor blades with a cambered profile, particularly as N increases.

The most common version of BEM as described in Chapter 7 uses the Prandtl tip loss factor to account for finite N, but it is inexact over the blade at low tip speed ratios [45]. Recently, helical vortex theory has been developed and implemented in BEM [46–48], particularly for the conventional three-bladed rotor, but this theory has not been applied to high-solidity windmills.

The other problem in BEMT analysis of windmills is their high σ which invalidates the approximation that blade elements behave as airfoils. In other words, blade element theory (BET) is strictly valid only in the limit that $\sigma \downarrow 0$. The two-dimensional equivalent of a high σ rotor is an infinite cascade of blade elements. Cascade theory has been deployed to account for blade-to-blade interaction in wind turbine performance analysis by BEM modification, e.g., [49–51]. Quamrul Islam and Sadrul Islam [49] demonstrated that modified BET using two-dimensional cascade gave better results than a standard BET for high solidity multibladed wind turbine rotors, though with a penalty in computational time [49]. Fagbenro *et al.* [52] studied high solidity effects on a circular-arc-bladed windmill rotor. Their computational study showed that solidity influences lift and drag, which agrees with the findings of Yan [53] even for low solidity rotors.

We now consider the BEM calculations of the power and torque of the model windmill described in John [54] and John *et al.* [55] that was a reproduction of that studied by Wegereef [43]. The experiments were done using identical blades at the same pitch but with the number of blades $N = 3$, 6, 12, and 24. Figures 6.9 and 6.10 show the conventional power (C_P) and torque (C_Q) coefficients, respectively, and the main results are summarized in Table 6.4 where λ_{opt} is the tip speed ratio for maximum power and λ_{run} is the runaway tip speed ratio that is the maximum operational value. Further details of the BEM calculations can be found in [54]. The theory is remarkably accurate and its main failing is to underpredict the torque at low λ by an amount that increases with increasing N. This is most likely a consequence of solidity modifying the blade element lift and drag. BEM accurately predicts $C_{P,\max}$, and, even more surprisingly, λ_{run}, presumably because it occurs at low λ whereas for electricity generating turbines

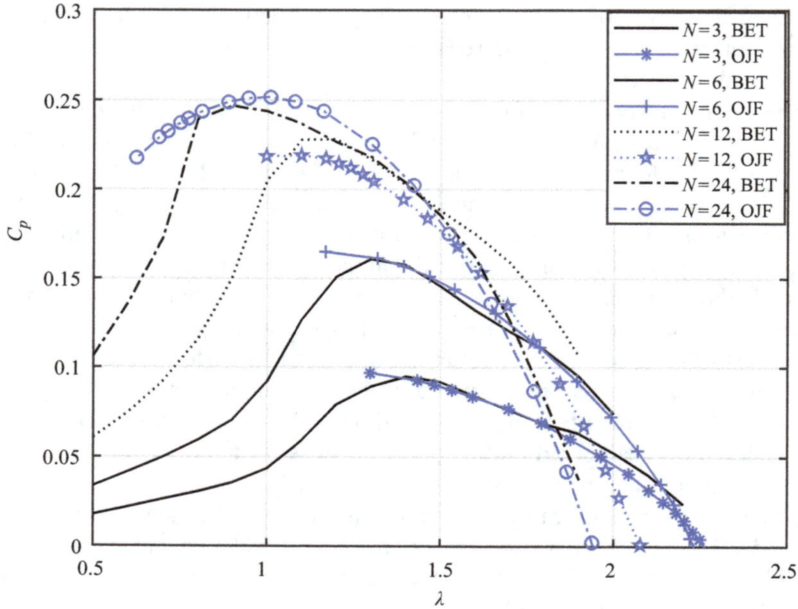

Figure 6.9 BET power coefficient compared to the measurements made in the open jet facility, (OJF) at TU Delft, the Netherlands

with λ_{opt} between 7 and 8, $\lambda_{run} \approx 15$ and well into the high thrust region where the momentum equation in BEM must be modified.

Figure 6.11 shows the present results and those of Sarkis and Pinilla [44] for the maximum C_P against λ along with the low-λ variation for Betz–Goldstein rotors by Wood [56] and Glauert [57]. Note that $C_P \rightarrow 16/27$ at large λ that is the Betz–Joukowsky limit. Sarkis and Pinilla [44] used a blade profile more like the CAAs used by John [54] and have higher lift/drag and therefore greater efficiency. The figure implies that it should be possible to increase the power production of windmills by a concerted effort at optimization that must, however, take account of the manufacturability of the optimum blades.

6.4 Starting behavior of windmills

It was noted in the introduction that the low-speed behavior of windmills is critical to achieve high overall effectiveness in water pumping. Good low-speed behavior is usually interpreted as the ability of the blades to rotate quickly as the wind speed increases from zero. This requirement is shared by small electricity generating turbines and results from the common fact that the blades of small turbines generally have no pitch adjustment. Pitching into the wind as done on large turbines to generate high starting torque is not possible.

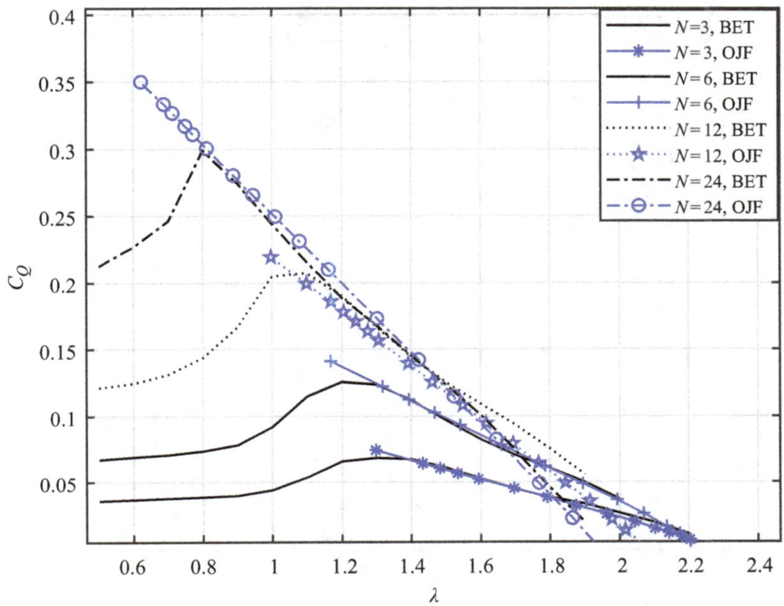

Figure 6.10 BET torque coefficient compared to the measurements made in the open jet facility, (OJF) at TU Delft, the Netherlands

Table 6.4 Comparison of measured and predicted torque and power coefficients of a windmill rotor

Rotor	$N = 3$		$N = 6$		$N = 12$		$N = 24$	
	OJF	**BET**	**OJF**	**BET**	**OJF**	**BET**	**OJF**	**BET**
$C_{P,max}$	0.097	0.095	0.165	0.161	0.219	0.228	0.252	0.247
λ_{opt}	1.297	1.4	1.168	1.3	1.099	1.1	1.008	0.9
$C_{Q,\max}$	0.075	0.069	0.141	0.126	0.2196	0.207	0.350	0.299
$C_{Q,\lambda=0}$		0.023		0.0392		0.095		0.21
λ_{run}	2.317	2.2	2.244	2	2.197	1.9	2.058	1.9

While many studies have focused on blade design for aerodynamic performance improvement, e.g., [58–61], only a few have examined the starting behavior. Ebert and Wood [62] examined on a 5-m diameter, two-bladed 5-kW wind turbine. The analysis showed that a complete starting sequence comprises two phases. The first is a long period of slow acceleration, in which the turbine accelerates slowly with the angle of attack decreasing from its high initial value at the stationary position until the blade reaches the region of the maximum lift/drag ratio. The second phase is characterized by a short period of rapid acceleration to the useful power extraction region, generating high torque. Thus, starting performance is

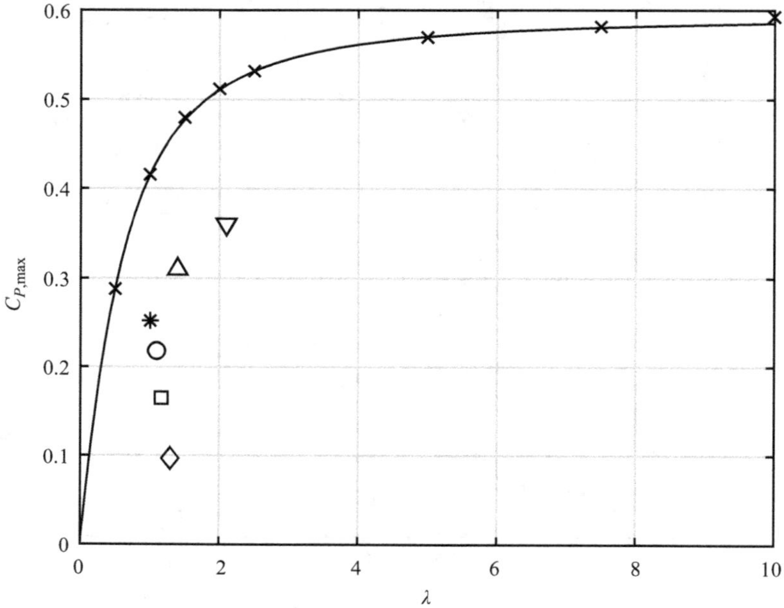

Figure 6.11 *Comparison of windmill maximum C_P with optimal Betz–Goldstein rotors from Figure 1 of Wood [56]: solid line, Glauert [57]; ×, Data from Sarkis and Pinilla [44]; Δ, from Figure 7; ∇, from Figure 11. Remaining data from Table 6.4: N = 3, ◇; N = 6, □; N = 12, ○; N = 24,*

dominated by high α, with the power extraction occurring at a low incidence and high lift to drag ratio (C_l/C_d). Mayer *et al.* [63] examined the effects of blade pitching angle on a similar small turbine as in [62]. Starting was observed for a range of pitch angles from 0° to 35° in steps of 5°. They found that the idling period decreased with an increasing pitch angle (reduced α) owing to the reducing α. By comparing measurement data with rotor acceleration prediction, a better agreement was found for higher pitch angles for the range of wind speeds investigated.

Further study on the starting performance was described by Wright and Wood [64]. They field-tested a three-bladed 2-m diameter turbine and showed from BEM that while the starting torque is generated largely at the blade root, the power extraction torque is produced at the tip region of the blade. They used approximate generic formula and a combination of interpolated airfoil performance data of different airfoils for estimating C_l and C_d at a high α due to the shortage of such data at the incidence angle (up to 90°) and at low *Re*. The authors noted the need to accurately characterize the drivetrain resistance of SWTs because their cogging torque increases with decreasing turbine size. Improvement in starting behavior can potentially increase energy yield by 15%–30% depending on the site [65], and up to 40% depending on the load [66]. Later studies by [67] concurred with the findings of [64] and assert that the idling period can be reduced by increasing the blade chord length and twist angles near the hub.

The aerodynamic torque must be sufficient to overcome the internal friction of the system [57]. The need to accurately characterized the internal friction or drivetrain resistance has been reported as early as the turn of the twentieth century. For example, Perry [68] performed some dynamometric experiments on a windmill that include the determination of starting torque and drivetrain resistance. In the early 1950s, a further attempt to measure the starting torque coefficients of windmills via a start and stop experiment was carried out by Sanuki [69], with Figure 6.12, showing the impact of blade number and varying wind speeds on the starting behavior of a Speedovane akin to a windmill. In the ensuing years, the research slowed due to growing alternatives to mechanical wind pumps. Currently, drivetrain resistance is becoming an increasingly critical issue in SWTs, particularly as it is an important design factor for low wind speed starting in HAWTs, [70–72] and VAWTs [73, 74]. Vaz *et al.* [70] investigated the effect of bearing friction on the drivetrain resistance of a small turbine, notably during its transition from the high static to low dynamic resistive torque in predicting the starting performance and runaway, defined as the maximum aerodynamically possible rotor speed for the no-load condition. As wind turbines must be operated below the runaway point to safely produce power, its estimation at the rotor design stage is important. The bearing frictional analysis showed that the static to dynamic resistive torque ratio is about 7:1, which is a significant variation compared to cogging torque (stationary and rotating) in permanent magnet generators. The authors found that starting wind speed is higher than the cut-in wind speed. They concluded that low wind turbine aerodynamics and accurate drivetrain resistance models have the capability to minimize the starting wind speed and broaden the operating range for optimal performance in rotor design. However, their study assumed no induction and was limited to a three-bladed rotor. We are unaware of any reported study, experimental or analytical, that has investigated the starting behavior of high solidity, CAA-rotors with C_l and C_d data at the appropriate Re and high α.

Figure 6.13 shows measurements from John [54] of the starting performance of his model windmill for the wind speed variation shown in the figure. No loading was applied to the rotor so the final angular velocity corresponds to λ_{run} in Table 6.4. The BEM calculations for $N = 3$, 6, 12 and 24 agree with the corresponding open jet flow (OJF) tunnel measurements. However, there are slight discrepancies. For $N = 3$, the region of acceleration shows significant discrepancy up until the steady-state region, probably because the resistive torque has not been adequately determined. For $N > 3$, the difference between the predicted and measured time-dependent rotor decreases in the transient phase of the starting sequence. This behavior is consistent with the error in the BEM calculations decreasing with increasing N as the aerodynamic torque increases much more than the resistance. Notice that when $N = 24$, a good agreement is achieved. However, relative to the experiment, the BEM under-predicted its final angular velocity. This is probably due to solidity not accounted for in the simulation, which acts to alter the lift and drag of the airfoil and hence the torque resulting in an increased ω of the

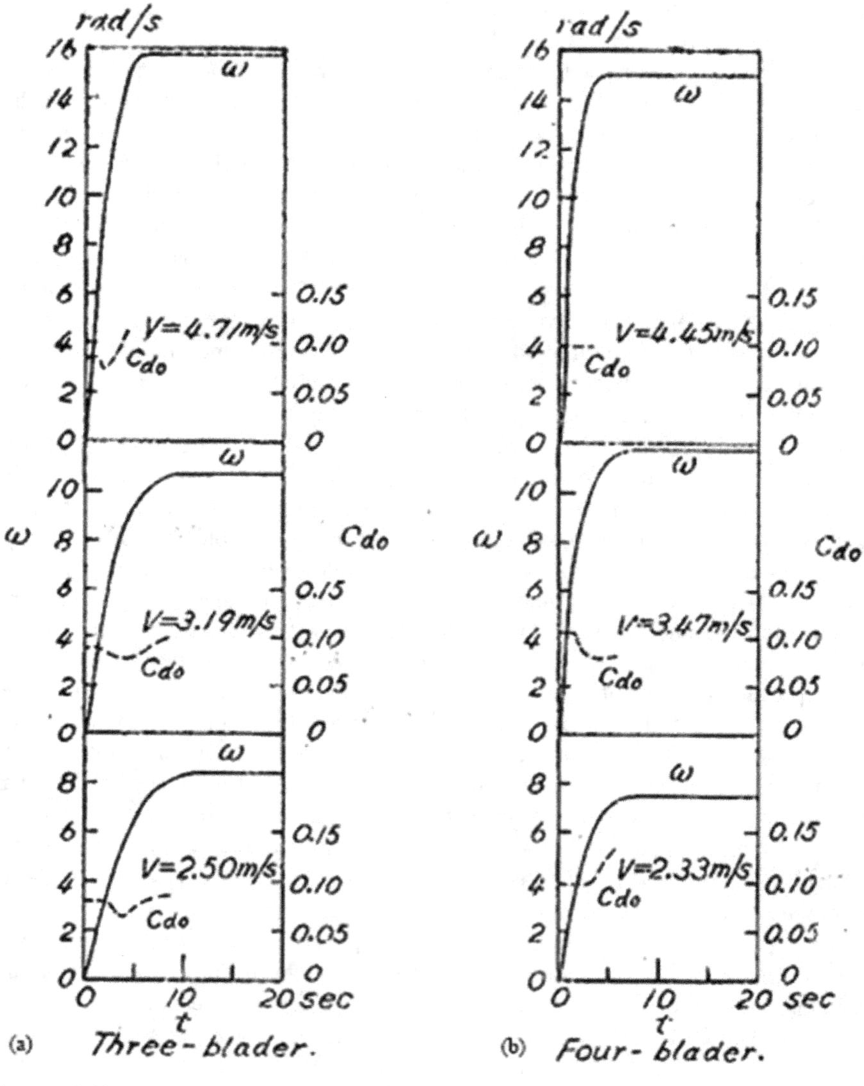

Figure 6.12 Starting of Speedovane in varying wind speeds and computed starting torque coefficients, (C_{d0}) [69]

experiment. Because the windmill was designed for low α_{opt} there is no period of slow acceleration as happens on other small turbines, see Figure 6.13.

6.4.1 Summary

Based on the literature review, the aerodynamics of a circular arc-bladed windmill with a significantly high blade number or otherwise solidity is still poorly

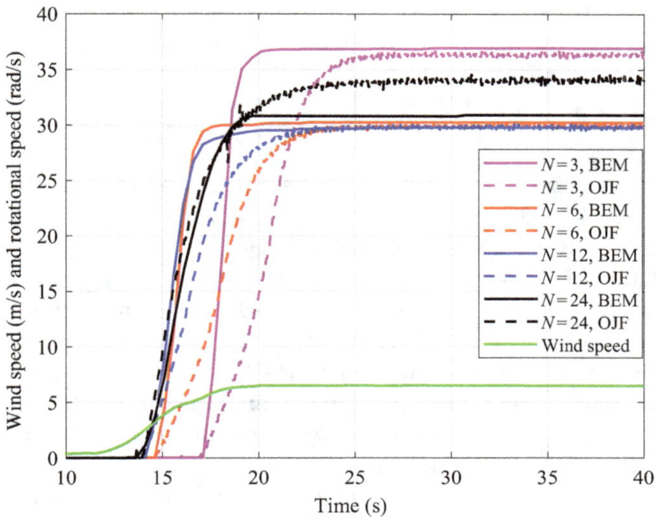

Figure 6.13 Comparison of rotational speeds with rotor experiment at constant wind speed

understood. Experimenting with a wide range of blade numbers and comparing the results with BEM (zero solidity) prediction will help determine if solidity effects are present and, if so, their significance at low *Re* operation.

The accuracy of BEM is dependent on the quality of the aerodynamic data of airfoil. Although there are low *Re* and high angles of attack airfoil performance data for camber airfoil, there is still a lack of data for certain geometrical characteristic combinations such as a high camber and thickness ratio, appropriate to a given application and *Re* requirement.

Further, vortex theory is not used in BEM, even though all blades in rotation shed helical vortices in their wake, which impact the velocity over the blades and the resultant forces acting on them. Application of a vortex-based tip loss factor for finite blade correction in a windmill with varying blade numbers will increase the understanding of the physics affecting windmill performance.

Starting is a critical operational issue for SWT, especially in low wind speed conditions, because the virtual lack of pitching mechanism and its drivetrain resistance can be significant. Accurate airfoil data at very low *Re* corresponding to the starting wind speed is required to model its starting behavior correctly. An attempt at modeling or experimentally investigating high solidity (high blade number) turbines has not been reported to the authors' knowledge.

Wind tunnel blockage significantly affects rotor test measurements. Most blockage studies have focused on low solidity, high tip speed ratio machines. While the aerodynamics of high solidity and low tip speed ratio are complicated because of mutual blade interaction, the combined effect of high solidity and high blockage on the rotor aerodynamic performance has not appeared in the open literature and therefore requires experimental investigation.

References

[1] Van Meel J and Smulders P. *Wind Pumping: A Handbook*. WTP101. The World Bank; 1989. Available from: http://documents1.worldbank.org/curated/en/119791468178164180/pdf/multi-page.pdf.

[2] Veldkamp D. *Recommended Practices for Testing Water-Pumping Windmills*. 100. World Bank; 1989. Available from: http://documents1.worldbank.org/curated/en/819831468764379771/pdf/multi-page.pdf.

[3] Kentfield J. *The Fundamentals of Wind-Driven Water Pumps*. Gordon and Breach Science Publishers. Amsterdam, the Netherlands; 1996.

[4] Hau E. *Wind Turbines: Fundamentals, Technologies, Application, Economics*. Springer Science & Business Media. Berlin; 2013.

[5] Spera DA. *Wind Turbine Technology: Fundamental Concepts of Wind Turbine Engineering*. vol. 3. ASME Press. New York; 1994.

[6] Fraenkel P, Crick F, and Cowley P. Wind power for pumping: The development of the ITP windpump. *Renewable Energy*. 1999;16(1–4):916–921.

[7] Handpump development – Neale consulting engineers; 2021. Available from: https://tribology.co.uk/poldaw-designs/handpump-development/ [Accessed: 19-August-2021].

[8] Harries M. Disseminating wind pumps in rural Kenya – meeting rural water needs using locally manufactured wind pumps. *Energy Policy*. 2002; 30 (11–12): 1087–1094.

[9] Rosato MA. *Small Wind Turbines for Electricity and Irrigation: Design and Construction*. CRC Press. Boca Raton, Florida; 2018.

[10] Pittleloud JD and Gsänger S. Small Wind World Report – Summary. World Wind Energy Association Publication; 2017. Available from: https://wwindea.org/wwea-released-latest-global-small-wind-statistics/ [Accessed: 18-August-2021].

[11] Jober Industries. Products; 2010. Available from: http://molinosjober.com/productos.htm [Accessed: 18-August-2021].

[12] Windpumps; 2020. Available from: https://aureka.com/windpumps/ [Accessed: 19-August-2021].

[13] WUWWA Programme. The United Nations World Water Development Report 2019: Leaving No One Behind. UNESCO Paris; 2019.

[14] Cace J, ter Horst RE, Syngellakis HK, *et al.* Urban Wind Turbines. Guidelines for Small Wind Turbines in the Built Environment; 2007. p. 1–41. Article online www.urbanwind.

[15] Kamkwamba W and Mealer B. The Boy Who Harnessed the Wind. Harper Collins. London; 2010.

[16] Energy from the wind; 2020. Available from: https://answers.practicalaction.org/our-resources/item/energy-from-the-wind/ [Accessed: 19-August-2021].

[17] Van Treuren KW. Small-scale wind turbine testing in wind tunnels under low Reynolds number conditions. *Journal of Energy Resources Technology*. 2015;137(5).

[18] Wood DH. *Small Wind Turbines: Analysis, Design and Application*. Springer. London; 2011.

[19] Argaw N, Foster R, and Ellis A. *Renewable Energy for Water Pumping Applications in Rural Villages; Period of Performance: April 1, 2001–September 1, 2001*. National Renewable Energy Laboratory (NREL). Golden, CO; 2003.

[20] Gopal C, Mohanraj M, Chandramohan P, *et al.* Renewable energy source water pumping systems: A literature review. *Renewable and Sustainable Energy Reviews*. 2013;25:351–370.

[21] Pinilla AE, Burton J, and Dunn P. Wind energy to water pumped: Conversion efficiency limits using single-acting lift pumps. In: *Proc. Conf. BWEA*, Cambridge University Press; 1984. p. 334–345.

[22] Dixon JC. Improving the mechanical load matching of wind energy converters. *Wind Energy Workshop*. 1979; 181–189.

[23] Bruining A. Aerodynamic Characteristics of a Curved Plate Airfoil Section at Reynolds Numbers 60000 and 100000 and Angles of Attack From −10 to +90 Degrees. Department of Aerospace Engineering, Delft University of Technology; 1979. Available from: http://repository.tudelft.nl/view/ir/uuid:6b92442a-01f7-4b7c-8d53-c4f10720ff3e/ [Accessed: 26-September-2018].

[24] Pandey M, Pandey K, and Ojha T. Aerodynamic characteristics of cambered steel plates in relation to their use in wind energy conversion systems. *Wind Engineering*. 1988;12:90–104.

[25] Okamoto M, Yasuda K, and Azuma A. Aerodynamic characteristics of the wings and body of a dragonfly. *Journal of Experimental Biology*. 1996;199 (2):281–294.

[26] Okamoto M and Azuma A. Experimental study on aerodynamic characteristics of unsteady wings at low Reynolds number. *AIAA Journal*. 2005;43 (12):2526–2536.

[27] Tezuka A, Sunada Y, and Rinoie K. Surface pressure distributions on 4% circular arc airfoil at low Reynolds number. *Journal of Aircraft*. 2008;45 (6):2164–2167.

[28] Okamoto M and Ebina K. Effectiveness of large-camber circular arc airfoil at very low Reynolds numbers. *Transactions of the Japan Society for Aeronautical and Space Sciences*. 2016;59(5):295–304.

[29] Flay RGJ, Piard A, and Bot P. Aerodynamics of a highly cambered circular arc aerofoil: experimental investigations. In: *International Conference on Innovation in High Performance Sailing Yachts*, Lorient, France; 2017. p. 150–162.

[30] Ackroyd JAD, Axcell BP, and Ruban AI. Early Developments of Modern Aerodynamics. *The Aeronautical Journal*. 2002; 106(1058): 202.

[31] Wallis R. *Wind Tunnel Tests on a Series of Circular Arc Plate Aerofoils*. Aeronautical Research Labs. Melbourne, Australia; 1946.

[32] Ikui T, Inoue M, and Kaneko K. Two-dimensional cascade performance of circular-arc blades (Circular arc blades two dimensional cascade performance test data for various cambers comparison with potential theory data). In: *Tokyo Joint International Gas Turbine Conference and Products Show*, Tokyo, Japan; 1971. p. 57–64.

[33] Mueller TJ and DeLaurier JD. Aerodynamics of small vehicles. *Annual Review of Fluid Mechanics*. 2003;35(1):89–111.

[34] Sunada S, Yasuda T, Yasuda K, *et al.* Comparison of wing characteristics at an ultralow Reynolds number. *Journal of Aircraft*. 2002;39(2):331–338.

[35] Mueller TJ. Aerodynamic measurements at low Reynolds numbers for fixed wing micro-air vehicles. In: *Development and operation of UAVS for military and civil applications*, Von Kaman Institute for Fluid Dynamics, Belgium; 2000.

[36] Bragg G and Schmidt W. Performance matching and optimization of wind powered water pumping systems. *Energy Conversion*. 1979;19(1):33–39.

[37] Nathan G. A simplified design method and wind-tunnel study of horizontal-axis windmills. *Journal of Wind Engineering and Industrial Aerodynamics*. 1980;6(3–4):189–205.

[38] Cleijne H, Smulders P, Verheij F, *et al.* Pump research by CWD: the influence of starting torque of single acting piston pumps on water pumping windmills. In: *European Wind Energy Association, Conference and Exhibition*. vol. 7; 1986.

[39] Manwell JF. Understanding wind energy for water pumping. In: *VITA Understanding technology series*, vol. 62; 1988.

[40] Rijs R and Smulders P. Blade element theory for performance analysis of slow running wind turbines. *Wind Engineering*. 1990;14:62–79.

[41] Rijs R, Jacobs P, and Smulders P. Parameter study of the performance of slow running rotors. *Journal of Wind Engineering and Industrial Aerodynamics*. 1992;39(1–3):95–103.

[42] Lysen EH. Introduction to Wind Energy, CWD 82-1. Amersfoort, The Netherlands: CWD, Consultancy Services Wind Energy Developing Countries; 1983.

[43] Wegereef E. Scale model of the ITDG 6 meter rotor wind tunnel tests for different bade setting angles. In: Report of Twente University of Technology, the Netherlands, windmill group. Report WM 076; 1984.

[44] Sarkis N and Pinilla AE. Experimental study on a model of a commercial wind pumping rotor. *Wind Engineering*. 2006;30(6):511–520.

[45] Wood D. Application of extended vortex theory for blade element analysis of horizontal-axis wind turbines. *Renewable Energy*. 2018;121:188–194.

[46] Branlard E and Gaunaa M. Development of new tip-loss corrections based on vortex theory and vortex methods. In: *The Science of Making Torque from Wind 2012: 4th Scientific Conference*; 2012.

[47] Wood D, Okulov V, and Bhattacharjee D. Direct calculation of wind turbine tip loss. *Renewable Energy*. 2016;95:269–276.

[48] Schmitz S and Maniaci DC. Methodology to determine a tip-loss factor for highly loaded wind turbines. *AIAA Journal*. 2017;55(2):341–351.

[49] Quamrul Islam M and Sadrul Islam AKM. The aerodynamic performance of a horizontal-axis wind turbine calculated by strip theory and cascade theory. *JSME International Journal Series B Fluids and Thermal Engineering*. 1994;37(4):871–877.

[50] Wood D. A cascade model of blade element interaction for wind turbines with unequal blades. *International Journal of Sustainable Energy*. 2016;35 (5):502–512.

[51] Jafari S, Kwok KC, Safaei F, *et al.* Aerodynamic analysis of a linear cascade wind turbine. *Wind Energy*. 2018;21(11):1141–1154.

[52] Fagbenro K, Mohamed M, and Wood D. Computational modeling of the aerodynamics of windmill blades at high solidity. *Energy for Sustainable Development*. 2014;22:13–20.

[53] Yan H. *Computational Modeling of Cascade Effect on Blade Elements with an Airfoil Profile*. Master's thesis, University of Calgary, Calgary; 2016.

[54] John IH. *Aerodynamic Performance and Starting Behavior of Multi-Bladed Water pumping Windmills*. PhD thesis, University of Calgary, Calgary; 2021.

[55] John IH, Vaz JR, and Wood D. Aerodynamic performance and blockage investigation of a cambered multi-bladed windmill. *Journal of Physics: Conference Series*. 2020;1618(4):042003.

[56] Wood D. Maximum wind turbine performance at low tip speed ratio. *Journal of Renewable and Sustainable Energy*. 2015;7(5):053126.

[57] Glauert H. Airplane propellers. In: *Aerodynamic Theory*. Springer. Berlin, Heidelberg; 1935. p. 169–360.

[58] Lanzafame R and Messina M. Optimal wind turbine design to maximize energy production. *Proceedings of the Institution of Mechanical Engineers, Part A: Journal of Power and Energy*. 2009;223(2):93–101.

[59] Lanzafame R, Mauro S, and Messina M. HAWT design and performance evaluation: improving the BEM theory mathematical models. *Energy Procedia*. 2015;82:172–179.

[60] Mesquita ALA and Alves AS. An improved approach for performance prediction of HAWT using the strip theory. *Wind Engineering*. 2000;24 (6):417–430.

[61] Yu D, Kwon H, and Kwon OJ. Performance enhancement of HAWT rotor blades by aerodynamic shape optimization. In: 50th AIAA Aerospace Sciences Meeting Including the New Horizons Forum and Aerospace Exposition; 2012. p. 1292.

[62] Ebert P and Wood D. Observations of the starting behaviour of a small horizontal-axis wind turbine. *Renewable Energy*. 1997;12(3):245–257.

[63] Mayer C, Bechly M, Hampsey M, *et al.* The starting behaviour of a small horizontal-axis wind turbine. *Renewable Energy*. 2001;22(1–3):411–417.

[64] Wright A and Wood D. The starting and low wind speed behaviour of a small horizontal axis wind turbine. *Journal of Wind Engineering and Industrial Aerodynamics*. 2004;92(14–15):1265–1279.

[65] Wichser C and Klink K. Low wind speed turbines and wind power potential in Minnesota, USA. *Renewable Energy*. 2008;33(8):1749–1758.

[66] Worasinchai S, Ingram GL, and Dominy RG. Effects of wind turbine starting capability on energy yield. *Journal of Engineering for Gas Turbines and Power*. 2012;134(4):042603.

[67] Khan S, Shah K, Khan H, *et al.* Observation of the starting and low speed behavior of small horizontal axis wind turbine. *Journal of Wind Energy*. 2014;2014.

[68] Perry TO. *Experiments with Windmills*. US Government Printing Office. Washington; 1899

[69] Sanuki M. Experiments on the start and stop of windmill-and cup-anemometers with particular reference to their over-estimation factors. *Papers in Meteorology and Geophysics*. 1952;3(1):41–53.

[70] Vaz JR, Wood DH, Bhattacharjee D, *et al.* Drivetrain resistance and starting performance of a small wind turbine. *Renewable Energy*. 2018;117:509–519.

[71] Farias GM, Galhardo MA, Vaz JR, *et al.* A steady-state based model applied to small wind turbines. *Journal of the Brazilian Society of Mechanical Sciences and Engineering*. 2019;41(5):209.

[72] Moreira JL, Mesquita AA, Araujo L, *et al.* Experimental investigation of drivetrain resistance applied to small wind turbines. *Renewable Energy*. 2020;153:324–333.

[73] Aso T, Seki K, and Ushiyama I. Study on the bearing resistance in a wind turbine generator system. *Wind Engineering*. 2015;39(1):113–128.

[74] Aso T, Aida T, and Seki K. Bearing resistance of wind turbine generator system. *Electrical Engineering in Japan*. 2016;194(2):64–73.

Chapter 7

Blade element analysis and design of horizontal-axis turbines

Jerson R.P. Vaz[1] *and David Wood*[2]

Horizontal-axis wind turbines (HAWTs) harness energy from the wind and horizontal-axis hydrokinetic turbines (HKTs) use the same basic principles in rivers, and tidal or marine currents. The value of wind turbines is well established. HKTs do not require large flooded areas or civil structures and so reduce environmental impacts. This chapter focuses on the analysis and design of HAWTs and HKTs using blade element momentum theory (BEMT). The theory is extended to deal with the major differences between wind and water operation—the possibility of cavitation in the latter—through the criterion of the local minimum pressure coefficient that is usually expressed as a cavitation number. We demonstrate the accuracy of BEMT and how it can be modified to include the effect of a diffuser surrounding the rotor as this modification has significant attraction for small turbines. Recent mathematical formulations to design diffuser-augmented turbines (DATs) are presented, and we consider BEMT design for maximum power and multidimensional optimization (MDO) when other objectives such as low noise and blade mass are also to be met.

7.1 Introduction

The aerodynamic analysis of wind turbines using BEMT is well established, e.g., Burton *et al.* [1], Wood [2], and Hansen [3]. Horizontal HKTs are less well known but there are substantial references in the literature, e.g., [4–6]. BEMT is a combination of the simple axial momentum and the blade element theory (BET) [2]. It is easy to formulate and code, and it usually executes quickly making it an ideal objective function for MDO that will be considered later. We also consider corrections for the finite number of blades and high rotor thrust [3]. For HKTs, a modification to include the effect of cavitation is necessary, in order to avoid damaging the blades [7, 8]. As in classical hydraulic turbines [9], cavitation is assumed to initiate on a blade section when the minimum local pressure falls below the vapor pressure of the fluid. Thus, the

[1]Faculty of Mechanical Engineering, Institute of Technology, Federal University of Pará, Brazil
[2]Department of Mechanical and Manufacturing Engineering, Schulich School of Engineering, University of Calgary, Canada

cavitation can be predicted by comparing the local minimum pressure coefficient with that pressure, usually in the form of the cavitation number as explained later. The chances of cavitation occurring increases toward the blade tip at low immersion depth, which is an important issue to be assessed on HKT design.

7.2 Horizontal-axis wind and hydrokinetic turbines

A basic HKT consists of a turbine rotor connected to a generator, often through a gearbox, as shown in Figure 7.1 in which the similarities to an HAWT are apparent. To use BEMT, it is necessary to assume that all components are infinitely rigid. Such a consideration is valid since the vibration modes of the system are assumed to be in frequency range higher than the operational one [10] and small blades tend to be more rigid than large ones. The drivetrain comprising the main shaft and its bearings, a gearbox if installed, and the generator bearings is important as its resistance can significantly increase the starting time of the turbine [11]. If a permanent magnet generator is used, the cogging torque contributes to the drivetrain resistance. This chapter concentrates on the design of rotors using BEMT analysis. BEMT is easily adapted to predict starting performance [2]. A more recent study that improves the modeling of the

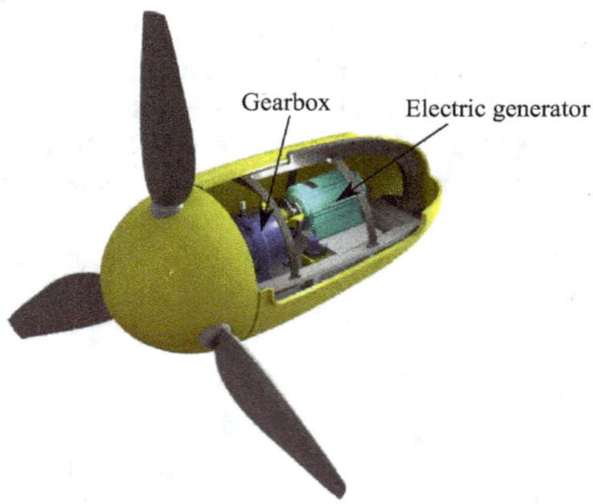

Figure 7.1 Illustrations of a hydrokinetic turbine [10]

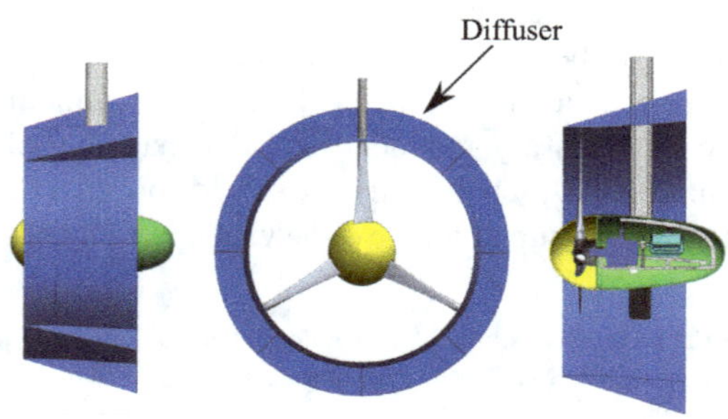

Figure 7.2 Illustrations of a wind or hydrokinetic turbine with diffuser [14]

drivetrain-resistive torque is in Vaz *et al.* [11]. A good balance between starting and power extraction is an example of MDO that we consider in Section 7.8.1.

Small diffuser-augmented wind turbines (DAWTs) are now on the market, Evans *et al.* [12], and DAHKTs are likely to follow, mainly because the diffuser increases the power output (Figure 7.2). The idea of adding a diffuser is simple but the additional aerodynamics are complex; for example, the very recent review of Bontempo and Manna [13] lists 125 references. In addition to increasing power, a diffuser also has the advantage of allowing the rotor to start rotating at lower speeds, and the fitting of screens to protect marine life.

7.3 Cavitation on hydrokinetic turbines

Cavitation must be avoided in the design of HKTs [7] because of its effects on performance and lifetime. During cavitation, liquid vaporizes almost instantly and forms a vapor bubble that alters the flow [15]. Figure 7.3 shows cavitation on a propeller. When a bubble implodes, the pressure on the blade surface increases, promoting erosion on the blade. The damage may eventually decrease lift and increase drag, leading to a reduction in turbine efficiency. Cavitation can be predicted by comparing the local pressure distribution with the cavitation number [16], which is classically defined as

$$\sigma = \frac{p_{atm} + \rho g h - p_v}{\frac{1}{2}\rho W^2}, \tag{7.1}$$

where p_{atm} is the atmospheric pressure, g is the gravitational acceleration, h is the distance between free surface and the radial position on the HKT blade, p_v is the vapor pressure, and $W = \sqrt{[V_0(1-a)]^2 + [\Omega r(1+a')]^2}$ is the relative velocity at each blade section. a and a' are the axial and tangential induction factors, respectively, which are introduced in the description of BEMT later. The angular velocity of the blades is Ω. Cavitation will occur if the local minimum pressure coefficient, $c_{p\min} = \min(c_p)$ (see Figure 7.4), is lower than σ. This coefficient can be used as a criterion to avoid cavitation:

$$\sigma + c_{p\min} \geq 0, \tag{7.2}$$

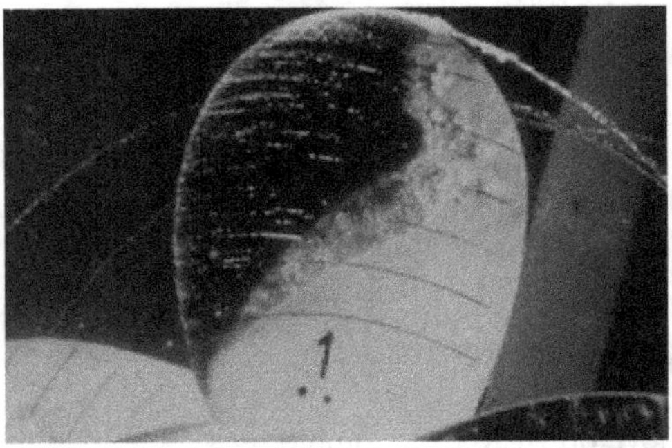

Figure 7.3 Cavitation on a marine propeller. Note the bubbles in the tip vortex [17]

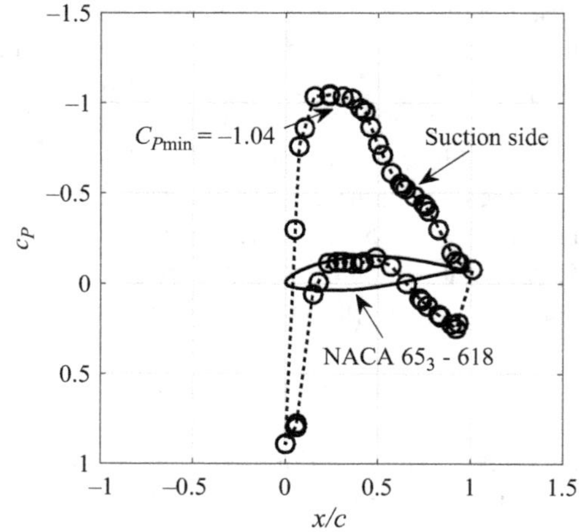

Figure 7.4 Typical minimum pressure coefficient around a rotor blade section near the tip [8]. x is the distance along the blade and c is the chord

where the pressure coefficient is defined as

$$c_P = \frac{p - p_0}{\frac{1}{2}\rho W^2}, \tag{7.3}$$

where p is the local pressure on the hydrofoil profile at the radial blade section, and p_0 is a reference pressure upstream of the rotor, given by $p_{atm} + \rho gh$. The usefulness of c_P comes from it being available for the airfoil comprising the blade section.

Cavitation is more likely near the tip of the top blade when it is closest to the water surface. Figure 7.5 illustrates the static pressure on a blade element (BE). Consequently, the velocity V_{CAV}, required to initiate cavitation, can be obtained combining (7.2) and (7.3). The expression for V_{CAV},

$$V_{CAV} = \sqrt{\frac{p_{atm} + \rho gh - p_v}{-\frac{1}{2}\rho C_{p\min}}}, \tag{7.4}$$

will be used later, through a correction in the rotor thrust coefficient, to avoid cavitation in the optimum design of HKT blades.

For the next sections of this chapter, the flow at each radial position r will be analyzed. This is the basis of BEMT, which describes the energy conversion by means of axial and angular momentum balances at each r.

7.4 Blade element momentum theory

In this section, we develop BEMT by combining simple momentum theory for the fluid flow and BET for the forces on the blades. These two classical theories are the

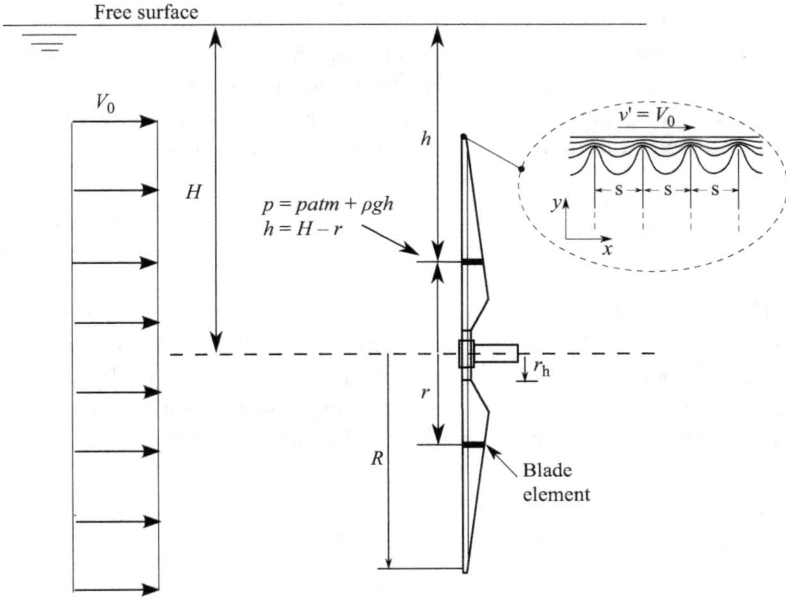

Figure 7.5 Illustration of the static pressure condition on a rotor blade section (adapted from [8])

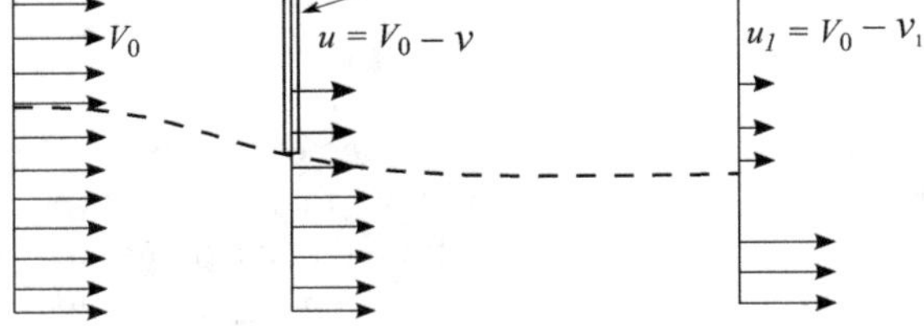

Figure 7.6 Classical actuator disk control volume [18]

simplest description of "bare" HAWTs and HAHKTs. The combination of the two theories in a computer program is usually done by iteration. Roughly, the current values of a and a' determined for the BEs are used to iterate new values using momentum theory. This continues until convergence. Typical BEMT algorithms are given in Section 7.8.

7.4.1 Axial momentum theory

Assume that a bare rotor, that is, the turbine has no diffuser, is represented as an ideal actuator disk, i.e., the fluid is frictionless and there is no rotational velocity in the wake. For one-dimensional analysis, the induced axial velocities u and u_1 at the disk and in the wake, respectively (see Figure 7.6), are written as

$$\begin{cases} V_0 - v = u \equiv (1-a)V_0 \\ V_0 - v_1 = u_1 \equiv (1-a_1)V_0, \end{cases} \tag{7.5}$$

where V_0 is the wind or water speed, and $v = aV_0$, $v_1 = 2aV_0$, and a is the axial induction factor at the rotor and $a_1 = v_1/V_0$ in the far-wake, the final flow region of no further development. For steady flow, the thrust (T) and power (P) in coefficient form are obtained by applying conservation of mass, momentum, and energy to the control volume shown by the dashed lines in Figure 7.6, yielding [2, 3]

$$C_T = \frac{T}{\frac{1}{2}\rho A V_0^2} = 1 - \left(\frac{u_1}{V_0}\right)^2, \tag{7.6}$$

and

$$C_P = \frac{P}{\frac{1}{2}\rho A V_0^3} = \frac{Tu}{\frac{1}{2}\rho A V_0^3} = \frac{u}{V_0}\left[1 - \left(\frac{u_1}{V_0}\right)^2\right], \tag{7.7}$$

where ρ is the fluid density that is about 800 times larger for water than air, giving HKTs a natural advantage in power production over HAWTs. Astute readers will have realized that a further assumption in deriving this equation is that the gauge pressure is zero in the far-wake. An alternative form of the thrust equation comes from first writing Bernoulli's equation between the undisturbed flow and the upstream face of the rotor and then between the rotor's downstream face and the far-wake. Assuming that the thrust on the rotor is given by the pressure difference, it follows that

$$u = \frac{1}{2}(V_0 + u_1) \quad \text{or} \quad a_1 = 2a. \tag{7.8}$$

Thus, the velocity at the rotor plane is the average of the undisturbed speed V_0 and the velocity in the far-wake u_1. It further follows that $C_T = 4a(1-a)$ and $C_P = 4a(1-a)^2$. By setting $dCp/da = 0$, we find the maximum power coefficient, $C_{p/\max} = 16/27$ occurs when $a_{opt} = 1/3$ for which $C_{Topt} = 8/9$ (Figure 7.7). This shows that the maximum energy extracted from the flow by a turbine is 59.3%, which is known in the literature as the Betz–Joukowsky limit [2]. These criteria can be used to design horizontal-axis turbines [6, 7, 18], but we emphasize that this is a one-dimensional design for maximum power only. There are, in practice, two very important corrections to this simple theory to account for (i) the finite number of blades on a real turbine and (ii) the high thrust because when $a > 0.5$, the far-wake velocity is negative, which is physically unappealing. Some corrections will be shown later.

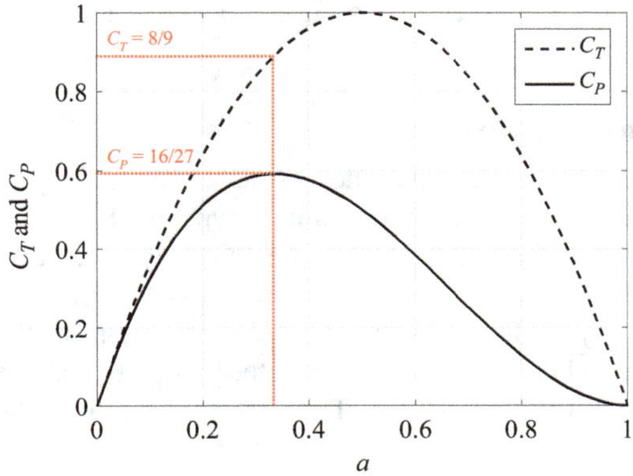

Figure 7.7 Thrust and power coefficients for an ideal bare turbine

7.4.2 Including the angular momentum

A turbine cannot produce power without causing wake rotation because the power extracted from the fluid is $Q\Omega$ where Q is torque on the rotor. In turn, Q must produce angular momentum in the fluid. Consider the differential torque at radius r, dQ, acting on an element of the rotor, not surprisingly called a "blade element," of radial extent dr. Assuming that there is no circumferential velocity upstream of the rotor, then all angular momentum wr must be produced at the blades. Taking an infinitesimal control volume of area $2\pi r dr$ that intersects the element and has its vertical face immediately downstream of the rotor, we write [19]

$$w = 2a'\Omega r, \tag{7.9}$$

which defines the tangential induction factor a'. The angular momentum balance is

$$dQ = 2\pi r^2 \rho u w dr = 4\pi r^3 \rho V_0 \Omega (1 - a) a' dr. \tag{7.10}$$

In coefficient form,

$$C_Q = \frac{dQ}{\frac{1}{2}\rho V_0^2 R dA} = \frac{4a'(1 - a)\Omega r^2}{V_0 R}. \tag{7.11}$$

The elemental power is $dP = \Omega dQ$ so that integrating (7.10) from 0 to the tip at radius R

$$P = 4\pi \rho \Omega^2 V_0 \int_0^R a'(1 - a) r^3 dr. \tag{7.12}$$

Hence, the power coefficient is

$$C_P = \frac{P}{\frac{1}{2}\rho V_0^3 A} = \frac{8}{\lambda^2} \int_0^\lambda a'(1 - a) x^3 dx, \tag{7.13}$$

where $\lambda = \Omega R/V_0$ is the tip-speed ratio, and $x = \Omega r/V_0$ is the local-speed ratio at radius r.

7.4.3 Blade element theory

The concept of a BE was introduced in the last subsection to account for any variation in Q and T (and other quantities) along a blade. A typical BE is shown in Figure 7.5. It is usually assumed that each BE behaves independently of all others so solving the sequence of BEs from, say, the hub to the tip, leads to a simple computer implementation of BET. Typically 30 BEs are used. This classical theory was initially proposed by Froude and Drzewiecki in the nineteenth century, and much of its further development was due to Glauert [20]. The early development focused on aircraft propellers that can be thought of as the converse of axial turbines.

Figure 7.5 illustrates a 2-bladed rotor of radius R, and hub radius r_h subjected to the uniform flow of V_0. There is nothing special about a 2-bladed rotor; BET is applicable to any number of blades, N.

The velocity diagram for a BE (Figure 7.8) is fundamental for the development of BET, since it shows the geometry of the velocity vectors at the element, in addition to the angles involved and the forces acting. The angle of attack is represented by α, the flow angle by ϕ and the twist angle by θ. These three angles are related by

$$\phi = \alpha + \theta. \tag{7.14}$$

The flow angle ϕ can be determined by

$$\phi = \tan^{-1}\left[\frac{(1-a)V_0}{(1+a')\Omega r}\right]. \tag{7.15}$$

Lift and drag forces are L and D, respectively, while normal and tangential forces are F_n and F_t, respectively. Note that F_n and F_t are written combining the

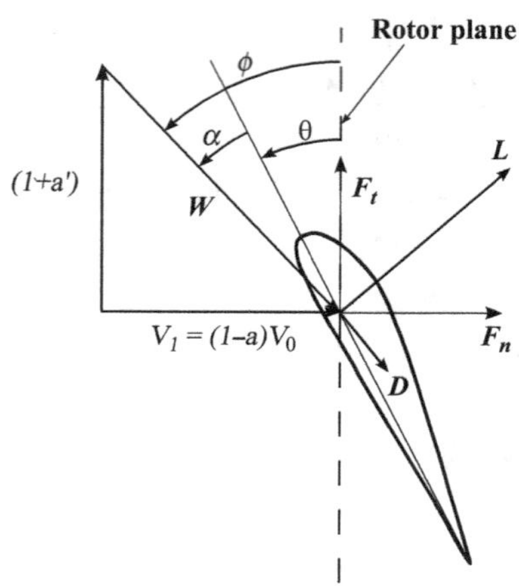

Figure 7.8 Blade element velocity diagram [21]

components of L and D, resulting in

$$F_n = L\cos\phi + D\sin\phi, \tag{7.16}$$

and

$$F_t = L\sin\phi - D\cos\phi. \tag{7.17}$$

We now come to the fundamental assumption of BET: the forces on any BE are those on an airfoil of the same cross-section at the same Reynolds number, $Re = Wc/\nu$, where ν is the kinematic viscosity of air or water, and α. Readers unfamiliar with the Reynolds number can safely ignore it as a detail introduced for completeness. This assumption is very powerful because it allows a vast amount of data on airfoil lift and drag to be used in turbine analysis, but there are limits as discussed later. The unit of lift and drag are force per length, defined as

$$L = \frac{1}{2}\rho W^2 c C_l, \tag{7.18}$$

$$D = \frac{1}{2}\rho W^2 c Cd, \tag{7.19}$$

substituting (7.18) and (7.19) into (7.16) and (7.17), yields

$$F_n = \frac{1}{2}\rho W^2 c(C_l\cos\phi + C_d\sin\phi), \tag{7.20}$$

$$F_t = \frac{1}{2}\rho W^2 c(C_l\sin\phi - C_d\cos\phi). \tag{7.21}$$

The normal and tangential force coefficients C_n and C_t are

$$C_n = \frac{F_n}{\frac{1}{2}\rho W^2 C} = C_l\cos\phi + C_d\sin\phi, \tag{7.22}$$

$$C_t = \frac{F_t}{\frac{1}{2}\rho W^2 C} = C_l\sin\phi - C_d\cos\phi. \tag{7.23}$$

Since there are N elements at each r, the thrust and torque on a given element of blade section are

$$dT = NF_n dr \quad \text{and} \quad dQ = rNF_t dr. \tag{7.24}$$

Note that the units of F_n and F_t are still force per length. Thus, substituting (7.20) and (7.21) into (7.24) gives

$$dT = \frac{1}{2}\rho NW^2 c(C_l\cos\phi + C_d\sin\phi)Ndr, \tag{7.25}$$

and

$$dQ = \frac{1}{2}\rho NW^2 c(C_l\sin\phi - C_d\cos\phi)rNdr. \tag{7.26}$$

Note that these two equations show that viscous effects in the form of drag, reduce dQ, and increase dT. It follows immediately that a BE will maximize its dP if the element operates at the α giving maximum L/D.

The local thrust and torque coefficients in terms of induction factors are given by

$$C_T = \frac{dT}{\frac{1}{2}\rho V_0^2 dA} = \sigma C_n \left(\frac{W}{V_0}\right)^2 = \sigma C_n \frac{(1-a)^2}{\sin^2\phi}, \tag{7.27}$$

and

$$C_Q = \frac{dQ}{\frac{1}{2}\rho V_0^2 R dA} = \sigma C_t \left(\frac{W}{V_0}\right)^2 \frac{r}{R} = \sigma C_t \frac{\Omega r^2}{V_0 R} \frac{(1-a)(1+a')}{\sin\phi\cos\phi}, \tag{7.28}$$

where $\sigma = Nc/(2\pi r)$ is called the local solidity. There should be no confusion with the use of σ in (7.1) as cavitation is not considered in this section. σ has the important interpretation that it measures the relative distance apart of BEs. The assumption of airfoil behavior as articulated earlier is strictly valid only in the limit $\sigma \downarrow 0$. Most HAWTs and HKTs, however, have low solidity so the airfoil assumption is reasonable. An interesting example of high–σ turbines where the airfoil assumption breaks down is given in Chapter 6.

Integrating (7.25) and (7.26) from r_c to R and rewriting them in terms of normal and tangential coefficients, the total thrust and torque of the rotor are

$$T = \frac{1}{2}\rho N \int_{r_c}^{R} W^2 c C_n dr, \tag{7.29}$$

$$Q = \frac{1}{2}\rho N \int_{r_c}^{R} W^2 c C_t r dr. \tag{7.30}$$

Therefore, the total power extracted from the fluid flow is given by

$$P = \Omega T = \frac{1}{2}\rho N \Omega \int_{r_c}^{R} W^2 c C_t r dr. \tag{7.31}$$

BEMT is the straightforward combination of the angular and axial momentum equations and BET. However, it is necessary to consider a correction for finite N. This correction will be described later. One of the most important uses of BEMT is determining formulations for a and a'. Thus, combining (7.6) and (7.27) for the axial induction factor, and (7.11) and (7.28) for the tangential one, yields

$$\frac{a}{1-a} = \frac{\sigma C_n}{4\sin^2\phi}, \tag{7.32}$$

$$\frac{a'}{1+a'} = \frac{\sigma C_t}{4\sin\phi\cos\phi}. \tag{7.33}$$

As mentioned before, these formulations need correction for a finite number of blades. Here, two important formulations are described: (i) the well-known Prandtl tip loss factor and (ii) the more recent "finite blade functions" described by Wood [22].

7.4.4 Prandtl tip loss factor

Figure 7.9 taken from Wood [2] shows BEMT calculations of the power extracted by a model 3-m diameter, 2-bladed wind turbine tested in a wind tunnel by Anderson *et al.* [23]. The tip loss corrections, as described in this section, typically reduce P by around 5% and so are the most important of the second-order aspects of BEMT.

Prandtl's tip loss factor, F_P, was originally defined as the ratio of the total bound circulation of the blades and the circulation of a rotor with an infinite number of blades. Alternatively, determining the BE lift and drag requires the velocities at the element, whereas the axial and angular momentum equations use the average velocities in the streamtube flowing over the element. As $N \rightarrow \infty$, the blade and average velocities coincide. The ratio of the blade to average velocity defines a finite blade function for each velocity component.

In many cases, BEMT calculations made with the Prandtl factor show good agreement with free-wake vortex theory and test data [24]. Prandtl's factor arises from an approximate solution for the potential flow a set of translating helicoidal surfaces representing the wake of each blade by the flow around the edges of a two-dimensional set of equally spaced semi-infinite laminae, as illustrated in Figure 7.5.

$$F_P = \frac{2}{\pi} \cos^{-1}\left(e^{-f}\right), \tag{7.34}$$

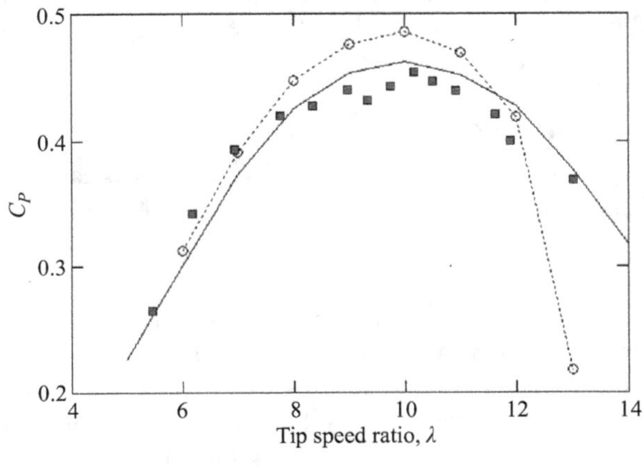

Figure 7.9 Accuracy of BEMT power coefficient with and without tip loss. Blue squares show the measurements, circles show BEMT without F_P, and solid line shows BEMT with F_P. Adapted, with permission, from Reference [2], Springer-Nature

where f to the blade tip is given by f_{tip}, which is

$$f_{tip} = \frac{B}{2} \frac{R - r}{r\sin(\phi)}, \tag{7.35}$$

to the blade root f is f_{root}, as

$$f_{root} = \frac{B}{2} \frac{r - r_h}{r_h\sin(\phi)}. \tag{7.36}$$

So, to be implemented in BEMT, Prandtl tip loss factor assumes the form

$$F_P = F_{Froot} F_{Ftip}. \tag{7.37}$$

Equation (7.34) is included in (7.6) and (7.11), resulting in

$$C_T = 4a(1 - a)F_P, \tag{7.38}$$

$$C_Q = \frac{4a'(1 - a)F_P \Omega r^2}{V_0 R}. \tag{7.39}$$

Combining (7.38) and (7.39) with (7.27) and (7.28), respectively, the formulations for axial and tangential induction factors become

$$\frac{a}{1 - a} = \frac{\sigma C_n}{4F_P \sin^2\phi}, \tag{7.40}$$

$$\frac{a'}{1 + a'} = \frac{\sigma C_t}{4F_P \sin\phi \cos\phi}. \tag{7.41}$$

The determination of a and a' is usually by fixed-point iteration. There have been several analyses of the suitable numerical methods for their solution, but the authors have found that this straightforward iteration using a relaxation factor works well in nearly all cases.

7.4.5 Finite blade functions

Finite blade functions are, in principle, more general than F_P because they are defined for the induced axial and circumferential velocities, a, and, a', respectively, as

$$F_u = a/a_b \quad \text{and} \quad F_w = a'/a'_b, \tag{7.42}$$

where the subscript b denotes a value at the blade and no subscript denotes a streamtube average. The standard approximation is $F_u = F_w = F_P$ as described in the previous subsection, where $F_P \leq 1$. For bare turbines, a difference between F_u and F_w was considered, but not used, by Shen *et al.* [25], while Wimshurst and Willden [26] determined F_u and F_w using a three-dimensional computational simulation. As $N \to \infty$, F_P, F_u, $F_w \to 1$ because there is no difference between a and a_b, or between a' and a'_b, for an actuator disk. Further, F_P goes to zero at the blade tip.

If the turbine has identical, equispaced blades that lie entirely in the radial direction, the difference between a and a_b and between a' and a'_b is due entirely to the helical vortices trailing from the junctions of the BEs. If the N vortices at radius t have pitch $p = dQ/dT$, Wood *et al.* [27] show that $w_b(r)$ is given by

$$\begin{aligned} w_b(r) \quad &\approx \quad \frac{N\Gamma(r)}{4\pi r} + \frac{1}{4\pi r(p^2+r^2)^{1/4}}\left(\int_0^r (p^2+t^2)^{1/4}\frac{\partial\Gamma}{\partial t}\right. \\ &\left[\frac{-N}{1-e^{\xi N}} + \frac{p}{24}\left(\frac{2p^2+9t^2}{(p^2+t^2)^{3/2}} - \frac{2p^2-3r^2}{(p^2+r^2)^{3/2}}\right)\log(1-e^{-\xi N})\right]dt \\ &\left. + \int_r^1 (p^2+t^2)^{1/4}\frac{\partial\Gamma}{\partial t}\left[\frac{N}{1-e^{-\xi N}} + \frac{p}{24}\left(\frac{2p^2+9t^2}{(p^2+t^2)^{3/2}} - \frac{2p^2-3r^2}{(p^2+r^2)^{3/2}}\right)\log(1-e^{\xi N})\right]dt\right) \end{aligned} \tag{7.43}$$

where

$$e^{\xi} = \frac{r\left(p+\sqrt{p^2+t^2}\right)\exp\left(\sqrt{1+(r/p)^2}\right)}{t\left(p+\sqrt{p^2+r^2}\right)}\exp\left(\sqrt{1+(t/p)^2}\right). \tag{7.44}$$

The equation for u_b is very similar, which reduces the computational cost of evaluation. The first term on the right is the streamtube average w so F_w depends on the remaining two integrals. Since the vortex geometry (p) depends on the BE thrust and torque, the calculation of F_u and F_w is done separately to the loop over all BEs to solve for a and a'. Their form is easy to implement in a BEMT program and their effect has been studied by Wood *et al.* [27] and Wood [22] for bare turbines. F_u and F_w are more accurate than F_P at low λ but the difference decreases at higher λ.

One important application of finite blade functions is to DATs. Vaz *et al.* [28] argued that F_P is then inappropriate because the diffuser ensures that a cannot approach zero as the tip is approached. That reference describes seemingly accurate BEMT calculations of a DAWT using F_u, F_w, and the theory described in the next section. It was found that the finite F_u at the blade tip as associated with a finite F_w.

7.5 Angular and axial momentum theory with a diffuser

BEMT as just described is now extended to include a diffuser. As reported in [21], to model a diffuser with losses, an approach similar to that used to determine duct flow in the presence of losses is required. The conditions previously assumed for an actuator disk are used here, where the fluid surrounding the rotor is frictionless and the rotational velocity component is ignored. The dashed lines in Figure 7.10 show the control volume used to analyze DAT performance.

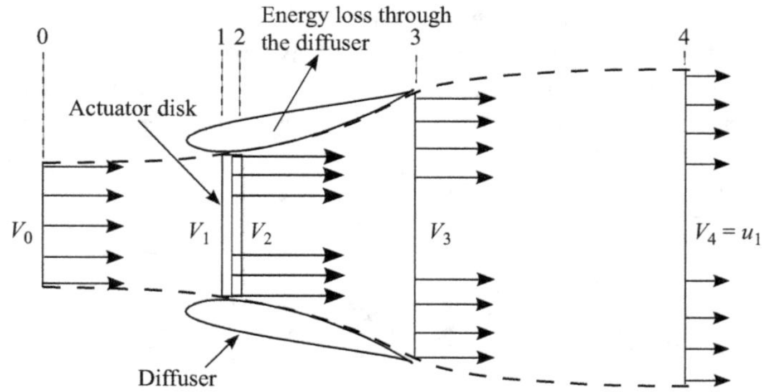

Figure 7.10 Simplified illustration of the flow velocities at the rotor plane and in the wake. The actuator disk models the rotor that lies between stations 1 and 2 and is surrounded by the diffuser [21]

Fletcher [29] gives the power coefficient, C_P, for a ducted turbine as

$$C_P = \varepsilon_1 \left[1 - \varepsilon_4^2 - (1 - \eta_d)\left(1 - \beta^2\right)\varepsilon_1^2\right], \tag{7.45}$$

where $\varepsilon_4 = V_4/V_0$ is the dimensionless far-wake velocity, V_4 is the velocity in the far-wake, $\varepsilon_1 = V_1/V_0$ is the velocity ratio, and $V_1 = V_2$ the velocity at the rotor plane. $\beta = A/A_3$, where A is assumed equal to the diffuser area at the rotor, A_3 is the cross-sectional area of the diffuser outlet, and η_d is the diffuser efficiency, defined as

$$\eta_d = \frac{p_3 - p_2}{\frac{1}{2}\rho\left(V_2^2 - V_3^2\right)}, \tag{7.46}$$

where p_2 and p_3 are the static pressures at the rotor plane and at the diffuser outlet, while V_2 and V_3 are the corresponding velocities, and ρ is the fluid density.

Applying the energy balance downstream to the diffuser exit assuming no losses gives for ε_4

$$\varepsilon_4^2 = \beta^2\varepsilon_1^2 + c_{p3}. \tag{7.47}$$

The pressure coefficient, c_{p3}, at the diffuser outlet is

$$c_{p3} = \frac{p_3 - p_0}{\frac{1}{2}\rho V_0^2}, \tag{7.48}$$

where p_0 is the static pressure in the freestream. Substituting (7.47) into (7.45) yields

$$C_P = \varepsilon_1\left(1 - c_{p3}\right) + \varepsilon_1^3\left[\eta_d\left(1 - \beta^2\right) - 1\right]. \tag{7.49}$$

Equation (7.49) shows that C_P depends on the back pressure caused by the diffuser, which also influences the far-wake velocity. Even for a DAT, the thrust is obtained by dividing the produced power by the local velocity at the rotor plane, as shown in [30, 31]. Thus, the losses through the diffuser (behaving as a duct) are $\triangle H_d = p_2 - p_3 + 1/2\rho(V_2^2 - V_3^2)$, which can be rewritten as

$$\Delta H_d = \frac{1}{2}\rho V_0^2 \varepsilon_1^2 (1-\beta^2)(1-\eta_d). \tag{7.50}$$

Applying the energy balance yields [32]

$$V_2 T = Q\left[\frac{1}{2}\rho(V_0^2 - V_4^2) - \Delta H_d\right], \tag{7.51}$$

where T is the thrust on the rotor, and $Q = V_2 A$ is the flow rate. Hence, C_T is given by

$$C_T = \frac{T}{\frac{1}{2}\rho A V_0^2} = 1 - \varepsilon_4^2 - (1-\eta_d)(1-\beta^2)\varepsilon_1^2. \tag{7.52}$$

As C_P, C_T, and c_{p3} depend on ε_4, an expression for V_4 needs to be formulated. For that, the formulation described by Sörensen [31] is applied. In this case, the momentum equation to the control volume shown in Figure 7.10 leads to

$$T + T_d = \rho A V_2 (V_0 - V_4), \tag{7.53}$$

where T_d is the thrust due to the diffuser. Substituting T from (7.51) into (7.53) yields

$$V_2 = \frac{1}{2}(V_0 + V_4) + \frac{T_d - A\Delta H_d}{\rho A (V_0 - V_4)}. \tag{7.54}$$

In dimensionless form,

$$\varepsilon_1 = \frac{1}{2}(1+\varepsilon_4) + \frac{C_{Td} - \varepsilon_1^2(1-\beta^2)(1-\eta_d)}{2(1-\varepsilon_4)}, \tag{7.55}$$

where C_{Td} is the diffuser thrust coefficient, which is important even if there were no losses in the diffuser ($\eta_d = 1$). Note further that (7.55) reduces to the model of a bare turbine if $C_{Td} = \varepsilon_1^2(1-\beta^2)(1-\eta_d)$ or $C_{Td} = 0$ and $\eta_d = 1$. Solving (7.55) in terms of ε_4 yields

$$\varepsilon_4 = \varepsilon_1 - \sqrt{(1-\varepsilon_1)^2 + C_{Td} - \varepsilon_1^2(1-\beta^2)(1-\eta_d)}. \tag{7.56}$$

Substituting (7.86) into (7.45), (7.47), and (7.52), gives

$$C_P = 2\varepsilon_1^2\left[1 - \varepsilon_1 + \sqrt{(1-\varepsilon_1)^2 + C_{Td} - \varepsilon_1^2(1-\beta^2)(1-\eta_d)} - \frac{C_{Td}}{2\varepsilon_1}\right], \tag{7.57}$$

$$\begin{aligned} c_{p3} = {} & \varepsilon_1^2(2-\beta^2) - 2\varepsilon_1\left[1 + \sqrt{(1-\varepsilon_1)^2 + C_{Td} - \varepsilon_1^2(1-\beta^2)(1-\eta_d)}\right] \\ & + 1 + C_{Td} - \varepsilon_1^2(1-\beta^2)(1-\eta_d), \end{aligned} \tag{7.58}$$

and

$$C_T = 2\varepsilon_1\left[1 - \varepsilon_1 + \sqrt{(1-\varepsilon_1)^2 + C_{Td} - \varepsilon_1^2(1-\beta^2)(1-\eta_d)} - \frac{C_{Td}}{2\varepsilon_1}\right]. \tag{7.59}$$

Note that (7.57) and (7.59) are the power and thrust coefficients, respectively, for a DAT. Another important expression is for the DAT far-wake velocity, (7.86), which is strongly dependent on the performance parameters of the diffuser. Figure 7.11 shows the effect of different area ratios, β, on the rotor trust coefficient, C_T.

Figure 7.12 shows the power coefficient in relation to ε_1. In this case, $\beta = 0.54$, $\eta_d = 1.0$ for part (a) and $\eta_d = 0.8$ for part (b). $C_{Td} = 0.2$ gives $C_{Popt} = 0.73$ (Figure 7.12(a)). Increasing η_d caused C_P to increase, clearly because the loss in the diffuser affects the turbine efficiency. Also, C_{Td} has a huge influence on C_P. This may be observed in both parts (a) and (b) of Figure 7.12, even for $\eta_d = 0.8$, in which $C_{Popt} = 0.61$. C_{Popt} exceeds the Betz–Joukowsky limit for $\eta_d = 0.8$ and $C_{Td} = 0.2$. However, it is worth noting that depending on the diffuser thrust, a DAT may or may not exceed the Betz–Joukowsky limit. A diffuser needs to be carefully

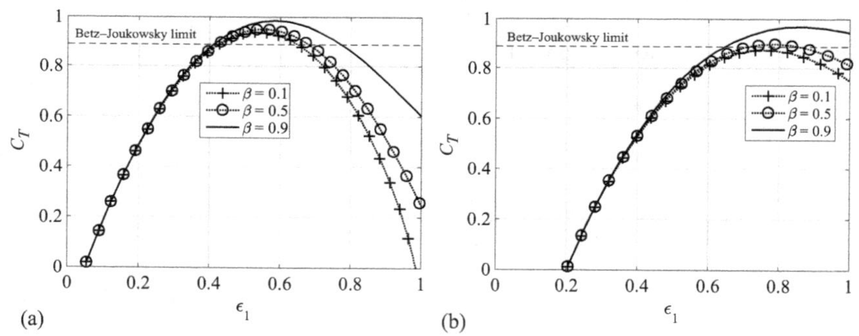

Figure 7.11 Rotor thrust coefficient as a function of ε_1: (a) $C_{Td} = 0.2$ and (b) $C_{Td} = 0.8$ [21]

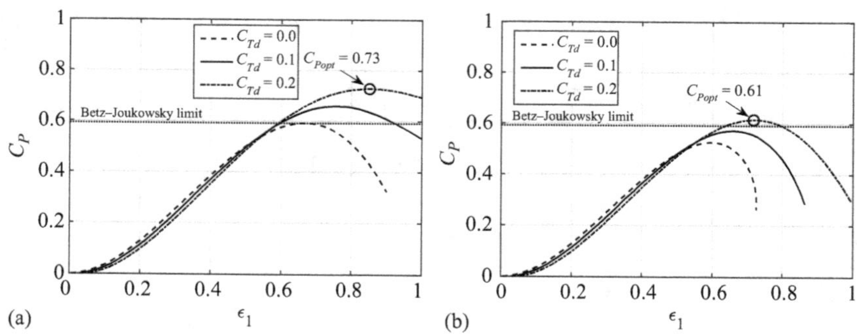

Figure 7.12 Power coefficient as a function of ε_1: (a) $\eta_d = 1$ and (b) $\eta_d = 0.8$ [21]

dimensioned to recover as much kinetic energy as possible. This result is further supported by Bontempo and Manna [33], who demonstrated through a semi-analytical approach that a diffuser must contribute significantly to the overall thrust in order to get a high-power coefficient. The diffuser thrust directly depends on the diffuser shape, and this interrelation requires evaluation of the local and global features of the flow throughout the diffuser. This interrelation is not assessed here, and in fact further analysis is necessary. However, the results are important for DAT analysis. For example, if $C_{Td} > 0$, C_P is probably increased. However, it is necessary to point out that $C_{Td} > 0$ does not cause C_P to exceed the Betz–Joukowsky limit.

The impact of η_d on C_P can also be assessed using the numerical results of Hansen *et al.* [30] for an actuator disk CFD model of the flow through a DAWT with $\eta_d = 0.83$ and $\beta = 0.54$. Figure 7.13 depicts the comparison between the BEMT analysis and [30]. For $\eta_d = 0.83$ the BEMT is in good agreement with the CFD results of [30]. Table 7.1 shows that the BEMT results for C_{Popt} and C_{Topt} have a maximum difference for C_{Topt} of 1.37%. C_{Popt} is reached when $C_{Topt} = 0.81$, which is close to the value found by CFD. No values of C_{Td} were reported in [30]. However, considering the high value obtained in [30] for C_{Popt}, high values for C_{Td} are also expected because C_P is directly proportional to C_{Td}. Thus, C_{Td} here is significantly larger than other studies. In this case, the BEMT gives $C_{Popt} = 0.94$ only if $C_{Td} = 1.12$.

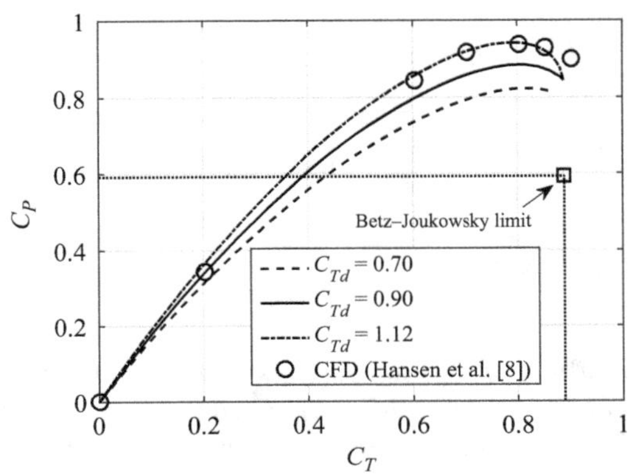

Figure 7.13 Power. coefficient as a function of the thrust coefficient [21]. Comparison with Hansen et al. [30]

Table 7.1 Comparison between BEMT [21] and CFD [30]

	Hansen *et al.* [30]	**BEMT**	**Difference (%)**
C_{Topt}	0.80	0.789	1.37
C_{Popt}	0.94	0.941	0.11

7.5.1 Correction for finite number of blades

In the absence of finite blade effects, the elemental thrust and torque coefficients on an annular control volume at radius r are

$$\frac{dC_T}{dr_*}+\frac{dC_Td}{dr_*}=4\varepsilon_1(1-\varepsilon_4)r_*, \tag{7.6}$$

and

$$\frac{dC_Q}{dr_*}=4\varepsilon_1 u_\theta r_*^2, \tag{7.61}$$

where $r_*=r/R$ and $u_\theta=2a'\lambda r_*$. To include F in (7.60) and (7.61), the axial velocity ratio ε_1 at the blade needs to be explicit. Assuming $\varepsilon_1=1-a$ and calculating ε_4 from (7.86) with $a=a_bF$ as the streamtube average, the result is

$$\varepsilon_4=1-a_bF-\sqrt{(a_bF)^2-(1-a_bF)^2(1-\beta^2)(1-\eta_d)+C_{Td}}. \tag{7.62}$$

In this equation, $F=F_u$ if finite blade functions are used, otherwise $F=F_P$. If the tangential finite blade function is used, $F=F_w$, otherwise $F=F_P$. Thus, (7.60) and (7.61) become

$$\frac{dC_T}{dr_*}+\frac{dC_Td}{dr_*}=4(1-a_b)r_*\times \left[a_bF+\sqrt{(a_bF)^2-(1-a_bF)^2(1-\beta^2)(1-\eta_d)+C_Td}\right], \tag{7.63}$$

$$\frac{dC_Q}{dr_*}=8a'_bF\lambda(1-a_b)r_*^3. \tag{7.64}$$

Note that for a bare turbine ($\eta_d=1$ and $C_{Td}=0$), (7.63) reduces to (5) of Wood and Okulov [34], while (7.64) matches their (6), when the missing λr is included in the latter. As for a bare turbine, the element power is obtained from $dP=\Omega dQ$. Thus, C_P is given by (7.13).

7.6 Blade element theory for diffuser-augmented turbines

The elemental thrust and torque coefficients, and tangential force coefficients C_n and C_t at any r are the same as for a bare turbine, see (16) and (17) of Vaz and Wood [19]. The flow angle, ϕ, is similarly obtained from the velocity diagram in Figure 7.8. Consequently, the thrust and torque coefficients can be found from (7.27) and (7.28). An extended formulation for the axial and tangential flow velocities can be obtained by combining (7.63) and (7.64) with (7.27) and (7.28), respectively, yielding

$$\frac{a_b}{1-a_b}+\frac{1}{1-a_b}\sqrt{a_b^2-\left(\frac{1}{F}-a_b\right)^2(1-\beta)^2(1-\eta_d)+\frac{C_{Td}}{F^2}} \\ -\frac{1}{4F(1-a_b)^2 r_*}\frac{dC_{Td}}{dr_*}=\frac{\sigma C_n}{2F\sin\phi^2}, \tag{7.65}$$

where $F = F_u$ or F_P and

$$\frac{a'_b}{1+a'_b}=\frac{\sigma C_t}{4F\sin\phi\cos\phi}, \tag{7.66}$$

where $F = F_w$ or F_P. Note that (7.65) is heavily dependent on C_{Td}, η_d, and β. The accuracy of this equation will be assessed later.

7.6.1 High thrust correction

As indicated at the end of Section 7.4.1, the thrust from axial momentum theory for a conventional turbine becomes unphysical at large a, and many corrections have been discussed in the literature. However, for DATs, an empirical correction similar to that of Buhl [35] for a bare turbine is proposed in [21]. A general quadratic equation for C_T is

$$C_T = b_0 + b_1\varepsilon_1 + b_2\varepsilon_1^2, \tag{7.67}$$

whose derivative with respect to ε_1 is

$$\frac{dC_T}{d\varepsilon_1} = b_1 + 2b_2\varepsilon_1, \tag{7.68}$$

where b_0, b_1, and b_2 are parameters determined using the following conditions: (i) for $\varepsilon_1 = 0$, $C_T = 2$, so that $b_0 = 2$. For $\varepsilon_1 = \varepsilon_{1opt}$, $C_T = C_{Topt}$ in both (7.52) and (7.67). Those conditions are obtained from Figure 7.14, whose experimental data is taken from Lock *et al.* [36] for a bare turbine, in which $\varepsilon_1 = 1 - a = 1 - Fa_b$. The optimum values, ε_{1opt} and C_{Topt}, are determined by maximizing C_P in (7.45) using (7.86), giving

$$6\varepsilon_{1opt}^3\left[\beta^2(1-\eta_d)+\eta_d\right]-C_{Td}\Delta \\ +4\varepsilon_{1opt}(1+C_Td+\Delta)-2\varepsilon_{1opt}^2(5+3\Delta)=0, \tag{7.69}$$

where

$$\Delta=\sqrt{1+C_{Td}+\varepsilon_{1opt}\left\{-2+\varepsilon_{1opt}\left[\beta^2(1-\eta_d)+\eta_d\right]\right\}}, \tag{7.70}$$

and then C_{Topt} is obtained by substituting ε_{1opt} from (7.69) into (7.52). Including F, (7.52) and its derivative become

$$C_T=\left[1-\varepsilon_4^2-(1-\eta_d)\left(1-\beta^2\right)\varepsilon_1^2\right]F, \tag{7.71}$$

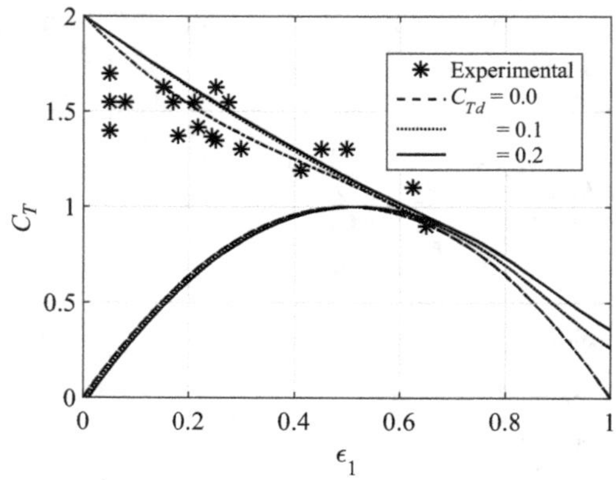

Figure 7.14 Thrust coefficient as a function of ε_1 for different values of C_{Td} [21] (experiment obtained from [36]

and

$$\frac{dC_T}{d\varepsilon_1} = \left[-2\varepsilon_4 \frac{d\varepsilon_4}{d\varepsilon_1} - (1-\eta_d)\left(1-\beta^2\right)2\varepsilon_1\right]F, \tag{7.72}$$

with

$$\frac{d\varepsilon_4}{d\varepsilon_1} = 1 - \frac{-2(1-\varepsilon_1) - 2\varepsilon_1(1-\eta_d)\left(1-\beta^2\right)}{2\sqrt{C_{Td} + (1-\varepsilon_1)^2 - \varepsilon_1^2(1-\eta_d)\left(1-\beta^2\right)}}. \tag{7.73}$$

Combining (7.67) and (7.68) with (7.71) and (7.72), respectively, and substituting the conditions stated earlier, gives for b_1 and b_2

$$2 + b_1\varepsilon_{1opt} + b_2\varepsilon_{1opt}^2 = \left[1 - \varepsilon_{4opt}^2 - \varepsilon_{1opt}^2(1-\eta_d)\left(1-\beta^2\right)\right]F, \tag{7.74}$$

and

$$b_1 + 2b_2\varepsilon_{1opt} = \left[-2\varepsilon_{4opt}\frac{d\varepsilon_{4opt}}{d\varepsilon_1} - 2\varepsilon_{1opt}(1-\eta_d)\left(1-\beta^2\right)\right]F. \tag{7.75}$$

Solving (7.74) and (7.75) for b_1 and b_2, noting that $b_0 = 2$, is a simple exercise in linear algebra.

The correction for C_T occurs when $\varepsilon_1 < \varepsilon_{1opt}$ and is calculated by the following procedure: if $\varepsilon_1 \geq \varepsilon_{1opt}$, then ε_1 is calculated using (7.65). Otherwise, ε_1 is calculated using

$$\left(b_2 - \frac{\sigma C_n}{\sin^2\phi}\right)\varepsilon_1^2 + b_1\varepsilon_1 + b_0 = 0. \tag{7.76}$$

7.7 The accuracy of blade element momentum theory

Figure 7.9 shows the typical accuracy of BEMT in calculating small wind turbine power and Figure 7.15, also taken from Wood [2], shows this for the thrust. It is worth noting that (i) the tip loss has little impact on the thrust, and (ii) the measured thrust is nearly always higher than the BEMT predictions. This increase can be caused by inaccurate estimates of the tower and nacelle drag and may not represent a shortcoming of BEMT.

Similar accuracy in using BEMT to predict the power and thrust of HKTs is shown in Figures 5 and 6 of Batten *et al.* [37] where the thrust also was slightly underpredicted in general. They also discuss the importance of considering cavitation in the choice of airfoils for the blades. Airfoil choice for small HAWTs is discussed in Wood [2]. For both applications, airfoils of thickness about 0.15*c* are preferred for the whole blade as opposed to the large wind turbine usage of very thick sections near the hub.

One important influence on the accuracy of BEMT results is the source of airfoil lift and drag data. That used for Figures 7.9 and 7.15 comes from wind tunnel tests as explained in [2]. Batten *et al.* [37] used the open-source software Xfoil to generate the data. We believe that extreme care needs to be applied in doing this. Xfoil can be accurate for the low *Re* airfoils used in small turbines, e.g., Morgado *et al.* [38] but it can give spectacularly incorrect results, as shown by Fig. 8 of Mack *et al.* [39]. The reason is that laminar separation bubbles can form on low *Re* airfoils and have major effects on the lift and drag as explained in [39]. Xfoil has very limited capability for dealing with these effects. In our experience, even the significantly more sophisticated transition/turbulence models can also be inaccurate. As a rule-of-thumb, any airfoil operating at $Re < 2 \times 10^5$ should be represented in BEMT by experimental data. This point is made forcefully by van Treuren [40].

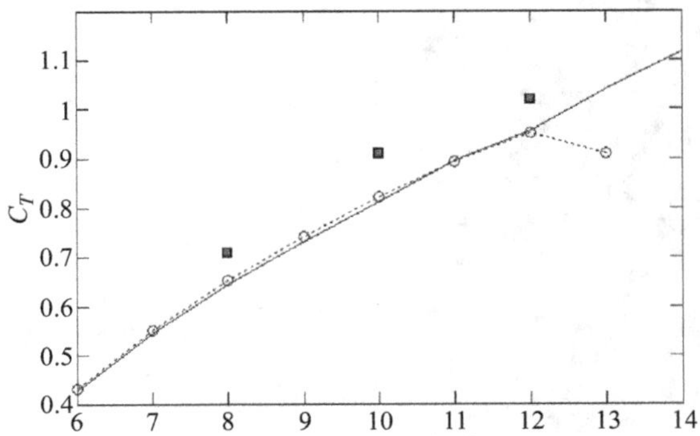

Figure 7.15 Accuracy of BEMT thrust coefficient from Wood [2]. Blue squares show the measurements, circles show BEMT without F_P, and solid line shows BEMT with F_P

There is much less experimental data for the performance of diffuser-augmented than for bare turbines. Vaz *et al.* [28] showed that BEMT using finite blade functions gave accurate power predictions for one operating condition, but so did the use of F_P. A more severe test of the two formulations is needed.

To evaluate the BEMT with all corrections previously stated, a comparison with the experiment of Ten Hoopen [41] is shown, which used a 3-bladed DAWT of a 1.5-m rotor diameter. The diffuser was a circular airfoil designed by National Aerospace Laboratory (NLR), as shown in Figure 7.16(a), where the suction side is pointed inward and the area ratio, β, is 0.5785. The blades and diffuser were designed using CFD [41]. The exit plane of the diffuser had a diameter of 2 m and was equipped with a Gurney flap of 0.04 m. Ten Hoopen [41] gave $C_T = 0.8$, $V_0 = 10$ m/s, and $\Omega = 75$ rad/s ($\approx$716.2 rpm). The diffuser Reynolds number was 6×10^5. Delta-shaped vortex generators were installed on the diffuser trailing edge in order to promote mixing. Figure 7.16(b) shows c and θ developed by NLR. The DAWT geometry and the experimental parameters are summarized in Table 7.2.

Table 7.3 presents the output power and torque for $V_0 = 10$ m/s. The BEMT is compared with experimental data from [41] and other methodologies available in the literature. It is worth noting that in [41] the diffuser efficiency is not reported; $\eta_d = 0.80$ gives the best fit for the BEMT. Note that the power output is very close to the experimental value: the error is 0.19% at the same Ω of 75 rad/s. The CFD results of Ten Hoopen [41] were obtained using the $k - \varepsilon$ turbulence model. The diffuser airfoil and details upon the code can be found in [41]. The computational $\Omega = 137$ rad/s, which is much larger than in the experiment.

Figure 7.17(a) shows the radial variation of ε_1 compared with the experiments [41] for 37 vortex generators on the diffuser. The influence of vortex generators occurs on C_{Td}. Thus, the measured ($C_{Td} = 0.1489$) was input to the BEMT analysis,

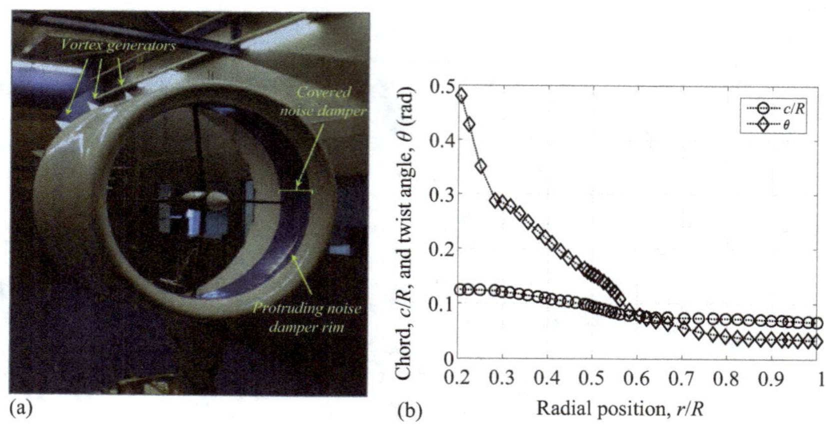

Figure 7.16 (a) DAWT equipped with vortex generators [41], and (b) chord and twist angle distributions as a function of the radial position [41]. From [21]

Table 7.2 Design parameters [21]

Parameters	Values
Turbine diameter	1.5 m
Hub diameter	0.3 m
Number of blades	3
Diffuser airfoil	NLR [41]
Blade airfoil	NACA 2207
Power output	531 W
Wind speed (V_0)	10 m/s
Air density (ρ) at 20°C	1.2 kg/m^3
Rotational speed	716.2 rpm

Table 7.3 Comparison between BEMT and experimental data (V_0 = 10 m/s)

	Angular speed (rad/s)	**Power output (W)**	**Torque (N m)**	**Thrust coefficient**
Experimental [41]	75	531	7.10	0.80
do Rio Vaz *et al.* [18]	75	526	6.10	–
BEMT	75	532	7.09	0.77
CFD [41]	137	545	4.00	–
CFD [41]	155	246	1.60	–

From [21].

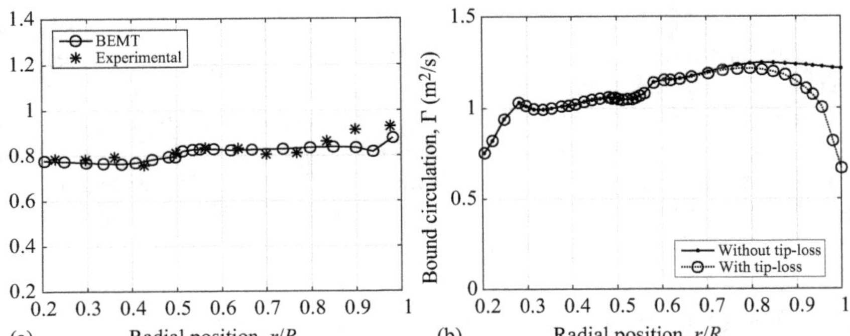

Figure 7.17 (a) Radial variation of ε_1 (experiment obtained from [41]) and (b) bound circulation as a function of the radial position. From [21]

in order to compare the calculated and measured velocity ratios. Note that ε_1 is in good agreement with the experimental value. Figure 7.17(b) shows the effect of the tip-loss, F_P, on the bound circulation ($\Gamma = 1/2W\, cC_l$) along the blade under same condition depicted in Table 7.3. Tip-loss has a considerable influence on DAWTs

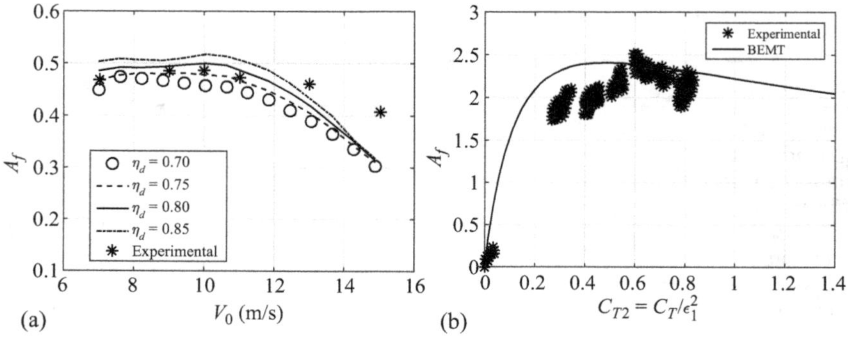

Figure 7.18 (a) Augmentation as a function of the freestream wind velocity (experiment obtained from [41]) and (b) augmentation as a function of C_{T2} (experiment obtained from [32]). From [21]

performance since the power output would be overestimated by 7% if $F_P = 1$ were used at every r. In the BEMT analysis, the principal modifications have occurred in the axial flow velocities and in the thrust coefficient for introducing the effect of C_{Td}, η_d, and β, whose modifications are responsible for the good behavior of this methodology.

The impact of C_{Td} and η_d on a DAWT performance is demonstrated through the augmentation factor, A_f, defined as the ratio of turbine total efficiency, $\eta_t = \eta_p \eta_g C_P$, to the Betz–Joukowsky limit; $A_f = 27\eta_t/16$, where η_p is the drivetrain efficiency and η_g is the generator efficiency. Figure 7.18(a) shows the results for A_f using the BEMT compared with the experiment of [41] in which $C_{Td} = 0.135$ was measured for a rotor with Gurney flap but no vortex generators. In this case, A_f was calculated using the measured value of C_{Td} for $\eta_d = \{0.70, 0.75, 0.80, 0.85\}$, $\eta_p = 85\%$, and $\eta_g = 74\%$ for a constant $\Omega = 716.2$ rpm. The result in Figure 7.18(a) suggests $\eta_d \approx 0.75$, and a strong dependence on η_d. Figure 7.18(b) shows a comparison with the experimental data obtained by Phillips [32], relating the augmentation factor with C_{T2}, defined as $C_{T2} = C_T/\varepsilon_1^2$, for the experimental values $\beta = 0.25$, $\eta_d = 81\%$, and $c_{p3} = -0.49$.

7.8 Design using blade element momentum theory

BEMT is an analysis tool that requires additional consideration before it can be used for design. This is illustrated by the Betz–Joukowsky limit that gives the value of $a = 1/3$ for maximum power but does not give any information about choosing N or the blade shape. Blade aerodynamic or hydrodynamic design usually starts by choosing the airfoil section. Small HAWTs normally use only one airfoil for the whole blade for two reasons [2]; briefly, thick airfoils do not perform well at low *Re* and much of the starting torque on a blade is generated near the hub. HKTs often have multiple airfoils, presumably because the two reasons are less applicable. The remaining part of the aerodynamic design consists of choosing the chord, c, and pitch, θ in Figure 7.8. Both can vary significantly along the blade.

By this stage of the chapter, the reader should appreciate that the BEMT equations are sufficiently complex that analytical design, say for maximum power extraction, is possible in only very limited situations. When the design is multi-dimensional, such as for high efficiency *and* low noise, numerical methods are always required.

In the following, bare turbines are discussed first. Then we consider HKT blade design in terms of the need for high power *and* the avoidance of cavitation using the cavitation velocity (7.4). Blade design for DATs is discussed last.

7.8.1 Bare wind turbines

The simplest optimization is to consider only maximization of power. If the BE drag is ignored along with F, then it is easy to derive θ and c for the chosen airfoil and λ by requiring $a = 1/3$ as the condition for maximum power. The resulting simple analytic equations for θ and c are (5.12a) and (5.12b) of [2].

Including drag and F prevents analytic optimization and some form of numerical method is needed. In addition, we consider MDO because power output, while always an essential objective, is only one of the possible objectives. Small turbines usually have no pitch adjustment so that starting can be a problem [2]. Obviously, there is no point designing a highly efficient blade if it loses power by taking too long to start. Further, turbine noise, which is generated mainly by the blades, may need to be minimized as might the blade mass that correlates with the cost of manufacture. Blade inertia also affects starting and so may also need to be minimized. Tip deflection may be a constraint to avoid contact with the tower, as could be the maximum chord for manufacturing reasons.

MDO can require substantial computer resources, especially when a high number of dimensions are required. For this reason the evaluation of the objective functions must be fast, which, in turn, means that BEMT is nearly always used for the aerodynamics.

Wood [2] provides programs for MDO of small HAWT blades. Sessarego and Wood [42] have updated them to include most of the objectives and constraints just listed. BEMT is used to determine power, which, as noted, must always be one of the objectives. Starting analysis uses a simplified version of BEMT to calculate the blade torque, from which the angular acceleration is calculated. It has been found that any resistive torque in the drivetrain can have a significant effect on starting. Using empirical models for the friction in bearings and other components, the resistive torque is subtracted from the aerodynamic torque before determining the angular acceleration, Vaz *et al.* [11].

Most MDO uses some form of genetic algorithm. An initial population of blades is produced randomly and rules provided for producing the next generation and deciding which members are best fit to survive. Reproduction continues until the population converges. The designer then examines the Pareto front for two-dimensional cases or the Pareto surface for higher dimensions. This is the surface formed by all blades (members of the final population) for which at least one objective is better than for every other member. The "best" design lies somewhere on that surface but the final choice of blade may be arbitrary. An example of designing a 5-kW blade for starting and power extraction is given by Wood [2]. Its starting performance as determined by field testing is shown in [42].

A bewildering number of possible MDO techniques are available, but it is likely that the technique used is not critical in blade design for reasons given later. In the freely available SWRDC code of [42], the multiple objectives are combined linearly to form a single objective function to determine blade fitness, and the Pareto front is determined using the very common NSGA-II algorithm. Sessarego and Wood [42] show optimal c and θ distributions for a 750-W blade optimized for starting and power extraction for five different blade materials.

One of the fascinating outcomes of MDO for blade design is that it is rarely an optimization; it is trade-off. A blade designed only for maximizing power, for example, will nearly always be the slowest to start, or the noisiest. To improve the time to start, it is necessary to reduce the maximum power, but it also has been found that the fractional decrease in starting time far exceeds the fractional decrease in power. This is the trade-off that leads to another rule-of-thumb: MDO is necessary to avoid worst case behavior of any objective not included in the optimization.

A further aspect of MDO is worth mentioning. It has been shown that it may be important to minimize blade inertia. This objective couples the blade aerodynamics and structure in a way that does not occur for large turbines, Pourrajabian *et al.* [43].

7.8.2 Bare hydrokinetic turbines

The optimization procedure here is based on [8]. This approach uses the simplified criterion to avoid cavitation described earlier together with the classical Glauert's optimization [20]. The cavitation criterion is that $W \leq V_{CAV}$. Correction is done by replacing W in the BEMT analysis by $W_{CAV} = (1 - f_S)V_{CAV}$, where f_S is an arbitrary safety factor between zero and 1 for the calculation of c. Replacing W by W_{CAV} in sections where $W \geq V_{CAV}$ leads to an increase in c, altering the pressure on the blade. The BEMT defines the thrust coefficient by (7.27), which gives c in terms of $(V_0/W)^2$ and V_{CAV}. For blade sections where the criterion is satisfied, it is then easy to find the ratio between the corrected and uncorrected c. In order to demonstrate this, start by using the uncorrected chord $c = c^{uc}$, which is defined by

$$c^{uc} = \frac{2\pi r}{N}\frac{C_T}{C_n}\left(\frac{V_0}{W}\right)^2. \tag{7.77}$$

To obtain the corrected formulation for the chord, it is necessary to replace W by W_{CAV}, yielding the equation for the corrected chord, c^{co}:

$$c^{co} = \frac{2\pi r}{B}\frac{C_T}{C_n}\left[\frac{V_0}{(1 - f_S)V_{CAV}}\right]^2, \tag{7.78}$$

where C_T is given by (7.38). Dividing (7.78) by (7.77) yields the ratio between c^{co} and c^{uc}, given by

$$c^{co} = c^{uc}\left[\frac{W}{(1 - f_S)V_{CAV}}\right]^2. \tag{7.79}$$

Note that cavitation causes $W \geq V_{CAV}$ so the term on the right will always be larger than 1, which increases c^{co}. This equation corrects the c to avoid cavitation. Equation (7.79) can be used in any optimization model. Here it is applied to the classical Glauert's optimization [20], which ensures that α has its optimum value along the span. The optimum angle of attack is assumed to be the one at maximum C_L/C_D. The optimum relationship between the local-speed ratio, $x = \Omega r/V_0$, and the axial induction factor, $a = a_{opt}$, is obtained by maximizing the power coefficient. A detailed description is reported in [3]. The optimum axial induction factor, a_{opt}, is

$$16a_{opt}^3 - 24a_{opt}^2 + a_{opt}(9 - 3x^2) - 1 + x^2 = 0. \tag{7.80}$$

It is noteworthy that (7.80) is assumed independent of F. This limitation is employed because F is variable with the axial induction factor a, making the optimization analysis rather complex. Studies on this were previously analyzed by Burton *et al.* [1] and further detailed by Clifton-Smith [44]. Thus, in the proposed optimization procedure, F is taken into account only in (7.38).

Having derived all the necessary equations to compute a given hydrokinetic blade geometry, it is possible to determine c and θ considering the possible occurrence of cavitation. The rotor optimization can be expressed as a function of a. The procedure to calculate the optimal chord c_{opt} and twist angle θ_{opt} for each radial section along the blade, starting at the hub, is described in the algorithm later, in which the iterative procedure for the calculation of optimum chord and twist angle at each blade section is shown.

Require: r, Ω, $C_L(\alpha_{opt})$, $C_D(\alpha_{opt})$ and V_0
Set initial values for a_{opt} and a'_{opt}. In this work $a_{opt} = 1/3$ and $a'_{\text{opt}} = 0$;
for $i = 1$ to Ns (Number of sections) **do**
While *error* $>$ *TOL* **do**
$iter = iter + 1$;
Compute ϕ_{opt} using (7.15) for a_{opt} and a'_{opt};
Compute C_n and C_t using (7.22) and (7.23), respectively, calculated for α_{opt} obtained from maximum C_L/C_D;
Compute C_T, using (7.38);
Compute $W = \sqrt{\left[V_0(1 - a_{opt})^2\right] + \left[\Omega r\left(1 + a'_{opt}\right)\right]^2}$;
Compute c^{uc}_{opt}, using (7.77) and $\theta_{opt} = \phi_{opt} - \alpha_{opt}$;
Compute V_{CAV}, using (7.4);
if $W > V_{CAV}$ **then**

Compute c_{opt}^{co}, using (7.79)
end if
Compute new a_{opt}, using (7.80), and new $a'_{opt} = \frac{1-3a_{opt}}{4a_{opt}-1}$;
Compute $error = |\phi_{opt}^{iter+1} - \phi_{opt}^{iter}|$.
end while
end for
Compute blade geometry.

To show the performance of this optimization procedure, the HKT described in [8] with the NACA $65_3 - 618$ hydrofoil is used, whose main parameters are listed in Table 7.4. H is the submergence of the turbine. The safety factor $f_S = 5\%$ is considered to assure that cavitation is prevented, which is really important for engineering purpose. In a real application, the parameters shown in Table 7.4 are input data to the algorithm, through which the rotor blade can be designed for a free-cavitation condition.

The hydrokinetic rotor geometry and hydrodynamic characteristics on each blade section are presented in Table 7.5, which shows the local Reynolds number based on the chord length and relative velocity ($Re_c = Wc/\nu$, where ν is the kinematic viscosity), the minimum pressure coefficient, and cavitation number or Thoma number as a function of the radial position. Note that $C_{P\min} + \sigma = -0.0693$ for $Ns = 7$ and greater, indicating that the criterion $C_{P\min} + \sigma \geq 0$ is not satisfied from $r \geq 3.66$ m.

Note in Figure 7.19(a) that cavitation occurs at approximately 70% of the span for the rotor when operating at 35 rpm. The correction on the blade chord is a consequence of the modification on the relative velocity, as shown in Figure 7.19 (b). This occurs due to W becoming higher than V_{CAV} for radial positions $r > 3.5$ m, where the term $[W/(1 - f_S)V_{CAV}]^2$ is greater than 1. In order to avoid cavitation, the proposed algorithm corrects W and assumes values always lower than V_{CAV}. A

Table 7.4 Design parameters used to design the HKT [8]

Parameters	Values
Turbine diameter (D)	10.0 m
Hub diameter	1.5 m
Number of blades	3
Current velocity (V_0)	2.5 m/s
Water density (ρ) at 25°C	997 kg/m^3
Submergence of the turbine (H)	6 m
P_{atm}	1×10^5 Pa
P_v	3.17×10^3 Pa
Gravity (g)	9.81 m/s^2
Safety factor	5%
Angular velocity	35 rpm

Table 7.5 Hydrokinetic rotor geometry and hydrodynamic characteristics on each blade section [8]

*Ns**	*r* (m)	*c* (m)	θ (°)	$Re_c \times 10^6$	$C_{P\min}$	σ	$C_{P\min} + \sigma$
1	0.9737	0.5637	20.34	2.9230	−1.140	14.5033	13.3633
2	1.4211	0.5273	14.09	3.3957	−1.142	8.0703	6.9283
3	1.8684	0.4273	10.37	3.4719	−1.142	4.9702	3.8282
4	2.3158	0.3552	7.94	3.5081	−1.143	3.3000	2.1570
5	2.7632	0.3028	6.24	3.5287	−1.143	2.3154	1.1724
6	3.2105	0.2632	4.99	3.5381	−1.143	1.6932	0.5502
7	3.6579	0.2583	4.04	3.5285	−1.144	1.0747	−0.0693
8	4.1053	0.2535	3.29	3.4527	−1.144	0.9891	−0.1549
9	4.5526	0.2620	2.68	3.0972	−1.145	0.7809	−0.3641
10	5.0000	0.1067	2.18	1.1105	−1.146	0.6264	−0.5196

**Ns* is the section number, usually called blade elements.

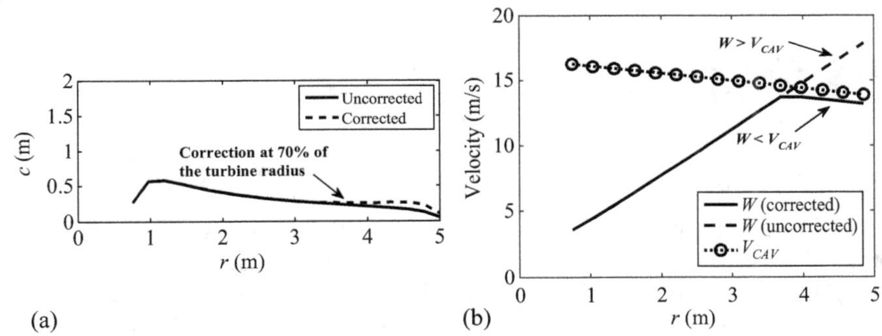

Figure 7.19 (a) Chord distribution along the blade and (b) relative and cavitation velocities as functions of the radial position [8]

comparison with CFD technique was made in [8], in order to show that the design approach can optimize an HKT free of cavitation. Figure 7.20 shows the region where cavitation occurs for the corrected rotor (Figure 7.20(a)) and uncorrected rotor (Figure 7.20(b)). Figure 7.20(c) clarifies the geometrical differences between the designs. The region where the blades were modified matches where cavitation occurs.

Optimization of HKTs using genetic algorithms has developed in parallel to those for HAWTs. A recent example is Zhu *et al.* [45]. Most studies have concentrated on maximizing power, presumably because objectives like starting time and noise are less important in HKTs. Similarly we are unaware of any HKT optimization that includes blade materials and structure. One noticeable absence from the list of objectives for both WTs and HKTs is blade fatigue. This is likely to change with the advent of simple models for blade fatigue, for which the work of Evans *et al.* [46] is a beginning.

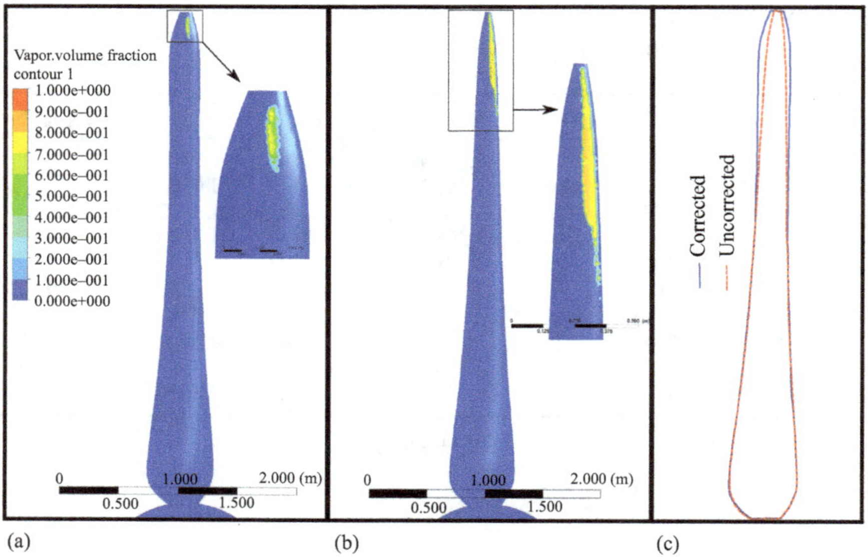

Figure 7.20 Cavitation in the rotors at 35 rpm: (a) corrected, (b) uncorrected, and (c) geometrical comparison [8]

7.8.3 For turbines with diffuser

The optimization procedure here is carried out without implementing any formal optimization process for the diffuser geometry. For the case of wind and hydrokinetic turbines, only the diffuser effect is considered, through basically three important parameters: η_d, β, and C_{Td}, as described in Section 7.5. These parameters come from the maximization of (7.45), which is reduced to (7.69). Once ε_{1opt} is determined, the far-wake velocity is obtained, ε_{4opt}, through (7.86). So, from the combination of (7.60) and (7.27), making $W = \varepsilon_{1opt}/\sin\phi$ from the velocity diagram in Figure 7.8, the optimum chord for each blade section is given by

$$c^{uc} = \frac{\pi}{NC_n W^2}\left[4\varepsilon_{1opt}(1-\varepsilon_{4opt})r_* - \frac{dC_Td}{dr_*}\right]. \tag{7.81}$$

As the diffuser increases the axial velocity at the rotor plane, consequently increasing the angle of attack, cavitation needs to be taken into account. Hence, under cavitation condition, $W \geq V_{CAV}$, (7.81) becomes

$$c^{co} = \frac{\pi}{BC_n[(1-f_S)V_{CAV}]^2}\left[4\varepsilon_{1opt}(1-\varepsilon_{4opt})r_* - \frac{dC_Td}{dr_*}\right]. \tag{7.82}$$

Dividing (7.82) by (7.81) results in the same equation as for a bare turbine (7.79). This expression shows that, even for shrouded turbine, the term $[W/((1-f_S)V_{CAV})]^2$ can be applied to correct the optimum chord distribution in order to avoid cavitation.

Once ε_{1opt} is obtained from (7.69), the optimized a_{opt} may easily be calculated through $a_{bopt} = 1 - \varepsilon_{1opt}$. a' as a function of ε_{1opt} is found using the conservation of energy, resulting in optimum element power:

$$dP_{opt} = \frac{1}{2}\rho V_0^3 \left[\varepsilon_{1opt}\left(1 - \varepsilon_{4opt}^2\right) - \varepsilon_{1opt}^2\left(1 - \beta^2\right)\left(1 - \eta_d\right)\right] dA. \tag{7.83}$$

Also, applying the angular momentum equation at a blade section,

$$dP_{opt} = 2\rho V_0 a_b^{'} \varepsilon_{1opt} \Omega^2 r^2 dA. \tag{7.84}$$

Equating (7.83) and (7.84) gives

$$a_{bopt}^{'} = \frac{2\varepsilon_{1opt}\left(1 - \varepsilon_{4opt}\right) - C_{Td}}{4x^2}, \tag{7.85}$$

with ε_{4opt} given by

$$\varepsilon_{4opt} = \varepsilon_{1opt} - \sqrt{\left(1 - \varepsilon_{1opt}\right)^2 + C_{Td} - \varepsilon_{1opt}^2\left(1 - \beta^2\right)\left(1 - \eta_d\right)}. \tag{7.86}$$

In this case, F_u is included in ε_{4opt} through $\varepsilon_{1opt} = 1 - a_{bopt}F_u$. Hence, the optimum flow angle, ϕ_{opt}, can be determined through (7.15). To compute a given DAHKT blade free of cavitation, the methodology is described in the following algorithm, in which the iterative procedure for the calculation of optimum c and θ at each radius is shown.

Require: r, Ω, $C_L(\alpha_{opt})$, $C_D(\alpha_{opt})$ and V_0
  for $i = 1$ to Ns (Number of sections) **do**
    while $error > TOL$ **do**
      $iter = iter + 1$
      Compute a_{opt} and $a_{opt}^{'}$ using (7.69) and (7.85), respectively;
      Compute ϕ_{opt} using (7.15) for a_{opt} and $a_{opt}^{'}$;
      Compute C_n and C_t using (7.22) and (7.23), respectively, calculated for α_{opt} obtained from maximum C_L/C_D;
      Compute the relative velocity, W;
      Compute c_{opt}^{uc}, using (7.81) and $\theta_{opt} = \phi_{opt} - \alpha_{opt}$;
      Compute V_{CAV}, using (7.4);
      if $W > V_{CAV}$, **then**
        Compute c_{opt}^{co}, using (7.81) or (7.79)
      end if
      Compute new a_{opt} and $a_{opt}^{'}$ using (7.69) and (7.85) again;
      Compute $error = |\phi_{opt}^{iter+1} - \phi_{opt}^{iter}|$..
    end while
  end for
Compute blade geometry.

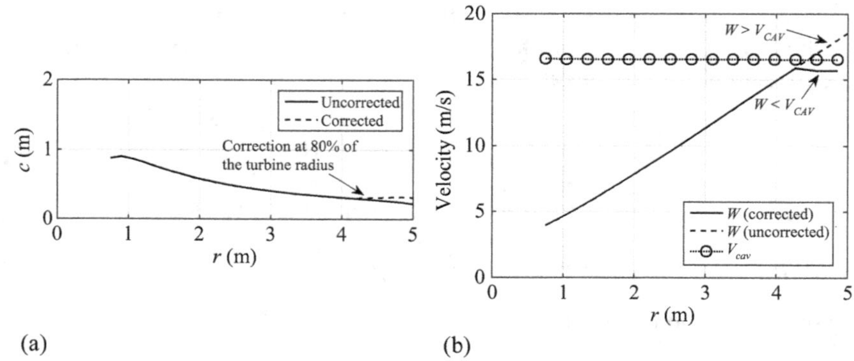

Figure 7.21 *(a) Chord distribution along the blade and (b) relative and cavitation velocities as functions of the radial position*

To evaluate the performance of the optimization procedure for a turbine with diffuser, we consider the design parameters in Table 7.4 for a diffuser with $\beta = 0.3587$, $\eta_d = 0.907$, and $C_{Td} = 0.456$. These values are only for the purpose of verifying the cavitation effect on a turbine with diffuser. Therefore, Figure 7.21(a) shows the chord distribution for both uncorrected and corrected models in relation to the cavitation effect. Note that cavitation occurs at approximately 80% of the blade length. When W becomes higher than V_{CAV}, the model imposes a correction in order to modify the relative velocity, as shown in Figure 7.21(b) for radial positions $r > 4.12$ m. In order to avoid cavitation, the approach corrects W and assumes values always lower than V_{CAV}.

7.9 Conclusions

BEMT for horizontal-axis turbines combines BET for the forces on the blade with conservation of axial and angular momentum in the air or water flowing over the element. In most cases the equations must be solved iteratively. The BE assumption, however, makes available to wind turbine designers and analysts a vast amount of lift and drag data on airfoils that can be used for the blades. It was pointed out that at the low Reynolds numbers that occur primarily on small wind turbines, it is dangerous to use lift and drag data obtained from computer simulation. Experimental data is usually much safer.

Although it was formulated well over 100 years old, BEMT is still useful and its usefulness has been increased by recent developments such as the development of finite blade functions and a cavitation calculation. Comparisons with experiments on wind and hydrokinetic turbines show that BEMT is reasonably accurate for the power but tends to slightly underpredict the thrust.

The extension of BEMT to turbines with a diffuser complicates the modeling because there is no easy way to include the diffuser performance. The limited experimental data for these turbines was shown to be in good agreement with BEMT simulations once the diffuser performance is modeled.

BEMT is an analysis tool but various ways were described to use it for blade design. First, the airfoil section (sections) for the blade is (are) chosen. Then BEMT can be used to determine the chord and pitch distribution along the blade to substantially complete the aerodynamic design. It was pointed out that blade design is often MDO or trade-off where power extraction must be balanced against other desirable objectives, including starting performance, noise minimization, blade mass, and inertia while satisfying constraints such as avoiding excessive tip deflection and cavitation. The fast evaluation of a BEMT makes it ideal for these calculations where the objective functions may have to be evaluated thousands of times in the context of a genetic algorithm optimization. Recently, BEMT has been combined with simple blade structural analysis to avoid the otherwise sequential nature of the design and to use the blade inertia in the optimization. It was suggested that future design will be able to incorporate fatigue analysis in a similar manner.

References

[1] Burton T, Sharpe D, Jenkins N, *et al. Wind Energy Handbook*. Wiley, Chicester; 2011.
[2] Wood D. *Small Wind Turbines*. Springer, London; 2011.
[3] Hansen MO. *Aerodynamics of Wind Turbines*. Routledge, Abingdon; 2015.
[4] Batten W, Bahaj A, Molland A, *et al.* Experimentally validated numerical method for the hydrodynamic design of horizontal axis tidal turbines. *Ocean Engineering*. 2007;34(7):1013–1020.
[5] Silva PA, Oliviera TF, Brasil Junior AC, *et al.* Numerical study of wake characteristics in a horizontal-axis hydrokinetic turbine. *Anais da Academia Brasileira de Ciências*. 2016;88(4):2441–2456.
[6] Silva PA, Vaz DAR, Britto V, *et al.* A new approach for the design of diffuser-augmented hydro turbines using the blade element momentum. *Energy Conversion and Management*. 2018;165:801–814.
[7] do Rio Vaz DA, Vaz JR, and Silva PA. An approach for the optimization of diffuser-augmented hydrokinetic blades free of cavitation. *Energy for Sustainable Development*. 2018;45:142–149.
[8] Silva PASF, Shinomiya LD, de Oliveira TF, *et al.* Analysis of cavitation for the optimized design of hydrokinetic turbines using BEM. *Applied Energy*. 2017;185:1281–1291.
[9] Raabe J. *Hydro Power: The Design, Use, and Function of Hydromechanical, Hydraulic, and Electrical Equipment*. VDI-Verlag Dusseldorf, Dusseldorf; 1985.
[10] Mesquita ALA, Mesquita ALA, Palheta FC, *et al.* A methodology for the transient behavior of horizontal axis hydrokinetic turbines. *Energy Conversion and Management*. 2014;87:1261–1268.
[11] Vaz JR, Wood DH, Bhattacharjee D, *et al.* Drivetrain resistance and starting performance of a small wind turbine. *Renewable Energy*. 2018;117:509–519.

[12] Evans S, Kesby J, Bradley J, *et al.* Commercialization of a diffuser augmented wind turbine for distributed generation. *Journal of Physics Conference Series*. 2020;1452:012014.
[13] Bontempo R and Manna M. Diffuser augmented wind turbines: review and assessment of theoretical models. *Applied Energy*. 2020;280:115867.
[14] Vaz JR, Mesquita AL, Mesquita ALA, *et al.* Powertrain assessment of wind and hydrokinetic turbines with diffusers. *Energy Conversion and Management*. 2019;195:1012–1021.
[15] Adhikari RC, Vaz J, and Wood D. Cavitation inception in crossflow hydro turbines. *Energies*. 2016;9(4):237.
[16] Sale D, Jonkman J, and Musial W. *Hydrodynamic Optimization Method and Design Code for Stall-Regulated Hydrokinetic Turbine Rotors*. National Renewable Energy Lab.(NREL), Golden, CO; 2009.
[17] Carlton J. *Marine Propellers and Propulsion*. Butterworth-Heinemann, Oxford; 2018.
[18] do Rio Vaz DATD, Mesquita ALA, Vaz JRP, *et al.* An extension of the blade element momentum method applied to diffuser augmented wind turbines. *Energy Conversion and Management*. 2014;87:1116–1123.
[19] Vaz JR and Wood DH. Aerodynamic optimization of the blades of diffuser-augmented wind turbines. *Energy Conversion and Management*. 2016;123:35–45.
[20] Glauert H. *Aerodynamic Theory: A General Review of Progress*, vol. IV, chapter Division L, Airplane Propellers. Dover Publications, Inc., New York, NY; 1963.
[21] Vaz JR and Wood DH. Effect of the diffuser efficiency on wind turbine performance. *Renewable Energy*. 2018;126:969–977.
[22] Wood D. Application of extended vortex theory for blade element analysis of horizontal-axis wind turbines. *Renewable Energy*. 2018;121:188–194.
[23] Anderson MB, Milborrow DJ, and Ross JN. *Performance and Wake Measurements on a 3 m Diameter Horizontal-Axis Wind Turbine. Comparison of Theory, Wind Tunnel and Field Test Data*; 1982.
[24] Wilson RE and Lissaman PB. *Applied Aerodynamics of Wind Power Machines*. National Science Foundation; 1974.
[25] Shen WZ, Mikkelsen R, Sörensen JN, *et al.* Tip loss corrections for wind turbine computations. *Wind Energy: An International Journal for Progress and Applications in Wind Power Conversion Technology*. 2005;8(4):457–475.
[26] Wimshurst A and Willden R. Analysis of a tip correction factor for horizontal axis turbines. *Wind Energy*. 2017;20(9):1515–1528.
[27] Wood D, Okulov V, and Bhattacharjee D. Direct calculation of wind turbine tip loss. *Renewable Energy*. 2016;95:269–276.
[28] Vaz JR, Okulov VL, and Wood DH. Finite blade functions and blade element optimization for diffuser-augmented wind turbines. *Renewable Energy*. 2020.
[29] Fletcher CA. Computational analysis of diffuser-augmented wind turbines. *Energy Conversion and Management*. 1981;21(3):175–183.

[30] Hansen MOL, Sörensen NN, and Flay R. Effect of placing a diffuser around a wind turbine. *Wind Energy: An International Journal for Progress and Applications in Wind Power Conversion Technology*. 2000;3(4):207–213.

[31] Sörensen JN. *General Momentum Theory for Horizontal Axis Wind Turbines*. vol. 4. Springer, Heidelberg; 2016.

[32] Phillips DG. *An investigation on Diffuser Augmented Wind Turbine Design*. PhD thesis, University of Auckland, Auckland, New Zealand; 2003.

[33] Bontempo R and Manna M. Effects of the duct thrust on the performance of ducted wind turbines. *Energy*. 2016;99:274–287.

[34] Wood D and Okulov V. Nonlinear blade element-momentum analysis of Betz-Goldstein rotors. *Renewable Energy*. 2017;107:542–549.

[35] Marshall L and Buhl Jr J. A new empirical relationship between thrust coefficient and induction factor for the turbulent windmill state. Technical Report No NREL/TP-500-36834. National Renewable Energy Laboratory, Golden, CO; 2005.

[36] Lock CNH, Bateman H, and Townend H. *An Extension of the Vortex Theory of Airscrews With Applications to Airscrews of Small Pitch, Including Experimental Results*. HM Stationery Office; 1926.

[37] Batten W, Bahaj A, Molland A, *et al.* The prediction of the hydrodynamic performance of marine current turbines. *Renewable Energy*. 2008;33 (5):1085–1096.

[38] Morgado J, Vizinho R, Silvestre M, *et al.* XFOIL vs CFD performance predictions for high lift low Reynolds number airfoils. *Aerospace Science and Technology*. 2016;52:207–214.

[39] Mack S, Brehm C, Heine B, *et al.* Experimental investigation of separation and separation control on a laminar airfoil. In: *4th Flow Control Conference*; 2008. p. 3766.

[40] Van Treuren KW. Small-scale wind turbine testing in wind tunnels under low Reynolds number conditions. *Journal of Energy Resources Technology*. 2015;137(5).

[41] Ten Hoopen P. An Experimental and Computational Investigation of a Diffuser Augmented Wind Turbine: with an application of vortex generators on the diffuser trailing edge. MSc thesis, TU Delft; 2009.

[42] Sessarego M and Wood D. Multi-dimensional optimization of small wind turbine blades. *Renewables: Wind, Water, and Solar*. 2015;2(1):1–11.

[43] Pourrajabian A, Dehghan M, Javed A, *et al.* Choosing an appropriate timber for a small wind turbine blade: a comparative study. *Renewable and Sustainable Energy Reviews*. 2019;100:1–8.

[44] Clifton-Smith M. Wind turbine blade optimisation with tip loss corrections. *Wind Engineering*. 2009;33(5):477–496.

[45] Zhu Fw, Ding L, Huang B, *et al.* Blade design and optimization of a horizontal axis tidal turbine. *Ocean Engineering*. 2020;195:106652.

[46] Evans S, Dana S, Clausen P, *et al.* A simple method for modelling fatigue spectra of small wind turbine blades. *Wind Energy*. 2020;24:549–557.

Chapter 8
Vortex-induced vibration-based energy harvesting

Arindam Banerjee[1] *and Kai He*[1]

The study of vortex-induced vibrations (VIV) can be attributed to many real-life scenarios such as the fluid flow in a heat exchanger, the vibration of structures, and the vibrations of both riser pipes and offshore structures [1]. The effects of VIV are detrimental and must be mitigated during the construction of civil, marine, and other engineering structures [2]. In the past, ignoring VIV has led to catastrophic failures: the collapse of the Tacoma Narrows Bridge in Washington in 1940 and the collapse of three (of eight) towers at Ferrybridge Power Station in the United Kingdom in 1965 are some well-documented examples of VIV. Several engineering solutions are available for mitigation purposes, including adding surface protrusions, shrouds, and near-wake stabilizers, which have been created to mitigate vibrations due to vortex shedding [3,4]. A relatively newer application is based on augmenting VIV to extract marine renewable energy (MRE). Pioneered by Bernitsas *et al.* at the University of Michigan [5–7], similar methods have been successfully used by several research groups [2,8]. The device in question is named VIVACE, an acronym for VIV for aquatic clean energy. The device was developed and patented by Bernitsas for converting the energy of slow-moving river/tidal/ocean currents to mechanical energy by inducing free vibration of cylinders. One advantage of VIV-based energy extraction is that it allows energy harvesting at low velocities (<1.0 m/s) below the cut-in speed for most tidal/hydrokinetic turbines. Thus, VIV-based energy extraction is feasible for shallow rivers and slow-flowing streams. A VIV-based energy harvesting device is easily constructed as it converts kinetic energy of a flowing stream directly to mechanical power through linear motions and does not use any rotating parts like turbines.

8.1 Physics of flow around cylinders

Fixed cylinders: A much-studied feature of flow around bluff bodies in general and around a circular cylinder, in particular, is that transition from laminar to

[1]Department of Mechanical Engineering and Mechanics, Lehigh University, Bethlehem, USA

turbulent flow occurs over a series of distinct transition states that takes place over an enormous range of *Re* [9–11]. The state of the flow disturbed by a cylinder may be fully laminar (L), in either of a series of three transitions (TrW, TrSL, and TrBL), or fully turbulent (T). The list of the flow states and adopted notations is given in Table 8.1 [3,4].

Typical transition states of flow around the circular cylinder are sketched in Figure 8.1(a)–(d) (adopted from [12]). The *first transition* state occurs in the wake (Figure 8.1(a)), where laminar vortices become turbulent due to three-dimensional distortions further downstream of the flow. The turbulence spreads upstream with an increase of *Re*; however, the free-shear layers surrounding the near wake always remain laminar. The *second transition* occurs in the free-shear layers (Figure 8.1(b)). The transition region gradually moves upstream toward the separation point as *Re* is increased. The *third transition* occurs around separation, seen in Figure 8.1(c), and produces the largest effect on the drag force as a result of a complicated interaction between the separation and transition before the boundary layer becomes fully turbulent. The final and *fourth transition* takes place in the boundary layer away from separation (Figure 8.1(d)). The transition region moves upstream with an increase in *Re*, and it ultimately leads to a merger of the transition and stagnation points [9]. Beyond the fourth transition state, all disturbed regions of the flow are fully turbulent. Experimental observations over the last century have revealed an enormous variety of regular and irregular flow regimes around circular cylinders. Zdravkovich [10] provides an extensive review of these flow patterns. The flow regimes and their subdivisions are expected to be confined in the fixed range of

Table 8.1 Description of flow regimes around a circular cylinder

State	**Regime**	***Re* no. range**	C_D
Laminar **(L)**	L1—no separation	0 to 4/5	↓
	L2—closed wake	4/5 to 30/48	↓
	L3—periodic wake	30/48 to 180/200	↑
Transition in wake **(TrW)**	TrW1—far wake	180/200 to 220/250	↑
	TrW2—near wake	220/250 to 350/400	↓
Transition in shear layer **(TrSL)**	TrSL1—lower	350/400 to 1–2 $\times 10^3$	↓
	TrSL2—intermediate	1–2 $\times 10^3$ to 2–4 $\times 10^4$	↑
	TrSL3—upper	2–4 $\times 10^4$ to 1–2 $\times 10^5$	↔
Transition in boundary layer **(TrBL)**	TrBL0—precritical	1–2 $\times 10^5$ to 3–3.4 $\times 10^5$	↓
	TrBL1—single bubble	3–3.4 $\times 10^5$ to 3.8–4 $\times 10^5$	↓↓
	TrBL2—two bubble	3.8–4 $\times 10^5$ to 0.5–1 $\times 10^6$	↓↓
	TrBL3—supercritical	0.5–1 $\times 10^6$ to 3.5–6 $\times 10^6$	↑
	TrBL4—post-critical	3.5–6 $\times 10^6$ to?	↔
Turbulent **(T)**	T1—invariable	? to ∞	↔
	T2—ultimate		?

Adapted, with permission, from References [9,10], Oxford Publishing Ltd.

↑ Indicates an increase, ↓ indicates a decrease, ↓↓ rapid decrease, ↔ indicates the same, and ? indicates not known.

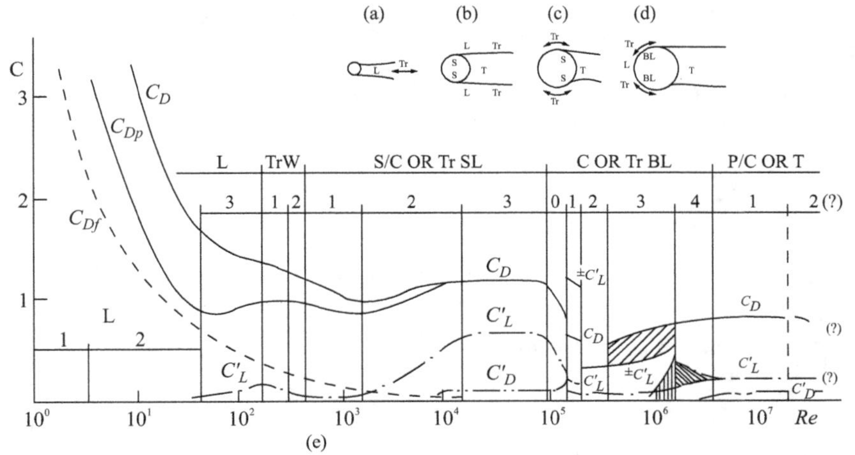

Figure 8.1 (a–d) Schematic of transition states (on top); (e) force coefficients versus Reynolds number. Adopted, with permission, from References [9,10], Oxford Publishing Ltd

Reynolds numbers (see Table 8.1). The variation of flow pattern in these regimes causes a continuous or discontinuous change of fluctuating, and time-averaged (mean) forces exerted on the cylinder. There are a total of six forces (or force coefficients) that act on the bluff body: the time-averaged and fluctuating drag and lift forces; denoted as C_D, C'_D, C_L, C'_L. The drag force is further subdivided into a skin friction drag (C_{Df}) caused due to viscous forces along the surface and a pressure drag (C_{Dp}), resulting due to the asymmetric pressure distribution on the upstream and downstream sides of the cylinder. The variation of all the six force coefficients acting on a nominally two-dimensional cylinder over a wide range of *Re* is presented in Figure 8.3(e) for a disturbance-free flow (vertical bars are used to represent the scatter in experimental data). A closer look at Figure 8.1(e) presents some important flow phenomena. The viscous friction coefficient (CDf) is substantial in the laminar state (L). It keeps decreasing with an increase in *Re* and becomes negligible starting from TrSL3. The pressure drag (C_{Dp}) shows a dependence on the different flow regimes. The least values of C_{Dp} were observed at the end of L2 (steady and closed near wake), at the transition from TrSL1 to TrSL2 (formation of long eddies), and in TrBL2 (formation of separation bubbles on both sides). An increase in fluctuating lift coefficient (C'_L) is seen throughout TrSL2, and the high values are maintained throughout TrSL3. The single-bubble regime, TrBL1, is marked by a discontinuous fall in C_D and the appearance of mean C_L. A discontinuous fall in both C_D and C_L is observed at the start of the two-bubble regimes, TrBL2. C'_L is always observed to be greater than C'_D.

Vibrating cylinders: The physics of VIV can be explained using a schematic diagram of an elastically mounted cylinder shown in Figure 8.2, first set up by Khalak and Williamson [13]. A rigid circular cylinder, free to oscillate in a direction transverse to the free stream, is immersed in a flow. The characteristic parameters of the setup are

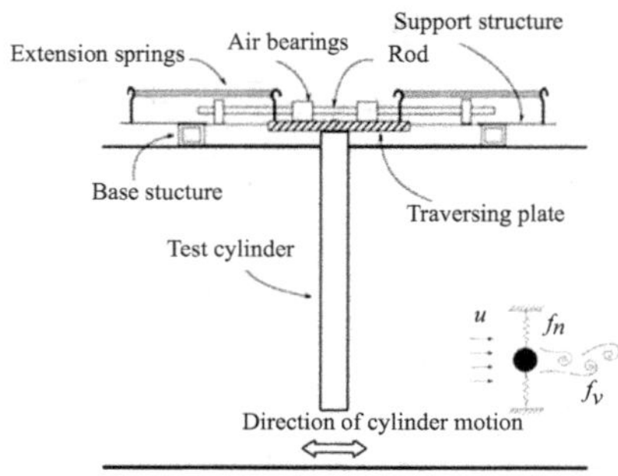

Figure 8.2 Schematic of a typical VIV experimental setup

Table 8.2 Nondimensional parameters used in the study of VIV

Parameters	Symbol	Definition
Amplitude ratio	A^*	A/D
Natural frequency	f_N	$(1/2\pi)\sqrt{k/(m+m_A)}$
Frequency ratio	f^*	f_{osc}/f_N
Drag coefficient	C_D	$2F_D/(\rho U^2 DL)$
Lift coefficient	C_L	$2F_L/(\rho U^2 DL)$
Mass ratio	m^*	$4m/(\pi\rho D^2 L)$
Reynolds number	Re	$\rho UD/\mu$
Strouhal number	St	$f_V D/U$
Reduced velocity	U^*	$U/(f_N D)$
Damping ratio	ξ	$c/2\sqrt{k(m+m_A)}$

Note: m is the mass of the oscillating system, m_a is the added mass, ρ is the density of water, D and L are the diameter and the length of the cylinder used, respectively, c is the damping constant, k is the stiffness of the spring, U is the free-stream velocity, f_n is the natural frequency of the system, A and f are the vibration amplitude and vibration frequency of the cylinder, ν is the kinematic viscosity of water, f_{vo} is the vortex-shedding frequency, and F_L and F_D are the lift and drag forces acting on the cylinder.

the oscillating mass of the system and the mass of the fluid displaced by the immersed segment of the cylinder. As the free-stream velocity increases, cylinder vibrations develop and reach amplitudes of one diameter over a limited range of velocities associated with a significant change in vibration frequency. Typically, the response profile of a cylinder undergoing VIV is categorized into three regimes, referred to as branches, based on the behavior of the nondimensional amplitude A^* in terms of the reduced velocity U^*. Table 8.2 provides a list of dimensionless variables used in studies of VIV.

Figure 8.3(a) is an illustration of the nondimensional amplitude (A^*) as a function of the reduced velocity (U^*) for low mass damping ($m^*\zeta = 0.015$) reported in experiments by the authors [14]. The initial branch is the regime where the vibrations

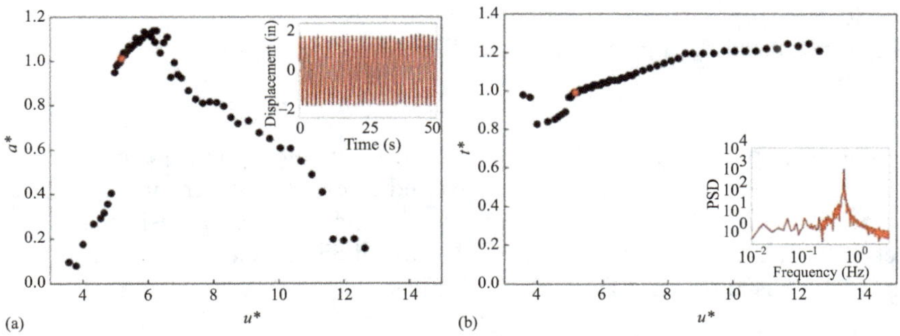

Figure 8.3 (a) Amplitude response (displacement trace at U = 5.01 as inset) and (b) frequency response (power spectral density at U* = 5.01 as inset) of a smooth circular cylinder at m*ζ = 0.015. Reproduced, with permission, from Reference [14], Elsevier*

begin to develop and was observed within $2.9 < U^* < 4.9$ followed by the upper branch in the range $5.0 < U^* < 6.3$, with A^* reaching a maximum of 1.14 and f^* remaining close to unity. In this upper branch, the vibration frequency tends to synchronize itself to the natural frequency of vibration in the water; a phenomenon commonly referred to as the lock-in. The lower branch was found to span a range of $6.4 < U^* < 11.3$ in which f^* increased to 1.2 and remained relatively constant for $U^* > 8.5$, while A^* was found to decrease with increasing reduced velocity. At the end of the lower branch, vibration dies down, and the region of negligible vibrations that follows is called the desynchronizing regime. A very similar amplitude response was also seen by Klamo *et al.* [15] at higher values of damping. The inset figures show the displacement trace (Figure 8.3(a)) and the corresponding frequency spectrum (Figure 8.3(b)) for the cylinder vibrating in the upper branch at $U^* = 5.1$. Steady vibrations with nearly constant amplitudes and a dominant peak in the frequency spectrum can be clearly identified.

8.2 Governing equations

Equations of motion: For a 1 degree of freedom elastically mounted rigid structure that is constrained to oscillate transversely to a stationary and uniform flow, the equation of motion can be described as a second-order linear oscillator as [16]

$$m\ddot{y} + c\dot{y} + ky = F_{fluid} \tag{8.1}$$

where y denotes the transverse displacement of the cylinder, c denotes the mechanical damping coefficient, k the system stiffness, and F_{fluid} is the force applied to the cylinder in a direction normal to the flow direction. The fluid force can be subdivided into two terms: a fluctuating lift force, $F_L = C_L \rho U^2 DL/2$, and an inviscid force, $F_{inviscid} = -m_A\ddot{y}$ due to added mass $m_A = \pi\rho D^2 L/4$ based on the mass of displaced fluid by the structure. Equation (8.1) can be rewritten as

$$(m + m_A)\ddot{y} + c\dot{y} + ky = F_L \tag{8.2}$$

For the purpose of energy harvesting, if the mechanical power generated by the transverse motion of the structure is converted to electric power through a piezoelectric transducer [17], we may consider artificially adding resistance in the electrical circuit, and (8.2) can be amended by adding an electromechanical force term, $-\theta V$, based on the VIV model of [16] as

$$(m + m_A)\ddot{y} + c\dot{y} + ky - \theta V = F_L \tag{8.3}$$

where V is the voltage across a load resistance R_L. The relation between voltage, displacement, and the capacitance (C_p) of the piezoelectric transducer can be described by a circuit equation [17]:

$$C_p\dot{V} + \frac{V}{R_L} + \theta\dot{y} = 0 \tag{8.4}$$

The electromechanical coupling term θ is a constant and relates the dynamic motion and electric output, thereby representing the amount of reaction force produced by a unit voltage. The introduction of θ to (8.3) adds an additional unknown (V); thus, (8.4) is essential to close the entire dynamic system. Several studies [17–20] have adopted this approach. The electromechanical coupling term θ is a constant and relates the dynamic motion and electric output, thereby representing the amount of reaction force produced by a unit voltage. The introduction of θ to (8.3) adds an additional unknown (V); thus, (8.4) is essential to close the entire dynamic system. Several studies [17–20] have adopted this approach.

Equations of mechanical power and energy-transfer ratio: The mechanical work (W_{mech}) can be evaluated over the cycle period T_{cyl} by multiplying the left-hand side of (8.2) with the oscillation velocity. The mechanical power (P_{mech}) generated is the mechanical work divided by the time period and can be written as [21]

$$P_{mech} = \frac{W_{mech}}{T_{cyl}} = \frac{1}{T_{cyl}}\int_0^{T_{cyl}} [(m + m_A)\ddot{y} + c\dot{y} + ky]\,\dot{y}\,dt \tag{8.5}$$

For the case of a sinusoidal response such that $y = A\sin(2\pi f_{cyl}t)$, only the term in phase with the velocity contributes a nonzero value to the integral. Thus, (8.5) can be written as

$$P_{mech} = \frac{1}{T_{cyl}}\int_0^{T_{cyl}} 4\pi(m + m_A)\dot{y}^2\,\zeta_{total}\,f_{n,water}dt \tag{8.6}$$

where ζ_{total} is the nondimensional damping coefficient and $f_{n,water}$ is the natural frequency of the system in water as is defined as

$$\zeta_{total} = \frac{c}{2\sqrt{(m+m_A)k}}; \quad f_{n,water} = \frac{1}{2\pi}\sqrt{\frac{k}{(m+m_A)}} \tag{8.7}$$

Upon integration, the total mechanical power that can be harnessed by the device can be expressed as

$$P_{mech} = 8\pi^3(m+m_A)\,\zeta_{PTO}(Af_{cyl})^2 f_{n,water} \tag{8.8}$$

As observed, the mechanical power is directly proportional to the power takeoff (PTO) damping coefficient (ζ_{PTO}). The component of P_{mech} associated with the structural damping is lost and cannot be recovered. The reader is also cautioned that (8.11) should not be interpreted as larger power output for a larger PTO damping. The amplitude A is inversely proportional to the damping coefficient, and f_{cyl} should be close to its natural frequency in water.

An efficiency or an energy-transfer ratio can be determined by using the conservation of momentum analysis and estimating the available power that can be harnessed from open flow by drawing an analogy to the Betz limit or coefficient (C_{Betz} = 16/27) that is considered the theoretical maximum power that can be extracted in a stream tube in the absence of viscous dissipation and wake rotation for an actuator disk-type model. The energy-transfer ratio (η_{mech}) is defined as [22]

$$\eta_{mech} = \frac{2P_{mech}}{C_{Betz}\rho(D+2A_{\max})LU^3} \tag{8.9}$$

For the piezoelectric transducer cases, part of the flow power is converted to electrical power [17]:

$$P_{elec} = \frac{V_{rms}^2}{R_L} \tag{8.10}$$

where V_{rms} is the root mean square value of voltage V across the load resistance R_L. V is the solution of (8.3) or (8.4). Similarly, η_{elec}, denoting the proportion of the available energy from an open flow that can be harnessed and converted to electricity, is expressed as

$$\eta_{elec} = \frac{2P_{elec}}{C_{Betz}\rho(D+2A_{\max})LU^3} \tag{8.11}$$

Along with the energy transfer ratio (η) defined in (8.9), the power quality is a key parameter for analyzing a VIV converter. The coefficient of variation (COV) was proposed for evaluating the dispersion in the instantaneous power ($P_{mech,i}$)

relative to its mean value $\overline{P}_{mech}$ [22] as

$$COV_p = \frac{1}{\overline{P}_{mech}} \sqrt{\frac{1}{N} \sum_{i=1}^{N} \left(P_{mech,i} - \overline{P}_{mech}\right)^2} \tag{8.12}$$

A small COV_p indicates that the mechanical power generated is relatively regular, and consequently, the VIV generator output remains steady with small dispersion.

8.3 Parameters that affect VIV

Mass-damping ratio: The mass-damping parameter has been shown to impact VIV amplitude response and lock-in regime. For cases of very low mass damping, Khalak and Williamson [23] showed that three distinct branches, initial, upper, and lower, are present in the amplitude response (see Figure 8.4). At low flow velocities, an initial branch consisting of low amplitude, sporadic vibrations are observed. As flow velocity increases and the vortex-shedding frequency approaches the natural frequency of the oscillating system, high-amplitude vibrations can occur with an oscillation frequency approximately equal to f_n, referred to as the upper branch. These amplitudes are typically on the order of one diameter for an elastically mounted cylinder. As flow velocity increases further, vortex-shedding frequency increases past the system's natural frequency resulting in a decrease in amplitude and decrease in the uniformity of vibration response. At high velocities, a smooth cylinder will only exhibit random, small-amplitude vibrations in what is called desynchronization. These observations contrast with the results by Feng, who used a system with very high mass damping. His experiments showed only an initial and lower branch response and no upper branch. The comparison of Williamson and Feng's results is presented in [23].

Klamo *et al.* [15] performed a set of experiments measuring the change in VIV response over a wide range of damping ratios. The response of the lowest and highest damping cases reflects the results presented by Williamson and Feng [23], respectively. As damping is increased, decreasing upper and lower branch amplitudes and shorter excitation regions are observed. Lee and Bernitsas [6] conducted a comprehensive study on the effects of varying damping at a constant spring stiffness and varying stiffness at constant damping. This was done by controlling the damping and stiffness of the system by means of a feedback loop, allowing them to easily test the independent effects of stiffness and damping in a systematic, controlled fashion [24]. They found that increases in damping for a given spring stiffness (Figure 8.5(a)) resulted in lower VIV amplitudes and a shorter range of excitation. The transition from the initial branch to the upper branch also becomes more gradual. At higher values of spring stiffness, increased damping is observed to push the peak upper branch amplitude to lower flow velocities, resulting in a more gradual transition to desynchronization. Figure 8.5(b) shows the variation in response with increasing stiffness at constant damping. As stiffness increases, synchronization is pushed to higher velocities, upper branches increase in length,

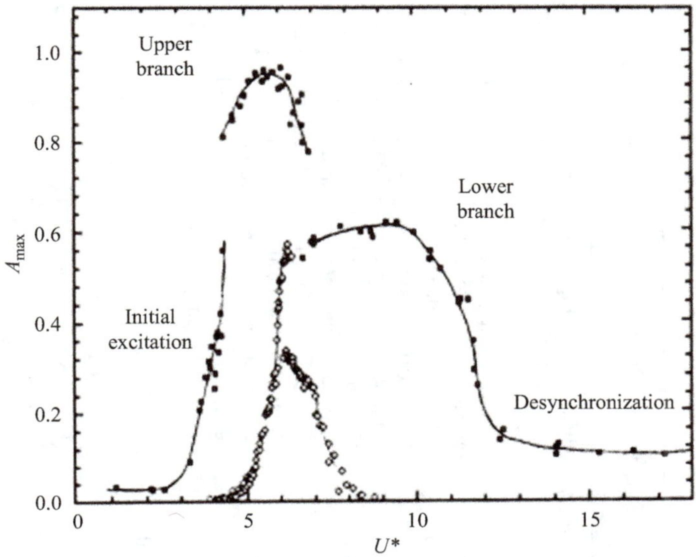

Figure 8.4 The amplitude response of circular cylinders. Adapted, with permission, from Reference [23], Elsevier

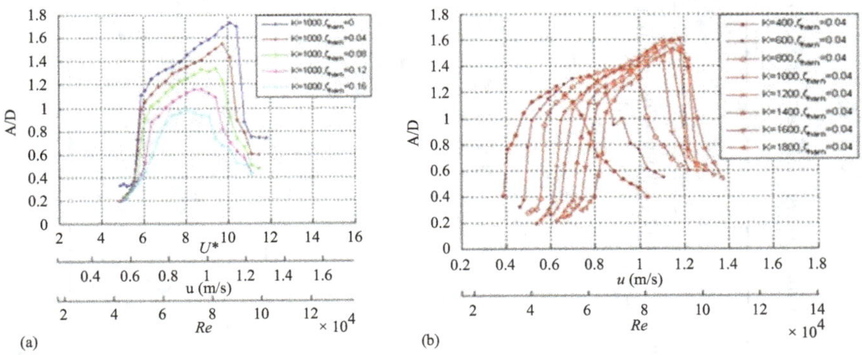

Figure 8.5 Amplitude response for (a) constant stiffness, varying damping and (b) constant damping, varying stiffness. Adopted, with permission, from Reference [6], Elsevier

and maximum amplitude experienced in the upper branch increases. The upper branch is followed almost immediately by desynchronization, occurring more rapidly with increasing spring stiffness. The authors noted that although increased stiffness pushes the response to higher Reynolds numbers, it occurs in approximately the same U^* range regardless of stiffness.

Passive turbulence control (PTC): Numerous attempts have been made to develop means that would mitigate vibrations due to vortex shedding. A comprehensive review of the various suppression methods tested was given by Zdravkovich [12].

Table 8.3 Effective added features in suppressing VIV. Adapted, with permission, from Reference [3], AIP Publishing

Phenomenological mechanism	Directional effectiveness	Added feature
Surface protrusion—affects separation lines and separated shear layers	Omnidirectional	Helical strakes (Figure 8.6(i)-(a)), helical wires (Figure 8.6(i)-(b)), rectangular plates forming a helix (Figure 8.6(i)-(c)), helical wires forming a herringbone pattern (Figure 8.6(i)-(d))
	Unidirectional	Four straight fins forming an X cross (OFigure 8.6(ii)-(a)), staggered straight wires (Figure 8.6(ii)-(b)), staggered rectangular fins (Figure 8.6(ii)-(c)), small spheres as turbulence promoters (Figure 8.6(ii)-(d))
Shrouds—affects the entrainment layers	Omnidirectional	Perforated shrouds with circular and square holes (Figure 8.6(iii)-(a)), fine mesh gauze used as a shroud (Figure 8.6(iii)-(b)), parallel axis rods forming a shroud (Figure 8.6(iii)-(c)), shroud reduced to four rods (Figure 8.6(iii)-(d)), shroud consisting of straight slats (Figure 8.6 (iii)-(e))
Near-wake stabilizers—affect switching of the confluence points	Unidirectional	Sawtooth fins (Figure 8.6(iv)-(a)), detached splitter plates (Figure 8.6(iv)-(b)), guiding plates (Figure 8.6(iv)-(c)), guiding vanes (Figure 8.6(iv)-(d)), base bleed into near wake (Figure 8.6(iv)-(e)), slit along with a cylinder (Figure 8.6(iv)-(f))

The features added to a cylinder to modify the flow were classified into three different categories, depending on the phenomenological mechanism that was employed. Surface protrusions were used to modify flow separation and the separated shear layers, shrouds were used to tweak the entrainment layer, and near-wake stabilizers were used to prevent the interaction of the entrainment layers. Table 8.3 presents the added features, successful in suppressing vibration, the phenomenological mechanism that they employ, and their directional effectiveness.

Figure 8.6 presents schematics of cylinders with the features discussed in Table 8.3. In the process of investigating suppression mechanisms, some surface protrusion-based features, or more appropriately, certain arrangements of some such features proved to be ineffective, as they ended up augmenting the oscillations. These are highlighted in Figure 8.7. Nakagawa *et al.* [25] carried out tests at supercritical Reynolds numbers ($1 \times 10^6 < Re < 6 \times 10^6$), with eight helical wires, wound around a flexibly mounted cylinder and observed amplitudes twice as high as those observed with the plain cylinder. Their experiments by replacing the eight helical wires with just one [26] also resulted in significant vibrations. Mahrenholtz and Bardowicks [27] achieved enhanced oscillations

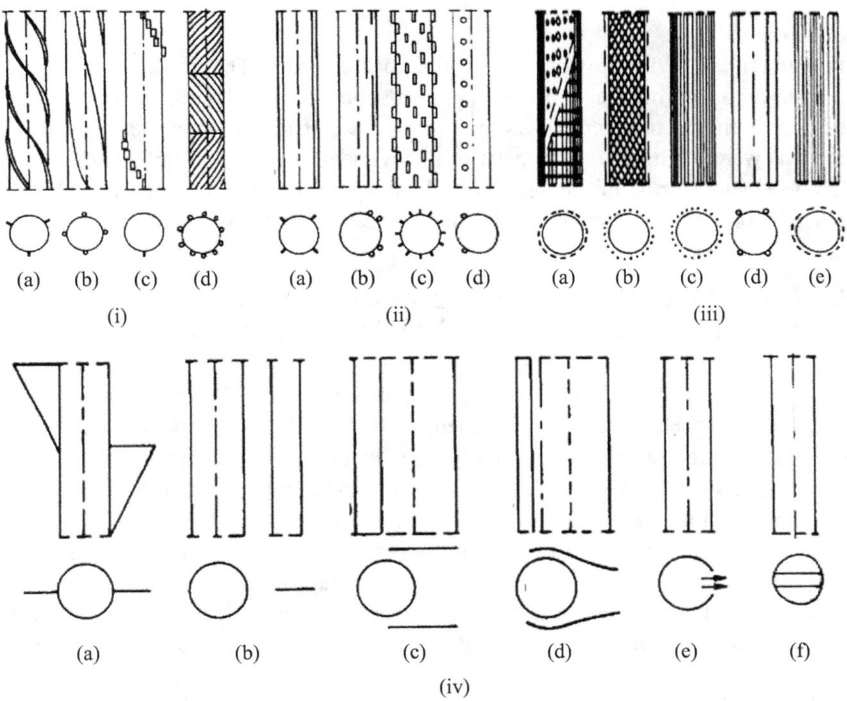

Figure 8.6 Techniques for VIV suppression: (i) omnidirectional surface protrusion, (ii) unidirectional surface protrusion, (iii) omnidirectional shroud-based techniques, and (iv) unidirectional near-wake stabilizers. Adopted, with permission, from Reference [12], Elsevier

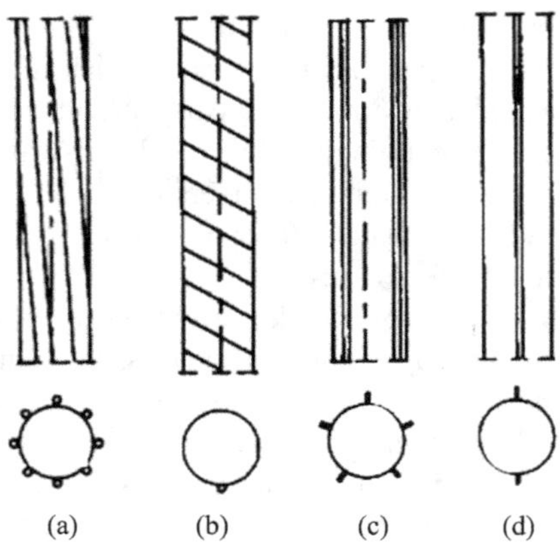

Figure 8.7 Surface protrusion-based mechanisms resulting in VIV enhancement. (a) eight helical wires, (b) single helical wire, (c) five straight fins, and, (d) symmetric side-fins. Adapted, with permission, from Reference [12], Elsevier

in the subcritical regime by fitting five fins to a cylinder. They observed maximum excitation for a symmetrical configuration. Gartshore *et al.* [28] performed experiments in both smooth and turbulent flows by attaching a pair of fins sidewise to a cylinder. These fins were 0.1 times the diameter of the cylinder in height. In turbulent flows, for reduced velocity (for definition see Table 8.2) ranging between 4 and 6, they observed increased vibrations, as high as five times that observed with a smooth cylinder.

8.4 Role of PTC in augmenting VIV in TrSL2 and TrSL3 regimes—transition from VIV to galloping instability

The practice of using sandpapers of various grits to enhance VIV in the TrSL3 regime was developed by Bernitsas. A comprehensive review of those works can be found in [29], a brief overview with some of the highlights is discussed in the section. The authors have used sandpapers and smooth strips [3,14] to explore enhancement methods in the TrSL2 regime. In general, sandpapers with various roughnesses and thicknesses were used on a circular cylinder. The placement angle was determined as the angle between the leading edge of the strip and the frontal stagnation point. The coverage area was defined by the angular coverage of the strip. The total height T is the sum of the height of roughness grit size k and the strip thickness p, as shown in Figure 8.8. Since the thickness of the strip is approximately equal to the boundary layer thickness [30], it trips the boundary layers and affects reattachment. The PTC can also introduce turbulence and consequently increases energy density in the downstream region.

The semicircle of the placement angle can be divided into six zones based on the amplitude responses, as shown in Figure 8.9. These include (1) weak suppression zone 1 (WS1), (2) hard galloping zone 1 (HG1), (3) soft or regular galloping

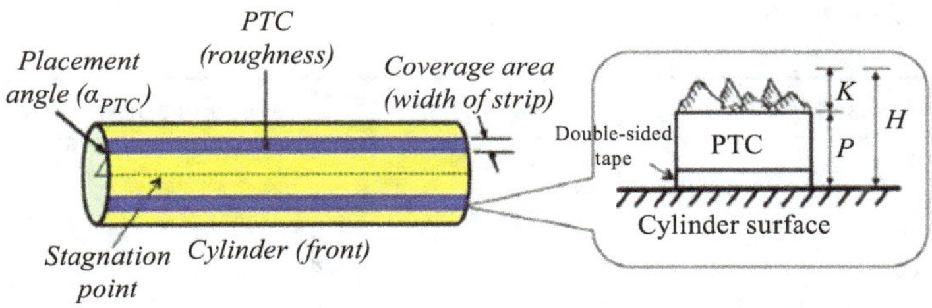

Figure 8.8 Selective distribution of surface roughness on a circular cylinder. Adapted, with permission, from Reference [31], Elsevier

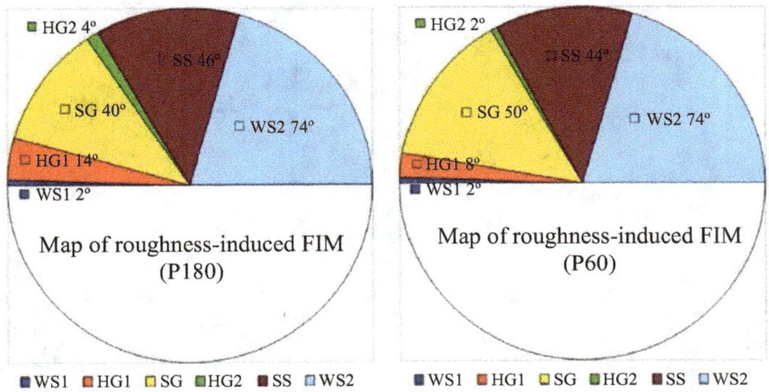

Figure 8.9 PTC-to-FIM maps for different sandpapers P180 and P60. Adapted, with permission, from Reference [31], Elsevier

zone (SG), (4) hard galloping zone 2 (HG2), (5) weak suppression zone 2 (WS2), and (6) strong suppression zone (SS). The PTC-to-FIM maps are shown to illustrate those divisions with P60 and P180 sandpapers and a coverage area of 16°. P60 sandpaper has a larger grit size *k*, as well as a larger total height T, compared with P180 (see Table 8.4 for particle size data). The two maps, however, exhibit minor variations in each zone, indicating that the amplitude responses seem dependence free on the types of sandpapers. Within WS1, WS2, and SS zones, FIM is significantly suppressed, characterized by the ignorable amplitude, a shrunk upper branch, and an earlier desynchronization. Conversely, PTC located at HG1, HG2, and SG zones can strengthen VIV or even initiate galloping.

The performance of PTC has been studied extensively by Bernitsas and other groups, and it has been concluded that the effectiveness of the PTC-to-FIM map relates to its robustness. The amplitude response is not sensitive to the PTC coverage width if the roughness strips are attached in their entirety inside a single zone (see Figure 8.10 (a)–(d)). As long as PTC covers the SS zone with or without coverage on SG or HG2 zones, the amplitude response tends to be strongly suppressed, suggesting that the SS zone is the most dominant. The SG zone has stronger effects than the WS1, HG1, and HG2 zones. When the four zones WS1, HG1, HG2, and SG are covered by PTC, VIV may transit to galloping as long as the SS zone was not covered. The width of coverage

Table 8.4 Sandpaper grit details

Strip used	Average grain size (μm)
P36	538
P60	269
P180	82
P320	46.2
Zero roughness	0

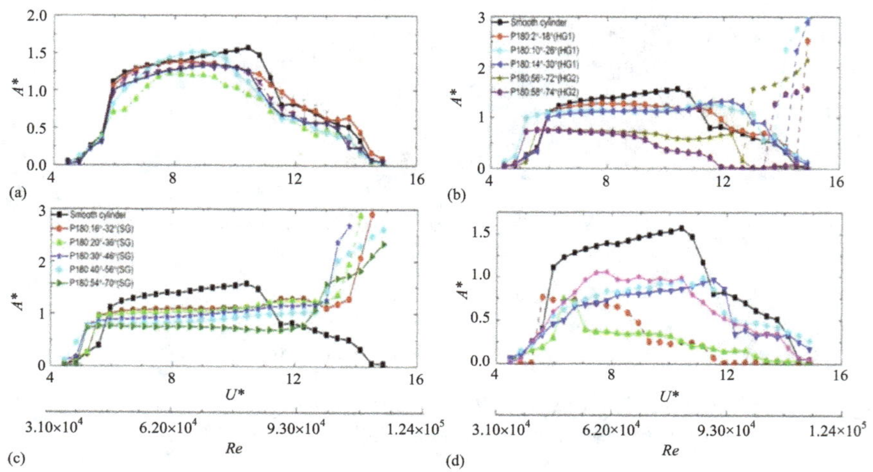

Figure 8.10 Amplitude response with P180 strips in (a) WS1, (b) HG1, (c) SG, and (d) SS zones. Adopted, with permission, from Reference [31,32], Elsevier

on the SG zone seems to significantly affect the lower branch. If PTC covers a small angle, VIV may directly transit to galloping, and no lower branch can be observed, while if it covers largely, the cylinder may undergo the lower branch.

8.5 Devices for marine and wind energy applications

In the last decade, there has been a significant uptake in research related to energy-harvesting concepts of flow-induced vibrations. These concepts are summarized in a recent review article [33]. The energy-harvesting mechanism and research progress are broadly focused on four types of flow-induced vibrations: VIV, galloping, flutter, and buffeting. In this section, we focus only on energy harvesting based on VIV by harnessing wind and MRE. We focus on two devices: the University of Michigan/ Vortex Hydro Inc. device, which is known as VIVACE for MRE generation, and Vortex Bladeless, a VIV-based method to harness wind energy without blades.

MRE generator—VIVACE: developed and patented by Bernitsas *et al.* at the University of Michigan [34–40]. The device (see Figure 8.11) converts the energy of river/tidal/ocean currents to mechanical energy by inducing free vibration of cylinders [36,41] and can be used in water bodies with current velocities as low as 0.25 m/s, as opposed to turbines that operate in currents with velocities >1.5 m/s. A VIV-based device is simple in construction and is modular, scalable, and robust and does not employ any rotating parts like turbines or propellers [36]. Up to three generations of VIVACE have been built and tested in the MRE Lab (Marine Renewable Energy laboratory) at the University of Michigan. The device is built to withstand environmental loads with low-maintenance requirements. Several technological developments have been undertaken that include (1) a virtual damping-spring device called V_{ck} system to replace the classical friction-elasticity system and (2) changeable mass cylinders are employed with a rail system, which can adjust the center-to-center distance between cylinders. The V_{ck} and changeable mass ratio results in the system's flexibility and controllability. Multiple cylinder configurations that are densely packed have been arranged to enhance VIV. Since the discovery and first patent in 2004, VIVACE has been scaled up to operate in different Reynolds number regimes discussed in Figure 8.12. Scale 1 (Lab Scale) prototype has been tested in MRE Lab to

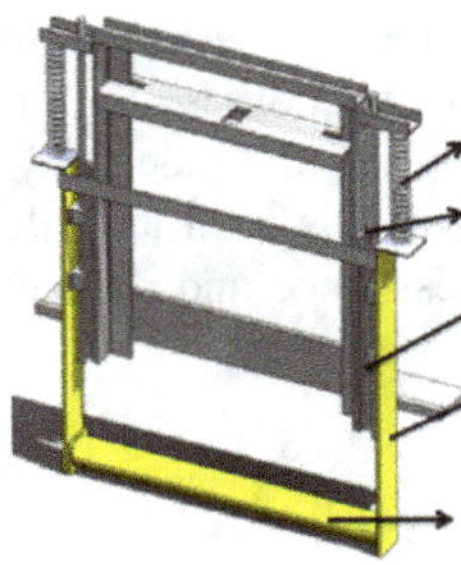

Figure 8.11 VIVACE (vortex-induced vibrations aquatic clean energy) for marine renewable energy generation developed by Bernitsas et al. at Vortex Hydro at the University of Michigan. Adopted, with permission, from Reference [32], Elsevier

Figure 8.12 VIVACE testing across different scales: single-cylinder configuration of (a) Scale 2 in the MRE Lab, (b) Scale 3 at St. Clair River, MN (2012); and multiple cylinder configurations of (c) Scale 4 device, at St. Clair River, MN (2016). Images from www.vortexhydro.com

show effectiveness in slow currents (~0.25 m/s) and produces up to 10 W of power and could be used for powering remote sensors, track spills, or fish populations, and charge unmanned underwater vehicles. Scale 2 prototype is a laboratory-scale device that produces ~200 W of power and operates in the TrSL3 regime. A configuration of densely packed cylinders allows significantly larger power density than wind farms or other MHK energy devices [29]. It is useful for military expeditions in remote locations or for generating power in remote communities. Larger devices (Scale 3) that operate in the TrBL regimes have been tested at different field sites in the United States and the Netherlands. A 10-m^3 device can produce 3.5 kW of power at flow speeds of ~1.5 m/s. Such devices can mostly cater to the river and coastal communities as well as powering remote infrastructure. The device could be further scaled up to operate in the post supercritical Reynolds number regime ($\geq$500,000). It is estimated that an 80-m^3 VIVACE device can produce up to 33 kW of power at flow speeds of ~1.5 m/s [29]. Like other technologies, the unit cost of power is expected to decrease with scale-up.

Wind energy generator—Vortex Bladeless: Vortex Bladeless (Madrid, Spain) has developed a VIV-based power generator that extracts energy from wind using VIV. A schematic of their device is shown in Figure 8.13. It consists of a vertical, slender, and cylinder-shaped wind generator. It is composed of a fixed part where the device is attached to a basement, and a flexible part which, acting as a cantilever, interacts freely with the fluid in an oscillation movement. The translational degree of freedom is restricted, while the rotational degree of freedom is not restricted. The top of the outer cylinder can oscillate transversely, driving the electricity generating device placed inside the cylinder. A reinforced polymer is used to build both the inner and outer structures resulting in good fatigue resistance and reduced power leak and friction.

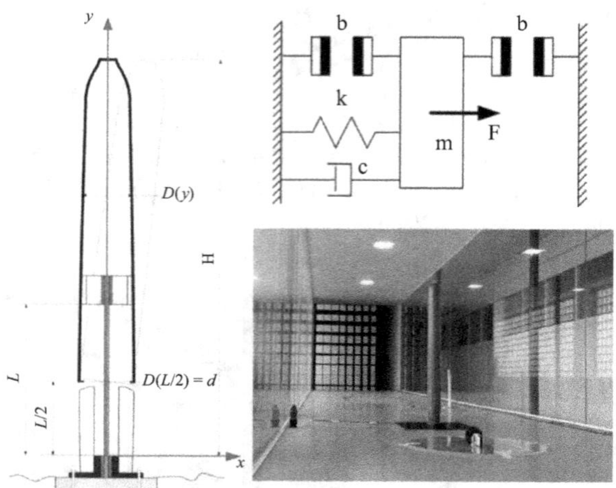

Figure 8.13 Right: schematic diagram of the wind power generator developed by Vortex Bladeless; top left: oscillator with a magnetic tuning system diagram; bottom left: the physical model [42]

Since the system is simple and free of multistage transmission, the device reliability increases, reducing operations and maintenance costs. The expected operational life span can be up to 20 years [42]. A 2.75-m tall bladeless Tacoma system is expected to produce ~100 W of power. However, due to its insignificant wake effects, an array of such devices could be closely packed, leading to larger energy density. The device is also modular and can be installed in complex landform or atmospheric conditions. Unlike the three-bladed wind turbine, there is no demand for expensive yaw mechanisms to turn the device to face the incoming wind. Besides, the device can adapt quickly to the approaching wind direction and speed. A tuning system in addition to the traditional spring-damping system is applied to achieve this function. A set of permanent magnets is used to adjust the system stiffness, resulting in an avoiding adjustment of mechanical parameters like mass-damping ratio and rigidity of the slender body. The tuning system works by capturing a wider wind speed range and extending the synchronization automatically. At the time of writing this chapter, Vortex Bladeless was focused on developing small vortex nano devices (85-cm high) and aims for distribution (mostly in Europe) toward the end of 2021.

8.6 Future scope

As the perceived need for sustainable development and renewable energy quickly rises, power extraction based on new techniques will usher in various new development opportunities. In the domain of VIV-based energy harvesting for wind and MRE researchers, further studies are required to develop technologies that can harness energy from multiple degrees of freedom excitations and use intelligent regulating elements (controls systems, artificial neural networks, and machine learning) to enhance efficiency and energy output from these devices. More field-scale deployment is required to boost investor confidence. Besides, coupling VIV-based systems with other energy harvesting technologies like wind turbines and photovoltaic devices will broaden the range of applications.

References

[1] Williamson, C.H.K. and R. Govardhan, Vortex-induced vibrations. *Annual Review of Fluid Mechanics*, 2004. 36(1): p. 413–455.

[2] Sun, W., S. Jo, and J. Seok, Development of the optimal bluff body for wind energy harvesting using the synergetic effect of coupled vortex-induced vibration and galloping phenomena. *International Journal of Mechanical Sciences*, 2019. 156: p. 435–445.

[3] Vinod, A. and A. Banerjee, Surface protrusion based mechanisms of augmenting energy extraction from vibrating cylinders at Reynolds number 3×10^3–3×10^4. *Journal of Renewable and Sustainable Energy*, 2014. 6(6): p. 063106.

[4] Zdravkovich, M.M., Review and classification of various aerodynamic and hydrodynamic means for suppressing vortex shedding. *Journal of Wind Engineering and Industrial Aerodynamics*, 1981. 7: p. 145–189.

[5] Bernitsas, M.M., K. Raghavan, Y. Ben-Simon, and E.M.H. Garcia, VIVACE (vortex-induced vibration aquatic clean energy): A new concept in generation of clean and renewable energy from fluid flow. *Journal of Offshore Mechanics and Arctic Engineering*, 2008. 130(4).
[6] Lee, J.H. and M.M. Bernitsas, High-damping, high-Reynolds VIV tests for energy harnessing using the VIVACE converter. *Ocean Engineering*, 2011. 38(16): p. 1697–1712.
[7] Raghavan, K. and M.M. Bernitsas, Experimental investigation of Reynolds number effect on vortex-induced vibration of rigid circular cylinder on elastic supports. *Ocean Engineering*, 2011. 38(5–6): p. 719–731.
[8] Hamlehdar, M., A. Kasaeian, and M.R. Safaei, Energy harvesting from fluid flow using piezoelectrics: A critical review. *Renewable Energy*, 2019. 143: p. 1826–1838.
[9] Zdravkovich, M.M., Conceptual overview of laminar and turbulent flows past smooth and rough circular-cylinders. *Journal of Wind Engineering and Industrial Aerodynamics*, 1990. 33: p. 53–62.
[10] Zdravkovich, M.M., *Flow Around Circular Cylinders—Volume 1: Fundamentals*. 1997, Oxford University Press: Oxford, UK. p. 672.
[11] Williamson, C.H.K., Vortex dynamics in the cylinder wake. *Annual Review of Fluid Mechanics*, 1996. 36: p. 413–455.
[12] Zdravkovich, M.M., Review and classification of various aerodynamic and hydrodynamic means for suppressing vortex shedding. *Journal of Wind Engineering and Industrial Aerodynamics*, 1981. 7: p. 145–189.
[13] Khalak, A. and C.H.K. Williamson, Motions, forces, and mode transitions in vortex-induced vibrations at low mass-damping. *Journal of Fluids and Structures*, 1999. 13: p. 813–851.
[14] Vinod, A., A. Auvil, and A. Banerjee, On passive control of transition to galloping of a circular cylinder undergoing vortex-induced vibration using thick strips. *Ocean Engineering*, 2018. 163: p. 223–231.
[15] Klamo, J., A. Leonard, and A. Roshko, The effects of damping on the amplitude and frequency response of a freely vibrating cylinder in cross-flow. *Journal of Fluids and Structures*, 2006. 22: p. 845–856.
[16] Facchinetti, M.L., E. de Langre, and F. Biolley, Coupling of structure and wake oscillators in vortex-induced vibrations. *Journal of Fluids and Structures*, 2004. 19(2): p. 123–140.
[17] Mehmood, A., A. Abdelkefi, M.R. Hajj, A.H. Nayfeh, I. Akhtar, and A.O. Nuhait, Piezoelectric energy harvesting from vortex-induced vibrations of circular cylinder. *Journal of Sound and Vibration*, 2013. 332(19): p. 4656–4667.
[18] Zhou, S. and J. Wang, Dual serial vortex-induced energy harvesting system for enhanced energy harvesting. *AIP Advances*, 2018. 8(7): p. 075221.
[19] Zhang, B., B. Song, Z. Mao, B. Li, and M. Gu, Hydrokinetic energy harnessing by spring-mounted oscillators in FIM with different cross-sections: From triangle to circle. *Energy*, 2019. 189: p. 116249.

[20] Barrero-Gil, A., S. Pindado, and S. Avila, Extracting energy from Vortex-Induced Vibrations: A parametric study. *Applied Mathematical Modelling*, 2012. 36(7): p. 3153–3160.

[21] Jadidi, P. and M. Zeinoddini, Influence of hard marine fouling on energy harvesting from vortex-induced vibrations of a single-cylinder. *Renewable Energy*, 2020. 152: p. 516–528.

[22] Sun, H., C. Ma, E.S. Kim, G. Nowakowski, E. Mauer, and M.M. Bernitsas, Hydrokinetic energy conversion by two rough tandem-cylinders in flow induced motions: Effect of spacing and stiffness. *Renewable Energy*, 2017. 107: p. 61–80.

[23] Khalak, A. and C.H.K. Williamson, Fluid forces and dynamics of a hydro-elastic structure with very low mass and damping. *Journal of Fluids and Structures*, 1997. 11(8): p. 973–982.

[24] Sun, H., E.S. Kim, M.P. Bernitsas, and M.M. Bernitsas, Virtual spring–damping system for flow-induced motion experiments. *Journal of Offshore Mechanics and Arctic Engineering*, 2015. 137(6): p. 061801.

[25] Nakagawa, K., T. Fujino, Y. Arita, Y. Ogata, and K. Masaki, An experimental investigation of aerodynamic instability of circular cylinders at supercritical Reynolds numbers. In *Proc. 9th Jpn, Congr. Appl. Mech.* 1959. Tokyo.

[26] Nakagawa, K., T. Fujino, Y. Arita, and K. Shima, An experimental study of aerodynamic devices for reducing wind-induced oscillatory tendencies of stack. In *Wind Effects on Buildings and Structures*. 1963, National Physics Lab: Teddington.

[27] Mahrenholtz, O. and H. Bardowicks, Aeroelastic problems of masts and chimneys. In *Proceedings 3rd Colloq. on Industrial Aerodynamics*. 1978. Aachen.

[28] Gartshore, I.S., J. Khanna, and S. Laccinle, The effectiveness of vortex spoilers on a circular cylinder in smooth and turbulent flow. In *Proceeding 5th International Conference on Wind Engineering*. 1978. Fort Collins, CO.

[29] Bernitsas, M.M., *Harvesting Energy by Flow Included Motions, in Springer Handbook of Ocean Engineering*, M.R. Dhanak and N.I. Xiros, Editors. 2016, Springer International Publishing: Cham. p. 1163–1244.

[30] Schlichting, H., *Boundary Layer Theory, 7th edn.* 1955, Springer: Berlin, Heidelberg.

[31] Park, H., R.A. Kumar, and M.M. Bernitsas, Enhancement of flow-induced motion of rigid circular cylinder on springs by localized surface roughness at $3 \times 10^4 \leq Re \leq 1.2 \times 10^5$. *Ocean Engineering*, 2013. 72: p. 403–415.

[32] Park, H., R. Ajith Kumar, and M.M. Bernitsas, Suppression of vortex-induced vibrations of rigid circular cylinder on springs by localized surface roughness at $3 \times 10^4 \leq Re \leq 1.2 \times 10^5$. *Ocean Engineering*, 2016. 111: p. 218–233.

[33] Wang, J., L. Geng, L. Ding, H. Zhu, and D. Yurchenko, The state-of-the-art review on energy harvesting from flow-induced vibrations. *Applied Energy*, 2020. 267: p. 114902.

[34] Bernitsas, M.M. and K. Raghavan, Reduction/suppression of VIV of circular cylinders through roughness distribution at $8 \times 10^3 < Re < 1.5 \times 105$. In

Proceedings of ASME 27th International Conference on Offshore Mechanics and Arctic Engineering. 2008. Estoril, Portugal.

[35] Park, H., M.M. Bernitsas, and E.S. Kim, Selective surface roughness to suppress flow induced motion of two circular cylinders at 30000 < Re < 120000. In *Proceedings of ASME 2013 32nd International Conference on Ocean, Offshore and Arctic Energy*. 2013. Nantes, France.

[36] Bernitsas, M.M. and K. Raghavan, VIVACE (vortex-induced vibration aquatic clean energy): A new concept in generation of clean and renewable energy from fluid flow. *Journal of Offshore Mechanics and Arctic Engineering*, 2008. 130(4): p. 041101.

[37] Chang, C.C., R.A. Kumar, and M.M. Bernitsas, VIV and galloping of single circular cylinder with surface roughness at $3.0 \times 10^4 <= \mathrm{Re} <= 1.2 \times 10^5$. *Journal of Ocean Engineering*, 2011. 38: p. 1713–1732.

[38] Bernitsas, M.M., K. Raghavan, *"Converter of Current, Tide, or Wave Energy." U.S. Provisional Patent Application*, United States Patent and Trademark Office Patent 7,493,759 B2. 2009.

[39] Bernitsas, M.M., K. Raghavan, *Enhancement of Vortex-Induced Forces and Motion Through Surface Roughness Control*, U.S.P.a.T. Office, Editor. 2011.

[40] Bernitsas, M.M., K. Raghavan, *Reduction/Suppression of Vortex-Induced Forces and Motion Through Surface Roughness Control*. 2014.

[41] Raghavan, K. and M.M. Bernitsas, Enhancement of high damping VIV through roughness distribution for energy harnessing at $8 \times 10^3 < \mathrm{Re} < 1.5 \times 10^5$. In *Proceedings of ASME 27th International Conference on Offshore Mechanics and Arctic Engineering*. 2008. Estoril, Portugal.

[42] Yáñez, D.J., *VIV Resonant Wind Generators*. 2018. Available from: https://vortexbladeless.com/technology-design/

Chapter 9

Field testing of a 5-kW horizontal-axis wind turbine

David Robert Bradney[1], Samuel Petersen Evans[1], Mariana Salles Pereira da Costa[1] and Philip Douglas Clausen[1]

Small wind turbines have been operating in a range of different forms for well over 300 years. Most of the earlier machines were drag-type machines with some still in use today; for example, the Southern Cross Windmill is still used to pump water across rural parts of Australia. The oil price shock of the 1970s saw 'large' horizontal-axis wind turbines (HAWTs) with an output power of 50–100 kW being developed and grid connected; today, these turbines are now considered moderately sized. On-going research and development have seen a steady increase in the size of these turbines and design power outputs with commiserate improvements in efficiency and reliability leading to a steady reduction in the levelized cost of producing power. Small wind turbines, turbines with a rated power output of less than 50 kW, unfortunately missed out on this early research and development funding due principally to the commercial interest in large machines. This chapter describes the work done to increase the knowledge of the performance of small high-efficiency HAWTs operating in turbulent wind flows. Findings show that the size of the tail fin has a strong influence on the yaw performance of the turbine. Gyroscopic loading from an operating yawing turbine can lead to potentially significant cyclical loading on blades and other components of the turbine. In turbulent flows, the starting performance of the turbine is important to help maximize energy capture. For a range of starting scenarios, the starting time of the turbine can be accurately modelled.

9.1 Introduction

Small HAWTs, defined by the International Electrotechnical Commission Standard IEC61400.2-2013 [1] as machines with power output of less than 50 kW and rotor

[1]School of Engineering, The University of Newcastle, Callaghan, Australia

swept area of less than 200 m^2, have not received the same level of research and development as their larger counterparts due to the significant commercial interest in high-performing large wind turbines. Small high-efficiency HAWTs, like their larger counterparts, have two or three blades mounted upwind of the tower, operate at variable speed and a high tip speed ratio but differ from their larger counterparts by being free yaw machines with the direction of the turbine governed by a tail fin and utilize over-speed protection mechanisms, including blade pitching, centrifugal or electrodynamic braking and furling. The design of small wind turbines cannot, therefore, be considered a scaled version of large turbines. HAWTs for commercial scale generation are sited in high-average low-turbulence wind speed sites whereas small HAWTs are mostly located close to their load and likely to be operating in highly turbulent wind regimes. As such, small wind turbines experience more complex additional forces due to the unsteady nature of their operating environment.

The small wind turbine group at The University of Newcastle, Australia has been investigating the performance of small high-efficiency HAWTs operating in unsteady flow regimes for close to 30 years. The group has significant experience at measuring the performance of an Aerogenesis high-efficiency 5-kW HAWT operating at a low-average wind speed and high-turbulence site and has used these results to validate the accuracy of modelling. In this chapter, the main features of the experimental turbine are described, the important aspects of the instrumentation on the turbine and data acquisition system are detailed and some of the more interesting findings are presented and discussed.

9.2 Description of the 5-kW horizontal-axis wind turbine

The 5-kW HAWT used for the experimental work described in this chapter was constructed and commissioned by Aerogenesis. The turbine has a nameplate generation capacity of 5 kW at a rated wind speed of 10.5 ms^{-1} and is a dual-bladed upwind HAWT. The twisted and tapered blades are nominally 2.5 m long and constructed from a glass-fibre reinforced polymer shell with a foam core. The blade utilizes an SD7062 profile on the working section with the radial variation of blade twist and chord length designed to have a cut-in wind speed of 3.5 ms^{-1} without unduly effecting the overall running performance. The turbine was designed as an IEC class III turbine with an average wind speed of 7.5 ms^{-1} (design speed of 10.5 ms^{-1}). The rotor has a design speed of 320 rpm and a design tip speed ratio of 8. The rotor drove a four-pole three-phase self-excited induction generator through an 8:1 two-stage helical reduction gearbox operating in reverse to maximize the efficiency of the generator. The turbine is controlled passively by a tail fin and has variable speed control to maintain the design tip speed ratio across a range of wind speeds without measuring the wind speed. The turbine is installed on top of an 18-m high-tapered octagonal cross-sectioned steel tower that is hinged halfway up the tower to allow easy lowering of the turbine to ground level for maintenance and instrumentation. Figure 9.1 shows a picture of the turbine on site.

Figure 9.1 The Aerogenesis 5-kW wind turbine at the Callaghan (University of Newcastle) site

9.3 Turbine instrumentation and data acquisition systems

To accurately model the dynamic response of a small operating wind turbine requires the instantaneous wind conditions to be known, and the response of the turbine's various interacting subsystems to be understood. Modelling is made more difficult when the turbine is operating in highly turbulent flow where rapid changes in wind speed and direction can result in unsteady aerodynamic behaviour over the blade's aerodynamic surfaces. To ascertain the veracity of each model requires appropriate experimental data of the performance of all the turbine's subsystems. This section describes the instrumentation used on the turbine and the important features of the data acquisition systems used to acquire response and performance data from the turbine.

Wind speed and direction were in the first instance measured using a Synchrotac 710 series cup anemometer and wind vane with both bolted to a frame attached to the turbine tower 3 m below the wind turbine's hub. Data from both instruments could be acquired at a rate of up to 5 Hz using a DT82E data logging system installed in a close by site hut and controlled by the site computer. This system was later replaced with three 3D ultrasonic WindMaster anemometers, manufactured by Gill Instruments, United Kingdom. The anemometers were mounted on a new frame 3 m below hub height and equispaced circumferentially at 120° around the tower on a radius of 1.2 m from the tower axis. Each anemometer can measure wind velocity in three orthogonal directions up to 30 Hz, with three

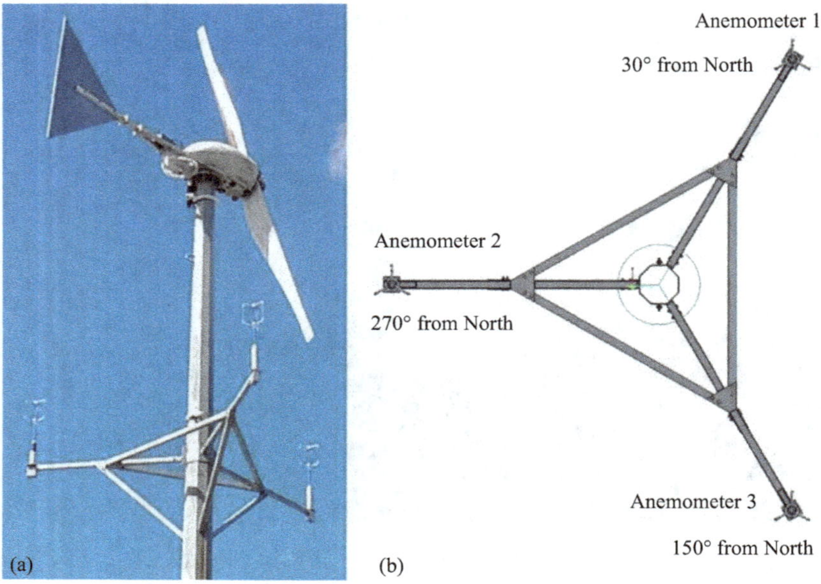

Figure 9.2 (a) Arrangement of ultrasonic anemometers on the wind turbine tower and (b) detailed of the positioning of the anemometers

anemometers allowing measurement redundancy if flow to one of the anemometers is affected by the tower shadow, see Figure 9.2.

To avoid using potentially noisy slip rings to transfer data from strain gauges attached to a blade's surface to a stationary computer for storage and subsequent analysis, the group developed a novel capacitive coupling data acquisition system [2,3]. This system was reasonably reliable but relatively slow and could only acquire data every 45° of blade rotation when the turbine was operating at its design rotational speed of 320 rpm. To measure more comprehensively the blade's dynamic response, signals from strain gauges attached to the blades' surface need to be acquired at around 500 Hz corresponding to every 4° of blade rotation at design operating speed. Seven-channel wireless V-link LXRS nodes capable of acquiring data at around 1,000 Hz were trialled, but at the time of the trials, each unit only had 4 MB of internal memory corresponding to 9 min of data when acquired at 512 Hz. Downloading these data wirelessly took some 20 min. In parallel with using the V-link data acquisition system, the group developed an Arduino Leonardo based system to acquire at 10-Hz turbine direction and turbine yaw rate data from a 9 degree of freedom Inertia Measurement Unit (IMU) from its inbuilt magnetometer and accelerometer, respectively. A Series 2 Xbee was used to control the Arduino system and to transfer acquired data back to a stationary computer. The versatility, cost, speed, and scalability of the Arduino-based data acquisition system led to it becoming the default data acquisition system. Due to the proximity of a high voltage transformer, the signals from the IMU magnetometer were affected by the transformer's magnetic field. Furthermore, a data acquisition rate of 10 Hz was found to be too low to accurately capture the yaw behaviour of the turbine.

Bradney *et al.* [4] described in some detail the important aspects of the data acquisition system used to acquire, in a comprehensive manner, performance data from the 5 kW wind turbine. Since publishing this article, the optical method to determine the yaw position of the turbine (a GoPro® viewing angular positions on the tower) was replaced with a 1:1 ratio 3D-printed gear coupling between the tower and the turbine that rotated a TT Electronics 6127 Non-contacting Hall effect single-turn position sensor. The analogue voltage from this sensor was acquired at 500 Hz using the same Arduino logging system as the tailfin load sensor. This new method of measuring turbine yaw position eliminated the time-consuming method of processing the images from the GoPro. Turbine yaw rate was determined from the yaw position data.

9.4 Site conditions and turbine performance

The wind profile at the wind turbine site at The University Newcastle, Australia is considerably unsteady, making it ideally suited for studying the effect of turbulence on the performance of small HAWTs. Wind speed and direction changes rapidly over the course of several seconds; Figure 9.3 shows this behaviour during a recorded period of 300 s. In this data set, the wind speed varies from 3.6 to 21 ms^{-1}, with changes in speed exceeding 10 ms^{-1} within 2 s. Site wind directions vary by almost 100°. Additional information on the results of the detailed wind data measurement campaigns can be found in [5].

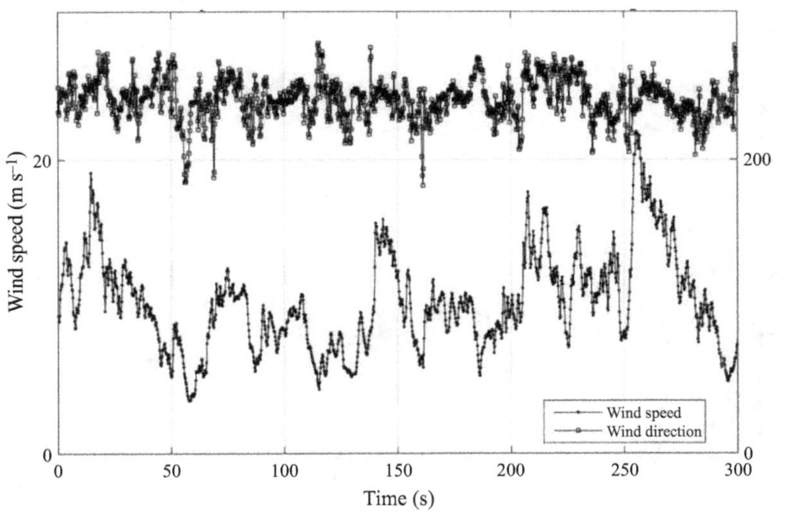

Figure 9.3 A 300-s sample of the Callaghan test site wind conditions, clearly showing its unsteady nature

For all wind speeds at the Callaghan test site, the 10-min turbulence intensity exceeds those stated by [1]; see Figure 9.4. Turbulence intensity, T_u, is defined by

$$T_u = \frac{\sqrt{\overline{u^2}}}{U} = \frac{1}{U}\left[\frac{1}{T_s}\int_0^{T_s} u^2 dt\right]^{1/2}, \tag{9.1}$$

where T_s is the sample time, and U and u are the mean and fluctuating velocities.

Figure 9.5(a) shows a Weibull wind speed distribution at this site, with turbine performance data in Figure 9.5(b)–(d) inclusive indicating substantial under-performance likely to be due to the highly turbulent nature of the site, significant yaw error and possibly some controller issues.

9.5 Tailfin size and turbine's dynamic response

As previously indicated, most small HAWTs use a tail fin to align the turbine with the direction of the incoming wind. With small wind turbines often operating in highly turbulent flow regimes, the tail fin's response needs to be sufficiently damped to prevent the turbine from following all instantaneous changes in wind direction yet sufficiently responsive to minimize the yaw error between the

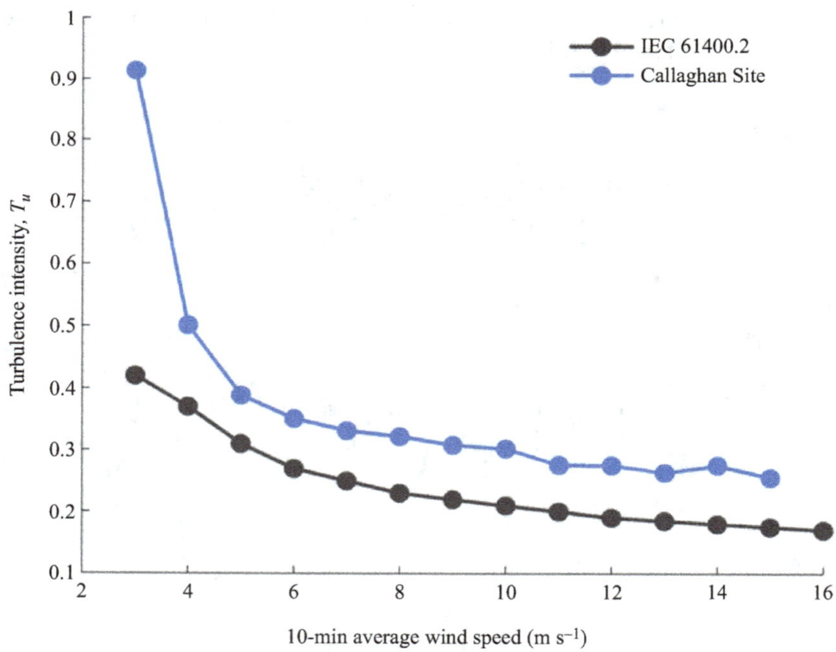

Figure 9.4 Comparison between the Callaghan site T_u and [1]

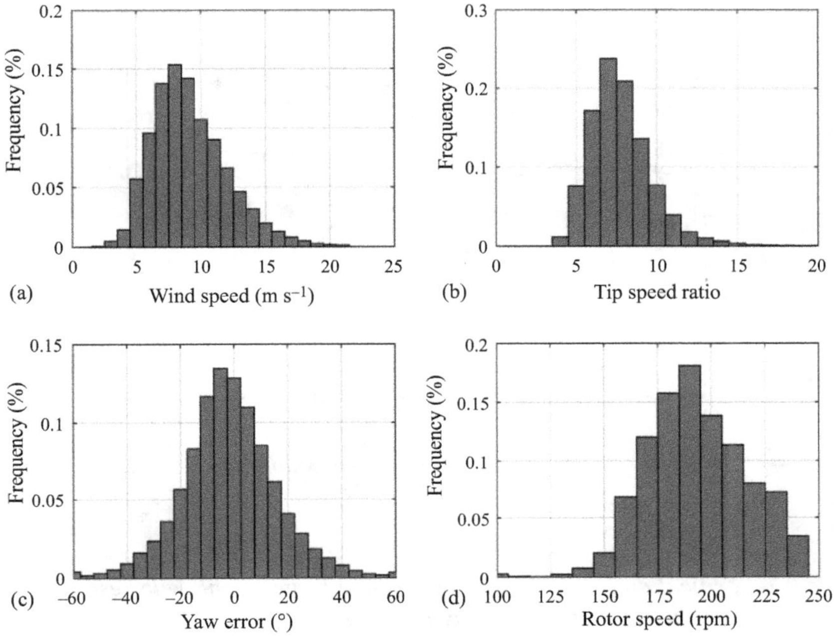

Figure 9.5 Site conditions and turbine performance with (a) showing a Weibull distribution, (b) operational tip speed ratio, (c) yaw error showing a minor bias of the nacelle to tend towards 0 to −5° yaw error and (d) rotor speed operating well below the rated design speed of 320 rpm

direction of the turbine and direction of the wind. The former is important as high yaw rates can lead to high gyroscopic loading on all turbine components – see Section 9.6, and the latter will ensure the yaw error, θ, has been minimized as the power extracted by the turbine is reduced by $\cos^2(\theta)$ [6,7].

The Aerogenesis wind turbine's original tail fin was triangular with a planform area of 0.255 m^2. During field testing, the turbine exhibited poor performance in following wind direction changes with extreme yaw errors of up to 180° regularly occurring. At low wind speeds, the turbine would 'wander' chaotically resulting in poor energy capture. With this tail fin, the wind turbine would often fail to start as it was unable to respond to wind direction changes causing the rotor to lose speed and, therefore, not to start. It is likely this tail fin is undersized for the turbulent wind conditions at the test site. A tail fin of the same triangular shape but twice the plan form area was tested against the original tail fin. Figure 9.6 and Table 9.1 give critical information and dimensions of the tail fins.

A series of experiments were undertaken to examine the effect of the tail fin size on the yaw performance of the wind turbine; Figure 9.7 shows the results of these experiments. More details of the experiments are available in [8].

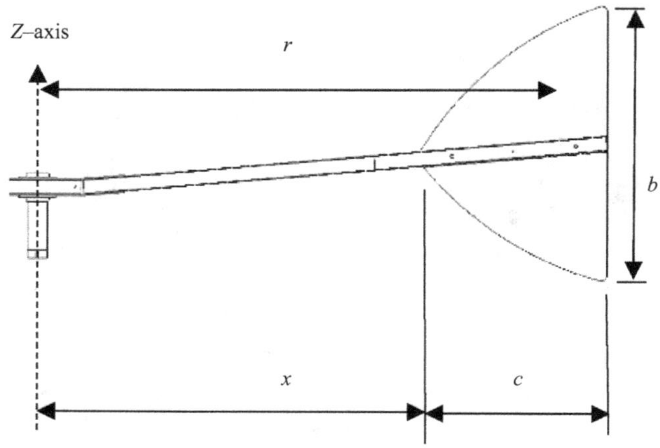

Figure 9.6 Tail fin parameters corresponding to Table 9.1

Table 9.1 Critical dimensions for the small and large tail fin with respect to the parameters indicated in Figure 9.6

Tail fin size	Material	*b* (m)	*c* (m)	*x* (m)	Thickness (mm)
Small	Steel	0.85	0.6	1.32	3
Large	Aluminium	1.54	0.85	1.32	3

The results presented in Figure 9.7 show the larger tail fin led to a mean yaw error of 15° more than the original tail fin for wind speeds less than 3.5 m s^{-1}. This is likely due to increased damping from the larger tail fin area. For winds speeds between 3.5 and 11 ms^{-1}, the larger tail fin has a smaller yaw error converging to a mean absolute error close to 8.5°. The wind turbine will lag changes in wind direction due to the inertia about the yaw axis ($I = 108$ kgm^2 – for the original tail fin), and so it is expected that it will always operate with some amount of mean yaw error. For the original tail fin, the average yaw error does not converge to the range of the larger tail fin until a wind speed of 10 ms^{-1}, meaning that the wind turbine is performing at a suboptimal level below this wind speed. As the wind turbine has a rated wind speed of 10.5 ms^{-1}, this poor convergence highlights the importance of an adequately sized tail fin. The standard deviation of the mean yaw error for the original tail fin at all wind speeds is approximately 50% higher than the larger tail fin, thereby leading to a lower power production.

Turbine yaw rate is also important to consider as high yaw rates can lead higher root bending moments and, hence, a reduction in the blade root bending

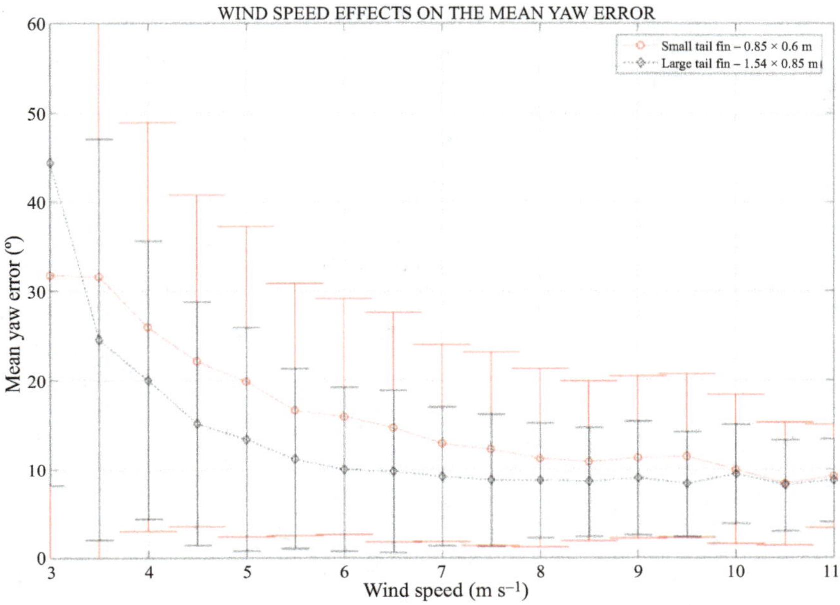

Figure 9.7 Mean absolute tail fin yaw error as a function of wind speed

fatigue life [9], due to gyroscopic moment M_x:

$$M_x = I_z \omega_y \psi_z, \tag{9.2}$$

where ψ_z is the yaw rate in rads^{-1}, ω_y is the rotational speed of the turbine in rads^{-1} and I_z is the total moment of inertia of the turbine about the yaw axis. It is, therefore, essential that any increase in responsiveness due to the size of the tailfin does not lead to higher yaw rates.

The results in Figure 9.8 show both tail fins operate at approximately the same mean yaw rate for all wind speeds up to 11 ms^{-1} with the larger tail fin's maximum yaw rate plateauing at 30°/s at a wind speed of 6.5 ms^{-1}. In contrast, the original smaller tail fin maximum yaw rate decreases with increasing wind speed. In both cases, these are less than the maximum yaw rate of 162°/s specified in equation 27 of [1].

Figure 9.9 shows a scatter plot of yaw rate plotted against yaw error producing a diamond shape with the points formed by the larger tail fin approximately half the size for that of the original tail fin. This effect has been previously observed by Wright [10]. This result indicates the larger tail fin reduces turbine yaw error and the yaw rate when compared to the smaller tail fin. Note that turbine yaw rate is defined relative to an earth-fixed coordinate system and yaw is the difference between the measured wind and turbine directions. In both cases, the highest yaw rates were experienced when the yaw error was close to zero degrees. This is likely

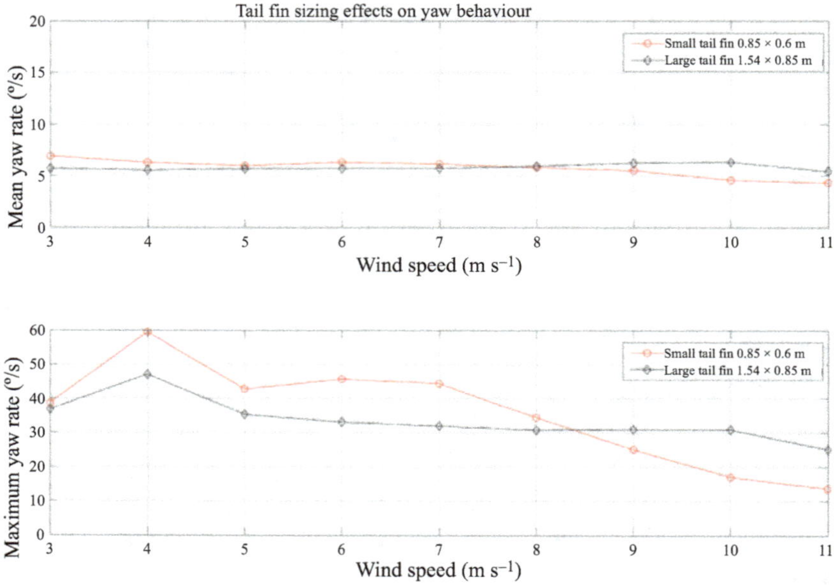

Figure 9.8 Mean and maximum yaw rates for both tail fin sizes

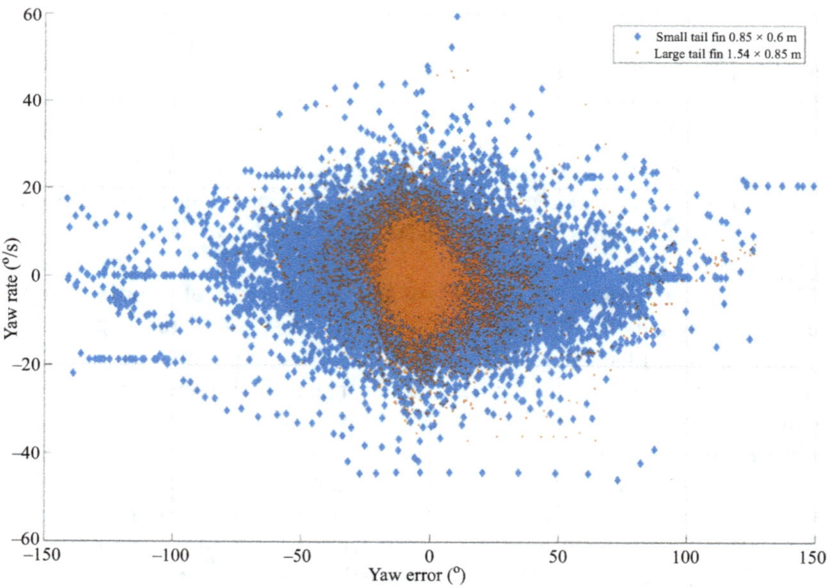

Figure 9.9 Yaw error plotted against yaw rate for both tail fins. Large blue diamond markers are for the small tail fin and the orange dots are for the large tail fin data set

due to a rapid yaw manoeuvre in which a zero-degree offset is temporarily overshot in the form of a decaying mathematical second-order ordinary differential equation.

So, what is the ideal sized tailfin for a small wind turbine? There is little published research on the size of tail fins for small HAWTs. The 'rule-of-thumb' for the area of a tail fin is about 5% of the rotor swept area [11]. As the tail fin provides a moment around the yaw axis, the boom length and rotor configuration are important variables in sizing the tail fin. Calvert [12] proposed small delta-wing tail fins and long boom arms whereas [13] minimized the offset between the yaw axis and rotor plane. Le Gourières [14] proposed (9.3) as an appropriate tail fin area; however, no references were given for this recommendation:

$$A_t = \frac{0.16 A_r E}{r}, \tag{9.3}$$

where A_t is the tail fin area, A_r is the swept rotor area, E is the distance between the rotor plane and the yaw axis, and r is the distance from the yaw axis to the centre of pressure of the tail fin. Based on (9.3), the 5-kW Aerogenesis wind turbine should have a tail fin area of $0.045A_r$, with the original tail fin is $0.017A_r$ and the larger one $0.034A_r$.

9.6 Blade response in unsteady wind flow

The dominant loads acting on the blades of small HAWTs during operation are aerodynamic – due to airflow interaction with the blades; centrifugal – from blade rotation about its main axis, and gyroscopic – from a rotating blade rotating around the yaw axis of the turbine [9,15]. The structural response of blades to both aerodynamic loading and centrifugal loading is well understood; however, the additional effect of gyroscopic loading on the blade structure is not as clearly understood. In addition, a small wind turbine operating in unsteady flow will experience varying yaw rates, causing fluctuating loading on the blades and potentially leading to blade structure fatigue issues.

A high-resolution finite element model of the Aerogenesis 5 kW wind turbine was built and solved for several load cases where the stationary blade was tip-loaded [16]. Predicted deflections and strains at key radial positions along the blade were mostly within 10% of measurements. Da Costa [16] developed a methodology to include the effects of gyroscopic loading in the blade finite element model; Table 9.2 shows details of turbine operating parameters for four different loading scenarios with Figures 9.10–9.13 inclusive showing the measured and predicted strains on gauges attached to the blade surface with signals acquired via the data acquisition system detailed in Section 9.3.

The most interesting observation from Figures 9.10 to 9.13 inclusive is the significantly lower strain on a blade as a response to load cases with positive yaw rate when compared to the load cases with negative yaw rate. The positive yaw rate results in the gyroscopic load acting in the opposite direction to the aerodynamic load, and as such, reducing blade streamwise deflection and, hence, bending strain

Table 9.2 Wind turbine experimental conditions

	Wind turbine experimental conditions				
	Rotor speed (rpm)	**Wind velocity (ms^{-1})**	**Yaw rate ($rads^{-1}$)**	**Yaw error (degrees)**	**Tip speed ratio**
Load case 1	62	5.3	0.16	26	3.11
Load case 2	69	6.1	−0.19	13	3.01
Load case 3	70	5.5	−0.14	19	3.31
Load case 4	66	5.8	0.15	16	2.97

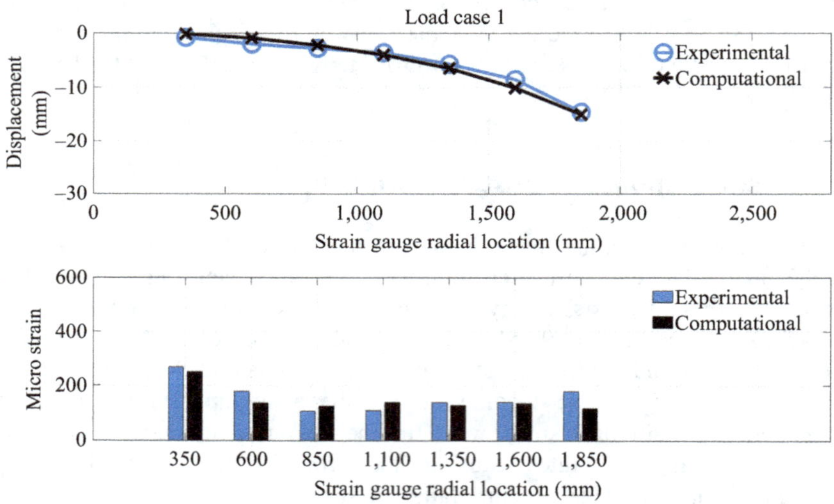

Figure 9.10 Blade response to load case 1, experimental measurements and computational predictions

during operation. The direction of the gyroscopic loading is a function of the azimuth position of the blade, and it will vary as the blade rotates.

The impact of the directional dependence of the gyroscopic load and its interaction with the aerodynamic load can be clearly seen in Figure 9.14. Here the blade is pointing vertically upwards. The figure shows the predicted blade tip deflection and strains at the strain gauge radial locations for the same operational conditions but with yaw rates of +0.5 and −0.5 $rads^{-1}$. Negative yaw led to approximately a 100% increase in blade tip direction compared to the positive yaw case with strains at the gauges increasing by over 100% compared to the positive yaw case.

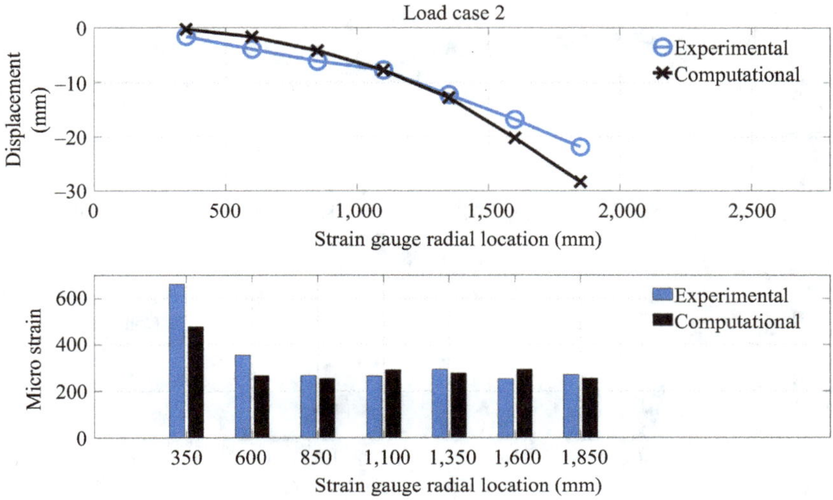

Figure 9.11 Blade response to load case 2, experimental measurements and computational predictions

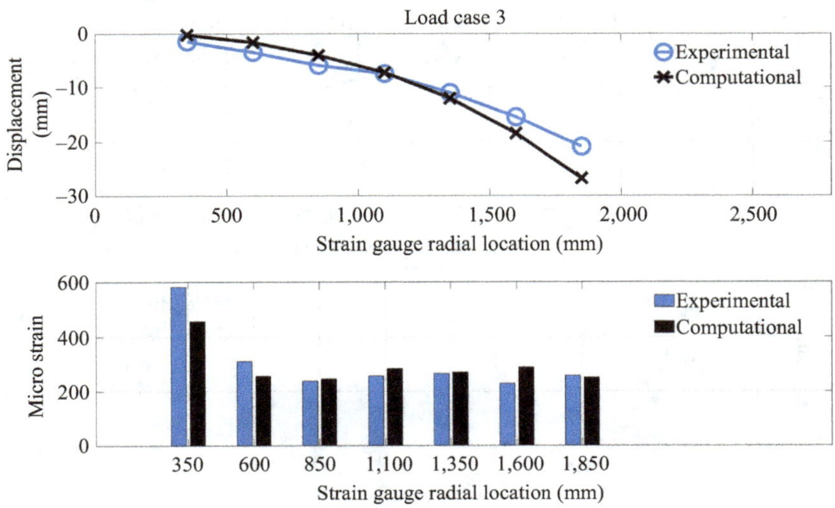

Figure 9.12 Blade response to load case 3, experimental measurements and computational predictions

To further explore the effects of gyroscopic loading and turbine operating conditions on blade straining, the effects of aerodynamic, centrifugal and gyroscopic loadings, capturing the maximum effects on the blade from gyroscopic loading for yaw in both directions were modelled for two operational conditions: (A) – turbine rotor speed of 320 rpm, wind velocity 10.5 ms^{-1},

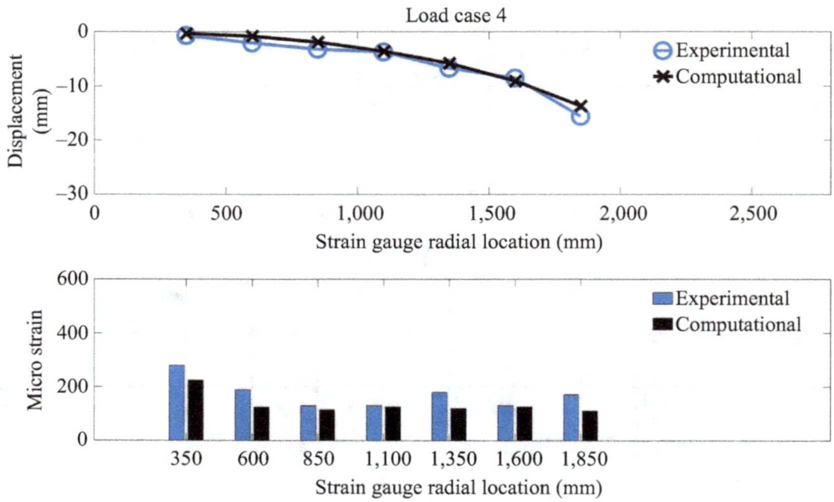

Figure 9.13 Blade response to load case 4, experimental measurements and computational predictions

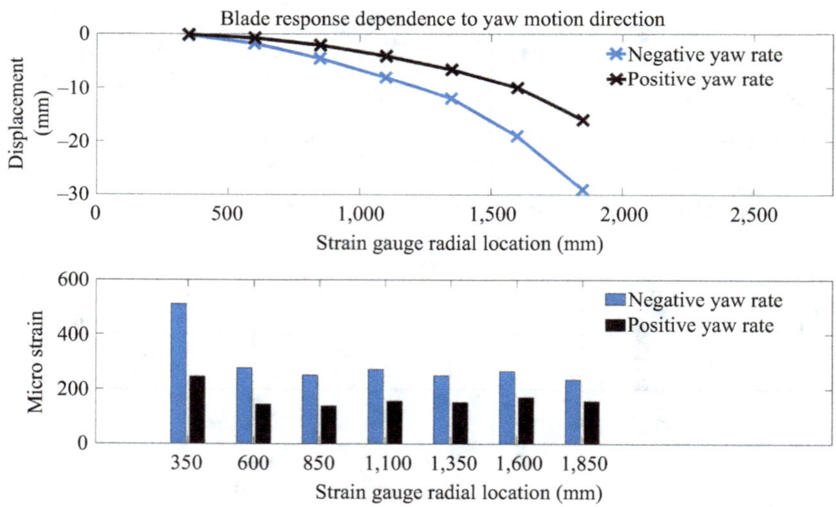

Figure 9.14 Computationally predicted response of blade model to opposite directions of turbine yaw motion

turbine yaw error 0°, blade tip speed ratio of 8 and yaw rate of 1 rads^{-1} and (B) – rotor speed of 214 rpm, wind velocity 8 ms^{-1}, turbine yaw error 0°, blade tip speed ratio of 7 and a yaw rate 1 rads^{-1}. Additional simulations were undertaken for operating cases (A) and (B) with yaw rate set to zero; cases (C) and (D), respectively.

Table 9.3 Maximum blade strains during turbine operation

	Maximum blade strain (μ strain)			
Blade position	**Operational conditions**			
	A	**B**	**C**	**D**
Upwards	300	144	517	335
Downwards	727	471	517	335

The results of these simulations, Table 9.3, indicate the cyclic nature of blade straining due to gyroscopic loading during turbine yaw. The results clearly indicate how gyroscopic loading could lead to an increase in fatigue loading experienced by the blades of small wind turbines operating in turbulent flows.

9.7 Turbine starting performance

The operating performance of any wind turbine is critical to its economic viability. For small wind turbines operating in turbulent flow, the turbine's starting performance is of equal importance. During turbine start-up, the turbine control system does not extract energy from the generator, and for a given rotational inertia of the blade rotor, drive train and generator, the turbine's starting performance is predominantly influenced by the following mechanisms: drive chain resistive torque; aerodynamic performance of the working section of the blade; the effect of high turbulence on the blade's aerodynamic performance and the effect of yaw error on the incoming wind velocity as seen by the blade. This section discusses the impact of each of the previous mechanisms on the turbine's starting performance.

The resistive torque of the gearbox used in the Aerogenesis wind turbine was measured at 1.9 Nm. An earlier higher cost gearbox had a quoted resistive toque of 0.9 Nm. The sensitivity of changes in resistive torque will be explored.

The aerodynamic starting performance of any aerofoil section is a function of the yaw error, yaw rate and the rotational speed of the turbine [9]. For rotational speeds close to zero, and relatively high yaw error, the angle of attack can be more than 90° for some sections along the blade. When the blade operates above the stall angle, the blade performs less like an aerofoil and more like a flat plate; Figures 9.15 and 9.16 show the lift and drag coefficient for a flat plate and an SD 7062 aerofoil section for angle of attacks between −180° and +180° for a number of Reynolds numbers [17]. As the rotational speed of the turbine increases, the angle of attack decreases and the performance of the aerofoil changes from that of a flat plate to an aerofoil section.

During turbine starting, there are two distinct phases, a slow ramp (aerofoil performing like a flat plate) followed by a rapid acceleration (normal aerofoil behaviour). As such, it is necessary to choose the modelling criteria to transition

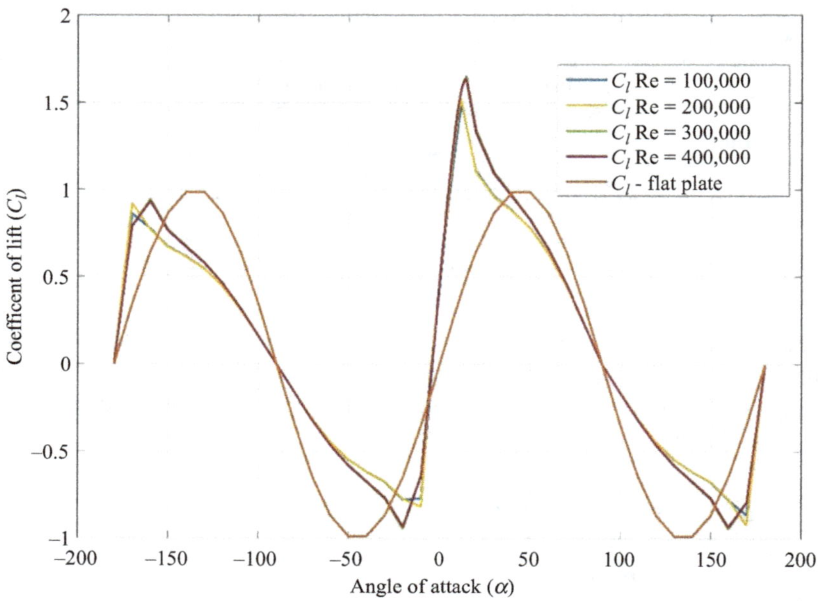

Figure 9.15 Comparison of the C_l curves for a flat plate and the SD 7062 aerofoil

from one phase to another. Previous studies have used the rotor speed to transition from one phase to the other whereas here each blade element is considered separately to decide when transition occurs between the two phases.

The primary effect of a turbine operating with a yaw error is a reduction of inlet velocity by a factor of $\cos\theta$, where θ is the yaw error. The effect of turbulence on the performance of a wind turbine is poorly understood. Following the work of [18,19], the C_l and C_d characteristics for the SD 7062 aerofoil have been modified to account for the effects of turbulence on the aerofoil characteristics. Figure 9.17 shows the original and estimated lift and drag characteristics for the SD 7062 profile for angle of attack between $-10°$ and $30°$ for a turbulence level of 30%.

During all starting simulations, the starting time is defined as the time taken for the rotor to reach the referred synchronous speed of the generator (175 rpm). This speed is indicated by the black horizontal dashed line in Figures 9.18–9.21 inclusive. All simulations used a combined rotational inertia of 14.01 kgm^2, and the wind speed was higher than the cut-in wind speed for all cases considered here.

9.7.1 Simulation 1: rapid wind gust

During starting, wind gusts may occur. Figure 9.18 shows a typical wind gust, where the wind speed changes from 4.5 to 8.5 ms^{-1} between $t = 21.42$ and 24.29 s, with a 3-ms^{-1} increase occurring in 1 s.

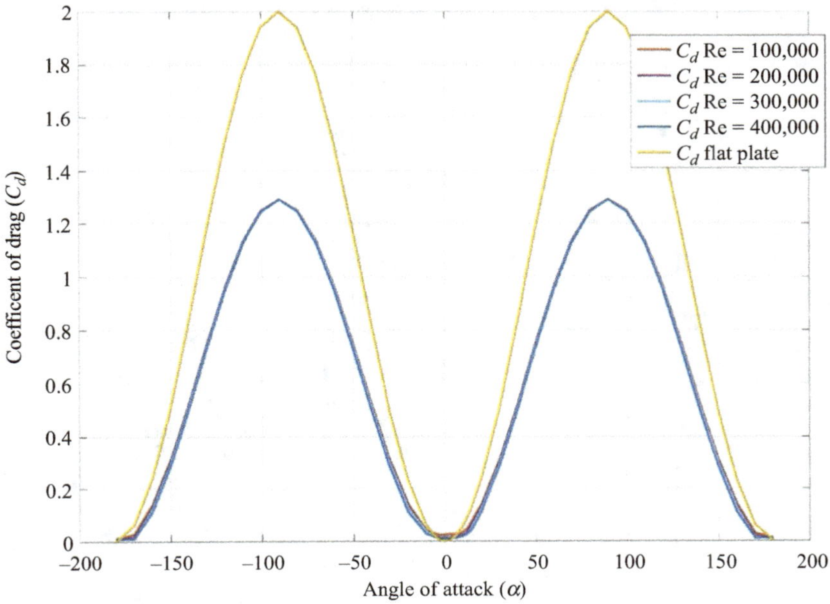

Figure 9.16 Comparison of the C_d coefficient curves for the SD 7062 aerofoil to a flat plate

The recorded starting time was 23.9 s. The simulated starting time using a resistive torque of 0.9 Nm, unmodified aerodynamic coefficients and no yaw effects, predicted 16.2 s (purple line). The inclusion of the higher resistive torque of 1.9 Nm increased the predicted starting time to 18.24 s (red line), an increase of 2 s, while the effects of yaw misalignment further delayed starting to 20.2 s (blue line).

The best performing simulation was the turbulence intensity adjusted C_l and C_d start. This sequence included yaw effects and 1.9-N m resistive torque, resulting in a starting time of 23.6 s, just 0.3 s slower than the wind turbine's measured starting time. The prediction of the phase 2 (aerodynamic) region, where the rapid acceleration occurred, is particularly accurate when using the modified C_l and C_d, with very similar acceleration rates. There is a slight under-prediction of the transition point at $t = 15$ s from phase 1 (high angle flat plate) to phase 2 (aerodynamic coefficient); however, its effect is very minimal.

9.7.2 Sequence 2: rapid wind deceleration

The second starting sequence, Figure 9.19, has a decrease in wind speed from 8.7 to 4.1 ms^{-1} between $t = 12.5$ and 16 s.

The recorded starting time was 22.5 s, and again it was found that resistive torques of 0.9 and 1.9 Nm underestimated the time taken to reach the synchronous

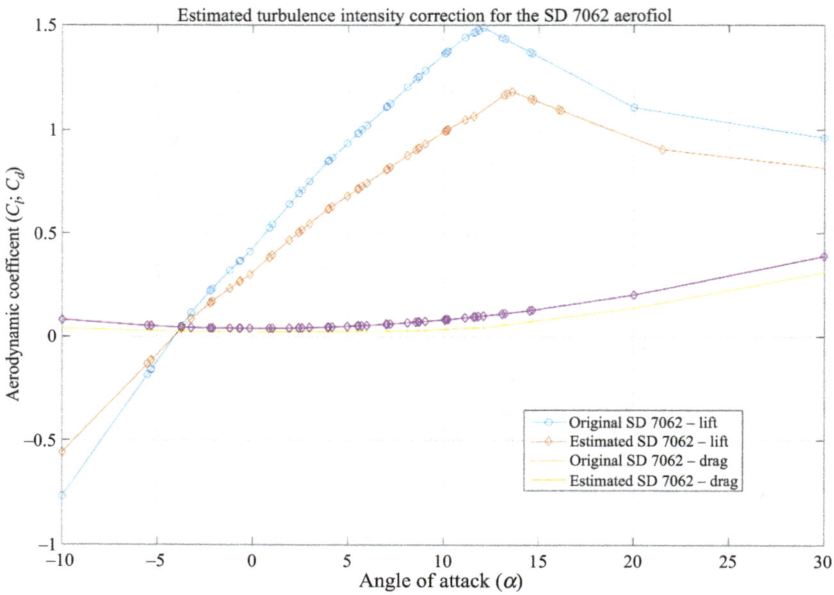

Figure 9.17 The SD 7062 C_l and C_d curves over the operating range of the wind turbine and the modified C_l and C_d curves to account for the effect of 30% turbulence level

speed, with estimates of 17 and 19 s, respectively (purple and red lines). Interestingly, this difference was again 2 s. Incorporation of the yaw error had little effect, and in this case, delayed the starting estimate to 19.45 s (blue line). However, as the yaw error is only 20° during the phase 1 section, this is not a surprising result. Using the modified aerodynamic data again showed excellent agreement with the measurements, estimating 22.63 s to reach the starting speed. There was a slight overestimate of the transition from phase 1 to phase 2 by approximately 1 s, but this did not adversely affect the results.

9.7.3 Sequence 3: ramping wind speed

The third sequence, Figure 9.20, shows a ramping wind series in which the wind progressively increases from 5 to 12.5 ms^{-1}. Yaw error remained within $\alpha \pm 30°$ range.

All four starting variations performed reasonably well in this case, presumably due to the favourable starting conditions. The recorded starting time was 25.6 s, with the best estimate again obtained using the modified aerodynamic coefficients, with a starting time of 25.6 s (green line). The yaw effects, the 1.9 Nm resistive torque and 0.9 Nm resistive torque methods simulated starting times of 24.9 (purple), 23.6 (red) and 21.8 s (blue), respectively. The change from 0.9 to 1.9 Nm again resulted in a 2-s time difference.

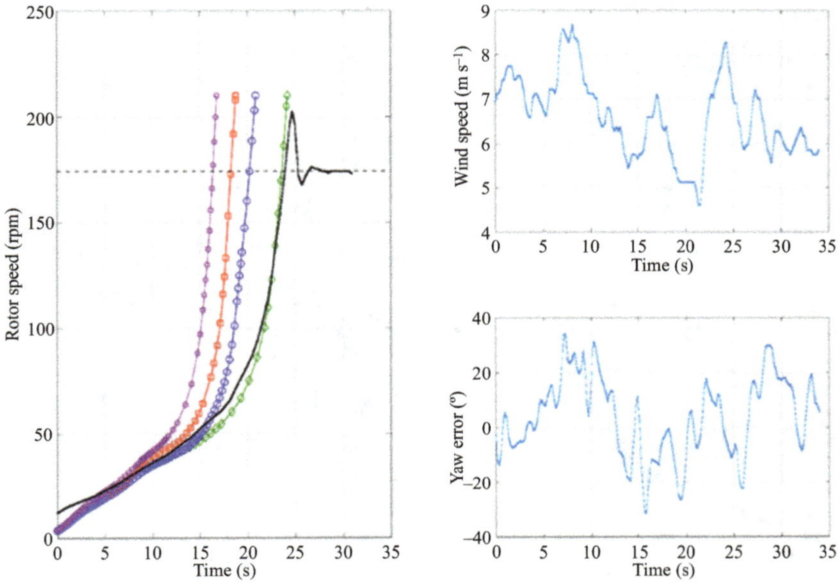

Figure 9.18 Starting sequence showing the effect of a rapid wind speed increase. Wind speed and yaw error for the time period is also shown

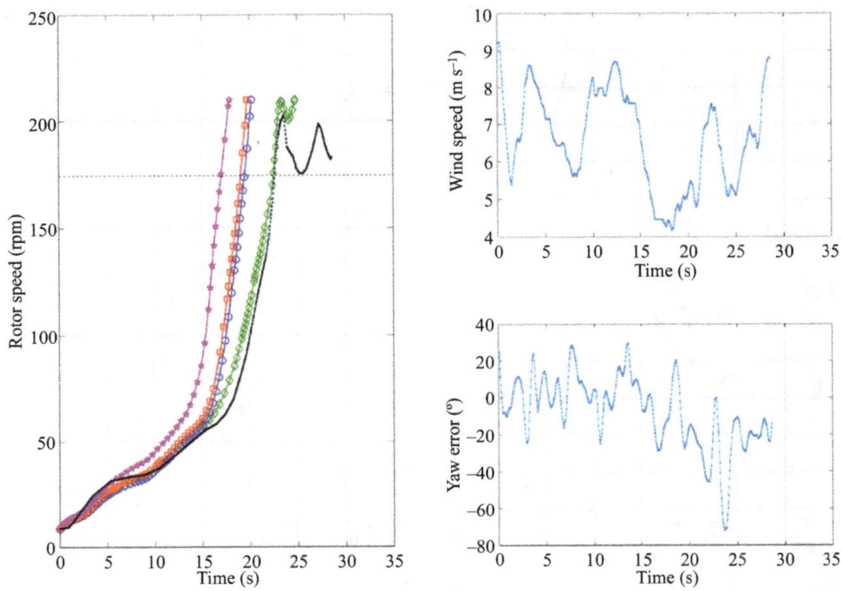

Figure 9.19 Wind turbine starting during a reduction in wind speed

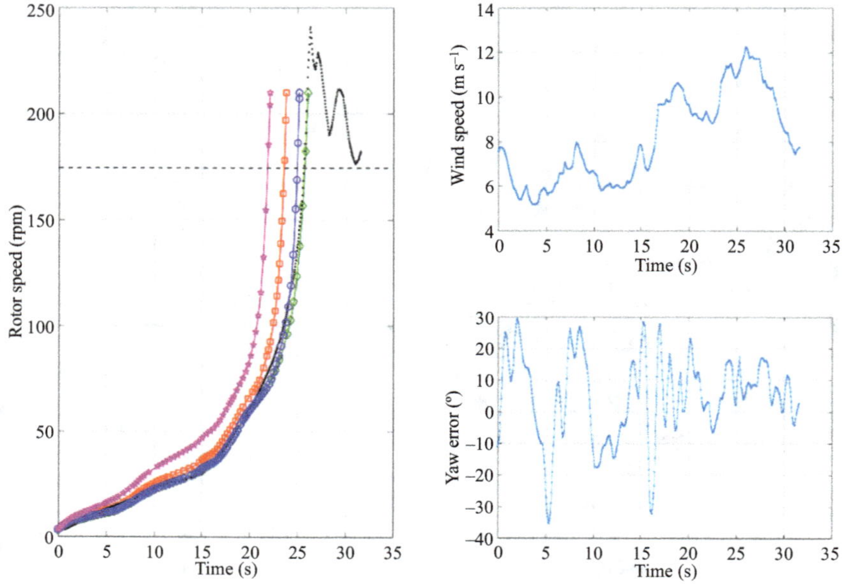

Figure 9.20 Starting sequence showing the impact of a 'ramping' wind speed

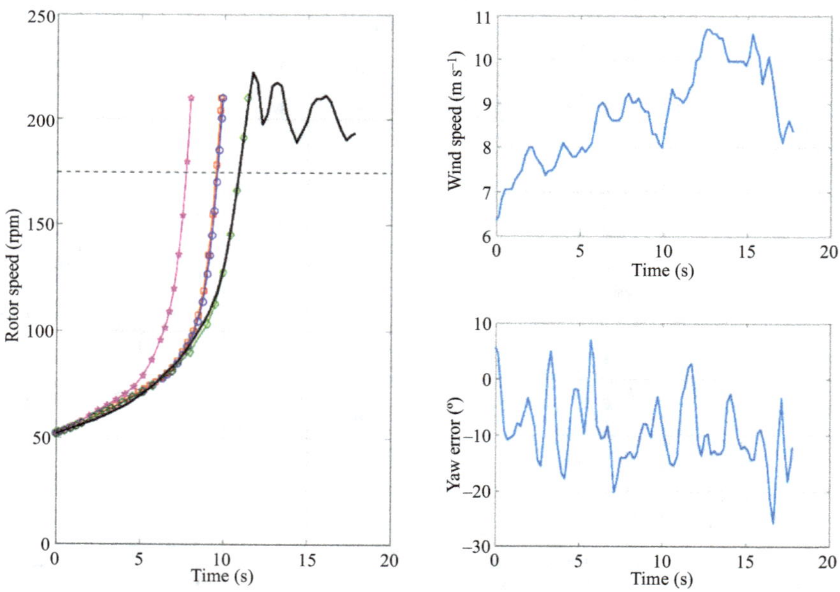

Figure 9.21 'High'-speed start, where the wind turbine begins with a rotational speed of 50 rpm

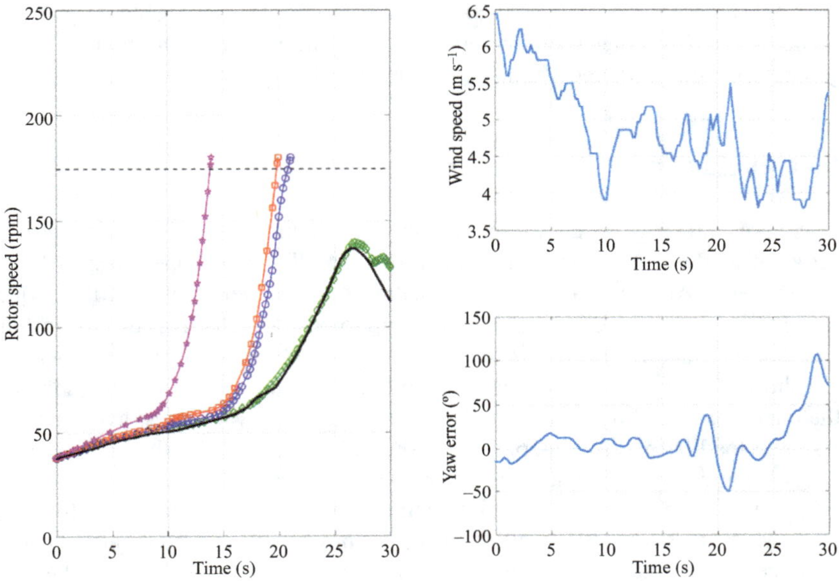

Figure 9.22 Failed starting sequence due to high-yaw misalignment

9.7.4 Sequence 4: high-speed start

As previously discussed, starting may be divided into two sections: the first when the blades are almost stationary and, therefore, have minimal rotational inertia, and the second when the blades are rotating before the starting event. Figure 9.21 shows measurements and predictions from the latter where the wind turbine blades were rotating at just over 50 rpm.

The recorded start time for this sequence was 10.89 s, significantly quicker than the 'slow' starts, owing to the prior rotational speed. A start sequence using the modified aerodynamic data shows excellent agreement with the measured starting time and follows the measured curve almost perfectly. In this sequence, the yaw error is minimal, and so the 1.9 Nm resistive torque and yaw misalignment cases show no significant difference, both starting in 9.6 s. In a similar trend to the previous starting cases, the 0.9 Nm resistive torque case was 2 s faster than the 1.9-N m case, with a starting time of 7.7 s.

9.7.5 Sequence 5: failed start

Not all wind turbine starts are successful, and high yaw misalignment events may result in a failed start. This sequence, Figure 9.22, presents a failed start in which the wind turbine yawed out of the wind with a maximum yaw error greater than 100°, only reaching a rotational speed of 137 rpm.

As can be seen, only the predictions using modified C_l and C_d data predicted the failed start. The other three methods were not impacted by the high-yaw misalignment as their predicted starting times were before the high-yaw event.

9.8 Conclusions

This chapter presents some of the interesting findings from detailed research undertaken to determine the performance of a small HAWT operating in highly turbulent flows. Arduino-based data acquisition systems were developed to acquire data from the operating turbine and were found to be cost-effective, robust, fast and can be highly customized to suit the specific application. Experimental measurements show that the size of a tail fin used on a small turbine is critical to responding to the change in wind directions whilst minimizing high yaw rates, avoiding high gyroscopic loading on all components on the turbine, especially the blades.

Gyroscopic loading is the result of an operating turbine rotating around its yaw axis, and a common phenomenon for a small wind turbine operating in turbulent wind flows. This loading cyclically adds and subtracts to the aerodynamic loading experienced by the blade of an operating wind turbine, thereby subjecting the blade to additional cyclical loading. This cyclical loading may lead to fatigue of the blade material.

The starting performance of a small wind turbine in highly turbulent flow is critical in its ability to capture energy. For a given rotor and drive chain rotational inertia, the starting performance is influenced by the drive train resistive torque; aerodynamic performance of the working section of the blade; the effect of high turbulence on the blade's aerodynamic performance; and the effect of yaw error on the incoming wind velocity as seen by the blade. For a range of starting scenarios, it was found when all influences were considered, the starting time of the turbine could be accurately modelled.

References

[1] International Electrotechnical Commission Standard IEC61400.2-2013, Wind turbines part 2. Design requirements for small wind turbines, SAI Global.

[2] Bechly, M. E. and Clausen, P. (2001). The dynamic response of a composite blade from a 5 kW wind turbine. Part 1: measured blade response, *Wind Engineering*, 25, 133–148.

[3] Wilson, S. and Clausen, P. (2007). Aspects of the dynamic response of a small wind turbine blade in highly turbulent flow. Part 1: measured blade response, *Wind Engineering*, 1, 1–16.

[4] Bradney, D., Evans, S., Chu, M., and Clausen, P. (2020). A low-cost, high-speed, multi-channel Arduino-based data acquisition system for wind turbine systems, *Wind Engineering*, 44, 509–518.

[5] Chu, M. and Clausen, P. (2014). The dynamic response of a small horizontal-axis wind turbine operating highly turbulent flow, In *Proc of 52nd annual conference: Australian Solar Energy Society*, Melbourne, 121–130.

[6] Maeda, T., Kamada, Y., Suzuki, J., and Fujioka, H. (2008). Rotor blade sectional performance under yawed inflow conditions, *Journal of Solar Energy Engineering*, 130(3), 031018.

[7] Pedersen, T. F. (2004). On wind turbine power performance measurements at inclined airflow, *Wind Energy*, 7(3), 163–176.

[8] Bradney, D., Evans, S., and Clausen, P. (2019). The effects of tail fin size on the yaw performance of small wind turbines operating in unsteady flow, In Wind Energy Exploitation in Urban Environment: TUrbWind 2018 Colloquium: Battisti L., Springer Nature, Cham, Switzerland, 55–70.

[9] Wood, D. (2011). Small wind turbines: analysis, design, and application. In *Green Energy and Technology*. Springer-Verlag, London.

[10] Wright, A. (2005). *Aspects of the Aerodynamics and Operation of a Small Horizontal Axis Wind Turbine*. PhD thesis, The University of Newcastle, Australia.

[11] WindyNation, *Sizing your wind turbine tail fin*. Available from: https://windynation.com/jzv/inf/wind-turbine-tail-fin-sizing-your-wind-turbine-tail [Accessed: 5-November-2016].

[12] Calvert, N. G. (1979). *Wind Power Principles – Their Application on the Small Scale*, United States.

[13] Kentfield, J. A. (1996). *Fundamentals/Wind-Driven Water*. Gordon and Breach Science Publishers, Amsterdam.

[14] Le Gourières, D. (1982). *Wind Power Plants: Theory and Design*, Pergamon Press, London.

[15] Wilson, S. V. R., Clausen, P. D., and Wood, D. H. (2008). Gyroscopic moments on SWT blades at high yaw rates, *Australian Journal of Mechanical Engineering*, 5(1), 1–8.

[16] Da Costa, M. (2020). Structural Optimisation and Response of Small Wind Turbine Blade, PhD thesis, The University of Newcastle, Australia.

[17] Selig, M. S. (1996). *UIUC Airfoil Data Site*. Department of Aeronautical and Astronautical Engineering, University of Illinois at Urbana-Champaign.

[18] Lyon, C. A., Broeren, A. P., Giguère, P., Gopalarathnam, A., and Selig, M. S. (1997). *Summary of Low-Speed Aerofoil Data* – Volume 3. Department of Aeronautical and Astronautical Engineering, University of Illinois at Urbana-Champaign.

[19] Devinant, P., Laverne, T., and Hureau, J. (2002). Experimental study of wind turbine airfoil aerodynamics in high turbulence, *Journal of Wind Engineering and Industrial Aerodynamics*, 90(6), 689–707.

Chapter 10
Aeroelastic modelling of a 5-kW horizontal axis wind turbine

Samuel Petersen Evans[1] *and Philip Douglas Clausen*[1]

Aeroelastic modelling is regularly used in the design and analysis of large wind turbines but has seen comparatively little use in the development of small wind turbines. Fewer studies still have considered experimental validation of model performance with field measurements. This study details the development of an aeroelastic model of a 5-kW horizontal-axis Aerogenesis turbine within the freely available software FAST (fatigue, aerodynamics, structures, and turbulence). The model includes rotor aerodynamics, blade and tower structural freedom, free-yaw tail fin dynamics, and a self-excited induction generator (SEIG) incorporating maximum power point tracking (MPPT) variable speed control. This model was compared to measured operating data to assess aerodynamic and structural performance. Blade loadings and deflections were simulated within 8% and 10%, respectively, at design conditions in the azimuthal domain. Simulated tail fin motions and yaw behaviour were bounded by measured field data. The simulated control system model was found to operate within the bounds of measured field data.

Areas for future model improvement are discussed in this chapter. The resulting aeroelastic model and methodologies presented here are expected to further the use of aeroelastic modelling in the small wind turbine field. Particular applications may include the verification of performance and design loads specified in IEC 61400.2-2013, the international standard for small wind turbines.

10.1 Introduction

10.1.1 Aeroelastic modelling

Aeroelastic modelling is a computational method used for predicting wind turbine performance and determining structural loads for component design. Aeroelasticity, in its simplest form, can be defined as a coupling between fluid forces and the resulting elastic and inertial response of a structure when acted upon by these forces.

[1]The University of Newcastle, Callaghan, Australia

For example, a wind turbine blade, when exposed to the wind, will naturally produce aerodynamic torque and thrust due to aerofoil lift and drag properties. These forces will act on the blade structure and will exhibit a structural response in the form of deflections and induced stresses. This simulation can then be executed iteratively, whereby any blade deflections, vibrations, or inertial effects will influence the local wind velocity, which in turn affect the aerodynamic loading.

Typically, this analysis technique is used to determine wind turbine rotor performance for a given wind velocity and input blade aerodynamic profile. Output parameters usually include the thrust and torque developed by the rotor, the aerodynamic power produced by the rotor, optimal rotor speed and tip speed ratio, and blade structural performance such as maximum tip deflection and developed stresses. Both steady-state and transient simulations can be undertaken to assess turbine performance across varying wind speeds and operational parameters. For the interested reader, a number of good resources are available which provide further information of this technique [1–5].

Aeroelastic modelling has seen widespread use in the large wind turbine field. The most notable example is the National Renewable Energy Laboratory (NREL) 5-MW wind turbine which is detailed in [6]. While this is purely a hypothetical turbine, it has been the subject of many studies of a wide range of topics from blade structural optimisation [7], offshore development and model validation [8], fatigue life assessments [9], and the integration of lidar with control systems to mitigate blade loads [10]. This has become a popular research topic likely most due to the fact that no physical manufacturing or testing is required, and that simulations can be executed simply using a desktop computer (i.e. there is no need for intensive computational resources when compared to detailed CFD simulations). Aeroelastic modelling is also used in by large wind turbine manufacturers for the design of wind turbine components (of which Bladed* is a noteworthy example). Aeroelastic software is an attractive design tool due to its low cost and resource requirements, whereby many designs or turbine configurations can be assessed in a relatively quick manner without the need for expensive and time-consuming prototyping and full-scale testing. Furthermore, codes like FAST are open source and approved for use in large turbine certification.

There are several potential issues with the popularity of aeroelastic modelling as computational tool. Like all other computational methods such as finite element (FE) analysis, computational fluid dynamics, or discrete element modelling, there is the risk of the user running 'black-box' simulations with very little understanding of underlying algorithms and theory resulting in incorrect implementation of simulations and erroneous interpretations of results. Another potential issue is the disproportionately low number of studies that focus on experimental validation of aeroelastic results. While this is understandable due to the complex nature of wind turbine field testing, large-scale wind tunnel testing, and difficulties in obtaining reliable experimental results, it is prudent that any potential users of this

*https:dnvgl.com/services/wind-turbine-design-software-bladed-3775

computational technique attempt to interpret simulations in line with existing experimental data where possible.

10.1.2 Modelling of small wind turbines

Comparatively few studies have attempted to both model and experimentally validate the aeroelastic behaviour of a small wind turbine. The authors are aware of five small wind turbine models of which published studies are available in the open literature. Operating parameters of these turbines are presented in Table 10.1. All models satisfy the definition of a small wind turbine within IEC 61400.2-2013, as they have a rotor-swept area less than 200 m^2. They also cover a range of operating speeds, rotor configurations, and passive yaw control methods that are representative of the many horizontal-axis small wind turbine designs available on the market. We note that IEC 61400.2-2013 does not apply to vertical-axis turbines.

Of note is work undertaken by the NREL, who have published studies of a 10-kW three-bladed furling turbine, modelled within FAST software and validated using data obtained from experimental measurements [11–14]. This site was classed as open terrain, with the turbine having several key operational differences to the 5-kW Aerogenesis wind turbine at the University of Newcastle site.

The SkyStream wind turbine has been recently used for the estimation of tower loads by the NREL [15]. The unsteady aerodynamics experiment (UAE) turbine and associated measurement campaigns are detailed in [16]. Details of the AOC-15/50 turbine can be found in [17].

Despite this research effort, techniques and methodologies specific to aeroelastic modelling of small wind turbines are seen to lag behind that of large wind turbines, especially with regards to the acquisition of operational data and validation with design loads specified in relevant standards [18].

10.1.3 Study aims

Few aeroelastic models of small wind turbines have been detailed in the literature, with even fewer subject to experimental validation. This is likely due to the proprietary nature of small wind turbine technology and the significant time investment and expert knowledge required to design and model a wind turbine system.

Table 10.1 Small wind turbine aeroelastic models detailed in the open literature

Model	Rated power (kW)	Rated speed (rpm)	No. of blades	Configuration	Tail fin
SkyStream	2.4	280	3	Downwind	No
Aerogenesis	5	320	2	Upwind	Yes
NREL SWRT	10	340	3	Up/Downwind	Yes
UAE VI	20	72	2	Up/Downwind	Yes
AOC-15/50	50	60	3	Upwind	No

To bridge this knowledge gap, the authors present a detailed methodology of the creation and validation of a 5-kW horizontal axis wind turbine. This turbine is the Aerogenesis wind turbine presented in Chapter 9. The intention is that this could be reproducible by other interested parties in small wind turbine analysis.

Developing and validating an aeroelastic model of a small wind turbine will assist in predicting performance, verifying loads via IEC 61400.2-2013, and furthering the capacity for improvements and innovation in the small wind turbine technology.

10.2 Methodology

10.2.1 FAST model overview

The computational model was constructed within the open-source software package FAST developed by NREL. This model incorporates blade degrees of freedom, tower structural behaviour, drive shaft freedom, passive yaw behaviour via a tail fin, and variable speed control for the SEIG.

To the best of the authors' knowledge, a model of this complexity has not been reported within the literature.

FAST allows for the following aspects of wind turbine aerodynamics, structural behaviour, and operational dynamics to be taken into account when modelling the 5-kW Aerogenesis turbine, including the following:

- Rotor aerodynamics (via a generalised dynamic wake (GDW) or blade element momentum (BEM) algorithm), including the following unsteady effects:
  - Time-varying inlet velocity (i.e. wind turbulence).
  - Wind shear due to roughness of the surrounding terrain.
  - Skew wake corrections due to yaw error.
  - Beddoes–Leishman dynamic stall correction to account for rapidly changing angle of attack and inlet wind conditions.
- Structural behaviour of the following elements via a Euler–Bernoulli/Rayleigh–Ritz beam method:
  - Blades (flapwise and edgewise degrees of freedom).
  - Tower (fore-aft and side-side degrees of freedom).
- Gearbox and drive train effects (drive train is considered rigid for modelling purposes).
- Yaw behaviour resulting from delta-wing tail fin, and nacelle rotational inertial effects.
- Generator operation and variable speed control via coupling with Simulink®.

It is important to note that all computational models have various simplifications and limitations which can potentially influence the results. FAST is not immune to this, with several potential shortcomings highlighted:

- No allowance for structural torsional degrees of freedom (unless FAST is implemented with the ADAMS preprocessor – considered beyond the scope of this project).

- No allowance for coupled modes in blade or tower bending, which could lead to high-order effects such as aerodynamic flutter or natural frequency excitation.
- No default setting for an SEIG and resulting variable speed control.
- Reduced accuracy with 'large' out-of-plane structural deflections (this is due to the use of Euler–Bernoulli beam theory, whereby small deflections are assumed, and, therefore, no fluid–structure interaction).
- Reduced aerodynamic accuracy with highly skewed inlet flow and high yaw misalignments (i.e. >20°).

Some of these issues can be mitigated, for example, by linking FAST with a Simulink model of the SEIG. One issue worth noting is FAST's performance at high skew angles, as site observations had shown that periods of operation can occur at high yaw errors (>20°), particularly in low wind conditions or during start-up when minimal yaw stability produced by the rotor is present. Results obtained during these periods of operation must be carefully considered during post-processing to ensure validity.

The 5-kW Aerogenesis operating parameters have been described in Chapter 9 and will not be repeated here in detail for brevity. The following freedom cases (ten in total) are allowed within FAST and will be used for the Aerogenesis turbine (Figure 10.1):

- First and second blade flapwise modal degree of freedoms, whereby the blades are considered rigidly fixed to the nacelle hub as per a cantilevered constraint (note that the edgewise mode will not be used as explained in the following section).
- First and second tower fore–aft (FA) and side–side (SS) freedom modes. Note that the tower is considered fixed to the ground as a cantilevered constraint. No foundational response will be considered.
- Rotor/tail-furl degree of freedom to facilitate tail fin actions.
- Yaw bearing rotational degree of freedom (to allow for the tail fin response).
- Generator azimuthal degree of freedom (i.e. variable speed control via Simulink).
- FAST allows for coupling between Simulink to allow for effects such as an active yaw drive or blade pitch control. While these Simulink models are not needed for stall regulated, passively controlled yaw machines, the control is necessary for production of generator power and shaft speed control. All other turbine components are considered to be rigid within FAST (i.e. drive train, gear box assembly, and nacelle housing).

10.2.2 Aerodynamic parameters

The FAST aerodynamic engine, AeroDyn, has a user-choice implementation of either the GDW model or BEM theory. Unsteady (i.e. time varying) aerodynamic effects that are accounted for include blade azimuthal variations in wind speed, fluctuations in inlet wind conditions, and yaw misalignment. Dynamic stall effects

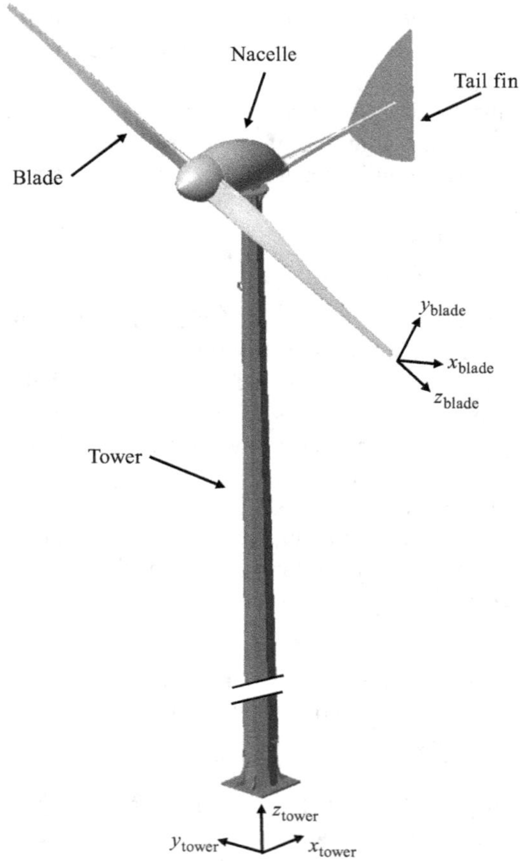

Figure 10.1 Aerogenesis 5-kW wind turbine schematic

are incorporated through use of the Beddoes–Leishman dynamic stall model [19]. Inputs for this subroutine are the 2D quasi-steady aerofoil properties of both the blade and tail fin (i.e. C_L and C_D curves) and dynamic stall parameters given by the normal force coefficient and the angle of attack at zero lift. Aerodynamic properties also input include drag properties of the tower section for tower dam/shadow calculations, tail fin lift and drag curves, and inlet wind conditions.

The aerodynamic profile of the blade consists of the SD7062 aerofoil section along the entire radial length. This aerofoil was specifically designed for low Reynolds number applications such as small wind turbine operation [20]. This aerofoil is designed to have a comparatively high thickness at 14% for improved blade structural considerations such as large bending moments that occur towards the blade root. The SD7062 aerofoil profile was chosen by Aerogenesis for the entirety of the blade due to the aforementioned properties and to simplify the manufacture of moulds for the blade. The coefficient of lift (C_L) and drag (C_D) with respect to angle of attack (a) are taken as inputs for the aerodynamic model within AeroDyn. FAST allows for the use of multiple Reynolds (*Re*) numbers and will perform a lookup function to determine the nearest Reynolds number (to avoid potential errors with a direct linear interpolation/extrapolation).

The SD7062 aerofoil lift and drag polars were obtained from experimental wind tunnel measurements undertaken in [20], and were input for $Re = 100{,}000$, 200,000, 300,000, and 400,000 for the linear range of operation to the onset of stall (i.e. approximately $a = -5°$ to 15°). As only a limited range of experimental data exists for this region of operation, corrections were to be made for a full 180° angle of attack. This correction is essential to make due to the fact that the linear range is easily exceeded either during rotor start-up, yaw misalignments, or highly turbulent inlet wind conditions. Anecdotal site evidence suggests that these events can occur regularly during operation and must, therefore, be accounted for. To extrapolate the existing aerofoil values, AirfoilPrep4 was used. This function uses the Viterna correction method, whereby further details can be found in [21]. This process was repeated for lift and drag data at all four Reynolds numbers, with the results for $Re = 400{,}000$ demonstrated in Figure 10.2.

10.2.3 Tail fin motions

In order to capture the yaw behaviour of the small wind turbine, the inclusion of a tail fin is necessary within FAST. This simple tail fin algorithm computes the lift

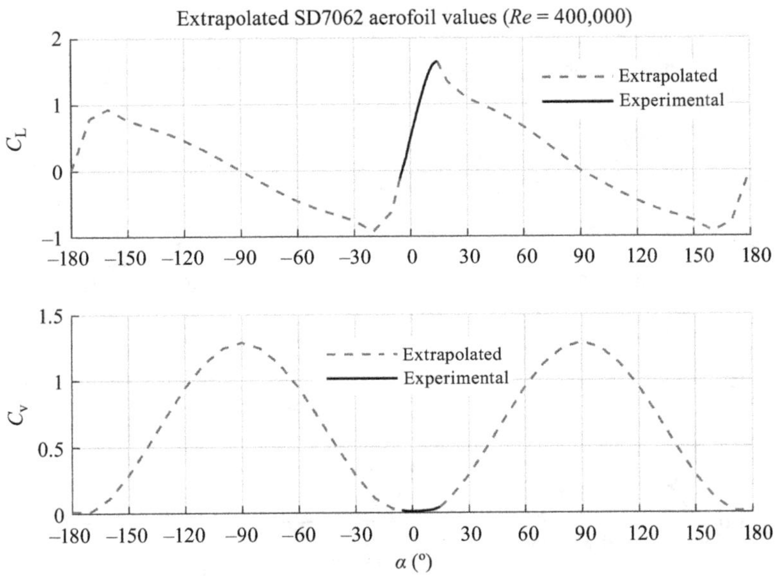

Figure 10.2 A comparison of experimentally acquired lift and drag data of extrapolated +/−180° values

and drag forces acting at the centre of pressure at the tail fin and determines the respective moment about the yaw bearing. The calculation includes the reduction in velocity at the tail fin due to downwind wake effects.

The tail fin utilised by the Aerogenesis wind turbine is a delta wing tail fin with a boom length, x, chord length, c, and span, b, of 1.32, 0.85, and 1.54 m, respectively, resulting in a total area of 0.65 m^2 and an aspect ratio ($AR = 2c/b$) of 1.10. Details of the tail fin configuration are provided in Figure 10.3. Aerodynamically, delta wings exhibit a similar linear increase in C_L with respect to angle of attack until the onset of stall at 30°. Delta wings are considered less sensitive to variations in Reynolds number compared with traditional aerofoils [13]. When determining lift and drag properties, the composite approximation (based on AR = 1.73 but encompassing delta wing tail fins of AR = 1.0–2.0) will be used as input for the aerodynamic tail fin properties, due to the limited experimental data of delta wings in the literature [22]. This tail fin input curve is shown in Figure 10.4.

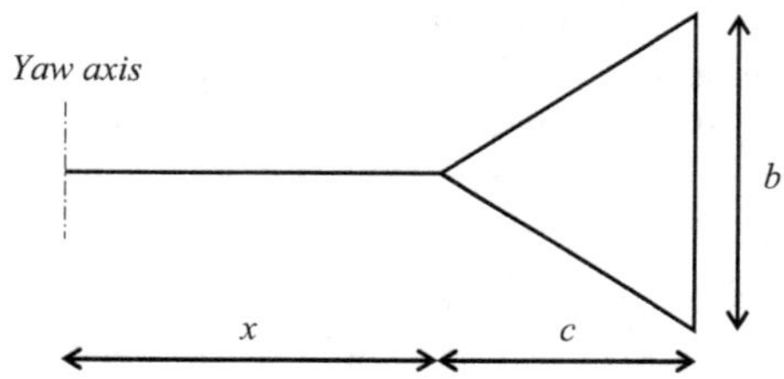

Figure 10.3 Geometry of a delta wing tail fin

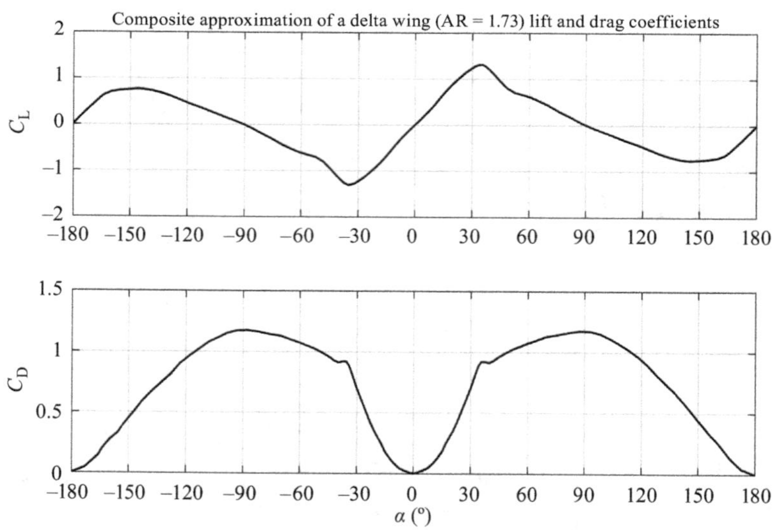

Figure 10.4 Composite approximation of the lift and drag of a delta wing reproduced from [18]

10.2.4 Blade structural properties

Structural properties of the blade are required so that blade deflections, and any resulting aerodynamic effects due to local blade velocities and mass inertial effects could be calculated within FAST. The blade model allows for the first two flapwise degrees of freedom and the first edgewise degree of freedom. As previously discussed, the 2.5 m blade of the Aerogenesis turbine is manufactured from glass fibre reinforced polymer (GFRP), which by nature has highly anisotropic stiffness properties. When considering blade elements as a series of 2D cross sections, the second moment of area (I_{xx} and I_{yy}) varies axially along the blade due to the chord, twist, and lay-up distribution. Due consideration must be taken when representing this blade model computationally either within a detailed FE package or via a more simple beam model. A fully anisotropic FE model of the blade was previously developed for blade calibration and testing purposes; however, an Euler–Bernoulli cantilevered beam model is required for use in FAST.

Within the FAST model, each 2D elemental section is attributed an equivalent stiffness, also known as the flexural rigidity constant, EI (Nm^2), principal stiffness angle, θ (degrees), and linear density, ρ (kgm^{-1}), based on the physical anisotropic GFRP property set. PreComp [23] was used to compute the equivalent values which can then be used in an Euler–Bernoulli beam model within FAST's structural routines. These results are shown for the 16 elements in Figure 10.5.

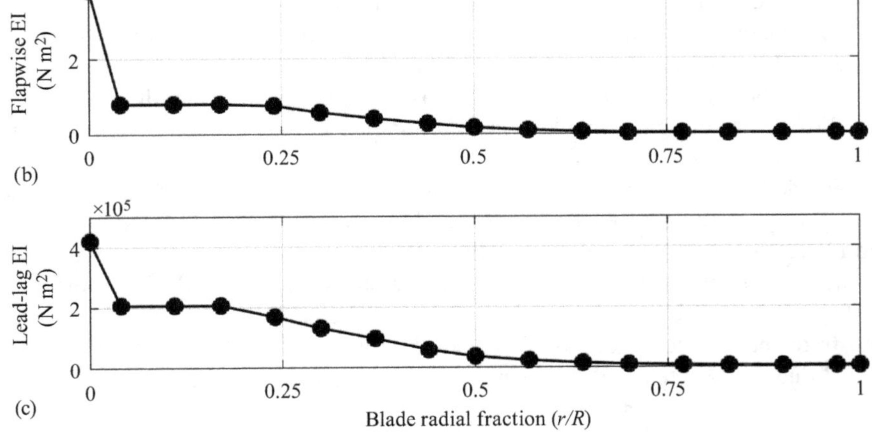

Figure 10.5 Radial distribution of blade elements (a), and respective stiffness in flapwise (b), and lead-lag directions (c)

Note that a larger chord length closer to the hub is required for good starting performance. One potential shortcoming of using PreComp is that each blade element is considered a closed, thin-walled section. This does not allow for the inclusion of the H80 structural foam ($E = 1.30$ GPa, $\rho = 100$ kgm^{-3}) which fills the shell section of the blade during manufacturing. While it is not expected to contribute significantly to the structural stiffness of the overall blade model (it was never intended to fulfil a structural role during the initial engineering design), it may contribute to the mass inertial effects present during centrifugal loading of the operating rotor. PreComp will, therefore, underpredict the net blade mass. The mass of each element will, therefore, be scaled with respect to the actual blade mass, resulting in a blade of equivalent mass as if the H80 structural foam was included. A secondary consideration of the H80 foam is its potential to resist compressive loading, and, therefore, prevent membrane buckling of the blade. Buckling is unlikely to occur during normal operation, and in any manner simulating this phenomenon is well beyond the capabilities of FAST.

FAST internally calculated the blade mass as 4.71 kg based on the equivalent densities determined within PreComp, with this underestimation attributed to the exclusion of the H80 structural foam. The blade fractional density was then scaled by a factor of 1.32 so that the final mass is equal to the measured mass of 6.20 kg. In Figure 10.6, the measured blade flapwise deflections for three calibration load cases were compared to the deflections calculated from the beam properties deduced in PreComp, in order to ensure good static agreement. The load case masses of 3.14, 5.23, and 7.23 kg were applied as point load at a distance of 50 mm inbound from the blade tip. Initially, the blade was understiff and deflecting at a mean value of 13% greater than that measured with the equivalent point loads. The stiffness values were, therefore, tuned by a factor of 1.15 (i.e. equal to 1/(1.0/0.13)), with resulting deflections presented in Figure 10.6. This tuning process is akin to what was undertaken in [24] where stiffness and damping parameters for an aero-elastic offshore wind turbine tension leg platform were iteratively tuned based on scale wind tunnel measurements. The need for scaling of the blade stiffness may be due to minor variations in the final composite lay up in the 'as-built' blade compared to the 'as-designed' blade. Also, the internal foam may have provided more structural stiffness than initially expected.

In addition to the sectional stiffness and mass properties, the normalised mode shapes are also required as inputs for the blade structural model. After computing the blade structural parameters, the flapwise natural frequencies were found via accelerometer measurements and natural frequencies determined via Strand7 FE package and are shown in Table 10.2. Due to difficulties in measurements of natural frequency modes, modes higher than the third could not be accurately determined due to expected high frequencies and are unlikely to effect the operational dynamics in any manner.

BModes6 (open-source structural analysis package) was also used to determine fundamental frequencies, with a rapid reduction in solution accuracy of the BModes function with increasing mode number likely attributed to the

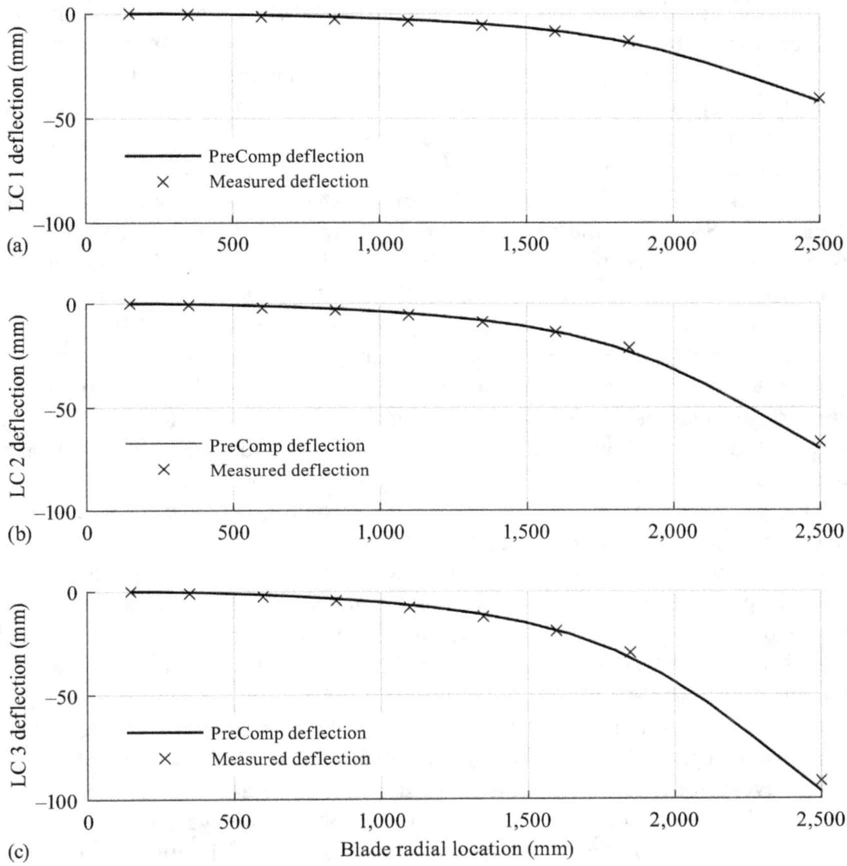

Figure 10.6 Comparison of three load cases; (a) 3.14 kg, (b) 5.23 kg, and (c) 7.23 kg, whereby point load was applied at a distance of 50 mm from the blade tip. Measured deflections compared with blade results from PreComp

Table 10.2 Comparison of the first three measured and predicted fundamental modes

Model	First mode (Hz)	Second mode (Hz)	Third mode (Hz)
Measured	7.41	17.62	45.78
Strand7 FE	8.09 (+9.18%)	19.65 (+11.52%) (+11.52%)	40.06 (−12.49%)
BModes	8.42 (+13.63%)	15.59 (−11.52%)	21.17 (−53.76%)

simplification of anisotropic stiffness/mass material property set into equivalent sectional values, and the lack of torsional coupling. The comparatively poorer performance of BModes raises questions as to whether it can accurately predict modes at rotational speeds – an important input for FAST, as rotational stiffness has been shown to impact modal response [25]. Strand7 was deemed the best method of finding rational mode shapes and frequencies, largely due to the impracticability of measuring rotational behaviour of a blade response independent of aerodynamic effects. BModes was, therefore, not used to determine rational modes due to the previously mentioned shortcomings.

The resulting natural frequencies and mode shapes (eigenmodes) were determined within Strand7 using a non-linear solve, with rotor speeds of 0, 320, and 400 rpm applied to correspond to the blade at design rotational speed, and an extreme overspeed case, respectively. A shifting of the mode shape was observed during an increase in rotor speed. Typically, this blade response will be dominated by the flapwise response, due to the comparatively stiff lead-lag direction. It is important to obtain correct values for mode shapes due to the FAST's approach for finding displacements. FAST allows for inputs of the first and second flapwise mode and the first lead-lag mode. No allowance is made for torsional modes or the coupling effect of mode shapes. Engineering judgement was used to determine the 'prominent component' of each mode, as the displacement of each mode is not simply confined to one plane. As expected, the first three modes were dominated by flapwise motions (expected as this is the 'weak' axis), with flapwise dominance observed up to the first ten modes calculated by Strand7. It is unlikely that any other modes (flapwise, lead-lag, torsional, or otherwise) beyond the third mode (>40.06 Hz) would be excited during normal operation, and for this reason lead-lag component was not used as input within FAST.

Upon solving for mode shapes in Strand7, the displacement vectors (eigenvectors) were represented as a sixth-order polynomial for input into FAST. Due to the cantilevered beam condition, the first two coefficients are zero as no translation or structural curvature occurs at the blade root. Polynomials were found at $\omega = 0$, 320, 400 rpm corresponding to the range of operating speeds likely to occur during variable speed operation. Another limitation of FAST is that it does not currently allow for the input of multiple rotational speeds with the rated speed $\omega = 320$ rpm chosen as a constant value for the blade polynomial, as this was deemed an acceptable trade-off between the rated speed of 320 rpm, high site turbulence, whereby the rotor may temporarily overspeed, and the underperformance of the turbine rotor control system. Care must, therefore, be taken in the interpretation of blade structural response at speeds much less than the rated speed (i.e. during start-up) or during high-gust events.

A sensitivity analysis of blade tip deflection with respect to rpm was undertaken within FAST using the aforementioned input polynomials at design conditions ($U = 10.5$ ms^{-1}, tip speed ratio = 8). The maximum variation between 400 and 0 rpm was found to be 7.7% showing that the blade tip displacement is moderately sensitive to rotational speed as implemented in FAST. The difference from the name-plate design speed of 320 to 0 rpm was only found to be 2.6%, which is not considered to be particularity significant. It was decided to use a blade corrected at the specified design speed of 320 rpm for the remainder of the FAST analyses.

10.2.5 Tower structural properties

In order to further understand the dynamic behaviour of the entire wind turbine system, it is necessary to include tower-related structural effects. There are several reasons for this, including possible interaction of the aerodynamic rotor response and tower structural response, potential for excitation of tower modes at various blade passing frequencies, and tower top translational effects on inlet wind velocities. Anecdotal site observations and preliminary qualitative experimental evidence suggest that these effects may be of significance, and any tower effects should, therefore, be captured within the aeroelastic model.

The Aerogenesis wind turbine tower consists of three tapered octagonal sections totalling 17,700 mm in height (nominally 18 m). The lower two sections are 6,500 mm long, while the top section is 5,700 mm high. The tower is constructed from hot-dipped galvanised structural steel (as per AS 4100-1998 'Steel structures') with a density, $\rho = 7{,}850\ \text{kgm}^{-3}$. The top section has a nominal plate thickness of 4 mm, while the bottom section has a nominal plate thickness of 5 mm. The tower has a top flange where the yaw bearing and turbine nacelle are connected. Another bottom flange and series of gusset plates are used to attach the tower to the reinforced concrete foundations. The foundations are considered to be rigid, so the bottom of the tower was modelled as a fixed cantilevered support. The tower is hinged through the use of a pin joint at approximately mid-height to allow for raising and lowering of the turbine to facilitate maintenance and experimental work. For the purposes of this model, the hinge will be considered rigid and will not be included in the FE model used to obtain structural properties and mode shapes. The structural properties were calculated using the aforementioned tower dimensions and properties.

An FE beam model of the tower was built within Strand7, with a total mass of 537.57 kg, excluding the turbine. The mass of the turbine was considered to be 170 kg from Aerogenesis design documentation and was included as a nodal point load to the top of the tower. The tower was then solved for gravitational effects of the turbine mass. This resulting linear static solve file was then used as initial conditions for the natural frequency solve to determine the first two mode shapes as sixth-order polynomials for input for FAST. The first two FA and SS modes were found at 1.04 and 4.84 Hz and 1.12 and 5.17 Hz, respectively (for interest, the third and fourth tower modes were calculated at 13.18 and 26.56 Hz).

10.2.6 Self-excited induction generator and variable speed control

Many studies which have utilised aeroelastic modelling typically oversimplify subsystems or treat them as a 'black box' [26]. There are a number of possible reasons for this, including user expertise and background knowledge (i.e. aerodynamicist vs control engineer), study focus and aims (blade loadings vs generator response and electrical output), and time frames which may not allow for detailed modelling of entire subsystems. The Aerogenesis wind turbine incorporates an SEIG and MPPT variable speed control, whereby the drive shaft and hence generator can operate across a range of rotor speeds. In theory, this allows for optimal power extraction, whereby the tip speed ratio is kept constant at (or close to) its design value across a range of wind speeds.

MPPT is a control methodology which aims to extract the maximum amount of power from a given inlet wind condition [27]. Recently, MPPT has seen much research effort, particularly with regards to doubly fed induction generators of which are implemented in large, off-shore wind turbines [26]. Practically, this is achieved in the Aerogenesis turbine by adjusting the generator load in such a way that the rotor is maintained at, or close to, its design tip speed ratio of 8. In practice, this task is somewhat difficult to implement as neither the rotor speed nor wind speed are measured via the turbine's control system. A further challenge is that the generator maximum power point does not necessarily match that of the blades due to the excitation capacitor deployed in the control system. While both rotational and wind inlet speeds are measured at the Callaghan site, both these measurements are for post-experimental analysis and are not inputs to the control system. As per Figure 10.7, there are several operational regimes as follows:

- Maintaining a higher than ideal optimal tip speed ratio from cut-in speed, U_C, to maximum rotor speed, U_B. This is due to compromises with generator selection detailed in [27].
- Rotor speed kept constant from U_B to rated speed, U_R.
- Rotor speed reduced from U_R until cut-out speed, U_F, whereby a mechanical brake (located at the rear of the high-speed shaft) is activated to stop the rotor.

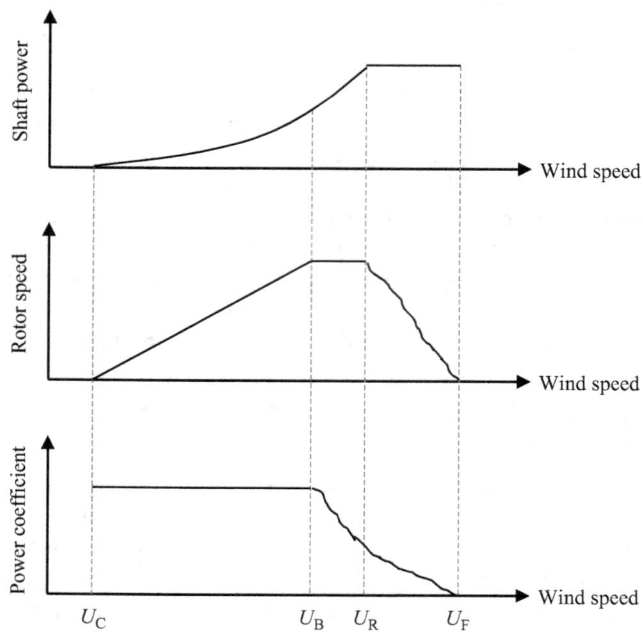

Figure 10.7 Optimal behaviour of the Aerogenesis turbine controller

The SEIG is essentially an induction motor operated in reverse as an induction generator (in this case, a 4-kW Donly induction motor), and provides a simple, reliable, and cost effective solution for generation of this scale [27]. Another advantage of SEIGs compared with other generators, such as a permanent magnet generator, is cogging torque – which is important for a start-up event. Three shunt capacitors are connected in a delta configuration to facilitate self-exciting and power production. Historically, SEIGs have had issues regarding output voltage and frequency control. These issues can be mitigated if the SEIG is then connected to a bridge rectifier and inverter. Another benefit is that the generator no longer has to run at a fixed speed – a key advantage for MPPT variable speed turbine operation. It should be noted that FAST does not natively support the use of an SEIG and associated MPPT variable speed control; however, it allows for interfacing with Simulink to control the following parameters:

- blade pitch control (i.e. for maximising starting performance or mitigating overspeed effects);
- yaw control (typically for use in large wind turbines with an active yaw drive);
- generator and variable speed control (for turbine generator systems that are different to the default generator options provided by FAST).

While the Aerogenesis turbine does not have blade pitch control or active yaw control, this interface with Simulink will be used to include the effects of the SEIG and MPPT variable speed control. A Simulink model of the Aerogenesis turbine generator and control system was developed at the University of Newcastle, Australia, by D.R. Bradney and has been provided for use in this study, where the reader is referred to [28] for further details on this model development. The influence of the variable speed control provided by the generator and control system impact on the rotor operation and, hence, blade loadings [29], it is therefore imperative to account for this affect. This model has been interfaced with FAST using the supplied 'OpenLoop' Simulink file, where the output parameters from FAST (i.e. high-speed shaft torque and rpm) are used as inputs for the SEIG model and then output and returned into FAST for execution.

10.3 Results and discussion

10.3.1 Initiation of FAST simulations

As with all computational models of physical systems, it is necessary to assess the performance of a given model with physical data, ideally from the actual system in question. While it is inherently difficult to accurately validate a complex system such as a wind turbine, this following section details an attempt to compare measured data with the simulated response of the Aerogenesis wind turbine in order to understand any potential shortcomings or limitations within the model. All models of this kind are likely to have some minor deficiencies due to limitations in algorithms such as aerodynamic or structural routines. For example, one potential shortcoming in the structural algorithm is that torsional motion (i.e. twisting) of the blade is ignored.

While it is unlikely that significant twisting will occur in operation, as small blades are inherently much stiffer than large blades, any minor torsion can vary the blade aerofoil angle of attack and, hence, aerodynamic response. Assumptions of this kind are often seen as an acceptable trade-off between computational resources and model accuracy.

The approach taken in validating the Aerogenesis wind turbine model with the experimental data is broadly based on assessing dynamic performance. Due to the unsteady site conditions, it is unlikely that the blade aerodynamic loads or tower structural vibrations will instantaneously match measured results. A 10-min time series simulation using an inlet wind series developed in TurbSim to match the conditions of the IEC standard was input into FAST with the turbine response output for comparison with site measurements. This wind series had a mean wind speed of 7.5 m s^{-1}, at a turbulence level of 24% (i.e. $I_{15} = 18\%$). This is seen as an attempt to reproduce the 'mean performance' of the Aerogenesis turbine within its intended design class. This simulated performance will then be compared to the measured data from campaigns detailed in Chapter 9.

Before undertaking full-scale simulations within FAST, the user guide recommends a sensitivity analysis be performed to ensure that an appropriate fixed-step numerical integration constant (dt) is used for both FAST and AeroDyn. It is important to ensure the correct step is used to produce time-efficient and computationally accurate results. While the default time step is provided at 0.0010 s, other values were investigated, with particular interest in 0.0020 s, as this corresponds to the Arduino data collection system sample rate of 500 Hz outlined in Chapter 9. This sensitivity analysis was performed on the C_P (power coefficient) with time steps from 0.02 to 0.005 s. It was noted that results of C_P did not vary beyond $dt = 0.0010$ s, with this time step chosen for future simulations.

10.3.2 Validation of dynamic rotor performance

Initially, the performance of the FAST Simulink MPPT controller was assessed across a range of wind speeds with respect to both tip speed ratio and rotor speed. It is imperative to understand how the controller performs in the aeroelastic model as the rotational velocity of the rotor can have a significant impact on the rotor aerodynamic response, power output, and structural loads. When assessing the simulated tip speed ratio to the measured tip speed ratio in Figure 10.8, results reveal a mean over-prediction of the simulated controller across all wind speeds. The overprediction at rated wind speed was found to be 11%, indicating that the modelled control system acts more aggressively in an attempt to keep the tip speed ratio higher at low wind speeds. A natural conclusion is that the 'as-designed' control scheme in the FAST model is different to the 'as-built' Aerogenesis controller. Knowledge of the PI control parameters could improve the accuracy of this model; however, they were not documented during commissioning of the controller, and the authors do not have access to them for the purposes of this study. Visually, the simulated data is bounded by the measured data, with the exception of several outliers at higher wind speeds. Further investigation revealed that rapid reductions in wind speed resulted in high instantaneous tip speed ratios due to the rotor inertia and response time of the controller.

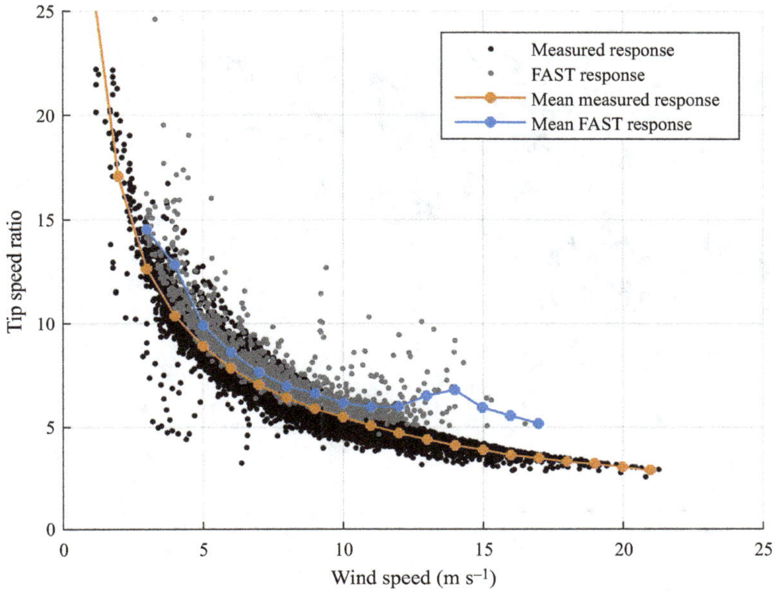

Figure 10.8 Tip speed ratio of the FAST model and the physical turbine. Both control curves exhibit a similar shape, with a FAST overprediction evident. Sample rate is 1 Hz

When investigating the performance in Figure 10.9, it can be seen that the rotor speed in the FAST model appears to be controlled more aggressively and confined to a much narrower band when compared to the measured response which shows a wider distribution of rotor speed with respect to inlet wind speed. As a consequence of the higher mean tip speed ratios in Figure 10.8, the mean simulated rotor speed is also higher for all wind speeds, with an increase of 8.5% at rated operating conditions. Of note is the significant spread of measured data points; for example, the maximum rotor speed of 240 rpm was measured from wind speeds of 4 to 21 m s^{-1}.

This indicates that the physical turbine control system has a significant amount of inertia that has not been represented in the FAST model. Reasons for this may be due to the fact that the Simulink generator model and MPPT curve are derived from steady-state idealised conditions, where variations in rotor speed are said to occur at discrete wind speed states. In practice, the wind turbine spends a significant portion of operating life in unsteady conditions due to fluctuating wind speeds. Implementing a variable speed control scheme at a highly turbulent site is a challenging endeavour. Despite the tendency of the simulated controller to be more aggressive when compared to the physical Aerogenesis controller, the simulated rotor speed is kept within the measured operating range.

The power curves of both the simulated and measured data sets are compared in Figure 10.10 to assess the electrical power that has been generated in both

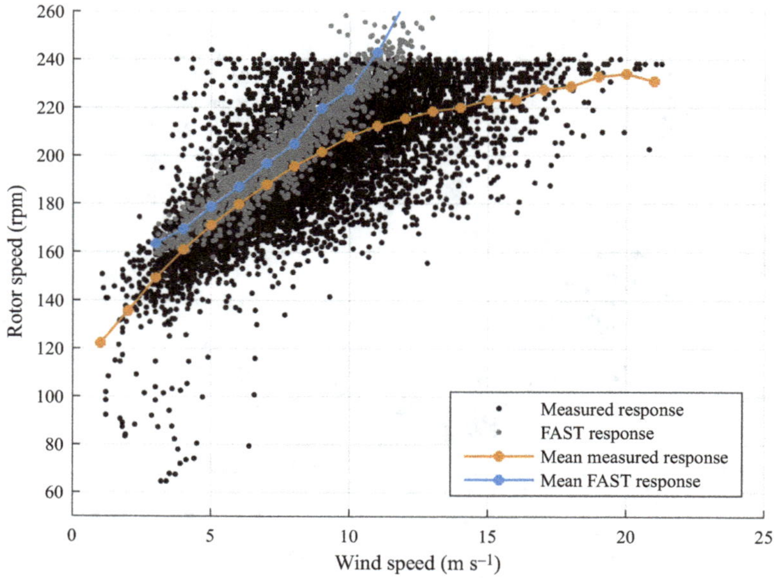

Figure 10.9 Comparison of measured and simulated rotor speed with respect to inlet wind speed. FAST appears to control the rotor response more aggressively. Sample rate is 1 Hz

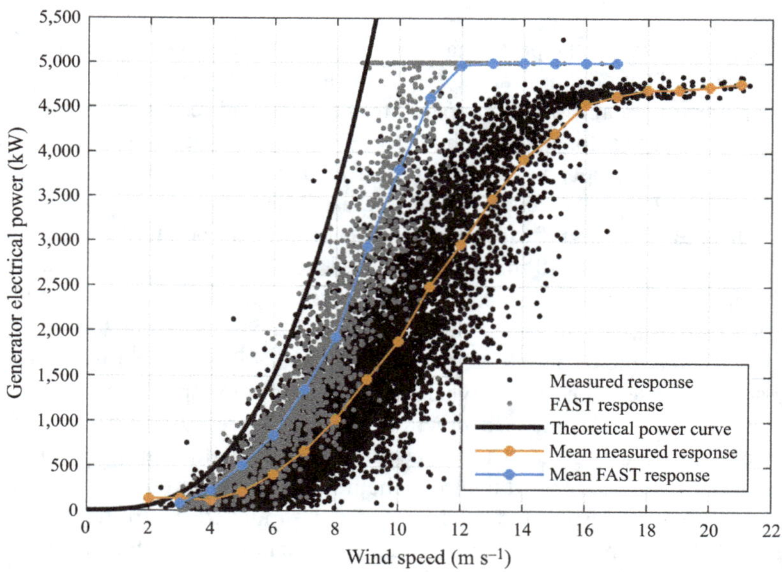

Figure 10.10 Measured and simulated power curve compared to the theoretical limit for this turbine. Sample rate is 1 Hz

instances. As expected, both methods show an underprediction of the theoretical power curve for an idealised Betz–Joukowsky rotor, with some outliers present, whereby electrical power have been generated in excess of the theoretical limit. This may be due to unsteady effects caused by rapid increases or decreases in the wind speed, whereby the generator controller does not instantaneously react to this which is manifest as a transient power spike. When considering the simulated wind set, FAST overpredicts the mean power that has been generated across the wind speed range, with a substantial increase of 50% found at rated conditions. This is not unexpected when interpreted in light of higher tip speed ratios and rotor speeds produced by the FAST controller. As this controller has to be developed from design documentation, it is consistent that the output electrical power is closer to the theoretical power curve than the physical controller which has been modified by an unknown amount post-installation.

Reasons for this overprediction in output power are likely to extend beyond discrepancies in controller schemes. The assumption is that aerodynamic loads calculated in the BEM algorithm are based on ideal steady-state operation and do not account for the unsteady effects of turbulence which has shown to derate the equivalent 2D quasi-steady performance of an aerofoil [30,31]. Furthermore, the BEMT assumption that the blade loads are continually in equilibrium with the wake is unlikely to be a valid assumption at turbulent sites where rapid aerodynamic changes occur before the wake can reach equilibrium. The FAST model also assumes that an ideal drive train is that in which small inefficiencies such as gearbox friction and resistive torque would slightly impact the conversion of torque developed at the rotor into generator electrical power. As a final consideration, blade surface condition due to wear and dirt accretion may reduce performance. FAST does not take these potential effects into account.

10.3.3 Validation of structural response

In Figure 10.11, the aeroelastic performance of the blade is compared to the measured response in the frequency domain via the use of a power spectral density analysis. This data was captured using a strain gauge data acquisition system operating at 500 Hz. Full details of this system can be found in [32]. When considering both blade flapwise bending moment and the structural deflection of the blade tip, reasonable agreement was found for the lower frequency behaviour of which most evident is the classic 1P blade response which occurs between 3 and 4 Hz on this variable speed machine. Beyond approximately 7 Hz, the agreement between measured and simulated signals tended to diverge, whereby the FAST model did not appear to capture some of the high-frequency phenomena which were present in the measured site data. Most obvious in the measured signal is the 3P blade response which is likely to be a harmonic of the one 1P response which occurs between 10 and 12 Hz. While ideally these higher order effects would be captured in the FAST model, these phenomena have a lower magnitude than the 1P response which is usually known to dominate blade response. Reasons for this 3P response are not immediately clear but may correspond with possible minor mass

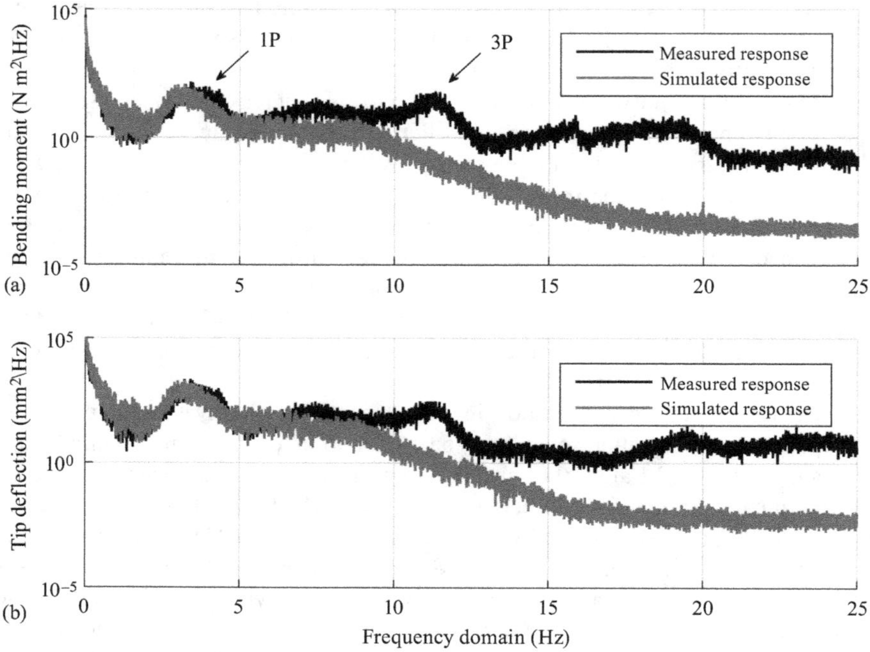

Figure 10.11 PSD of blade flapwise bending moment (a) and tip deflection (b)

or aerodynamic imbalances in the rotor, with these observed in the measured tower accelerometer response. A possible line of investigation is the added mass of the strain gauge lead wires and copper tape shielding, which was estimated at 242 g. This is equivalent to 4% of the blades mass and equates to a rotor eccentricity of 0.0137 m as per IEC 61400.2. Despite the fact that this is greater than the suggested value of 0.0125 m for this rotor, no provision is given for quantifying these effects on resonance.

To assess the equivalent steady-state response of the blade at the design wind speed for the Aerogenesis turbine, azimuthal averaging was implemented. The philosophy behind this technique is that while it is difficult to instantaneously compare blade response between modelled and measured data in the time domain, averaging of the response in the rotational domain will effectively reduce the impacts of any short-term high-frequency phenomena and reveal long-term behaviour which is more indicative of mean steady-state operation. Figure 10.12 details the results of azimuthal averaging between 0° and 360° at a mean wind speed of 10.5 ms^{-1}, whereby yaw error and yaw rate are limited to between 2.5 and 5.0°/s, respectively, to mitigate any unsteady effects of yaw behaviour. The simulated flapwise bending moment found to be accurate to within 7.7% of the measured data, with the tip deflection was found to be accurate to measurements within 9.9%. When comparing the results of the blade response in the azimuthal domain, the lack

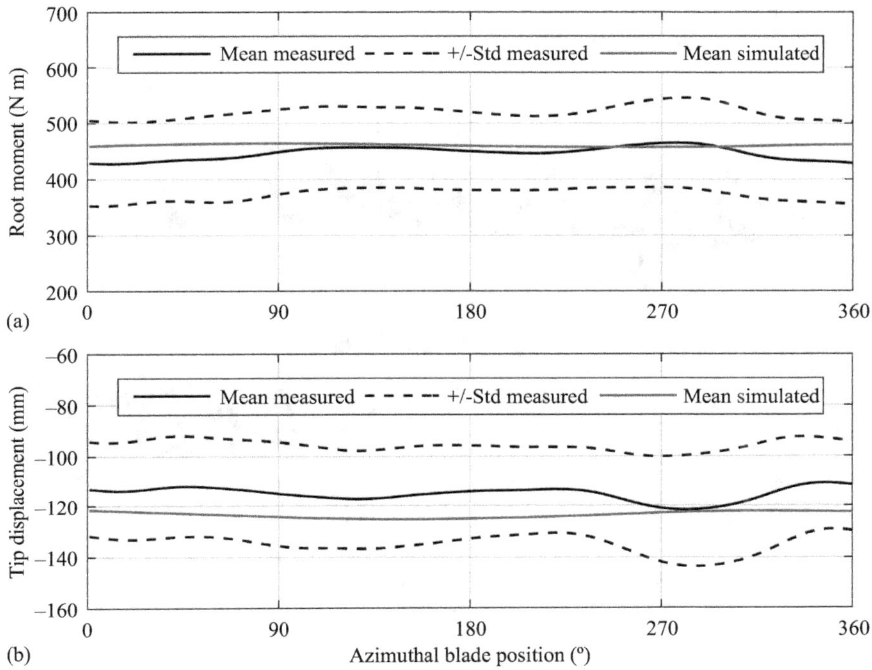

Figure 10.12 Azimuthally averaged response of measured and simulated blade response at design wind conditions. Aerodynamic moment is predicted to within 7.7% (a), and tip deflection within 9.9% (b)

of higher order effects which were revealed in the frequency domain in Figure 10.11 once again became apparent. While it is obvious that these high frequency perturbations are not accurately captured in the aeroelastic model, these perturbations are very minor in magnitude and are unlikely to contribute significantly to effects such as fatigue loading.

The net dynamic acceleration response of the wind turbine tower top was compared in Figure 10.13 to the measured response. Data was transformed to the frequency domain by a power spectral density (PSD) analysis. Immediately obvious are the first two tower modes in both the measured and simulated response, which have been captured well by FAST in terms of both magnitude and frequency of occurrence. Beyond approximately 7 Hz, the simulation tended to diverge from the measured response. This may be due to the fact that higher order tower modes and structural response cannot be accounted for in FAST. Harmonics of the rotor per-cycle response (e.g. 1P–3P) are evident in the measured response. As previously stated, these are likely to be caused by minor imbalances in the rotor due to mass or aerodynamic effects which are difficult to completely eliminate during rotor manufacture and installation. Nevertheless, the fundamental dynamic response of the tower has been accounted for with regards to natural frequency excitation and modal response.

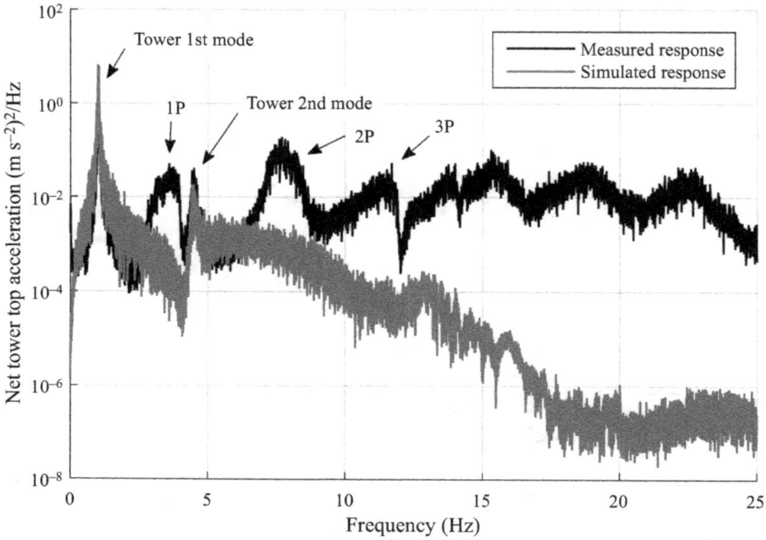

Figure 10.13 Results of FAST simulations and measured data of the net tower acceleration response. Very good agreement is found for the tower fundamental modes

10.3.4 Validation of dynamic yaw behaviour

The simulated yaw behaviour was assessed with respect to the measured data in the form of a scatter plot of yaw rate and yaw error in Figure 10.14. This is in line with techniques used in [33] to assess yaw dynamics of small wind turbines. In both cases, it was observed that lower yaw rates corresponded to higher yaw errors, and that higher yaw rates occurred at lower yaw errors – possibly due to the fact that the rotor instantaneously passes through a zero yaw error as it overshoots the inlet wind direction during a corrective yaw action. The simulated yaw behaviour was bounded by the site measurements, indicating that the simulated yaw dynamics acts within the physical range of the system as measured. Some degree of possible bias was observed in that more of the simulated data points are present in the first and third quadrant of the scatter plot. While the effect appears to be minor, it may represent some form of 'handedness' in the yaw behaviour or could be due to interactions of the aerodynamic wake and tail fin. Further investigation is required.

10.3.5 Discussion

Upon comparing simulated turbine response to the measured turbine response, the simulated turbine dynamics showed broad agreement with field measurements. The tip speed ratio, rotor rpm, and resulting electrical power output tended to be higher than the measured results, despite being kept within the measured operational range. Reasons for this controller discrepancy may be due to the fact that only

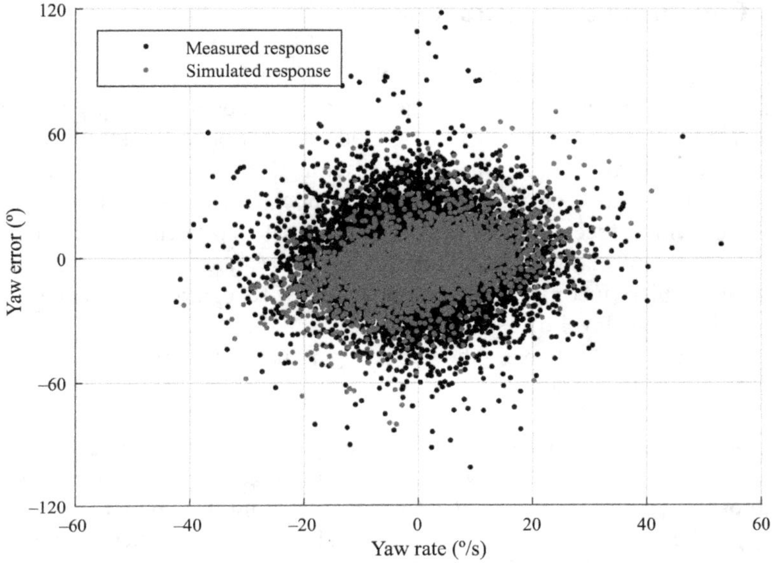

Figure 10.14 Comparison of yaw rate and yaw error for both measured and simulated response. Sample rate is 1 Hz

design documentation is available for the controller, which has since been subject to undocumented modifications. Another consideration is that the Simulink controller was based on steady-state parameters, whereas in reality, a majority of operating life occurs in unsteady inflow. Yaw control via the means of a tail fin showed good agreement in FAST, whereby the both yaw errors and respective yaw rates were bounded by the measured response.

Structurally the blade showed good behaviour compared to the measured response, whereby the azimuthal 'steady-state' equivalent flapwise bending moment was found to be within 7.7% of the measured blade load for design conditions. In a similar manner, the simulated tip deflection was found to be within 9.9% of the measured response. In the frequency domain, the simulated blade response corresponded well to operational frequencies <7 Hz with a dominant once-per-cycle effect evident in both spectra. Agreement deteriorated at higher operational frequencies, whereby a 3P response was evident in the measured data but not apparent in the simulated response. Reasons for this could be the fact that higher order structural effects such as blade torsion were not modelled in FAST, and high frequency aerodynamic effects, such as operation at high yaw errors (whereby the modified BEM theory has a limited ability to accurately capture this), or minor rotor aerodynamic and mass imbalances (which were not modelled in FAST). The simulated tower dynamic response showed very good agreement to the measured response for the first and second fundamental modes. However, higher order effects such as 1P–3P vibrations induced by the rotor did not appear in the spectra of the tower response.

10.4 Conclusions and future work

In this chapter, an aeroelastic model of the Aerogenesis 5-kW wind turbine has been developed and compared to field test data. The model is comprehensive and includes blade and tail fin aerodynamics, blade and tower structural response, and variable speed control from an SEIG. The resulting FAST files are available in the public domain for non-commercial use, see Appendix A. Verification of model performance with respect to measured operational data has been undertaken. To the best of the authors' knowledge, this is the first time that a small, passively controlled wind turbine (incorporating variable speed SEIG control) has been modelled within FAST to this level of detail.

In concluding this chapter, the following remarks are made by the authors regarding aeroelastic modelling:

- Aeroelastic modelling has seen a substantially disproportionate use in large commercial-scale wind turbines, compared to the small wind turbine field. The authors' suggests several reasons for this, including the following:
    - While the aeroelastic software (FAST) is free and open source, it has taken a substantial investment of time and engineering experience to develop a small wind turbine model to an appropriate level of detail. These engineering and time resources may be out of reach of many small wind developers. The authors' note that as a compromise, structural components, yaw control and generator control can be considered rigid, fixed, and fixed respectively. This would allow for an acceptable 'initial' estimation of loads and performance for steady-state operation; therefore, possibly subverting the conservative IEC 61400.2-2013 simplified load model.
    - A broad engineering knowledge base is required in several disciplines, including aerodynamics, structural dynamics, GFRP composites, and generator control theory.
    - The lack of a GUI and reliance on detailed text-based input files may provide an initial learning-curve barrier which potentially limits usability.
    - Commercially available aeroelastic software such as Bladed are likely to be prohibitively expensive and may not model specific small wind turbine features (i.e. tail fin motions and SEIG control).

- Generator response and variable speed control is important to capture the time-varying behaviour of the entire wind turbine system. Previous wind turbine models have considered subsystems such as the speed control to be a 'black box', which would not allow for the analysis of events such as a rapid increase in tip speed ratio caused by a rapid drop in wind speed and a lagging rotor speed due to inertia and generator response effects.
- Blade mass centrifugal effects are important when considering blade structural inputs for small wind turbines which may operate at rotor speeds well in excess of 300 rpm. The user should be cautious of events where the operating rpm

exceeds the blade structural input rpm. This could result in large non-linear deflections and failed FAST execution.

- Acquiring operational data for model verification can be a complex task, particularly for variable speed turbines operating at highly turbulent sites where sustained steady-state operation may not be observed. Future work should focus on more accurately modelling the controller behaviour to unsteady inflow conditions.

Despite these difficulties, a complete aeroelastic model is nevertheless a useful tool for the wind turbine engineer to validate and assess a wide range of both steady-state and unsteady operational regimes.

Future work may be associated with including torsional modes for the blade structural response, implementation of aerodynamic routines, for instance, when the rotor is at very large yaw errors, a comprehensive tail fin model, and investigating the utility of the GDW model for modelling unsteady turbine behaviour. A migration to the recently released OpenFAST may facilitate this with improved structural routines and the ability to develop a custom yaw routine based on a passive tail fin mechanism.

As small wind turbines have a large array of configurations, it is natural that effort should be placed in the development and verification of models beyond the classic 'Danish' three-bladed HAWT. For example, as turbines with a larger number of blades, vertical-axis rotor configurations, or a diffuser-augmented wind turbine. Verification of aeroelastic models of a VAWT would be of significant value to the small wind turbine field. Of note is QBlade that has the capacity to model such rotors.

Finally, from a commercial aspect, the DWEA SMART Wind Roadmap[†] and a recent NREL report[‡] show substantial opportunities still remain for the small wind market. Higher adoption can be achieved with increased system reliability and reduced levelised cost of energy. Aeroelastic modelling may be a key tool in unlocking some of these potential benefits, by knowing loads to a higher degree of accuracy, compared to simplistic load equations, prior to physical prototyping.

Appendix

The FAST input files for the 5-kW Aerogenesis turbine model developed in this study are available at http://hdl.handle.net/1959.13/1349817, or from the corresponding author upon request.

At the time of writing, the current release of FAST (v8.12.00a-bjj) did not support tail fin aerodynamics and furling behaviour. The latest version of FAST that supported tail fin motion (v7.02.00d-bjj) was used in this study. It is recommended that any interested user first familiarises themselves with FAST software available at: https://nwtc.nrel.gov/FAST.

[†]https://distributedwind.org/wp-content/uploads/2016/05/SMART-Wind-Roadmap.pdf
[‡]https:nrel.gov/docs/fy17osti/67337.pdf

References

[1] M. Hansen, J. Sørensen, S. Voutsinas, N. Sørensen, and H. Madsen. State of the art in wind turbine aerodynamics and aeroelasticity. *Progress in Aerospace Sciences*, 42(4):285–330, 2006.

[2] F. Rasmussen, M. Hansen, K. Thomsen, *et al.* Present status of aeroelasticity of wind turbines. *Wind Energy*, 6(3):213–228, 2003.

[3] J.M. Jonkman and M.L. Buhl. *FAST user's guide*. NREL (National Renewable Energy Laboratory), Golden, CO (United States), 2005.

[4] P.J. Moriarty and A.C. Hansen. *AeroDyn theory manual*. NREL (National Renewable Energy Laboratory), Golden, CO (United States), 2005.

[5] M.L. Buhl and A. Manjock. *A comparison of wind turbine aeroelastic codes used for certification*. NREL (National Renewable Energy Laboratory), Golden, CO (United States), 2006.

[6] J.M. Jonkman, S. Butterfield, W. Musial, and G. Scott. *Definition of a 5-MW reference wind turbine for offshore system development*. NREL (National Renewable Energy Laboratory), Golden, CO (United States), 2009.

[7] M. Sessarego, K.R. Dixon, D.E. Rival, and D.H. Wood. A hybrid multi-objective evolutionary algorithm for wind-turbine blade optimization. *Engineering Optimization*, 47(8):1043–1062, 2015.

[8] A.J. Coulling, A.J. Goupee, A.N. Robertson, J.M. Jonkman, and H.J. Dagher. Validation of a FAST semi-submersible floating wind turbine numerical model with DeepCwind test data. *Journal of Renewable and Sustainable Energy*, 5(2):023116, 2013.

[9] S. Lee, M.J. Churchfield, P.J. Moriarty, J. Jonkman, and J. Michalakes. A numerical study of atmospheric and wake turbulence impacts on wind turbine fatigue loadings. *Journal of Solar Energy Engineering*, 135 (3):031001, 2013.

[10] E. Tofighi, D. Schlipf, and C.M. Kellett. Nonlinear model predictive controller design for extreme load mitigation in transition operation region in wind turbines. *In 2015 IEEE Conference on Control Applications (CCA)*, pages 1167–1172. IEEE, 2015.

[11] D. Corbus and M. Meadors. Small wind research turbine: Final report. *Technical report*. NREL (National Renewable Energy Laboratory), Golden, CO (United States), 2005.

[12] J.M. Jonkman and A.C. Hansen. *Development and validation of an aeroelastic model of a small furling wind turbine*. NREL (National Renewable Energy Laboratory), Golden, CO (United States), 13, 2005.

[13] D. Corbus and D. Prascher. *Analysis and comparison of test results from the small wind research turbine test project*. NREL (National Renewable Energy Laboratory), Golden, CO (United States), 2005.

[14] D. Corbus, A.C. Hansen, and J. Minnema. Effect of blade torsion on modeling results for the small wind research turbine (SWRT). *Journal of Solar Energy Engineering*, 128(4):481–486, 2006.

[15] S. Dana, R. Damiani, and J. van Dam. Validation of simplified load equations through loads measurement and modeling of a small horizontal-axis wind turbine tower. *Technical report*. NREL (National Renewable Energy Laboratory), Golden, CO (United States), 2015.

[16] M. Hand, D. Simms, L. Fingersh, *et al.* Unsteady aerodynamics experiment phase vi: wind tunnel test con figurations and available data campaigns. *Tech. rep.* National Renewable Energy Lab., Golden, CO (United States), 2001.

[17] J. Tangler. Advance wind turbine design studies. *Tech. rep.* National Renewable Energy Lab., Golden, CO (United States), 1994.

[18] D.H. Wood. *Small wind turbines: analysis, design, and application*. Springer, Springer-Verlag London, 2011.

[19] J.G. Leishman and T.S. Beddoes. A semiempirical model for dynamic stall. *Journal of the American Helicopter Society*, 34(3):3–17, 1989.

[20] C.A. Lyon, A.P. Broeren, P. Giguere, A. Gopalarathnam, and M.S. Selig. *Summary of low-speed airfoil data: Volume 3*. SolarTech Publications, Virginia Beach, VA (United States), 1997.

[21] P.J. Moriarty and A.C Hansen. *AeroDyn theory manual*. NREL (National Renewable Energy Laboratory), Golden, CO (United States), 2005.

[22] A.K. Wright. Aspects of the aerodynamics and operation of a small horizontal axis wind turbine. PhD thesis, School of Engineering, The University of Newcastle, Australia, 2005.

[23] G.S. Bir. User's guide to PreComp (pre-processor for computing composite blade properties). *Technical report*. NREL (National Renewable Energy Laboratory), Golden, CO (United States), 2006.

[24] G. Stewart, M. Lackner, A. Robertson, J. Jonkman, and A. Goupee. Calibration and validation of a FAST floating wind turbine model of the DeepCwind scaled tension-leg platform. *Technical report*. NREL (National Renewable Energy Laboratory), Golden, CO (United States), 2012.

[25] D.M. Eggleston and F. Stoddard. *Wind turbine engineering design*. Van Nostrand Reinhold Co. Inc., New York, NY (United States), 1987.

[26] M. Singh, E. Muljadi, J. Jonkman, V. Gevorgian, I. Girsang, and J. Dhupia. Simulation for wind turbine generators–with FAST and MATLAB-Simulink modules. *Technical report*. NREL (National Renewable Energy Laboratory), Golden, CO (United States), 2014.

[27] N. Batt and C. Coates. Maximising output power of self-excited induction generators for small wind turbines. In *2012 XXth International Conference on Electrical Machines*, pages 2105–2111. IEEE, 2012.

[28] D.R. Bradney. Measured and predicted performance of a small wind turbine operating in unsteady flow. PhD thesis, School of Engineering, The University of Newcastle, Australia, 2017.

[29] Y. Vidal, L. Acho, N. Luo, M. Zapateiro, and F. Pozo. Power control design for variable speed wind turbines. *Energies*, 5(8):3033–3050, 2012.

[30] Ph. Devinant, T. Laverne, and J. Hureau. Experimental study of wind-turbine airfoil aerodynamics in high turbulence. *Journal of Wind Engineering and Industrial Aerodynamics*, 90(6):689–707, 2002.

[31] N. Cao, D.S.K. Ting, and R. Carriveau. The performance of a high-lift airfoil in turbulent wind. *Wind Engineering*, 35(2):179–196, 2011.

[32] D. Bradney, S.P. Evans, M. Chu, and P.D. Clausen. A low-cost, high-speed, multi-channel Arduino-based data acquisition system for wind turbine systems. *Wind Engineering*, 44(5):509–518, 2020.

[33] S.V.R. Wilson and P.D. Clausen. Aspects of the dynamic response of a small wind turbine blade in highly turbulent flow: part 1 measured blade response. *Wind Engineering*, 31(1):1–16, 2007.

Chapter 11
The very low head turbine—a new hydro approach

Paul G. Kemp[1]

The very low head (VLH) turbine is a relatively new application of long-standing hydroelectric power technology using recently available modern technological advances. These advances have enabled a fundamental design shift that substantially improved all efficiency, cost and performance in the low-head hydro regime.

This chapter describes the conception of the VLH turbine design, the basic technology design improvements and the key adaptations used to make it suitable for use in cold climate regions followed by a case study of the first plant installation in Canada.

The VLH turbine uses a variable-speed, permanent magnet generator (PMG), which is rare in the hydro industry, as most hydro turbine units are fixed speed, synchronized or induction generator systems. For VLH applications where the upstream water level to downstream water level is in the order of a 2–5-m drop, the economics of the plants are very difficult. This is due to the large volume of water required to generate only a few hundred kilowatts of power. To solve this, the VLH was conceived to be installed at existing water control facilities, either adjacent to or within, to minimize civil works, reduce environmental impact and shorten regulatory approvals. The VLH itself is designed to provide multiple functions, over and above producing hydropower, including a water control gate, flood passage, debris rack, fish passage, water level control weir and rural electrical grid support.

11.1 VLH conceptualization

Deployment opportunities abound with an estimated 80,000 existing low-head dams and water control structures in Canada and the United States that do not have hydropower deployed at them [1] (Figure 11.1). These unpowered dams, weirs and locks could be developed to provide over 70 TWh of renewable energy generation per year.

[1]Canadian Projects Limited, Calgary, Canada

Figure 11.1 *Typical low-head water control structure found in Canada that does not have power production facilities. Many of the estimated 80,000 structures like this in North America could become commercially viable renewable energy sources*

This hydro technology development story begins in 2001–04 when preliminary engineering work started on a large hydro scheme called Dunvegan, located on the Peace River in Alberta. This project came up with an innovative turbine idea utilizing 40—2.5-MW small bulb turbines across the 300-m wide river. A new turbine technology was created for this application with a collaboration of a leading European turbine manufacturer, the project engineer and project developer. This turbine idea was developed by applying new wind turbine generator technology at the time to hydro, which utilized PMGs that could operate at variable speeds to improve their efficiency, reduce their size and provide several electrical operation advantages. The turbine's powerhouse structure was designed to be submerged such that it acted as a weir so overtopping flow could occur to meet extreme flood passage requirements. It took over 10 years to achieve the project approval for Dunvegan but by then it was not built due to changed power market conditions and new ownership.

In concert with this, in 2004, several VLH sites were being looked at on the Otonabee River in Ontario that comprised old lock dams in the Trent-Severn Waterway National Park. However, initial feasibility studies indicated that these were marginally economic sites based on using traditional low-head turbines with large powerhouses and canal works.

Around this time, the original Dunvegan turbine lead designer had semiretired and was granted permission from his former employer to explore this idea further to develop a small turbine (0.5 MW) using some of the previous Dunvegan learnings with some support of his Canadian colleagues in Quebec. However, this time the focus was applied to the abundant small unpowered low-head sites such as those on the Otonabee. Research funding was tough in Canada, so the development of the first VLH turbine mostly occurred in France (albeit still with some NRCan support) and pilot VLH turbines were deployed and then commercialized. There are now more than 50 VLH turbines deployed in Europe.

Interest for this technology continued in Canada, so in 2008 discussion took place to adapt the VLH design for cold climate use. A new developer, Coastal Hydropower Corporation, secured funding from NRCan to do a VLH Turbine Cold Climate Adaptation study that was undertaken by Canadian Projects Limited (CPL), and it identified several design improvements and modifications to make the VLH design viable in Canada's freezing rivers.

So, the initial VLH concept was first identified and conceived in Canada then developed and refined in France with some Canadian funding then upgraded to adapt to cold climate use so it could be brought back to Canada for its first deployment in North America.

11.2 VLH technology

For some time now, waterpower projects from sites with less than 5 m of head have not been commercially viable in many regions. However, a new approach for deployment of an innovative low-head hydroelectric turbine system was developed to address this challenge. The resulting VLH turbine technology allows the deployment of VLH turbines within or adjacent to existing dams, spillways, weirs and canal structures at significant civil works savings. High power efficiencies and capacity factors are achieved in head ranges between 2 and 5 m.

11.2.1 Low-head hydro benefits

The key characteristics of this innovative capture of river power compared to other conventional and renewable energy alternatives are important to appreciate in order to assess the long-term potential of low-head turbine opportunities:

- Predicable power—Energy production from river flow is very predictable with high seasonal reliability and has been an electricity mainstay in Canada for more than 100 years. In comparison, wind and solar energy are a lot less predictable day to day.
- Existing infrastructure support—Many old dams and weirs require upgrade and ongoing operational funding to safely maintain their functionality. With funding for existing infrastructure being difficult, adding a VLH plant can change a cost burden into a revenue asset to provide much needed long-term sustainable infrastructure support to these vital water management facilities.
- Dense energy—Water is about 830 times denser than air; therefore, water turbine runners are much smaller than wind turbine rotors of comparable power output. The capital costs of water turbines are somewhat less but balance-of-plant costs can be much higher. Operations and maintenance costs are much lower with hydro compared to wind and thermal facilities.

- Small footprint—Since water turbines are much smaller and may be placed closer together compared to wind and solar plants, the environmental "footprint" of a run-of-river hydro project is relatively small for its output.
- Low visual impact—These small hydro systems are mounted below the water surface; therefore, they minimize "visual impact" compared to what exists with conventional hydro powerhouses, or thermal plants, or many wind and solar plants.
- Fish friendly—The VLH turbines operate at a very low speed with large water passages between its blades, therefore, are fish friendly. Fish can swim through the turbines without injury as has been confirmed by numerous fish mitigation studies in Europe and in Canada.
- Scalable deployment—The VLH facilities are scalable in size to meet the requirements of a single small community or industry and by adding more units can provide larger scale electricity supply to the regional power grid, depending on river size.
- Carbon credits—This technology substantially reduces the consumption of fossil fuels and the generation of pollutant by-products; thus, the reduction of Green House Gas emissions allows the plant owners to accrue additional income through carbon tax credits and similar incentive mechanisms.
- Easily expandable and re-deployable—Unlike conventional hydro projects that require upfront construction of a specific facility size at a specific location, VLH turbines can be added to an existing facility as needed to meet expanded production requirements and redeployed at new location as needs change.

These attributes make the VLH turbine attractive for many communities, industries and regions particularly in bolstering investment support for existing water management infrastructure.

11.2.2 VLH turbine technology

The VLH is an integrated generating set that has a Kaplan runner with 8 adjustable blades, a fixed distributor with 18 fixed wicket gates with flat bars inserted in between to also act as a trashrack (Figure 11.2).

The turbine has a PMG that is directly coupled to the runner shaft and an automatic trashrack cleaner (rotating wiper) mounted on the distributor. The runner of the VLH is relatively large and as such, the speed of the water flow through the runner is quite slow, which eliminates the need for complicated inlet and outlet water passages and the associated expensive civil works typically required for energy recovery. The low rotational speed of the runner and its low flow velocity make the VLH a fish-friendly turbine and less prone to debris blockage.

The flap gate on the turbine crest serves three functions: (1) allows water level and flow control, (2) passes floating debris from the river over the top of the turbine and (3) cushions forces from ice sheet formation in winter.

The VLH is entirely submersible allowing a near silent operation (the sound of cascading water) and has minimal visual impact. It is mounted at an angle to maximize a runner diameter for a given river depth and is removed from the sluice

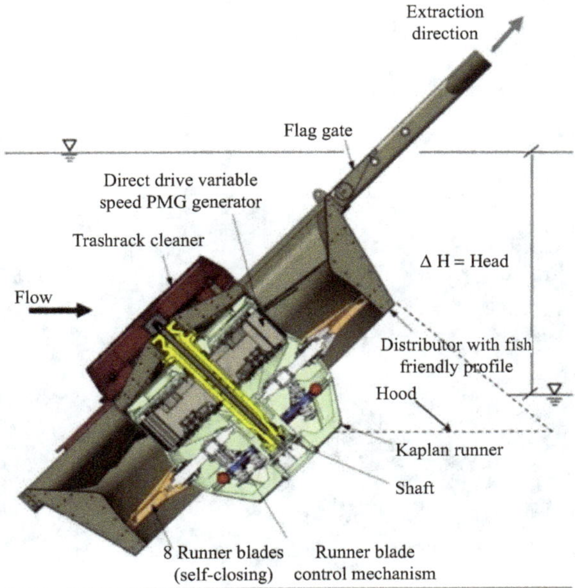

Figure 11.2 The very low head (VLH) turbine components with the turbine blades connected to the central shaft and a PMG generator (illustration from NRCan website https:nrcan.gc.ca/science-and-data/funding-partnerships/funding-opportunities/current-investments/wasdell-falls-hydro-power-project/16155)

passageway by a lifting system so the unit can be raised for maintenance or to attain the full flow capacity of the sluice during high flood flow events (Figure 11.3).

The runner blades are adjustable and self-closing to each other allowing automatic upstream water level regulation and eliminating the need for a separate upstream gate to shut down the unit or control headwater levels. The unit can also operate in discharge mode (when not generating electricity) to maximize flow through the sluiceway without lifting it up. In addition, if desired, the VLH can operate in isolated mode when not connected to the grid once extra electrical controls are added for off-grid power supply to support remote locations.

The VLH turbine generating set is double regulated in a nontraditional way with adjustable blades and variable speed as opposed to fixed speed and variable wicket gates. This characteristic allows operation on sites where the head varies with changes in flow. Thus, the variable speed capability of the VLH turbine assures a stable and efficient operation under moderate variable head. This innovation effectively provides a double regulated design by varying turbine speed as opposed to the traditional more complex method of adjustable mechanical wicket gates. To facilitate the variable speed nature of the turbine, the VLH system includes a frequency converter to enable delivery of 60-Hz synchronous power to the electrical grid.

Figure 11.3 The very low head (VLH) turbine installed on its lifting system that was developed as part of the Cold-Climate Adaptation Package (lifting system developed by Canadian Projects Limited, Calgary, AB and built by Mecan Hydro of Granby, QC)

The VLH's extremely low rotational speed (35 rpm), a large runner diameter, very low water velocity (less than 2 m/s) together with other technical features make it a fish-friendly turbine that allows downstream fish migration through the turbine runner itself. Extensive efficiency testing and operational data from several years of installations have verified a high turbine efficiency of about 86% that is similar to traditional larger, more mechanically complex, hydro turbines.

Unlike most hydro turbines that are custom designed for specific sites, the VLH turbine is a non-site-specific design that results in significant manufacturing economies and reduced construction costs. There are a set of model ranges that were selected to fit the majority of low-head sites for common existing structure depths and flows with a reasonable power output per unit. The standardized turbine range consists of 5 model sizes with runner diameters varying from 3.5 to 5.6 m and generator output ranging between 200 and 700 kW per unit for head ranging from 1.5 to 4.5 m and operational flows from 10 to 30 m^3/s per turbine.

The rapid deployment and minimal modifications needed to accommodate this technology make the VLH turbine an attractive upgrade to existing water control structures that otherwise have had no commercially viable means to generate renewable energy and yield valuable revenue. This technology opens an opportunity to existing structure owners to offset operation and maintenance costs of their essential water control structures by converting existing capital cost burdens to revenue generators at minimal risk or imposition to their existing operations.

The feasibility of hydro on structures of this nature has been difficult due to long approval times and marginal economics for some time now. However, the nature of this innovative fish-friendly, low-impact design enables relatively straightforward environmental assessment for approval by regulatory agencies. Further, once approvals and agreements are in place, the deployment system developed by Canadian Projects enables hydro installation of a multiunit VLH plant in only a few months, thus reducing deployment timeframe from years to months and diminishing the development risks.

With state-of-the-art electrical controls, these plants operate in full remote mode and, when necessary, can lift themselves fully out of the way at any time to allow floods or debris to pass. The modular system design allows staged implementation, so units can be added or removed to suit site requirements and changing needs.

This innovative system is well suited to garner waterpower energy from a yet-untapped, low-head water resource in Canada and the United States by adding new energy opportunities to hundreds, if not thousands, of existing low-head water control structures.

11.2.3 VLH hydro plant components

A typical VLH hydro plant consists of the following components:

11.2.3.1 VLH structure

The VLH structure is an existing or an adjacent new concrete structure typically consisting of a bottom sill, pier walls and top deck. The piers form water passage bays that have gates or stoplogs between them to control the upstream water level and allow flow to pass through each bay. The VLH is installed in one or several of these bays to replace the existing gates or stoplogs with the turbine, while maintaining the same water control functionality.

11.2.3.2 Bridge deck

The bridge deck structure provides equipment and vehicle access for operations and maintenance, including bridge access across the dam. The deck is complete with guardrails and light standards.

11.2.3.3 Stoplogs

Stoplogs are the multiple interlocking type with a spreader bar lifting device and are required to allow the extraction of the turbine-generator for maintenance. Each bay has stoplog guides upstream and downstream, so the turbine bay can be dewatered in the event that a turbine cannot be lifted normally or for structure inspection and maintenance in the dry. One set of stoplogs (both upstream and downstream) is provided so that one bay can be dewatered. The structure usually can only accommodate one bay dewatered at a time to prevent unnecessary uplift forces.

11.2.3.4 Grizzly racks

Removable grizzly racks (coarse trashracks) are installed in the upstream stoplog guides in each turbine bay. Grizzly racks prevent large debris, such as large branches and trees, from reaching the turbine fine trashrack. Grizzly racks are cleaned manually by the operator using a pike pole or rake. If necessary, a small crane is used to remove larger debris. The grizzly racks are fitted with a rigging attachment for removal using a stoplog spreader bar.

11.2.3.5 Turbine-generators

Turbines-generators are typically identical VLH turbines that are Kaplan-type units with eight hydraulically operated and eccentric axial, fail-closed blades that are mounted inside of distributor housing with fixed guide vanes. The generators are directly driven by the turbines and are located inside the turbine runner hubs. Generator excitation is by rotating permanent magnets mounted on the turbine hub with the stator in the core of the generator on the fixed shaft. The stator is cooled by the water flowing around the external face of the hub with rotor fan blades circulating the air inside the hub. A separate dehumidified air system in the control building pressurizes the hub via air lines to prevent water ingress complementing the efforts to completely seal the hub and protect its electrical components.

The turbines are capable of operating in speed-no-load conditions (runaway) for sufficient time to maintain plant flow and/or control the headwater level while the flow is transferred to the dam's bypass gates after trip stops.

The turbine-generators are mounted on removable support frames that include the turbine extraction hoist systems and the crest flap gates. Turbine parts exposed to atmosphere on their downstream side during operation are protected from icing by a heavy vulcanized-rubber curtain.

The noise generated by the VLH turbines is negligible, essentially no more than the sound of falling water similar to that of any dam gateway or stoplog bay.

11.2.3.6 Frequency converter

To address the fact that the VLH turbine operates at variable speeds, the VLH system includes a frequency converter that adjusts the power generated to the grid frequency and to the required power factor. The frequency converter includes the combination of a rectifier that converts the asynchronous output of the VLH generator into DC energy, and an inverter that then converts that DC energy into 60-Hz synchronous energy as required to deliver to the grid. Like the PMG's, large frequency converters were originally developed for wind turbines and at a smaller scale for variable frequency drive motors.

11.3 Cold climate adaptation

In order to deploy the VLH turbine in Canada, a Cold Climate Adaptation Study* was completed in October 2010 that was funded by Natural Resources Canada (NRCan) and undertaken by CPL. The study focused on approaches to upgrade and modify the prototype models being deployed in France [2].

*Canadian Projects Limited 2010 VLH turbine cold-climate adaptation study.

The critical cold climate modifications that were adopted include the following:

- Insulated and heated turbine housing.
- Linear extraction system using hydraulic rams with a robust HSS frame and HDPE glides as opposed to the tilt hinge system used in Europe.
- Heavy-duty bar grizzly made of an ovalled heavy-wall pipe to form hydrodynamic shaped bars with robust strength to resist ice flow impact and debris blockage.
- Larger yielding crest flap gate that is designed to absorb and deflect ice expansion to prevent ice force transfer to the turbine housing.
- Ice-phobic coatings on exposed surfaces to reduce ice adherence.
- Modified outlet hood with heavy vulcanized rubber cladding to resist ice formation and impingement on the turbine discharge hood and to maintain water seal integrity of the discharge flow.
- Heat tracing, steam ports and integrated air bubblers in critical areas to prevent ad-freezing of frazil ice and spray on moving components.
- Robust flexible-armored umbilical festoon cable management system for power and control cables as well as for the hydraulic and pneumatic piping.
- Weathertight Sea-Can control building to house power converter, auxiliaries and control system equipment nearby that are shop-built and tested (Figure 11.4).

All these Cold Climate Adaptation Measures have been deployed with some field modification and fine-tuning to optimize their effectiveness. These systems have now been successfully in operation for over 5 years in central Canada year-round with minimal downtime due to cold climate impacts.

Figure 11.4 *Cold Climate Operations at the Wasdell VLH Hydro plant. Left and right turbines in operation, center turbine raised for maintenance during low water season. Note the armored control and power cable festoon arrangement to accommodate turbine lifting and ice protection*

11.4 VLH hydraulic studies

There is a variety of existing weirs, spillways and other structures on which the VLH can be considered for deployment. The effect of structure geometry on performance is important to understand as this can markedly affect the operation and economics of VLH installation. Understanding these affects is critical in assessing suitable sites across North America for VLH deployment and predicting turbine performance.

To gain this understanding for various geometric and hydraulic conditions, a hydraulic modeling study was performed at the University of Calgary with the Institute of Sustainable Energy, Environment and Economy (ISEEE) [3]. The study quantified the effects of an upstream obstruction, such as an outcrop, low weir or drop, on turbine performance that are often found on existing structures. Figure 11.5 shows the scale model built to study VLH hydraulic performance.

The study was funded by Natural Science and Engineering Research Council of Canada and CPL of Calgary. The 2-year investigation included both numeric modeling and physical modeling using the water tunnel shown in Figure 11.6, at ISEEE's experimental hydraulics laboratory.

The study utilized a backward-facing step geometry, as shown in Figure 11.7, upstream of the turbine as the effective shape of an upstream obstacle or drop that are sometimes present at existing low-head hydraulic structures in Canada. The open-channel water tunnel enabled a detailed observation of the effects of the step at varying distances upstream of the turbine and time-resolved particle image velocimetry was used downstream of the backward-facing step. Power was determined through torque and power measurements.

Based on the measurements, a drop in efficiency of up to 7% was observed with the introduction of an obstruction upstream of the turbine at a distance of half the turbine diameter. Unsettling was the result that comparable efficiency impacts were observed with the obstruction positioned a full runner diameter upstream. These results suggest that a noticeable loss in efficiency was a result of the highly

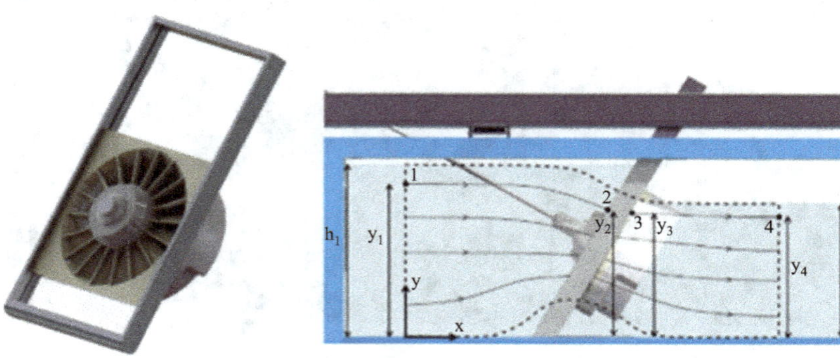

Figure 11.5 The 1/15th scale VLH model used for the University of Calgary Hydraulic Model Studies in 2013–14 and a schematic of the experimental setup of the VLH model with flow lines [3]

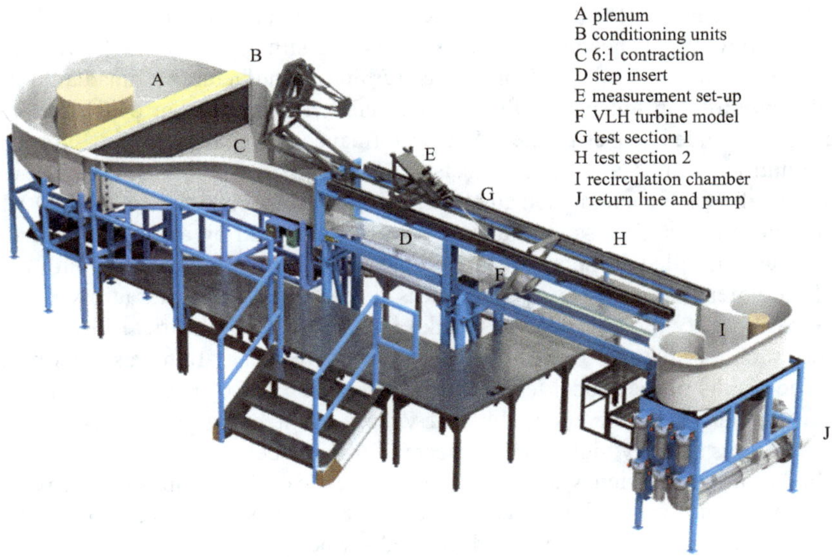

Figure 11.6 Water tunnel used for VLH hydraulic model testing at the University of Calgary, [3]

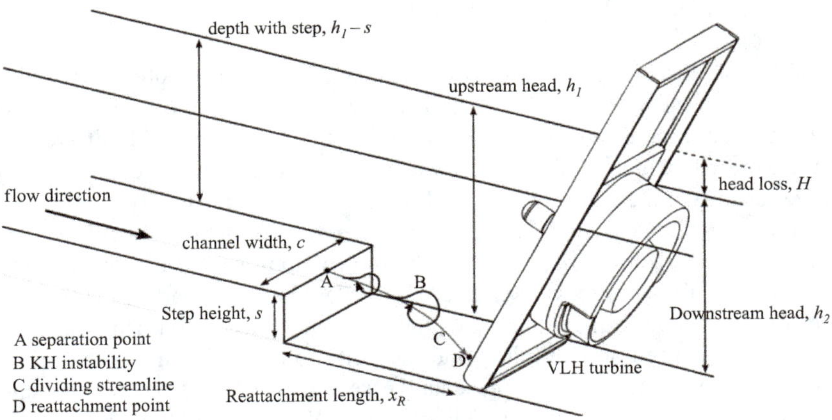

Figure 11.7 Backward-facing step model setup and streamline [3]

nonuniform velocity profile entering the turbine. The study provided valuable input in the selection of turbine positioning within existing structures or consideration of structure profile modifications, since such reduction in efficiency could have detrimental economic impact on VLH deployment.

The laboratory results were correlated to measured full-scale turbine efficiency curves to develop comprehensive performance characteristics. Prior to the initiation of the Wasdell Falls VLH project (see the Case Study in Section 11.6), no full-scale testing of flow versus power over the expected operating range of the VLH turbine had been conducted independent of the turbine manufacturer. Canadian stakeholders required an independently verified detailed measurement of the turbine's performance index to verify the VLH performance being reported in Europe. In collaboration with Canadian Projects, the turbine manufacturer along with OEL—Hydrosys Inc. conducted an efficiency test† of an existing Model 4000 VLH turbine installed at Terrasson, France. Flow was recorded by an array of acoustic flow meters installed at the upstream end of the concrete forebay along with water level measurements, and power output was recorded by the plant's utility revenue meter. The utility meter was used to take into account all electrical losses of the facility such that the true overall water-to-wire power generation was determined.

The results of the model tests indicated that the VLH efficiency matched closely with the manufacturers' published values. As a bonus, the measured field efficiency was actually higher than expected for flows below turbine median flows. For higher flows, efficiency dropped slightly below expected values but on average was only insignificantly lower than expected. For flows in excess of the turbine design flow, the turbine efficiency decreased further as expected.

Overall, the hydraulic modeling study provided invaluable insight on the performance characteristics under adverse operating conditions and independently verified the VLH's full-scale efficiency over its full operating range.

11.5 Canadian VLH fish studies

Several fish studies [4–6] have been undertaken in Europe since 2008 when the first VLH prototype was installed. These studies have been done on a frequent basis as new models or variants were deployed or different river systems with different fish species were encountered. The fish passage and mortality study results are quite positive leading to the claim of the VLH turbine to be "fish friendly."

Similarly, following the first VLH deployment in North America and as part of the environmental commitment made in the project approval documents submitted to the agencies, fish studies were planned and undertaken in Canada as well. A two-part study [7] was performed in 2019 with the first part focused on fish passage and the other part focused on fish injury and mortality to determine the overall risk of the VLH turbine to fish. Five representative species were selected for the study: Largemouth Bass, Smallmouth Bass, Rock Bass, Walleye and Northern Pike.

For the fish entrainment part, the fish were monitored with acoustic telemetry for a year. Specifically, the fish use of the forebay areas upstream of the VLH turbines was studied where it was found that entrainment (fish passage) through the VLH turbines did not occur over the course of the entire year. Further findings were that the forebay usage was not species specific, as well as the forebay usage was somewhat limited. It is suggested that the rather sterile rock cut walls and bedrock bed with little food or hiding locations could be causing sparse fish usage of the forebay.

†OEL-Hydrosys Inc. 2011 VLH performance testing summary report.

For the fish injury and mortality due to entrainment part, the selected fish were experimentally introduced into the turbines via plastic pipe sluices and then recaptured downstream. Body sizes varied from 17 to 69 cm. Only one fish was killed representing 1.2% of the sample; otherwise turbine strike abrasion related injuries were the most common finding.

The results of this study suggest that entrainment events by the VLH turbines are rare on the species studied and that entrainment by the VLH turbine has minimal effects on the fish with very low mortality occurring.

Therefore, the Canadian fish study further reinforces the "fish-friendly" claim in that the risk of the VLH turbine to the studied fish species and life stages is low.

11.6 Case study: Wasdell Falls VLH Hydro Plant

The Wasdell Falls VLH Hydro Plant is situated on the Severn River approximately 6-km North of Washago, Ontario, Canada at the site of the existing Wasdell Falls Dam and an abandoned Ontario Hydropower generating plant.

11.6.1 Introduction

The Wasdell Falls Dam is a 7-m high by 43-m long concrete gravity dam that was originally constructed in 1913 and 1914 simultaneously with the original historic hydroelectric plant as shown in Figure 11.8.

Figure 11.8 *Aerial view of the Original 800 kW Wasdell Falls Hydro Plant on the left and the still operating dam on the right; both were built in 1913/14 by the Ontario Hydro Electric Power Commission. Arrows indicate where a tragic boating accident occurred in 1942. (From Toronto Star Archives compiled by Steve Stanton in 2018 with credit to "Severn River" by James T. Angus, Severn Publications, Orillia, 1995.) The inset photos are of the old hydro powerhouse during construction and the turbine hall prior to being decommissioned in 1955*

The original 800-kW hydroelectric plant was the first generating station constructed by the Hydro-Electric Power Commission of Ontario (now known as Ontario Hydro). The plant operated as a separate unit serving the northern district of Ontario County and western portion of Victoria until 1924 then it became part of the Georgian Bay System. The plant was decommissioned in 1955, as larger hydro developments displaced it. The powerhouse was demolished in 1973 and only a portion of the concrete foundation remains as it forms part of the water retaining structure of the existing dam. The dam is now owned by the province under the Ministry of Natural Resources (MNR).

Wasdell Falls Limited Partnership, the project developer and CPL, the Project Manager and Engineer developed the Project. The VLH plant consists of a new approach channel leading to a new structure to contain three VLH turbines. The new hydro plant structure is located adjacent to the existing Wasdell Falls Dam and the old decommissioned powerhouse foundation. A new small control building and transformer yard is located 60 m away from the new hydro plant. The VLH plant generates 1,650 kW of power and is connected to the Hydro One 44-kV distribution system via a 4.1-km transmission interconnection line, most of which is buried.

The plant is designed and operated in accordance with the required federal, provincial and municipal regulations and approval conditions. Key approvals and conditions relevant to the project design are as follows:

- Permit to Take Water—Ministry of Environment (MOE)
- Connection Impact Assessment—Hydro One Networks Inc.
- Connection Cost Agreement—Hydro One Networks Inc.
- Site Lease—MNR
- Navigable Waters Protection Act Approval—Transport Canada
- Fisheries Act Authorization—Department of Fisheries and Oceans
- Certificate of Approval—MOE
- Lakes and Rivers Improvement Act (LRIA) Approval, Sec. 23.1—MNR
- LIRA Technical Bulletins and Best Management Practices—MNR
- Construction Environmental Management Plan—MOE

11.6.2 Site conditions

Wasdell Falls is fed from the Severn River, a part of the Trent Severn Waterway (TSW). The Severn River is fed by Lake Couchiching and the much larger Lake Simcoe just upstream of Wasdell Falls. Lake Couchiching is regulated by six dams controlled by Parks Canada as part of the TSW. The Black River joins the Severn River just upstream of Wasdell Falls. It is unregulated and is prone to high and rapidly changing flows compared to the Severn River exiting Lake Couchiching.

The project hydrology was developed by combining data from Water Survey of Canada hydrometric stations on the Severn and Black Rivers, for a combined 40-year period of record from 1965 to 2004.

The existing works include the Wasdell Falls Dam, a few small control buildings, and an old powerhouse foundation. The existing water control structure has 7 bays: 5 bays have manually operated mechanical gates with 3 stoplogs on each, 1 bay has 12 stoplogs, and the larger center bay has 4 stoplogs over an Ogee spillway, the bottom stoplog has been fastened to the weir crest and is not removable.

Wasdell Falls Dam is currently operated to maintain headwater levels high enough for recreational purposes during low flows and low enough to minimize potential flooding in the community of Washago during high flow events. The operations at Wasdell Falls Dam are coordinated with the TSW to maintain water levels at the downstream of Lower Dam D (Washago Mill).

Investigations and studies were completed and used to characterize the site geological conditions. For the project, the excavation of approach and tailrace channels was mainly in bedrock and minimal excavation or stability issues were encountered. The bedrock-bearing capacity and shear resistance were quite adequate to support the new VLH structure in the new approach channel.

A road and parking area already existed at the dam, with adequate soil-bearing capacity for construction equipment and for the final access road and yard area on the west side of the structure. The access area between the Wasdell Falls Dam and the new structure is located partly over the old powerhouse foundation containing the old turbine pit, intakes and draft tubes that were previously filled with granular materials and covered with a concrete slab. It was discovered that parts of the old turbines were also still in place. This plethora of abandoned works is one of the key reasons why building adjacent to this area was a wise choice. The bearing capacity of the old structure was assessed during construction to verify that it was adequate for equipment access to the existing dam for construction and operations.

The riparian land at the site is owned by the MNR, while the riparian land upstream and downstream of the site is privately owned. The plant did not change the current operating regime or water levels. Only a small section of the shoreline in the immediate vicinity of the existing dam was affected where the new approach and tailrace channels were located.

No significant aquatic or shoreline habitats were identified at the proposed channel inlet/outlet locations. No significant archeological artifacts were identified at the site.

A butternut tree assessment was carried out in support of the project that identified two butternut trees that met retention guidelines protecting them from harm under the Environmental Site Assessment. A protective 25-m radius buffer from each butternut was established to prevent disturbance. However, with the assistance of the MNR biologist, some work associated with the upgrade of the existing site access road was allowed near the buffer.

Regional economic influences were considered in the design along with sociological conditions, such as nearby landowners interests and requirements, including water wells; current waterway commercial and recreational uses and impacts of the development on these current uses; nearby municipalities; current water management plans in the region, including operational rules and requirements. Specific design elements include a provision of a canoe portage and boat launch on the west side of the VLH structure and relocation of the historical hydro plant monument placard to facilitate better access and viewing.

The river, both upstream and downstream of the plant site, is used predominantly for local recreational and water supply purposes. Navigational use by large watercraft is generally directed along the Trent-Severn Canal that parallels the river system offset some 5 km away. Several landowners along the riverbank in proximity to the hydro plant draw their domestic water supply from the river. A few municipalities, including Washago and Orillia, operate water treatment plants in the region.

11.6.3 Project design

The project is designed for a useful life of 100 years and a design flood of 1:1,000 years. All structures, components and buildings are designed to meet the following requirements:

- Maximum Design Earthquake—Canadian Dam Association, Dam Safety Guidelines
- Seismic Loading—Provincial and Municipal Building Codes, Post Disaster Seismic Performance Category
- Waterway Signage—MNR Lakes and Rivers Improvement Act, Public Safety Around Dams Best Management Practices
- Electrical Safety Signage—Canadian Electrical Code and Ontario Electrical Safety Code
- Control Building—Ontario Building Code, Exposure Classification "A"

The project was designed by CPL of Calgary with support from its Barrie, Ontario office for field investigations and construction management. Standard design tools, guides, manuals, codes and industry practices were used for analysis and design on this project as appropriate for the project.

Since this was the first demonstration project of this technology in North America and was under cold climate conditions, many critical systems needed to be reviewed for this project application that made the design task quite challenging.

Converting the design from the European 50-Hz system to the North American 60-Hz system proved to be a challenge as well. Even though the driver systems being used were available in both European and North American standards, it was quickly determined that there were several technical nuances to this swap over that took several months to resolve. Further to this, the utility was not well versed in PMG Generator and Frequency Converter System characteristics for interconnection, so discussion on plants transformer yard protection and the upgrades to the Hydro One receiving substation was an educational process that took almost a year.

Another critical design challenge was attaining agreement on the water management operating regime for such an old and historic site. Even though the intent was not to change the site operations, there was significant uncertainty as to what the actual operating rules were and if the levels in the documents corresponded to the actual levels being used on site under various operating conditions. It turned out that the operations documents were not correct, so a process of verification and acceptance was required to properly establish the rule curves, and the water management documents were revised accordingly. This process took more than a year.

11.6.4 Wasdell Falls Hydro Plant components and construction

The Wasdell Falls VLH Hydro Plant components and some aspects of their construction are as follows:

Siteworks—The project involves new and existing siteworks, including roads, yards and parking, drainage, borrow and disposal areas, reclamation, historic works and works for recreational use. Site layout and arrangement considered the aesthetic presentation of the project from various key recreational vantage points. The existing heritage monument was relocated to the west side of the structure to remain readily accessible to the public. An existing portage trail to allow people with small watercraft such as canoes or kayaks to bypass the project site was maintained. The canoe portage landings and trail were kept in their original locations and some shoreline enhancement was provided on the downstream side where the existing rock shore was steep, to improve boater access. The existing access road from Wasdell Falls Road was upgraded to provide site access. The yard and parking area is treated near the control building to limit the amount of dust drawn into the frequency converters cooling fans. The parking area acts as an emergency overflow spillway in the case of an extreme flood event beyond the design flood. The headrace and tailrace channels edges are lined with road barriers for safety. To the extent possible, excavated materials were reused at site to backfill the structures and rough grade the transformer yard, control building pad and parking area. Areas affected by earthworks were reseeded and replanted (Figure 11.9).

Headrace and tailrace channels—New headrace and tailrace channels were constructed by drilling and blasting the bedrock to accommodate the new turbine structure. The design of the headrace and tailrace channels is optimized for cost and headloss. The design minimizes the channel footprint and avoids impacting the upstream wetlands area. The channels are designed to withstand free discharge through the turbine bays during extreme flood events. Whereas the channel velocities and depths vary with river flow, the normal design condition is the maximum plant flow (without flap gate flow) and lowest minimum release flow at the dam, which is when water levels are lowest downstream and consequently the channel velocities highest, thereby resulting in the most headloss. The headrace channel velocity is designed to promote the formation of a stable ice sheet to minimize frazil ice conditions in the proximity of the turbines. A debris boom is placed at the approach channel inlet to divert floating debris and ice to pass over the existing dam as before and minimize the amount of debris at the turbine structure. This boom also acts as a barrier for boaters in addition to the safety cable. The channel excavation in bedrock is adequate to resist erosion during normal and flood conditions. Side slopes in overburden are armored with appropriately sized riprap as required to prevent slope erosion. Sequencing of construction was done in such a way that the majority of the work was carried out in the dry except for the upstream cofferdam plug removal (Figure 11.10).

VLH structure —There are three VLH turbine generators installed within turbine frames mounted to the new concrete VLH structure. The structure location was

Figure 11.9 Aerial view during construction of the new 3-bay VLH structure on the left with its approach channel and the existing Wasdell Falls Dam on the right with the historic powerhouse foundation remnants in-between. Note the new control building being installed on the far-left side on the raised berm

positioned in accordance with the approvals to comply with the advisory conditions and changes dictated by the Minister of Natural Resources. The location was selected to optimize hydraulic conditions, facilitate construction, limit the channel excavation and facilitate access and installation of the turbines. The VLH structure includes a bridge deck, stoplogs and grizzly racks for debris management as described in Section 11.2.2. The VLH structure is cast-in-place concrete with the nose and tail of the pier protection formed by prefabricated steel liners that also acted as concrete end-forms during construction. Q-deck was used to construct the concrete structure deck. Trays for electrical cabling and mechanical hoses and tubing are along the downstream side of the structure deck. Commercial low intensity, low-height and down-directional floodlights are provided for work lights and aesthetics to minimize lighting impacts on neighboring residences (Figure 11.11).

Existing Wasdell Dam upgrades—The Wasdell Falls Dam electrical gate operators were modified and automated. Existing gate seals on Wasdell Falls Dam were repaired. Gates 5 and 6 were modified to allow full opening of the gates with two new electric hoists for automatic operation. The new hoists do not restrict the movement of the two existing hoists across the entire dam length. The existing hoists can be used as redundant lifts for the automated gates in the case of failure of

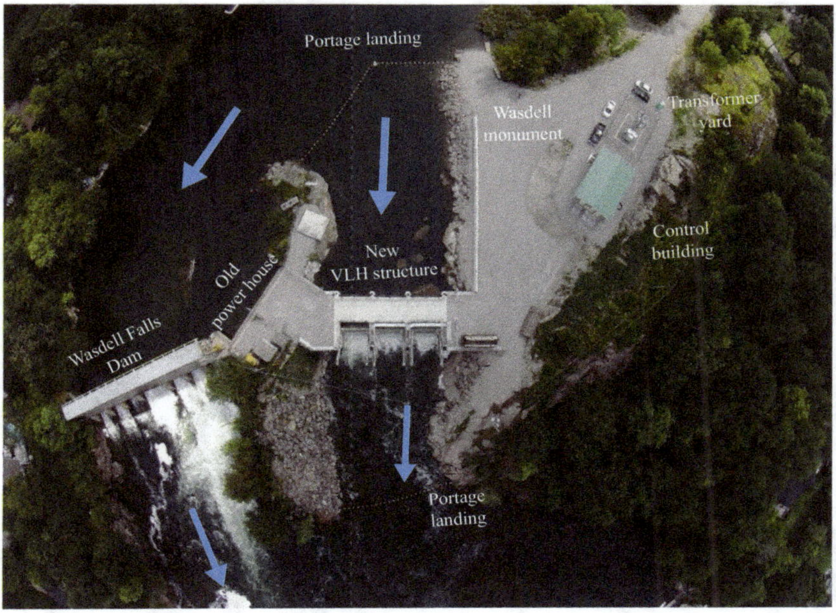

Figure 11.10 Aerial photo of the completed Wasdell Falls VLH Hydro Plant. Original Wasdell Falls Dam on left, the old powerhouse foundation in middle and adjacent the new VLH structure with new headrace and tailrace channels. Note the control building and transformer yard on the right. River flow is top to bottom

the new hoists. The new hoists are automatically controlled by the plant control system and include manual operation override capability directly at the dam. The local programmable logic controller (PLC)-based controller provides flexibility to implement additional functionality should it be required. Power for the new hoists is from the existing power distribution at the dam with the existing diesel generator still available to provide backup power. Heat tracing was installed on the automated gates to prevent ice up in winter.

Safety systems—A safety cable is installed across the headrace channel inlet similar to the one installed at the adjacent dam. Sirens and/or warning lights were considered at the Wasdell Falls Dam and at the turbine/generator structure for rapid changes in water level that may create a public safety hazard but this was determined to not be necessary by the dam owner (MNR). Fencing and gates are installed to limit access to hazardous equipment. The VLH structure deck remains accessible for pedestrians with physical barriers installed at the entry to limit unauthorized vehicle access. The portage route is identified with signs, and enhancements were added to improve access to the river. Video cameras are installed to provide remote surveillance of key areas of the plant.

Figure 11.11 VLH turbines installed in the structure looking from downstream. The left-side turbine is in raised position; center turbine is shut down and raised slightly for boat inspection with stoplogs placed upstream; and the right-side turbine operating with some spill over its flap gate

Control building—The turbines have frequency converters for independent operation to maximize plant capacity and switchgear installed in the control building. The control building is constructed using standard "Sea-Can" shipping containers connected side-by-side. The external finishes are insulated metal panels colored to compliment the surrounding cottage environment on all sides, including covering the container double doors. The building is installed above maximum extreme flood water elevation. There is one main entry into the control room with three emergency exits at the opposite end of the building. It does not have potable water or sanitary service as those are available nearby. Interior finishes are painted sheet metal ceiling and walls capable of supporting the necessary wall mounted electrical and mechanical equipment, and the floors are rubber matted. Control, electrical and mechanical equipment are all located in separate rooms and laid out to have easy access for maintenance. Alarm and security systems are provided (Figure 11.12).

Control system—The plant has an automatic PLC-based control system with human machine interface, as well as a control and data-monitoring system. There is a supervisory control panel and separate control panels for each turbine-generator unit. The plant is continuously monitored by the control system and fully remotely

Figure 11.12 Control building and transformer yard on raised berm adjacent to the VLH structure. The control building comprises four Sea-Can containers built off-site then modified with cladding and a metal roof to emulate neighboring cottages

monitored by the operator. With the significant energy being handled through the frequency converters and the relatively small footprint of the control building, a powerful HVAC design was required for ventilation and temperature management that automatically adjusts for varying ambient temperature conditions based on the units power conversion.

Transformer yard—The power is stepped up from 500 V to 44 kV via a pad-mount transformer in the transformer yard adjacent to the control building. The transformer yard includes the main power step-up transformer, high-voltage switching and disconnect equipment, voltage and current transformers (VTs and CTs) for electrical system protection relays, grounding, fencing and signage. The transformer yard design is a typical open-air-type design as used in Ontario but its footprint was minimized by careful configuration and using a limited amount of equipment to create a compact accessible design. Major maintenance and servicing is done by equipment located outside the transformer yards fence. The main power step-up transformer is color matched to the control building. The surrounding ground resistivity grid is based on the site conditions measured during construction.

Interconnection line—The power generated is tied into the nearby existing 44-kV transmission line via a 4.1-km line consisting of 3.5 km of buried powerline, 0.6-km aerial wood pole line and a pole-mounted meter at the point of common coupling (PCC) near the highway. The 44-kV interconnection line extends from the plant to the PCC at Coopers Falls Road and then to a point of connection on the Hydro One M2 Feeder, with the nearest Hydro One Supply Station being the Orillia Transformer Station. The power metering is pole mounted at the PCC. The interconnection line construction was a joint project of Hydro One and Wasdell Falls Power Corp.

11.6.5 VLH Hydro Plant operation

The VLH Hydro Plant operates as a "run-of-river" facility, whereby the headpond level upstream of the dam is maintained at a relatively constant level. The headpond is not used to store water, so it does not affect the river flow passing the dam site, which is consistent with the existing dam operation rules (Figure 11.13).

11.6.5.1 Operational approach

Naturally occurring river flow that was previously passed at the dam is now discharged through the hydro plant, up to the hydro plant capacity, with the exception of the minimum dam release flows. Flow above the plant capacity passes over the VLH turbine flap gates and/or through the Wasdell Falls Dam by removing stoplogs and operating gates. This follows the same operating protocols as was done previously at the dam in accordance with the Wasdell Falls Dam Safety and Emergency Preparedness Study.

Figure 11.13 Completed Wasdell Falls VLH Hydro Plant with all turbines installed and in their raised position just prior to commissioning

Figure 11.14 Completed Wasdell Falls VLH Hydro Plant with all turbines operating during mild winter conditions

Automation of two gates on the Wasdell Falls Dam was done to allow automatic flow transfer between the dam and the plant for plant start-up and shutdown sequences to maintain continuous passage of the river flow and consequent maintenance of upstream and downstream water levels.

The Wasdell plant does not coordinate its output with other plants, as it is not dispatchable or operable under islanded conditions. The plant inherently provides grid support to the area, but it does not actively alter its output for this purpose (Figure 11.14).

An operator conducts routine inspections on a weekly basis regardless of plant operating status. The water levels upstream of the dam and the turbine-generator structure are continuously monitored by the control systems, and it has redundant sensors with error checking logic and alarming. Discharge through the hydro plant and dam is adjusted to maintain water levels at the Washago Mill Dam according to the previously established rule curve. The water level at the Wasdell plant is maintained to maximize head on the turbines and limit the risk of unnecessary spill.

Minimum flow is maintained through the existing dam to maintain flow over the broken rock fish habitat located about 50-m downstream of the Wasdell Falls Dam at the boulder weir on the northeast shore. The minimum flow is released through gate 1 on the dam during the prescribed (Walleye) fish spawning period, manually controlled by the plant operator based on the water temperature reported by a temperature transducer located just upstream of the dam.

11.6.5.2 Control modes

The VLH turbines are normally operated in "level control" mode, whereby headwater level transducers located upstream on the VLH structure continuously input the headpond water level to the control system. The control system then automatically increases or decreases the plant flow to maintain the headpond level at the required set point. This is done by adjusting the VLH turbine blade opening and rotational speed.

During turbine operation, the automated dam gates are set, so they always have enough remaining discharge capacity to pass the current turbine flow in the event of an unplanned plant shutdown. The number of units in operation and the flow distribution between units are adjusted to optimize energy production efficiency and accommodate the flow available for operation.

Bypass mode occurs when flow is required to bypass the turbines, as spill, this happens when

1. the river flow exceeds total plant capacity and minimum release flow combined, or
2. some or all of the turbines are unavailable for operation, or
3. river flow is less than the minimum required for continuous turbine operation and minimum release flow combined.

Flow is spilled initially by the turbine crest gates and then at the dam by automated gate operation and then by manual gate and stoplog operation, similar to the existing dam-operating protocol. When flows begin to exceed plant capacity, the headpond level is controlled by modulating the turbine crest gates.

For start-up mode, the plant is started by the operator from the control building or via remote computer access to the control system. The automated dam gates 5 and 6 automatically transfer flow between the plant and the dam during plant start-up or shutdown. During start-up, all available flow, less the minimum release flow, is transferred to the turbines until plant capacity has been reached. During plant shutdown, the plant flow is transferred back to the dam. If a problem (fault) occurs during start-up, operation or stop sequences, critical shutdown functions cause a lockout and require an operator to visit the site to correct the problem and restart the unit(s). Key alarms and shutdowns are included in the design, including high and low water levels, gate operation errors and electrical abnormalities.

For debris-passing mode, the turbines are designed to pass floating organic and small-to-medium size woody debris by spilling over the unit. The crest gate is automatically lowered briefly, on adjustable preset timing, to spill floating debris at a frequency dependent on the quantity of incoming debris. Large floating debris is stopped by the grizzlies upstream of the unit and is removed manually by the operator with a pike pole and/or a small truck-mounted crane.

The plant units are designed to be operated continuously unless there is a shutdown of one or more units due to water availability. Regular maintenance is done during low flow periods on one unit at time so that the other units can maintain production with the available flow.

11.6.5.3 Controls, auxiliary systems and communications

All key auxiliaries are designed to provide at least one level of redundancy, with emergency power to provide key functions for shutdown of the plant. Turbine mechanical controls are provided by one hydraulic power unit (HPU) per turbine. HPUs have hydraulic accumulators with sufficient capacity to close the turbine blades twice and for two full open/close cycles of the crest flap gates (Figure 11.15).

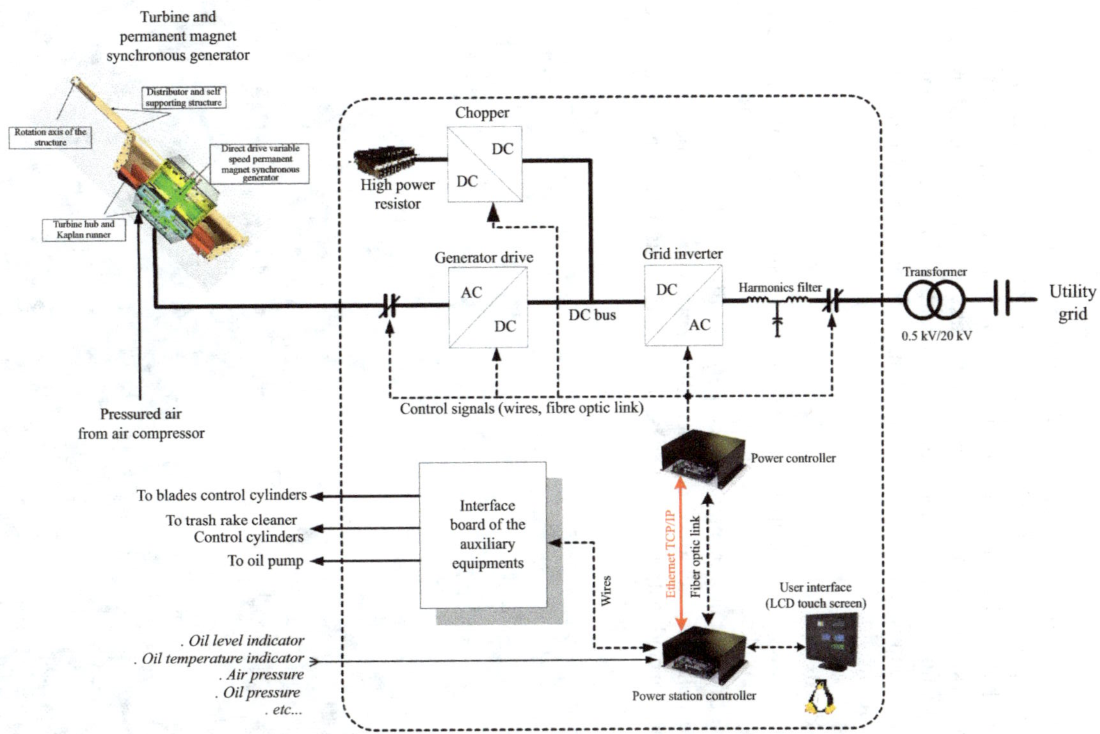

Figure 11.15 VLH Plant Electrical and Controls System showing the water to wire scheme. This AC/DC/AC conversation system uses back-to-back driver and inverter technology to convert the 12-Hz output from the turbines to 60-Hz and 500-V output to the transformer to step up to 44 kV (VLH diagram from Wasdell Falls VLH Operations Manual, MJ2 Technologies)

There is one air pressurization unit per turbine, each with one compressor-receiver tank, dryer, pressure regulator and safety relief valve. Receiver tanks have sufficient capacity to maintain hub pressurization for 48 h without recharge. Power for the extraction of the turbine frames is by a separate dedicated HPU for all units with accumulator backup. All auxiliary systems have individual starters and disconnects.

The primary source of station electrical power for the hydro plant is by the AC power from the plant generator(s) after frequency conversion. The dam operations continue to be powered as they were previously, which is by supply from the Hydro One system backed up by an onsite diesel generator requiring manual transfer.

A SCADA link for interconnection communications protection and metering is provided by a dedicated telephone leased line. Internet connection and phone connection are provided at the plant.

11.6.5.4 Plant operations

The plant is operated by one part-time operator who visits the site five times a week for a brief system check, and there is also a backup operator to provide assistance from time to time for larger maintenance periods (Figure 11.16).

The Wasdell Plant achieved commercial operation in December 2015 and has attained the expected output efficiencies. The VLH units have had some internal

Figure 11.16 The operating Wasdell Falls VLH Hydro Plant on the left and the original Wasdell Falls Dam on the right with the new automated center gates to coordinate flow passage

mechanical issues needing repairs that the manufacturer is working to address. Wasdell Falls Power Corp. has a contract with the utility to buy power for $130/MWh for 40 years under the Ontario Feed-In-Tariff program for renewable energy.

11.7 The art of engineering design

While the Wasdell Hydro project provides a good example of the deployment of new renewable energy technology, it also provides interesting insight into the "art of engineering design." Through a careful consideration of all aspects of engineering design ranging from environmental impacts and existing site conditions to detailed analysis of all required operating conditions, the design engineers delivered a project that optimized all aspects from footprint and site aesthetics to efficiency and performance resulting in an overall elegant engineering solution.

The critical aspects of engineering design for renewable energy facilities should not be overlooked. It is as much an art as it is a science, one that is tried to various degrees in other industries but is epitomized in this industry given its unique drivers for success; the renewables industry strives to a higher standard through all of its participants.

This industry is founded on environmental stewardship, gaining societal acceptance and inspiring community involvement. The investors, project developers, the government agencies, the stakeholders and indeed the practitioners of the technical design, supply and construction of these facilities take enormous pride in being able to do things differently, with a softer touch, with a balanced outlook and a "can do" spirit that is fulfilling to be part of.

In the case study on Wasdell Falls, the breadths of all the factors, both big and small, that were integrated into a relatively small renewable energy development are presented here to give some perspective of their multitude. This is not unique; as a matter of fact, this is the norm in how renewable developments are properly executed. It is not good enough to have a better proverbial "mouse trap," and it is imperative that the technology is intimately fit into the ecosystem, society and the local community while respecting all of the physical and engineering challenges. This is the "art of engineering design" that enables the renewable energy industry to excel. Each new development is striving to be best, but it can always be done better and simpler; innovation will not happen unless it is tried and tried again.

11.8 VLH summary

Small hydro continues to have significant potential for the development of renewable energy in Canada. The VLH turbine technology allows economic, low environmental impact, construction of existing low-head facilities by minimizing the civil works typically associated with hydropower projects.

VLH technology has demonstrated its promise through the successful installation of over 50 units in Europe and 3 units in Canada. With thousands of existing unpowered structures within North America, there exists great potential for VLH hydro deployment within Canada and the United States.

For successful deployment of this emerging technology, a knowledge base must be garnered to identify specific applicable sites. Scoping studies with the engagement of government agencies, universities, industrial proponents, stakeholders and consultants would advance opportunities to apply this technology throughout North America.

The VLH's low impact environmental and aesthetic design has achieved societal acceptance and public approval at numerous sites, including historic waterways. VLH Fish Study work has demonstrated unprecedented "fish friendliness" for a variety of fish species at numerous operating installations and garnered agency recognition as safe to operate in fish bearing rivers.

The Cold Climate Adaptation Study identified the critical issues and modification solutions for deploying the VLH European design in Canada's harsh northern climate to achieve year-round operation and confident performance in ice covered rivers with over 5 years of successful operation.

Comprehensive hydraulic studies have investigated the effect of upstream obstacles on VLH performance and established deployment criteria for real-word application. This knowledge, along with independent onsite full-scale efficiency testing in Europe, provides confidence in site selection and predicable performance that is needed to facilitate development investment.

To meet North American electrical requirements, modification of the frequency converters, controls and protection systems from the European 50 Hz to North American 60-Hz system has been accomplished and is now well understood.

The VLH turbine technology and deployment design has integrated the critical aspects required: from hydraulic considerations of project sites, to robust design for the harsh conditions of Canadian waters, to the protection and control of associated systems and finally to social and regulatory acceptance. Each element of the technology now has a strong foundation in research and development with specific site knowledge that has resulted in an innovative deployable turbine technology. This accomplishment provides a new renewable energy supply with commendable environmental and economic benefits particularly for rural and remote regions.

The knowledge gained has resulted in the successful first North American deployment of the VLH technology in Canada at Wasdell Falls. The ability to integrate all aspects of VLH deployment as outlined here has created a holistic design that bolsters the achievements that the VLH technology has attained in Europe.

This small hydro technology story demonstrates the art of engineering design as it has been applied over the past two decades of renewable energy development in Canada and the many global connections it fosters.

References

[1] Kemp P, Williams C, Sasseville R, and Anderson N. Very Low Head Turbine Deployment in Canada, *27th IAHR Symposium on Hydraulic Machinery and Systems (IAHR 2014) IOP Publishing IOP Conf. Series: Earth and Environmental Science 22* (2014) 062005. doi:10.1088/1755-1315/22/6/062005.

[2] Northwest Hydraulic Consultants Ltd. Andres D, 2010. Very low head hydropower installations: ice related design considerations.

[3] Fernando J and Rival D, 2014. Characterizing the influence of upstream obstacles on very low head water-turbine performance. *Journal of Hydraulic Research* 52.

[4] Lagarrigue T, Voegtle B, and Lascaux J, 2008. Tests for evaluating the injuries suffered by downstream-migrating salmonid juveniles and silver eels in their transiting through the VLH turbogenerator unit installed on the Tarn River in Millau. *MJ2 Technologies Fish Friendliness® Tests.*

[5] Lagarrigue T and Frey A, 2010. Tests for evaluating the injuries suffered by downstream-migrating eels in their transiting through the spherical discharge ring VLH turbine generator unit installed on the Moselle River in Frouard. *MJ2 Technologies Fish friendliness® Tests.*

[6] ECOGEA, 2013. Preliminary report on testing fish mortality during downstream migration through the VLH in Millau, at La Glacière. *MJ2 Technologies Fish friendliness® Tests.*

[7] Tuononen E, 2019. Fish Community Interactions with Very Low Head (VLH) Turbine Technology, MSc Thesis, Carleton University.

Chapter 12

Development and experience with a vertical-axis hydrokinetic power generation system

Clayton Bear[1] *and Michael Bear*[1]

This chapter chronicles New Energy Corporation Inc.'s development and experience with its vertical-axis cross-flow hydrokinetic power generation system. New Energy has taken this concept from initial trials to full commercialization.

The vertical-axis turbine is suited to a wide variety of applications, including rivers, canals, industrial outflows, and tidal streams. The turbines can be installed on a floating platform or fixed in position. A distinguishing feature, the vertical orientation, means that the generator and power conditioning equipment can be situated above the water permitting the use of conventional components. These systems can be configured as complete water-to-wire solutions for both remote stand-alone and microgrid applications, and centralized grid-connected power generation applications in single unit or array configurations. The systems are quiet, unobtrusive, produce zero emissions, and have a low impact on aquatic life. To date, system sizes of 5–25 kW have been developed with plans to increase to 250 kW.

New Energy has conducted a long-term development program primarily focused on building and field testing its systems in a variety of operating conditions and application types. Initial turbine rotor designs and control concepts were tested in the outflow of a wastewater treatment plant and then on a purpose built self-propelled platform in open water. The test program then moved to a hydro plant supply channel, testing throughout the winter in subzero temperatures, to evaluate and refine system designs in preparation for first deployments.

New Energy has deployed its systems in various locations around the world in a variety of applications. These range from an installation in a small stream high in the Himalayas to power a village, to powering a school in Myanmar, to producing grid power from a canal in India. The installations have presented some unique challenges, including transporting hardware to a location with a 3-day walk from the nearest road, finding a local partner in geographically remote locations to provide long-term service and support, and dealing with different cultures and political situations.

[1]New Energy Corporation Inc., Calgary, Canada

12.1 Introduction

An introduction to vertical-axis turbines is given in Chapter 4. This chapter focuses on the development and experience with New Energy Corporation Inc.'s EnCurrent and EnviroGen Hydrokinetic Power Generation systems. Some of the unique challenges of installation are highlighted.

12.2 History

The first attempts at harnessing water as a source of energy was through waterwheels used for grinding, pumping, and driving various types of equipment. Some accounts suggest that waterwheels were first used as far back as 4,000 years ago. These waterwheels used either the elevation change across the wheel to turn it as in the case of an overshot waterwheel, akin to today's hydro turbines, or used the velocity of the water to turn the wheel as in an undershot or vertical-axis waterwheel. The undershot and vertical-axis waterwheels were based on the principle of drag, with the force of the water pushing buckets or paddles along causing the wheel to turn. Generally, this was a highly inefficient means of energy extraction.

The concept of a vertical-axis turbine using the principle of lift instead of drag was first developed and patented by the French inventor Georges Darrieus in 1931 for a vertical-axis wind turbine. This principle is essentially the same whether the turbine is used in wind or in water. In the early 1980s, the concept was adapted to water through a program led by Barry Davis with support from the National Research Council and Natural Resources Canada as well as private investors. During this program, a series of prototypes were built and tested. These prototypes experienced varying levels of success, but all proved the technology was viable. However, the concept was never taken to a fully commercial state.

New Energy Corporation Inc. was formed in late 2003 with the intent to complete the commercialization of the vertical-axis hydrokinetic turbine and begin commercial production and sale. New Energy's turbine designs leverage the R&D previously undertaken and the products build on the designs and recommendations from this program, adding enhancements to improve the performance and operational characteristics of the turbine.

12.3 Technology

As mentioned previously, New Energy's vertical-axis hydrokinetic turbine is based on the Darrieus wind turbine concept. The main difference between New Energy's design and the original Darrieus concept is that New Energy's turbine employs straight blades as opposed to the characteristic eggbeater design of the Darrieus turbine.

In New Energy's design shown in Figure 12.1, a series of aerodynamically shaped blades are mounted parallel to the vertical shaft in a concentric and equi-spaced arrangement. The individual hydrofoils are connected via support arms to a

Figure 12.1 5-kW rotor and generator

central shaft that then transmits the torque to a generator or other power transfer device. In the turbine's simplest form, the blades are mounted rigidly to the arms at an optimum pitch angle designed to maximize power extraction efficiency in the desired range of flows. In this configuration, the turbine rotor design is mechanically simple and robust. Variations in this design include variable pitch blades such as those used in the Kobold turbine [1,3], or blades that are angled to reduce torque variations over a blade cycle such as those used in the Gorlov turbine [2–4]. Each configuration has advantages and disadvantages, but all work relatively well in their intended applications.

The shape of the hydrofoils is a key element in generating the torque that is ultimately converted to rotational energy. The hydrofoil (airfoil) shapes used in New Energy's turbines are based on shapes that were created and cataloged by the National Advisory Committee for Aeronautics to give developers of various aerodynamic (and hydrodynamic) devices the tools they needed to help them select specific desired performance characteristics for a specific application [5].

The basic principles of vertical-axis turbines are described in Chapter 4. There it is shown that the hydrofoil shapes are not only optimized for specific lift characteristics but can also be used to minimize drag that may seem intuitive but is an important consideration for any component exposed to a fluid stream, not just the hydrofoils.

Depending on the flow regime and other characteristics of the application, the turbine can have a reasonable number of blades in combination with an appropriate chord length and blade profile. Typically, three to five blades are used. New Energy has settled on a four-blade design that is thought to be a good compromise between efficiency, complexity, and torque pulsations. A three-blade design may be slightly more efficient but has higher torque pulsations, while a five-blade design may have a lower efficiency but may be better suited to restricted flow applications.

The vertical-axis configuration is also well suited to extract energy from flows such as rivers, tidal streams, and human-made channels since the rotor diameter to height aspect ratio can be tailored to maximize the swept cross-sectional area of the stream. Most rivers are shallow and wide making a rotor with a 2:1 aspect ratio more practical.

12.4 Applications

New Energy's turbine can be applied to any application where a moving stream of water with sufficient volume exists. This includes industrial outflows, irrigation canals, rivers, and tidal streams. The two broad applications for hydrokinetic power generation systems are "free stream" that relies only on kinetic head in the flow and "restricted flow" where a small potential head can be developed across the turbine.

In free stream applications as shown in Figure 12.2, water moves freely around the turbine so no potential head can be developed; however, energy can be extracted from the kinetic head of the flow passing through the turbine rotor. A trade-off exists between the amount of flow that will freely flow through the turbine, and the amount of resistance to that flow through the energy extraction efforts from the hydrofoils. The Betz–Joukowsky limit demonstrates the optimal point beyond which additional resistance reduces the flow available for energy extraction to a greater degree than the extraction per volume of water.

In fast moving flows, the turbine is simply deployed with an open rotor, whereas in slower water it may be necessary to restrict the flow to increase the kinetic energy at the rotor as illustrated in Figure 12.3. This enables consideration of a much larger number of sites for the installation of this technology. Note that these systems whether free stream or restricted flow can be installed as floating or bottom mounted units.

For restricted flows, potential head up to about 1.0 m can be utilized for energy extraction above which cavitation becomes an issue and other conventional low-head hydro technologies may become more suitable. However, closely spaced turbines can produce similar effects. The peak efficiency in this style of installation is not governed by the Betz–Joukowsky limit but falls somewhere between that of a more traditional hydro installation and that of a free stream installation. Due to the low energy density of the water, drive system losses are a significantly larger component of the total energy thus restricting hydraulic efficiencies to less than

Figure 12.2 5-kW floating system

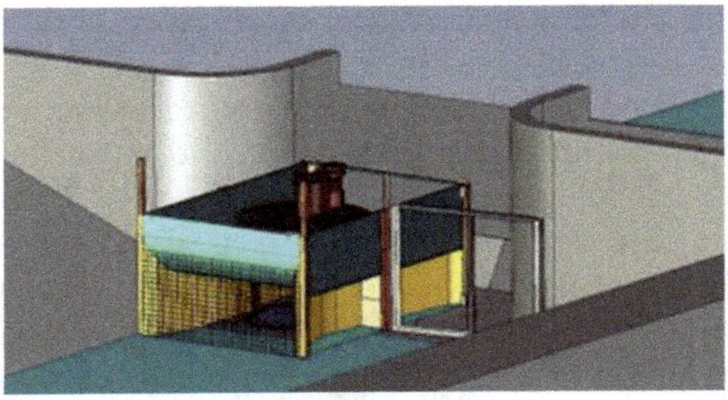

Figure 12.3 Ducted turbine concept in restricted flow application

60% for current designs. Restricted flow applications are found in existing low-head dams, lower velocity flows in irrigation systems, and industrial outflows as well as river systems where room exists to back the flow up to a small extent.

Industrial outflows can include municipal sewage treatment plants, thermal power plants, pulp mills, desalination plants, and other plants that use a high volume of water. Energy in the outflows is normally lost as the water leaves the plant, but some of this energy can be recouped and fed back into the plant or used for a specific local process. Irrigation canals are typically slow-moving streams with occasional drop points. Effort is spent to dissipate energy at these drop points to minimize bank erosion, but instead this energy can be extracted at these points and supplied to local loads. These applications are generally small depending on the volume of water flowing through the canal but can be large in quantity of sites given the number of canals in North America and overseas in countries such as India.

The output from the turbine does not need to be limited to the generation of electricity. It can also be used to directly pump water. While this has not been done to date with New Energy's hydrokinetic systems, there is a need to pump water for irrigation, especially in the developing world.

12.5 System configuration

The vertical-axis cross-flow design makes it natural to have the electrical equipment, specifically the generator, above the water line. In most natural waterways, this typically means floating the powertrain on a small platform or barge and anchoring to the bottom of the waterway. Installation, monitoring, servicing, and retrieval are all much simpler than if the equipment was mounted in the bottom of the waterway. As well, the highest water velocities are generally in the top third of the water column. Deployment and removal can be done with the turbine rotated out of the water simplifying transport to that of moving a small barge. Floating the system also allows it to rise and fall with the water surface.

In smaller streams where the flow must be concentrated to ensure sufficient depth to cover the rotor, a fixed arrangement can be constructed to hold the rotor in position. This is also true of human-made channels especially if concrete supports are available.

In large rivers, arrays of turbines can be installed and arranged to maximize energy extraction. The turbine is also ideally suited to application in tidal streams or estuaries since operation is independent of flow direction. It can be incorporated in a bridge structure or in narrow channels between islands where the channel can be spanned, or it can be deployed on its own. Access to mechanical and electrical equipment is relatively easy since most of the equipment is above water, and on-site equipment is available for the removal of components. In all cases, the impact on the environment, fish, and navigation must be considered.

New Energy's power generation systems can operate as complete water-to-wire solutions and can be configured to operate in stand-alone mode, as part of a microgrid system (either as primary or supplementary) such as a diesel-generator-based system for remote communities, or they can be connected to the main centralized grid. Aside from offering a clean environmentally responsible and economical source of power, the distinguishing feature with using water as the power source is that it is constant 24 h a day in river applications and in tidal applications is extremely predictable. Remote communities and stand-alone installations can take advantage of this feature in assessing and meeting capacity needs, know that it is a reliable energy source for critical applications such as telecommunications for distance learning and remote healthcare, or refrigeration for food or medicine.

For stand-alone applications, the grid-tie inverter needs a grid forming signal with which to synchronize. This is handled with a battery charger/grid forming inverter. A typical configuration is shown in Figure 12.4. The batteries then can be used to store excess energy and can be used as an additional energy supply for times of high usage, effectively doubling the maximum output of the system until such time that the stored battery energy is depleted. The batteries can then be recharged during times of low usage. With other forms of renewable energy, there is no certainty about when additional power can be generated, so the energy in the batteries must be used judiciously until the next opportunity to generate power.

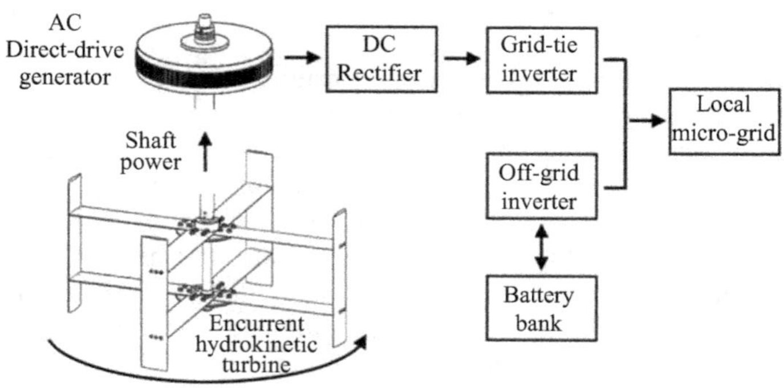

Figure 12.4 System block diagram

With grid-connected installations (microgrid or main), grid-tie inverters are simply connected to the grid.

With the variability in water speed, optimal rotational speed of a fixed geometry turbine varies with it. Thus, the output from the generator will be an uncontrolled variable frequency and variable voltage signal. To turn this into useful electricity, it is necessary to first rectify the generator output and then provide this DC output to an inverter that, in turn, will output a controlled AC signal at the required voltage and frequency.

12.6 Development program

New Energy's development program has been focused on identifying important aspects of the turbine, optimizing performance, and understanding system and operational characteristics required to bring a robust yet highly functional system to the market. The development program consists of three components: theoretical modeling, model verification through experimental testing, and longer term field testing.

12.6.1 Early days

Theoretical modeling was initially done on the basic turbine design using code originally developed for the wind industry and subsequently adapted for water. The computer performance model employed predicts the performance of a ducted turbine resulting in quantification of the power, discharge, and extraction-efficiency coefficients. This was an iterative numerical approximation based on a dual stream tube analysis (one passing through the turbine and the other through the space between the turbine and ductwork) based on an analysis derived from first principles. This code gave a good first approximation of the expected performance of the turbine under normal conditions. However, it was not designed to consider the effects of cavitation on performance. As well there can be no comparison of different ducting arrangements with this code as it relies on empirical inputs for any duct configuration. Effects of different ducts must be tested empirically. Even so this modeling exercise was valuable in establishing basic performance and design criteria for initial prototype rotors.

12.6.2 Novel approach

To validate the initial prototypes, a novel approach was used for experimental testing. A development platform and a test turbine system were built on the basis of the state of the art at that point of time. The test platform, shown in Figure 12.5, consisted of a 4.5 m×10 m pontoon boat driven by a 250-hp outboard motor, customized so that a turbine could be mounted between the pontoons and moved at a precise speed through still water. A 2.5 m×6 m open center section allowed the installation of various turbines and test fixtures. The initial test prototype consisted of a 1.0-m diameter×0.5-m high rotor capable of producing 2.5-kW shaft power. It was connected to a variable speed pulley and step up gearbox arrangement driving a synchronous generator. Power generated was dissipated through a

Figure 12.5 Pontoon boat test platform

Figure 12.6 Development turbine

resistive load bank. The development rotor in Figure 12.6 has adjustable rotor hydrofoil pitch angles and configurations combined with the ability to mount various fixtures allowing us to test and quantify a matrix of key variables relating to turbine performance and operation for ducted and un-ducted systems.

As well, not only did this test platform give us the ability to accurately measure and control our speed through the water, as can be done in a tow tank, but also we had the ability to conduct long test runs ensuring that steady state was achieved at each operating point, as can be done with a test flume. Overall, the test program validated performance predictions for the initial prototypes and allowed us to move on to field trials.

Initial performance testing was followed by tests of the surrounding infrastructure, including the evaluation of drivetrain, electrical generation components, control schemes, and the impact of ducting arrangements on the performance and energy extraction capabilities of the technology.

The field testing program supported by Natural Resources Canada consisted of building and deploying between twelve and eighteen 5- and 25-kW turbines in a variety of field applications and configurations, closely monitoring the performance of these systems, and adopting any operational improvements as a result of this experience.

The first system developed was a constant speed design employed in a stand-alone application using a constant speed synchronous generator with a load bank and load controller. The turbine developed for this application was a plate design with a rated capacity of 5 kW. Speed control was achieved with a Thompson and Howe load controller widely used in stand-alone micro-hydro applications. The controller used a load priority methodology. Speed control was achieved by modulating any excess generated power through a resistive load, dissipating this energy as heat. The system shown in Figures 12.7 and 12.8 was installed in the effluent outflow of the Bonnybrook Wastewater Treatment Plant in Calgary, Alberta.

Figure 12.7 Ducted 5-kW turbine

Figure 12.8 Installation at Bonnybrook Wastewater Treatment Plant

The Bonnybrook installation proved out the ability to operate the turbine in a constant speed configuration. Power generation was in the range of 3–4 kW during design flow conditions utilizing only part of the effluent flow stream. However, during off-design flow conditions, which occurred twice during a daily cycle, power generation would rapidly drop to zero since the turbine operating point would migrate down the front side of the performance curve. This is one of the drawbacks of operating as a constant speed system in a variable flow regime.

The original turbine design consisted of a heavy aluminum top plate with the hydrofoils welded around the periphery, and the bottoms welded to an annular ring. The top plate was mounted to the drive shaft. This design proved difficult to manufacture and created dynamic alignment issues with the flow and as a result was replaced with a center shaft design using radial arms to support the hydrofoils. Figure 12.9 shows the difference in these two design approaches.

The center shaft design was tested in an irrigation canal over the better part of an irrigation season (Figure 12.10). Of special interest was the impact of weeds and other debris in the water. Weeds did not pose a large issue with the turbine, but they did catch on other obstructions near the water surface such as a flow meter installed as part of the data collection system. The irrigation canal water carried significant silt and dirt. After 1,200 h of operation, the blades were inspected and found to have experienced some polishing due to abrasion. This was considered an easily manageable issue with the use of coatings or surface hardening.

It was apparent from the Bonnybrook testing that applications with highly variable flows required a variable speed turbine system allowing operation at peak efficiency for a wide range of flow conditions. To this end, a variable speed system was designed and constructed capable of operating under a wide range of flow conditions. This system used a high efficiency permanent magnet generator and gearbox connected to a battery charger/inverter.

Figure 12.9 Center shaft vs plate. (The five blades shown were initially chosen because it was thought that four blades would increase torque pulsation magnitude due to symmetry. This was later determined not to be a significant issue.)

Figure 12.10 Installation in Western Irrigation District canal

Figure 12.11 Prototype 5-kW system

12.7 Initial prototypes

The first prototype system was initially tested on the pontoon boat test platform and functioned as predicted. However, cogging torque from the permanent magnet generator combined with the high step-up ratio of the gearbox (75–1,800 rpm) resulted in excessive breakaway torques for the turbine. Once operating, this was not an issue but it meant that the turbine was not self-starting under most flow conditions, needing to be started externally. This could be done electronically but was temporarily addressed with a simple hex head on the top of the shaft to briefly spin the rotor to get it going. The initial prototype system shown in Figure 12.11 was installed on the Yukon River in Alaska.

12.7.1 Pointe du Bois test program

In the fall of 2007, New Energy initiated a new testing and development program in Pointe du Bois, Manitoba. This test program was designed to test the equipment under extreme environmental and real field, full system conditions.

Pointe Du Bois is in Eastern Manitoba on the Winnipeg River. The test site was set up by the University of Manitoba with support from Manitoba Hydro. The actual site was a supply channel upstream of the Manitoba Hydro Pointe du Bois dam and hydro plant. The University of Manitoba installed anchors in the bottom of the channel designed for mooring hydrokinetic equipment. New Energy provided a test barge to be used as a platform for testing equipment. Manitoba Hydro also provided permission to connect to the local power line, thus providing access to the grid.

The test program was designed to provide a location for long-term testing of a complete system in real field conditions, including severe weather, ice, and debris. This approach was successful in that it not only confirmed performance of the equipment, but it also identified other operational issues, resulting in solutions to some of those issues.

One of the characteristics of hydrokinetic turbines is the large drag forces associated with the extraction of energy. It stands to reason that lift and drag have an association given one is 90° to the other. Effectively, this drag is roughly located at the centroid of the rotor cross section parallel to the flow. The drag creates a moment arm from this point to the point at which the boat or barge is moored (ignoring any angle off horizontal for the mooring line). The torque created by this moment pushes the front of the boat into the water. This can be compensated by using a long hull to maximize the counteracting buoyancy moment, or the effective mooring point can be lined up with the center of drag. New Energy's solution, shown in Figure 12.12, was to take this one step further and mount the entire drivetrain on a pivot while holding the pivoting assembly above and below the torque center [6]. This decouples the drag from the boat, reducing the buoyancy requirements to simply supporting the weight of the system, and thus reducing the size of the boat required.

The control system used a look-up table based on a theoretical power output curve to schedule the output power for a given water velocity. If an attempt was made to draw too much power, the turbine would stop rotating (an unstable situation), so the table was set up to draw slightly less power than available. However, in an unrelated effort to reduce costs by using different materials and profiles for support arms, system performance was observed to be noticeably worse with less aerodynamic-shaped arms. Clearly, the rotational drag had a significant impact on rotor performance, so a program was initiated to minimize this parasitic drag. As this drag was reduced, performance steadily improved.

As previously mentioned, it was thought that with a three-stage gearbox and high-speed synchronous motor, the starting torque was just too high for self-starting, but with the performance improvements this issue disappeared (Figure 12.13). The by-product of this program was consistent self-starting capabilities once the drag was minimized. Problem solved.

Figure 12.12 Mooring arm rigging and pivot support

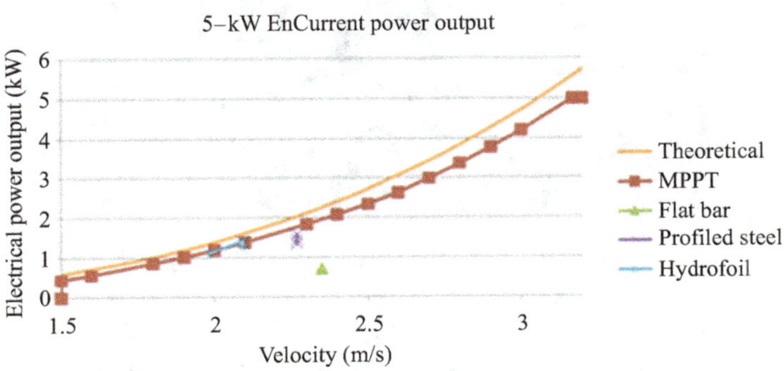

Figure 12.13 Theoretical performance, look-up table (MPPT) points, and actual performance curves with data points for various support arm configurations

Ice was a major issue to which we had to contend. While the fast-flowing channel did not freeze, with temperatures as low as −39°C, anything exposed to the air was obviously below freezing so water splashing on the boat or any of the equipment immediately froze as can be seen in Figure 12.14. Ice buildup was an

Figure 12.14 Test barge showing severe icing during cold weather testing

Figure 12.15 Testing the 25-kW system

ongoing issue, especially with temperatures below −15°C. Weekly efforts to break off as much ice as possible were undertaken during the coldest part of the winter. Interestingly, the turbine rotor remained relatively ice free during this time that even with super cooled water, operation could continue with a fully submerged system provided the water surface was not pierced.

During the spring, testing was expanded to include a 25-kW system and continued for the remainder of the year (Figure 12.15). Testing was conducted to validate the 25-kW design, pivoting mooring system, self-starting capabilities with the larger rotor and various support arm profiles, and the power conditioning hardware interconnection.

12.8 First-generation designs

First-generation 5- and 25-kW systems, branded the EnCurrent product line, were based on the development work demonstrated at the Pointe du Bois test site. The drivetrain for these systems consisted of an overhung rotor supported with conventional bearings above the water line, a multistage step-up gearbox, coupled to a conventional high-speed generator (Figure 12.16). Safety systems included a disk brake attached to the generator and as a last resort the entire rotor assembly could be rotated out of the water until horizontal (transport mode). The power electronics were generally configured for grid-tie or microgrid connection.

While the systems operated well in this configuration and with these features, it was believed to be highly desirable to eliminate the gearbox from the design from simplicity, reliability, and environmental perspectives—simplicity in that a multistage gearbox adds considerable complexity to the drivetrain; reliability in that gearboxes are a significant failure point; and environmental in that the gearbox is full of oil and tends to leak around the shaft seal potentially causing an environmental issue.

The wind industry had been considering alternatives in their drive systems due to gearbox failures. Low-speed generators that are direct drive or include a single-stage speed increaser have been developed for wind applications and were close to meeting the requirements for generators in New Energy's EnCurrent turbines, but operate at slightly different speeds and lack functionality specific to the water current turbine. As a result, New Energy embarked on a development program to develop a high-efficiency low-speed generator for use with its turbines.

12.8.1 New bearing concept

Additionally, having to use conventional bearings above the water line in an application with extremely high thrust loads meant that the rotor shaft and support housing were required to be correspondently large and heavy to handle the bending

Figure 12.16 First-generation 25-kW system

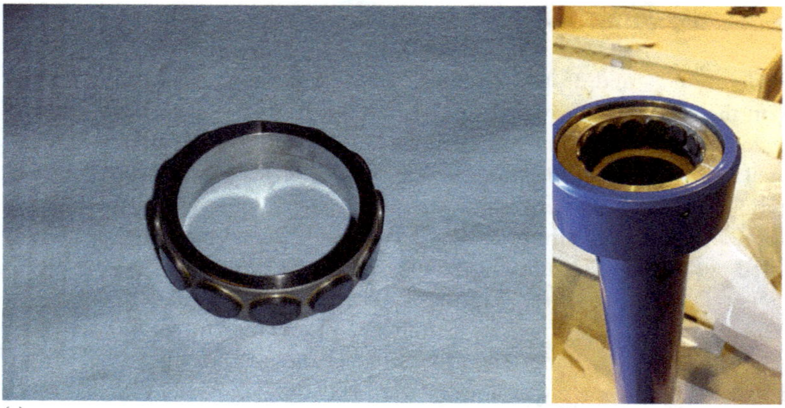

(a) (b)

Figure 12.17 Diamond button bearing rotor and stator

loads. It would have been beneficial to move the shaft support bearing down as low as possible to reduce the bending moment arm, but no bearing technology existed that could withstand the high loads, low speed, and submerged fluid environment without an expensive sealing system.

A novel approach to this problem was to develop a bearing that could operate below the water line that was impervious to impurities in the water such as mud and silt and needed no external lubrication other than the water itself. The solution was to develop a bearing using diamond buttons (used in drilling bits and down hole applications) as the wear surface, shown in Figure 12.17 [7,8]. Diamond on diamond has a low coefficient of friction, is extremely hard, sees almost all other materials as a lubricant, and can operate in an environment with contaminants. As such, a diamond-button-based radial bearing was developed that could be installed below the water line and exposed directly to the water regardless of mud, silt, or other contaminants in the water.

12.9 Second-generation designs

New Energy's second-generation products, branded EnviroGen, as shown in Figure 12.18, incorporated the learning from the initial deployments of the first-generation systems. Improvements include incorporation of the diamond bearing, elimination of the gearbox, development of a direct drive generator, and improvements in the power conditioning hardware.

Since the diamond bearing can be fully submerged and in fact requires submergence to pull heat away from the faces, it was placed just above the upper rotor support arms. This dramatically reduced the bending moment and, in turn, reduced the required size and weight of the rotor shaft. With no need for lubrication and the long wear life of the bearing surfaces, the long-term maintenance requirements were reduced, improving reliability of the drivetrain.

Figure 12.18 Second-generation 25-kW system with direct drive generator

Eliminating the gearbox and developing a purpose-built low-speed direct drive generator meant that the turbine shaft could be connected directly to the generator without the need for an exceptionally large coupling (necessary due to the low speed). The trade-off to this approach is that the low-speed generator must then be correspondingly larger and heavier.

The base technology for power conversion between alternating current and direct current is widely available due to the increase in clean energy applications. Solar PV modules produce DC current, so this must be inverted for grid connection or even stand-alone systems that require AC power, so solar inverters are quite common. The primary difference between a solar inverter and an inverter required for hydrokinetic power generation is in the setup of parameters, primarily how maximum power point tracking is done. Using "hill climbing" algorithms that test the operating points and select the optimum power extraction level works well for a noninertial system, because there are no moving parts. With an inertial system, this algorithm will work provided care is taken to ensure parameters are chosen to maintain stability and smooth operation. Otherwise passive techniques such as look-up tables can be used to closely approximate maximum power curves.

Since hydrokinetic systems produce wild AC with varying voltage and frequency, this must first we rectified before it is suitable as an input to these inverters.

With the dramatic increase in the development and sales of electric vehicles, bidirectional power management systems are becoming increasingly common and available. An electric vehicle accelerates by taking DC energy from batteries, inverting it, and driving a variable speed motor that turns the wheels, but when the brakes are applied, this same motor turns into a variable speed generator, output of which must then be rectified to recharge the batteries. This is like the conversion process that occurs with a hydrokinetic system: variable frequency AC power is generated, then rectified, and then inverted at the appropriate frequency and voltage.

Technology and availability of products for energy conversion is still rapidly evolving.

12.9.1 Impact on aquatic life

With the cross-flow turbine rotor design, the hydrofoils are mounted parallel to the shaft, so the entire blade moves at the same speed as opposed to horizontal propeller-type rotors where the speed of the blade is a function of the distance from the shaft. Combined with the relatively slow tip speed ratio, the cross-flow turbine design is not difficult for fish to avoid.

To confirm this is in fact the case, New Energy has extensively monitored its turbines for impact on aquatic life. An independent study initiated by the Electric Power Research Institute and supported by Natural Resources Canada was conducted by the S.O. Conte Anadromous Fish Research Center located in Turner Falls, MA, on juvenile Atlantic Salmon and adult American Shad to determine the impact of a 5-kW New Energy turbine on fish having to traverse a restricted channel. Figure 12.19 shows the turbine installed and in operation in a very tight channel.

The fish were instrumented with transmitters, released into the channel, and behavior monitored. Even though this was an extreme situation with the fish having to pass close to or through the turbine, no mortality or injuries were observed [9].

12.10 Tidal deployments

New Energy has deployed and demonstrated the turbine technology in a variety of permanent field installations, with 5- and 25-kW deployments in controlled rivers, uncontrolled rivers, small streams, canals, and estuaries. Our freshwater demonstration program provided us with valuable field experience needed to manage the

Figure 12.19 5-kW EnCurrent turbine installed and operating in the test flume where fish were released and monitored

Figure 12.20 25-kW system being secured to a bridge pier in an estuary with provision to rise and fall with the tide

risks associated with moving into a tidal application. Since the technology is very adaptable, the next logical step was to build a system for a tidal application, with the first system, shown in Figure 12.20, installed in an estuary.

Tidal sites differ from freshwater sites in that the tides flow in one direction and then reverse and flow in the opposite direction twice a day (called a semidiurnal tide) in most locations around the world but can also occur once a day (diurnal tides) or even mixed. Regardless, the characteristics of a tidal flow are that the velocity of the water is continuously variable, the water surface height is continuously variable, and the flow is bidirectional. Along with the corrosive properties of salt water, these characteristics present unique challenges for hydrokinetic systems.

The vertical-axis cross-flow design has the advantage of being directionally independent. The rotor does not care what direction the water is flowing, it will always turn in the same direction. Thus, the rotor does not need to orient itself to the flow. Mooring the vertical-axis turbine in a tidal application can be done by either holding the system in place regardless of the flow direction requiring anchoring and mooring in two directions, allowing the system to pivot about a single anchor point, or securing it to a fixed object such as a bridge pier, regardless of method, provision must be made to handle the changes in surface level.

The continuously variable flow velocity presents a challenge for power conditioning in that the system cannot be tuned for a specific velocity as in some freshwater applications. A detuned system will have a low efficiency and thus low-power output because of a poor power factor. Active rectification is required to ensure reasonable efficiency through the range of tidal flow velocities. This development work is ongoing.

12.11 Summary

The vertical-axis cross-flow turbine has shown to be a versatile and effective hydrokinetic power generation technology. The rotor configuration allows it to be deployed in a broad range of application types and situations. The floating arrangement means easy deployment, retrieval, and servicing. The vertical-axis cross-flow design is independent of flow direction eliminating the need for yaw control in bidirectional tidal flows. The relatively simple design, although highly engineered, results in a lower capital cost of the equipment as well as lower installation costs.

New Energy continues to advance the state of the art from the hydraulic, control system, and physical perspectives, especially with deploying systems in tidal applications generally located in much more energetic flow regimes. Hydraulic modeling work has been conducted using supercomputing capabilities to determine the interaction between turbines [10] with future work planned to model the impact of turbulence, impact of wake structures from the relative positioning of multiple units, cavitation at higher water velocities, and optimizing the design of the rotor. As mentioned in the previous section, variable flows present a control system challenge to maintaining efficient operation through an entire range of flows with a variable speed rotor design so active rectification must be incorporated to maintain a reasonable power factor throughout the range. From a physical perspective, bidirectional currents and varying water surface levels require further development of anchors, mooring lines, flotation, and electrical cable lines to shore. This work continues, with the mantra that has always been part of New Energy's culture—to get equipment in the field as soon as possible to really evaluate what works and what does not work.

12.12 Case studies

The following case studies describe the experiences New Energy has had in deploying hydrokinetic power generation systems in geographically remote locations around the world, and some of the challenges encountered with these installations.

12.12.1 Ringmo, Nepal

The Himalayan Mountains include some of the most rugged terrain and harshest climate conditions anywhere on the Earth. Total 19 major rivers and a vast number of tributaries flow through the Himalayas. Villages line these rivers with many being largely inaccessible by road and often do not have access to electricity. These communities are truly "off the grid." Some have access to solar arrays, or small wind turbines, but with most located at the bottom of steep valleys, the sun is limited, and wind turbines must be mounted high up on ridges, presenting difficult logistic challenges. While micro-hydro exists in some locations, not all villages are located near waterfalls suitable for generating hydro power.* Ringmo (Figure 12.21) is one of those locations.

*Many small hydropower systems in Nepal use a canal beside the river to produce a head of around 10 m. This is labor intensive but avoids needing a waterfall.

Figure 12.21 Project site—Ringmo

Ringmo, Nepal, is a village located high in the Himalayas in the Dolpa region of Northern Nepal, along the banks of a small stream that flows out of Phoksundo Lake. There are no roads leading to the village. Access from the nearest motorized transportation is by way of a multiday hike through the mountains. All materials must be carried in by hand, or on the backs of pack animals.

With no existing electricity, and limited options for generating power, New Energy was approached by Himalaya Currents and the World Wildlife Fund to see if it would be possible to generate electricity for the village from the stream flowing alongside the village. After visiting Ringmo, it was determined that with some channeling of the river flow, it would be possible to install a small hydrokinetic power generation system. However, given the remoteness of the village, logistics in getting a system into this location would be a challenge. New Energy's second-generation 5-kW EnviroGen Power Generation System rigidly mounted was chosen as the best size and configuration for the site with the lightest overall weight.

12.12.1.1 Transportation

Getting the equipment to Ringmo meant transporting all components by air to the nearest landing strip where they were then carried by hand or on pack animals (Figure 12.22). With no roads, this trip took several days on trails and over passes. It was clearly important to minimize weight of individual components and be mindful of the weight of the heaviest components. This factored into the selection of the 5-kW EnviroGen system which is a direct drive system with no heavy speed reducing gearbox.

12.12.1.2 Site preparation

Since the stream was limited in size, it was decided that a weir would be built across the stream bed with water channeled through one narrow opening slightly wider than the turbine rotor. The weir would also serve to moderate the flow through the channel by allowing spillage over the weir during times when the flow in the stream was high. The weir was built out of rock filled wire cages (gabions)

Figure 12.22 Transporting turbine components to Ringmo

Figure 12.23 Site preparation, Ringmo

along with a support structure to support the turbine rotor housing. Local village residents were extremely helpful in building the weir structure (Figure 12.23).

A control room was built using traditional methods and materials of stone, mud, and timbers (Figure 12.24). This room housed the power electronics, including inverter, battery charger/grid forming inverter, batteries, and switch gear.

(a)

(b)

Figure 12.24 (a) Control building and (b) power conditioning equipment

Figure 12.25 Two views of the installation of turbine in gabion structure

12.12.1.3 Installation

Installation included assembling the turbine system on site (Figure 12.25), which could all be done with hand tools mounting into the support structure, running the electrical cable to the control room, and installing the power electronics and battery storage system.

12.12.1.4 Operation

With 40 dwellings in the village, each house was allotted 50 W of energy usage on average leaving some additional capacity for other uses and to ensure that during peak times there will be enough capacity available to avoid a brownout. No one wants the power to go out in the middle of their favorite show!

12.12.1.5 Learning

Installing a hydrokinetic power generation system in an extremely remote location has its challenges, especially with the logistics of getting the equipment to the site. It is also essential to have the proper tools available for the assembly and installation, and for the unexpected that invariably occurs because it is a long way back to the hardware store. Of equal importance is that there are resources available locally or at least regionally to provide help if something goes wrong. Something as simple as a reset switch or a failed battery can make the difference in whether the system is generating power or not. The more remote the location, the more important it is to have local service and support capabilities.

In the end, this installation successfully demonstrated that reliable continuous power can be provided to the most remote locations with just a stream flowing by the community. As with traditional micro-hydro energy from flowing water can provide energy 24 h a day that can then be used for critical applications such as telecommunications for distance learning, and remote healthcare as well as for the refrigeration of food and medicine. New Energy's EnviroGen turbine is a key building block to power Ringmo's nascent micro economy.

12.12.2 Hpa-An, Myanmar

Myanmar has one of the lowest rates of electrification in Asia with about 50% of households connected to the public grid as of 2015, with even lower rates in rural communities. At best, some of these communities may have power for a few hours a day. It is unlikely that many of these communities will connect to a centralized grid in the foreseeable future so they must rely on local sources of energy. The many rivers in Myanmar offer an abundance of hydropower potential, but in locations where conditions are not conducive to hydro dams where the terrain is flatter, hydrokinetic power generation can be developed. It is estimated that as many as 31,000 communities are not connected to the main utility grid and as a result have insufficient access to electricity.

New Energy was asked to provide a stand-alone power generation system that could provide continuous electricity to a school located near Hpa-An, Kayin State, Myanmar. The school had access to a diesel generator, but it would only run for a couple hours in the evening due to the cost of fuel. The school was located near the bank of the Thanlyin River that had a reasonably high flow velocity, so it was well suited for hydrokinetic power generation.

12.12.2.1 Equipment selection

The estimated maximum load was not expected to exceed 10 kW so two floating second-generation 5 kW systems were selected for the project. These units are also small enough such that logistic issues and access to heavy equipment (potentially an issue) for deployment would not be required.

12.12.2.2 Transportation and assembly

The systems were crated unassembled and shipped by air to Yangon, Myanmar, and then transferred to site by ground transport.

The school grounds next to the river provided a convenient location for assembling the units. Curiosity about what this project was all about and enthusiasm for the prospect of having electricity in the school rooms and surrounding buildings meant that there was no shortage of help to assemble the boats and other components. Assembly of these systems is not difficult requiring a few hand tools and very little heavy lifting (Figure 12.26).

Figure 12.26 Assembling the units

12.12.2.3 Site preparation and installation

Site preparation was minimal with a couple exceptions. A ramp was required to be carved into the riverbank so that the units could be launched into the water from the staging area (Figure 12.27). This was accomplished with hand shovels and a bit of elbow grease.

Moving the units to the water was expected to be a challenge, but while New Energy personnel were devising a plan on how best to move the units, the local people who were assisting in the assembly, or just watching with interest, picked up the entire system, boat, and all and moved it down the river bank by hand (Figure 12.28).

Anchor placement was another challenge. Hydrokinetic systems create a surprisingly high drag force and, therefore, require a heavy anchor to ensure their position is held. To be conservative, the anchors were oversized, but the downside was that they were very heavy and difficult to move into position in the soft mud

Figure 12.27 Ramp carved through the river bank for deployment

Figure 12.28 Moving the 5-kW system by hand

Figure 12.29 Moving the heavy anchor to the river bank for placement in the river

and silt of the riverbank. They were, however, moved into position by boat without the need for additional external lifting equipment. A separate Danforth anchor, shown in Figure 12.29, was provided for each system.

The units, Figure 12.30 were then moored to the anchors with cables that were supported with buoys.

A separate building was not required for the power electronics and battery storage system as a convenient location was found on the side of the main school room. A panel was attached to the wall of the school under an overhang where all power conditioning and control equipment could be mounted with the batteries situated below. The electrical cable was run overhead from the units to shore and then to the power electronics location.

12.12.2.4 Operation

The system was configured with a battery charger/grid forming inverter that simulated a grid for the Aurora inverters (required to synchronize the inverters), charged the batteries when required, and could also output AC power in addition to the output from the units (Figure 12.31). This meant that during peak usage times,

Figure 12.30 The 5-kW units in the river

Figure 12.31 Power Conditioning equipment for two 5-kW units

as much as 20 kW of power could be provided while charge remained available in the batteries.

If the electrical load being consumed was less than that being produced, the batteries would charge until fully charged at which time output from the units would be reduced.

12.12.2.5 Challenges

Components such as the anchors and steel cables were purchased locally. With the anchors, this was not a problem because they are just big hunks of metal. The steel cables were more of a challenge because it is difficult to confirm material specifications from a local supplier to which you have no previous experience. Despite assurances that the cables met the specifications, the cables purchased did not last long before corroding, and failing. Luckily the two units were tied together so neither was lost, but an expensive repair was required to bring both units back online.

12.12.2.6 Considerations

The challenge with all geographically distant installations is that a local individual must be trained to take charge in maintaining the equipment. Further, this individual must be motivated to continue maintaining it. Without a cluster of installations in the region, it is hard to justify supporting the individual to a reasonable degree and maintenance may eventually become an issue. Communication and cultural differences can also make it difficult to maintain the support. A change in the political landscape can also cause a loss of a contact and the associated support.

These are all variables that need to be considered when considering entering a new market.

References

[1] Calcagno, G., Salvatore, F., Greco, L., Moroso, A. and Eriksson, H., Experimental and numerical investigation of an innovative technology for marine current exploitation: the kobold turbine, *Proceedings of The Sixteenth International Offshore and Polar Engineering Conference Vol. 1*, 2006, p. 323–330.

[2] Gorban, A.N., Gorlov, A.M. and Silantyev V.M., Limits of the turbine efficiency for free fluid flow. *Journal of Energy Resources Technology*, 123, 2001, p.311–317.

[3] Bedard, R., Previsic, M., Siddiqui, O., Hagerman, G. and Robinson, M., EPRI Survey and Characterization Tidal in Stream Energy Conversion (TISEC) Devices, November 9, 2005.

[4] Verdant Power LLC., and GCK Technology Inc., Amesbury Tidal Energy Project: Integration of the Gorlov Helical Turbine into an Optimized Hardware/Software System Platform Final Report, April 30, 2005, Rev 07.26.05.

[5] Allen, B., "NACA Airfoils", 2021. Available from: http://nasa.gov/image-feature/langley/100/nacaairfoils.

[6] Bear, C., Wessner, C. and Ginter, V. *Torque neutralizing turbine mooring system*, 2009. US Patent 8,466,574.

[7] Tessier, L. and Doyle, J. *PDC bearing for use in a fluid environment*, 2010. US Patent 8,613,554.

[8] Tessier, L. and Schock, N. *Self aligning shaft assembly*, 2015. US Patent 9,939,011.

[9] Castro-Santos, T. and Haro, A., Survival and Behavioral Effects of Exposure to a Hydrokinetic Turbine on Juvenile Atlantic Salmon and Adult American Shad, Coastal and Estuarine Research Federation, July 11, 2013.

[10] Korobenko, A., Mohamed, B., Bear, M. and Bear, C., Performance analysis of two vertical-axis hydrokinetic turbines using variational multiscale method. *Computers & Fluids*, 200, 2020, p.104432.

Chapter 13

SWTs for arctic applications: powering autonomous rovers

Morten Hedelykke Dietz Fuglsang[1] *and Robert Flemming Mikkelsen*[2]

In arctic regions, small wind turbines (SWTs) fill an important niche for powering weather and research stations as well as telemetry in the dark winter months. This chapter revolves around another, rather unusual, use-case: using SWTs for powering autonomous rovers. Specifically, rovers are designed to roam the Greenland ice sheet, with the purpose of monitoring ice sheet conditions during the dark winter months, when PV is not an option. The chapter is based on a study assessing the feasibility of powering a rover with a micro wind turbine, considering weather and wind conditions of a specific location on the ice sheet. It includes a study of local conditions and derived functional requirements, a review of SWTs with arctic references compared against established selection criteria, and lastly, a short field test of a selected turbine on the ice sheet mounted at a very low hub height and the results thereof. While the study was preliminary, we demonstrate that harvesting wind energy at very low hub heights is possible on the Greenland ice sheet due to low turbulence: that maximum thrust must be considered for rover stability; and that one off-the-shelf SWT performs significantly below manufacturer specifications, in part attributed to cold temperature effects.

13.1 Introduction

Monitoring the Greenland ice sheet constitutes a significant source of data for climate researchers. During the summer months, manned research stations and affiliated projects, including ice core drilling, together with satellite imagery, contribute with a significant amount of data. The dark winter months, however, yield significantly less data, as manned missions are difficult due to extreme cold, and satellite observations are limited. A research group at the Niels Bohr Institute, Copenhagen University, is working to overcome this challenge: by developing and

[1]Department of Engineering Technology and Didactics, Technical University of Denmark, Denmark
[2]Department of Wind Energy, Technical University of Denmark, Denmark

deploying an array of autonomous rovers on the Greenland ice sheet, aiming to dramatically increase the amount of in-situ observations [1].

As a power source for the rover during the winter period, a micro wind turbine is considered. Using a wind turbine for power production on a moving vehicle presents an unusual set of design constraints, as balancing hub height and loading transferred to the vehicle becomes paramount. The study described in this chapter concentrates on the following questions:

- Is an off-the-shelf horizontal axis micro wind turbine mounted at a 2-m hub height likely to meet demands for power production for the rover in the winter period at the specific location?
- Does the loading from the turbine at a 2-m hub height risk tipping the rover at maximum expected wind gusts, when the turbine is free-yawing and the rover is standing still on a horizontal surface?
- Related to the previous two sub-questions, what is an appropriate hub height at which to mount the turbine?

To address the previous questions, we first look at the local conditions and wind resources to establish a set of functional requirements for a turbine. Thereafter, a review of SWTs with arctic references is presented. Lastly, a field test and associated test results are shown and discussed, before conclusions are described.

13.2 Location and local conditions

While the rovers are envisioned to roam the entire ice sheet, only one specific location was considered for this study: the East Greenland Ice-core Project (East- GRIP) [2] research based on the eastern Greenland ice sheet, latitude 75.6247N, longitude 35.9748W (75°37′28.9″N, 35°58′29.3″W), location as shown in Figure 13.1. Figure 13.2 illustrates the very flat terrain of the camp site, with a mean altitude of 2,660 m.

Weather and wind were assessed with data from the PROMICE Automatic Weather Station [3] located at EastGRIP, using data from May 2016 to January 2019. All measurements are 1-h average values. These are based on instantaneous measurements every 10 min, except for wind speed measurements that are based on 10-min averages [4]. Wind speeds are measured at roughly 2.1–2.6 m above surface (depending on time of year, due to snow precipitation). Figure 13.3(a) illustrates the seasonal variations of temperature and wind speeds, while the wind rose (b) reveals a remarkably steady wind direction at the site.

One of the unique local conditions of the ice sheet is the surface roughness, as can be seen in Figure 13.2. A power law fit of the wind speed profile was approximated from cup anemometer readings [5] from EastGRIP at 0.5-, 1- and 2-m heights above the surface. The power law exponent a was calculated from 1- to 2-m data and found to be $\alpha = 0.10$. The exponent is used for scaling of wind speed and profile of the "winter period" of the year where the rover is reliant on the wind resource for electricity, defined as from 15 October to 28/29 February. Key values from this period are presented in Table 13.1. Note the very low humidity.

Figure 13.1 EastGRIP location in Greenland (Map data: Google, Data SIO, NOAA, US Navy, NGA, GEBCO, Landsat/Copernicus, IBCAO, US Geological Survey, 2021)

Figure 13.2 Terrain around EGP (Photo: East Greenland Ice-core Project, www.eastgrip.org)

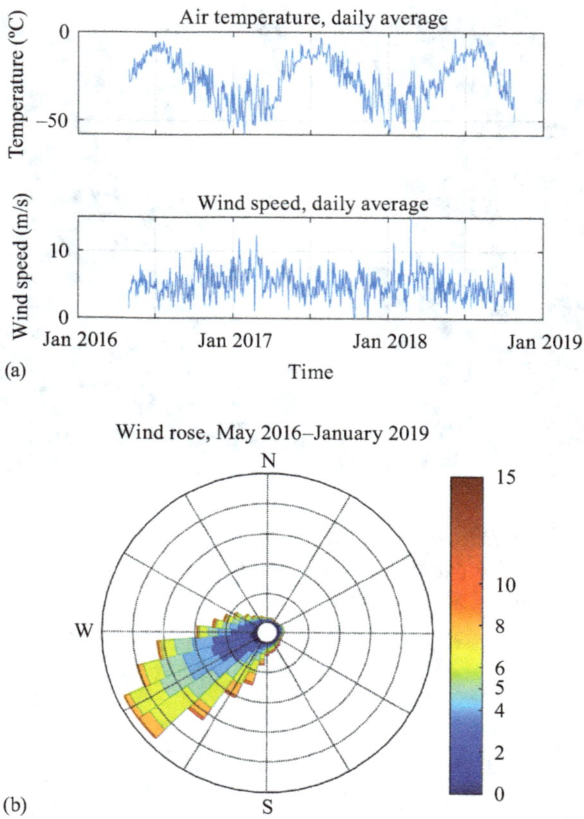

Figure 13.3 Overview of temperature and wind conditions at EastGRIP: (a) EastGRIP temperature and wind speeds, daily averages across seasons, (b) wind rose indicating stability of wind direction and speeds

Table 13.1 Key values of weather and wind in the winter period from 15 October to 28/29 February

Parameter	Min	Max	Average
Temperature (°C)	−60.6	−12.6	−37.9
Specific humidity (g/kg)	0.01	1.8	0.3
Pressure (Pa)	67,924	73,909	71,122
Wind speed @ 2.16 m avg. above ground (m/s)	0	21.3	5.7
Wind speed scaled to 2 m above ground (m/s)	0	21.2	5.7
Wind speed scaled to 1.5 m above ground (m/s)	0	20.6	5.5
Average air density (kg/m^3)	1.11	0.99	1.05

Table 13.2 Weibull parameters and power densities for the winter period

Hub height (m)	Scale (–)	Shape (–)	Power density (W/m^2)
2	6.34	2.27	161
1.5	6.16	2.27	144.9

Table 13.3 Wind turbine functional requirements

Parameter	Criteria
Minimum avg. power production, winter	522 Wh daily (22 W for 24 h)
Survival wind speed	35 m/s
Loading	Rover must not tip
Minimum operating temperature	−65°C
Cut-in wind speed at low temperatures	Start at 5.5 m/s at –65°C
Icing	None

The maximum likelihood Weibull parameters and associated power density are determined for the winter period, for the scaled wind speeds summarized in Table 13.2. The Weibull parameters serve for assessing statistical power production.

13.3 Turbine functional requirements

Based on the local conditions outlined above, along with rover functional requirements, the functional requirements for the wind turbine were decided on. They are summed up in Table 13.3. Both thrust and gyroscopic forces are evaluated for the loading requirements. Note that icing conditions are not considered a functional requirement, based on the low humidity during the winter period as well as anecdotal evidence from several years of missions to EastGRIP. Survival wind speed is determined from a maximum recorded wind gust of 27.5 m/s during the period of weather data, with a safety factor added.

13.4 Selection of a turbine with prior arctic references

Ten micro wind turbines were reviewed, with the purpose of identifying suitable turbines for use on the rover. Most of these had arctic references – meaning, one or more mentions of them being used in arctic or Antarctic conditions – while others were added for the sake of comparison, particularly the Proven P600. The turbines were a mix of HAWT and VAWT. The basis of the review was manufacturer specifications along with user reviews. The turbines were consequently evaluated against each other on the following six criteria:

- Average theoretical power production on site, based on manufacturer-based power curves and derived power coefficients.

- Low-temperature design. Factors include bearing and lubricants, sealing, material used particularly for blades and electronic components.
- Robustness and build quality. Factors considered include overspeed protection, amount of moving parts, supposed quality of materials used and manufacturer warranty and recommended maintenance.
- Loading transmitted to rover, including mass of the turbine along with thrust. Thrust data were only available for a few of the turbines. For the rest, the evaluation was based on assessing rotor solidity and overspeed mechanisms.
- Credibility of manufacturer, evaluated based on years in business, units sold, assessment of quality of technical information published and customer reviews.
- Price.

A weighted sum model was used. Results are summarized in Table 13.4. The review concluded the following:

- Theoretically, several commercially available micro wind turbines can generate an average of 25 W or more on the ice sheet during the winter months.
- Vertical axis turbines scored lower. A big drawback for all VAWT models reviewed was their mass per watt.
- There is not one of the reviewed turbines that is a perfect fit for the wind and weather conditions, nor for fitting on a rover. Some excel on some criteria but score lower in others, i.e. some of them are robust and engineered for the cold but have a high mass and/or rotor thrust and sub-optimal power curve for the wind resource available.

Table 13.4 Turbine weighted ranking

Criteria weight	**Power 25%**	**Low temp. 20%**	**Robust. 20%**	**Load 20%**	**Cred. 10%**	**Price 5%**	**Total 100%**
Superwind 350	0.75	0.6	1	1	0.5	0.15	4
Rutland FM910	0.75	0.4	0.6	0.8	0.3	0.25	3.1
Eclectic Energy D400	1	0.4	0.8	0.4	0.3	0.15	3.05
APRS World WT10	0.75	1	1	0.6	0.5	0.05	3.9
Leading Edge LE-300	1.25	0.6	0.6	0.6	0.4	0.2	3.65
Proven P600	0.75	0.6	0.8	0.8	0.3	N/A	3.25
Hi-VAWT DS-300	0.75	0.4	0.6	0.6	0.2	0.05	2.6
Forgen V30 Antarctic	0.25	0.8	1	0.4	0.4	0.2	3.05
Leading Edge LE-v150	0.5	0.6	0.8	0.4	0.4	0.2	2.9
WindSide WS-0,30	0.5	1	1	0.2	0.5	N/A	3.2

The review concluded that other than designing a wind turbine from scratch, the following four turbines could be modified to better fit the conditions. The four turbines are illustrated in Figure 13.4.

- Superwind 350 [6]: Very well engineered. Does not seem optimized for below −40°C but could be fine, particularly with bearing lubrication replaced. The exotic mechanical pitching system secures low rotor thrust but does depend on more moving parts. This may or not prove a problem. It is not optimized for the lower wind speeds, but this could potentially be solved with new blades – in this case, however, careful attention should be paid to engineering the blades to fit with the pitching system.
- APRIS World WT10 [7]: Very robust and optimized for the cold. Like the SW350, it is not optimized for lower wind speeds, but this could be solved with new blades. Is very likely to survive without issue in the conditions, but in higher winds, rotor thrust will be significant and this could be an issue for the rover.

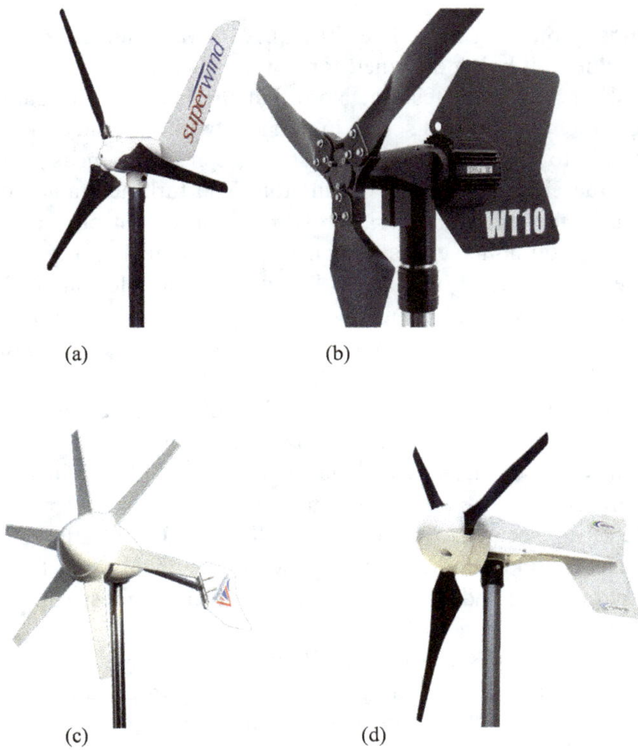

Figure 13.4 Four of the top reviewed turbines: (a) Superwind 350, (b) APRS World WT10, (c) Rutland FM910-4, (d) Leading Edge LE-300. Images courtesy of the manufacturers

- Rutland FM910 [8]: Medium build quality, but cheap, and optimized for lower wind speeds. Of the top candidates, this is the only turbine with a furling system to reduce thrust in higher wind speeds. It is uncertain whether the thrust might rise again beyond a certain wind speed. The furling mechanism will add a gyroscopic moment that must be taken into account. Bearing lubrication should be replaced, and possibly seals and nacelle sealing too. The rotor hub and blade assembly could be modified to be without the heavy flywheel, and possibly with fewer and different blades in another material.
- Leading Edge LE-300 [9]: Good build quality, reasonably priced and optimized for lower wind speeds. Its rated survival wind speed is quite low at 27 m/s. The reason for the latter is unknown and would be important to ascertain. It could be the blade material. High rotor thrust in high winds could be an issue, and if so, retrofitting a furling system might be an option.

13.5 Description of the field test

Of the four most promising candidates, the FM910 was selected for field testing. This was both due to it being designed for low wind speeds and because one was available on short notice for testing. Note that the bearing lubrication was not replaced before the test. The test set-up included a two-axis load cell mounted just below the nacelle to measure turbine thrust, and sensors to measure current and voltage of turbine electrical power production. The turbine was connected to a 12-V lead-acid battery through a TriStar TS-45 charge controller connected to a power resistor for dumping excess electricity. For data logging, a low power microcontroller-based set-up was used. Thrust was sampled at 50 Hz, and all other values at 5 Hz. Average values were recorded every 5 s. Furthermore, the necessary values for computing maximum, minimum and standard deviation thrust for every 5-s recording was logged. Prior to the field test, the load cell and other sensors and electronics were calibrated at different temperatures, −18°C being the lowest.

Mechanically, all electronics and the battery were placed in an aluminium cargo box, with sealed cable entries. The turbine and load cell were mounted to a telescope pole set-up secured with guy wires, allowing for adjusting hub height by 0.5 m during a test period. Before mounting the turbine on the site, all the rest of the equipment was mounted and the load cell was zeroed. The test equipment on the field test site is shown in Figure 13.5.

The field test at EastGRIP was carried out from 26 May to 24 June 2019. The turbine was mounted at a 2-m hub height until 10 June. On 10 June, the turbine was lowered to a 1.5-m hub height, where it stayed until the end of the test. From the 27 to 28 May, a loose connection on a screw terminal on one of the full bridges of the load cell resulted in no-thrust measurements in this period. Furthermore, on the 28 May, a software update fixed an error in recording minimum and maximum thrust events, such that these were recorded from 28 May and on.

Figure 13.5 The field test set-up installed and running by EastGRIP on the ice sheet

Wind and weather data were acquired from PROMICE Automatic Weather Station at EastGRIP, roughly 100 m from the turbine test site. The dataset consists of 1-h average values.

Prior to the field test, a 5-h test run of the set-up was conducted at the DTU Risø Campus wind turbine test site in Denmark. The turbine was mounted at a 2.4-m hub height. This test run offers some data for comparing the thrust to power relationship in a different temperature regime than the EastGRIP test.

13.6 Results from the field test

An overview of the test results from the EastGRIP test is given in Figure 13.6, based on 1-h average data. Of note is that in some parts of the period, wind speeds were very low, resulting in low power and thrust but also an unstable thrust angle. For the remainder of the test, power and thrust follow wind speeds, as do the thrust direction.

An overview of daily power production is given in Figure 13.7, indicating the higher wind speeds in the end of the testing period. Key values from the test are summarized in Table 13.5.

13.6.1 Turbulence, yaw error and maximum expected thrust

Turbulence and yaw error are investigated by assessing a high thrust event for each of the 2- and 1.5-m periods, occurring on 9 and 23 June, respectively. Figure 13.8(a) illustrates the 2-m period, and (b) the 1.5-m period. In the time interval of both the 2- and 1.5-m thrust event, the average thrust angle remains stable. This indicates low turbulence and associated yaw error for both hub

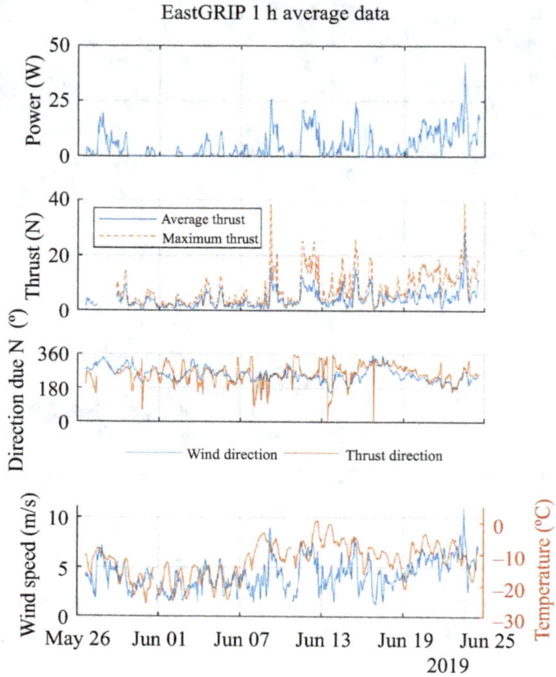

Figure 13.6 EastGRIP 1-h average data for the complete test period

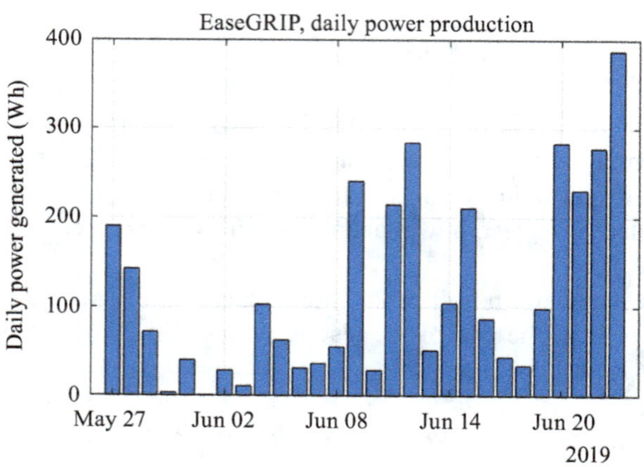

Figure 13.7 Daily power production

Table 13.5 EastGRIP test key data

	2-m test period		**1.5-m test period**	
	Average	**Maximum**	**Average**	**Maximum**
Power (W)	2.9	36.4	7.5	35.3
Thrust (N)	3.1	62.1	5.8	38
Wind speed @ scaled hub height (m/s)	3.7	8.8	5	10.8
Temperature (°C)	−15.4	–	−9.2	–
Air density (kg/m^3)	0.99	–	0.97	–
Pressure (hPa)	730.6	–	730.7	–

Note: Note that maximum values of power and thrust are from 1-min average data; the rest is from 1-h average data.

heights, a conclusion that is supported by observation of yaw error of the wind turbine on site, and general observations of turbulence at the EastGRIP site. This is further supported by comparing to high thrust event data and observations during the Risø test, where the average thrust angle was found to vary significantly following a spike in thrust.

The variations in maximum thrust follow overall power production and, while the maximum thrust angle varies, it seems stable during peaks of maximum thrust, indicating that the high thrust is due to gusts. It should also be noted that maximum thrust and associated thrust angle is one data point only during the sampling period (5 s), and some variance is expected on this account. Note that as the turbine tested is equipped with a furling system, it is likely that the yaw error is greater compared to a turbine without a furling mechanism.

Lastly, the maximum thrust for both intervals is more than twice the average thrust. This must be taken into account when considering the thrust at maximum wind speeds and rover stability.

13.6.2 Thrust vs power

The thrust–power relationship for the duration of the EastGRIP test is shown in Figure 13.9(a) and for the Risø test in (b). Each circle represents a 10-min average for the EastGRIP test and a 1-min average for the Risø test. A fairly linear trend is observed for both observations although with more scatter in low winds/low power production at EastGRIP. As the equipment used for the two tests was identical, these figures give an indication of low-temperature effects on the system, considering that the difference in mean temperature between the two tests was 24.3°C (11.7°C mean during Risø test, −12.6°C mean during the EastGRIP test). The result of low temperatures at the EastGRIP test is a greater thrust required per watt of power, mostly likely due to friction in bearings and internal battery resistance. For the rover case, a temperature of − 12.6°C is relatively warm compared to the average −37.9°C temperature of the winter period, which likely would result in an increased thrust per unit of power.

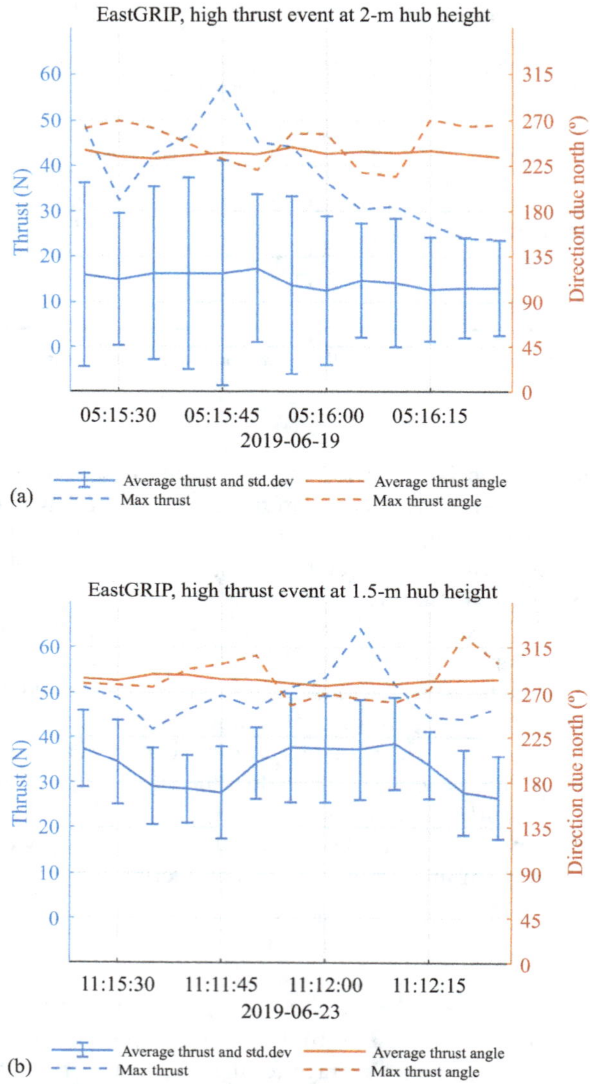

Figure 13.8 EastGRIP high thrust events at (a) 2- and (b) 1.5-m hub heights

13.6.3 Coefficient of performance

Figure 13.10 compares the power performance of the FM910 turbine and system during the EastGRIP test against manufacturer specifications. The power coefficient is calculated with 1-h average data for the whole test period, with wind speeds scaled for each hub height. Measured density from the weather station is used for

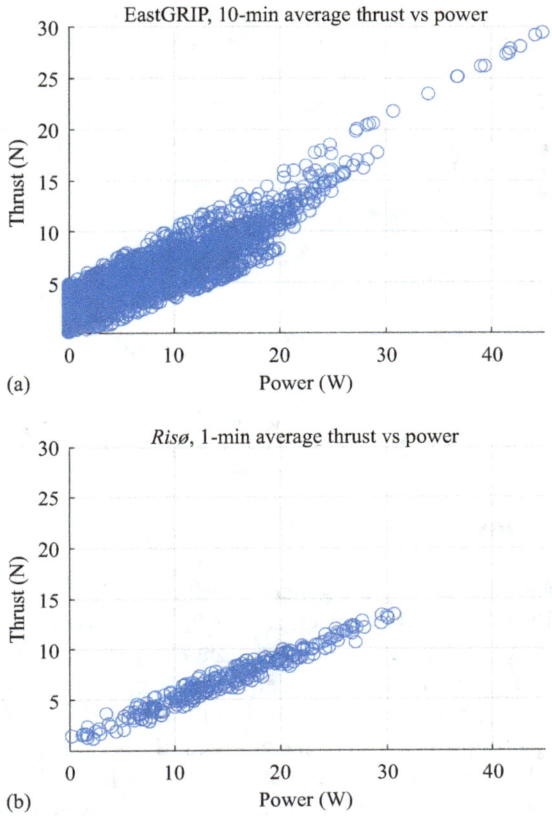

Figure 13.9 Thrust–power relation of (a) EastGRIP field test, from 10-min average data and (b) Risø test run, from 1-min average data.

computing the power coefficient. With the range of wind speeds occurring during the test period, it is not possible to construct a complete performance curve for the conditions. Furthermore, the density used to compute the normalized power coefficients varies; hence, they cannot be directly compared. It does although indicate that the total efficiency is significantly lower than the manufacturer specifications in general. A site-specific scaling factor, used for calculating statistical power production, was set to 50%. With this along with scaled statistical wind speeds and average density for the winter period, the FM910 was found to generate a daily average of 290 Wh at 1.5-m hub height and 307 Wh at 2 m. This translates to 55.5% and 58.9% respectively, of the rovers power requirement. With this in mind the FM910 is not likely to meet the presently considered power requirements for the rover. A larger swept area turbine appears to be needed to meet the power requirements, as well as appropriate measures for low temperature operation, particularly bearings and lubrication.

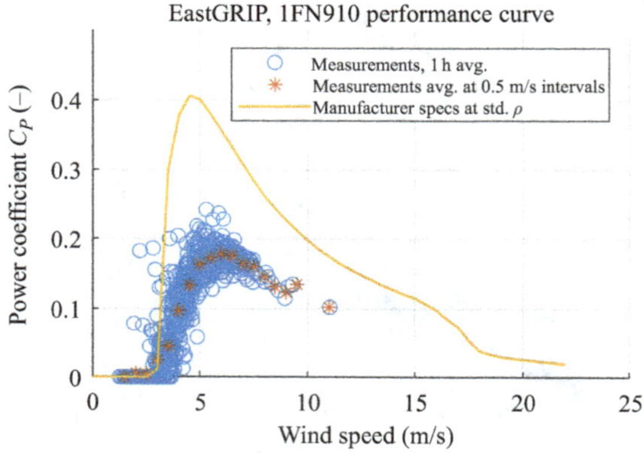

Figure 13.10 Performance curve based on EastGRIP field test

13.7 Conclusions

This study focused on assessing the general feasibility of powering a rover with a wind turbine on the Greenland ice sheet. An off-the-shelf SWT was field tested by EastGRIP on the ice sheet at very low hub heights (2 and 1.5 m) for just short of a month, and a 6-h test run at 2.4-m hub height in Risø, Denmark before then. By comparing thrust magnitude and direction from the Risø test with the EastGRIP test, the thrust direction was found to be significantly more stable, pointing to an unusually low turbulence that can be explained by the low surface roughness of the ice sheet. No significant differences were observed between 1.5- and 2-m hub heights, suggesting that 1.5-m hub height is viable. This is preferable due to stability, as highlighted by calculations not covered in this chapter (29% increase in safety factor against tipping for a specific set of conditions). Low-temperature effects on the turbine were illustrated by comparing thrust as a function of power for Risø and EastGRIP tests, not surprisingly finding an increase in thrust per produced watt. A turbine performance curve was computed and was found to be significantly lower than as specified by the manufacturer, also considering the low density of air at the EastGRIP site. From this, a site- and condition-specific power scaling factor of 50% was determined, and the projected daily average power production was calculated for both hub heights. This was found to be 58.9% and 55.5% of the projected required 522 Wh daily average, at 2 and 1.5 m, respectively.

The results indicate that powering a rover with a wind turbine on the ice sheet is realistic and within reach, but additional long-term testing is necessary before a turbine powered rover can explore the dark winter on the East Greenland ice sheet.

References

[1] Hvidberg CS. (2018). Arctic Rover project description [online]; [cited 2021 April 8]. Available from: https://veluxfoundations.dk/en/about/projects-grantedt /0060X00000U9keUQAR. Project funded by VILLUM FONDEN.

[2] Physics of Ice Climate and Earth, Copenhagen University. (2016). EastGRIP project website [online]; [cited 2021 April 8]. Available from: https://eastgrip.nbi.ku.dk/.

[3] Fausto RS, van As D. (2019). Programme for monitoring of the Greenland ice sheet (PROMICE): Automatic weather station data. Version: v03, Dataset published via Geological Survey of Denmark and Greenland. [dataset]; [cited 2021 April 8]. Available from: https://doi.org/10.22008/promice/data/aws.

[4] Programme for Monitoring of the Greenland Ice Sheet. (2019). Automatic weather station documentation [online]; [cited 2019 June]. Available from: http://promice.dk/PROMICEnreadnme.pdf.

[5] Madsen, M.V., Steen-Larsen, H.C., Hörhold, M., Box, J., Berben, S.M.P., Capron, E., *et al.* (2019). Evidence of isotopic fractionation during vapor exchange between the atmosphere and the snow surface in Greenland. *Journal of Geophysical Research: Atmospheres*, 124(6), 2932–2945. Available from: http://doi.org/10.1029/2018JD029619.

[6] Superwind. Superwind 350 official product page [online]; [cited 2019 March]. Available from: https://superwind.com/superwind-350-353/.

[7] APRS World. APRS WT10 official product page [online]; [cited 2019 March]. Available from: http://aprsworld.com/wtaprs/WT10/.

[8] Marlec. Rutland FM910-4 official product page [online]; [cited 2019 March]. Available from: https://marlec.co.uk/product/rutland-fm910-4-furlmatic-windcharger/.

[9] Leading Edge. Leading Edge LE-300 official product page [online]; [cited 2019 March]. Available from: https://leadingedgepower.com/le-300-wind-turbine-standard-1013010.html.

Chapter 14
Commercialisation of a small diffuser-augmented wind turbine for microgrids

Samuel Petersen Evans[1,2], *James B. Bradley*[2] *and Joss E. Kesby*[1,2]

Small-scale distributed wind generation faces challenges in being cost competitive due to recent advances in solar photovoltaic (PV) and battery storage technology. Reductions in levelised cost of energy (LCOE) can be achieved by improvements in aerodynamic efficiency, generator controller design, or reducing cost of manufacture. In this chapter, we present a case study detailing the commercialisation of a novel 200-W high-efficiency diffuser-augmented wind turbine (DAWT).

We highlight the particular challenges to be overcome when designing the electrical control system for interface with microgrid architecture. Results include increased rotor efficiency, bespoke controller design, and the novel use of manufacturing processes. Findings and conclusions are of direct interest to small wind turbine designers as they seek to develop innovative methods of reducing LCOE.

14.1 Introduction

Small wind turbines are defined as having a rotor area less than 200 m^2 and power output less than 50 kW [1]. Turbines in this class are less technologically mature than utility-scale wind turbine technology. They have been documented to suffer from poor realised power output, structural failures, and a higher LCOE [2,3]. More recently, the significant reduction in cost of solar PV modules and battery storage systems has driven the need to reduce LCOE of small wind energy [4]. Small wind turbines typically need feed in tariffs (FITs) to become cost competitive [5]. Despite these challenges, small wind turbines still have a part to play in increasing the renewable penetration and reducing system cost of hybrid microgrids [6].

Aside from the use of FITs, one obvious way of reducing small wind turbine LCOE is by improving the system efficiency. This efficiency has a theoretical upper limit of 59.3%, also known as the Betz limit. Its derivation is covered in

[1]School of Engineering (Mechanical), The University of Newcastle, Newcastle, New South Wales, Australia
[2]Diffuse Energy Pty Ltd, Newcastle, New South Wales, Australia

significant detail in [7]. DAWTs hold potential to increase power output and have been the subject of previous investigation [8]. In this embodiment, the performance of the turbine is augmented by the use of a cylindrical diffuser or shroud that surrounds the rotor plane of a horizontal-axis wind turbine (HAWT) as per Figure 14.1. This cylindrical shroud usually has an aerofoil-shaped cross-section. The lift generated by the shroud increases the mass flow rate across the rotor plane and hence power coefficient. A DAWT with a given exit area can, therefore, generate more energy than an equivalent-sized HAWT rotor.

Studies have revealed other advantages of DAWTs, including improved performance in turbulent conditions and lower susceptibility to a reduction in power output due to yaw errors [9]. They are often quieter due to the diffuser's alteration of the blade tip vortices and physical inhibition of sound wave propagation [8] and are safer as the blades are encapsulated in the event of failure or when placed in proximity to human interaction.

There has been renewed interest in the application of DAWTs [10–12]. Recent focus has consisted of improving the design of the rotors with respect to the inlet wind profile at the rotor, optimising the location of the rotor in the throat of the diffuser, and the interaction of the actuator disk within the DAWT. There exist published methods for optimising the rotor design, where this usually involves the use of computational fluid dynamics (CFD). The need to reduce small wind turbine

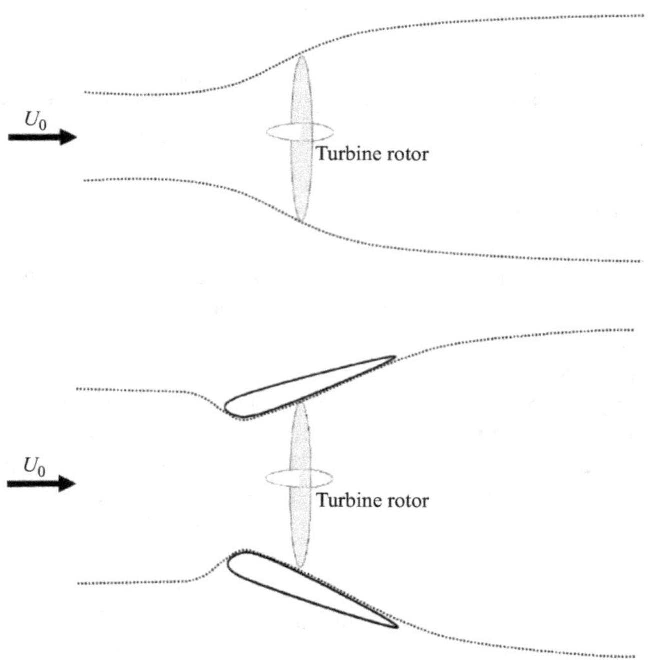

Figure 14.1 Diagrams of the stream tube intercepted by the rotor of a HAWT (top) and a DAWT (bottom)

LCOE and the simultaneous reduction in cost of computing power has undoubtedly driven the renewed academic and commercial pursuit of DAWTs.

Despite these performance gains, many designers have claimed to have exceeded the Betz limit. Often these claims are misleading as the rotor area of a DAWT is erroneously used in calculations that show an 'exceedance of the Betz limit', when in reality the larger exit area of the diffuser is the more correct metric. Commercial failures of DAWTs such as the Vortec7 and Donqui most likely contribute to their bad public perception. Despite these failures, it is encouraging to see commercialisation efforts undertaken by Ducted Wind Turbines, Halo Energy, Bluepower, and Wind Lens in developing their technological readiness level towards a commercially viable product.

This lack of publicly available commercialisation knowledge has been raised by organisations such as the NREL, and Distributed Wind Energy Association. Specifically, a major finding of the DEWA SMART Wind Roadmap [13] is the lack of publicly available best practice data from operational small wind turbines. A recent report by the NREL concluded that new financing and business models are required to deliver unmet needs for high penetration renewables with small wind generation [14]. It is clear that technical and commercial outcomes go hand-in-hand to ensure successful product-market fit.

14.2 Aims

The study documented in this chapter is aimed at reducing the LCOE of small wind generation by the use of an aerodynamically shaped diffuser to improve HAWT performance. A detailed case study describing the commercialisation process will be presented and discussed. Commercialisation of technology in any field is a challenging endeavour. The aim of this chapter is to present a case study and add to the literature of shared experiences and lessons learnt. The authors intend that this would contribute to the commercialisation effort of small wind technology and thereby realise reductions in LCOE for small wind or hybrid distributed energy systems.

14.3 Methodology

14.3.1 Turbine design

The initial stage of turbine development is deciding on the rotor design and hence power output. Once these critical details are 'locked in', engineering effort can be applied to designing and developing the balance of the system, including component sizing, material selection, and manufacturing techniques. This step is known as design for manufacture or DFM. Once DFM is complete, a bill of materials can be produced which forms the basis of vendor selection for manufacturing said components.

Complicating the design further, there is also no specific guidance for shrouded or DAWTs in the small wind turbine design standard IEC 61400.2. This standard is limited to open bladed HAWTs. Despite this, a deft designer could appropriate this standard for initial determination of design loads. This lack of a relevant standard may lead to the bad market perception of small-scale wind, extensive customer interviews by the authors revealing that noise, safety concerns (either real or perceived), and cost were major barriers to adoption.

In previous studies, a small 35 W DAWT ($U = 10.5$ m/s) was designed using a CFD methodology. This wind turbine had a diameter of 430 mm and was printed from PETG material. This 3D-printed wind turbine then underwent aerodynamic testing in a wind tunnel at the University of Newcastle, Australia. The CFD design methodology was validated using data obtained from the wind tunnel measurements [15]. Sufficient confidence in the design methodology and diffuser augmentation has been developed by the authors to allow for scale to a larger design. Larger designs usually result in an increase in coefficient of performance due to the reduced effects of hub losses, etc. with respect to the total rotor area.

A 200 W DAWT ($Cp = 0.42$, $U = 10.5$ m/s) has been developed as per previous work by the authors in [15]. This represents an appreciable improvement compared to existing market products such as the Rutland 914i (0.26), Superwind SW1250 (0.30), and the Leading Edge LE-300 (0.35). Despite the promising theoretical design, there is a significant gap between research and commercial application. This commercialisation process is rarely discussed in the open literature. In Australia, this landscape is changing with the creation of programs such as the CSIRO ON Program. This is Australia's National Science and Technology Accelerator and is focused on driving and supporting the commercialisation of publicly funded research. Energy-specific accelerator programs such as A-Labs run by the Australian Renewable Energy Agency (ARENA) assists in navigating the challenges of translating academic output into a commercial product. Outputs of ARENA-funded projects are published in the public domain* for the wider benefit of the renewable energy industry both nationally and internationally.

The commercialisation journey is extensive for any piece of technology and it is envisaged that this manuscript would serve as a case study with learnings applicable to others within academic organisations who wish to commercialise their research.

14.3.2 Manufacturing techniques and materials

Small wind turbines are constructed from a myriad of different materials and manufacturing techniques. Common materials include fibreglass, carbon fibre, timber, sheet metal, and injection-moulded polymers [16,17]. All have various trade-offs in terms of stiffness, cost, and manufacturability. One other challenge is that lower production run of units is undertaken, which reduces the possibility of savings for high volume orders. This means that the amortisation of expensive capital items such as moulds are less favourable, thus increasing the overall LCOE.

The nature of the wind turbine blades and diffuser makes them challenging to produce. For example, both have a varying continuously smooth surface that tapers to a trailing edge. This surface is critical to the aerodynamic operation and, therefore, must be machined to a very high geometrical tolerance. The masses (and

*https://arena.gov.au/projects

hence inertia) of the blades and diffuser are required to be kept to a minimum as the net mass has ramifications to the support structures, while a high rotor inertia has negative effects on starting performance and gyroscopic loadings. The blades are also subject to fatigue loading during operation, especially in turbulent wind [18–20]. They are known to be fatigue critical by design, experiencing up to 10^7 cyclical loads during an expected life of 20 years; an order of magnitude more cycles than experienced by large-scale wind turbines [21]. In summary, the blades and diffuser have three main manufacturing requirements; high stiffness to mass ratio, accurate geometrical construction, and high resistance to fatigue loading.

The choice of material affects the manufacturing technique, and vice versa, so it is necessary to consider both the material choice and manufacturing method in parallel. While a model was 3D printed using a PETG material for testing, it was never intended as a material for manufacture in large quantities. The obvious starting point for the manufacture of the 400-mm long blades is the application of fibre-reinforced polymer composites such as carbon fibre and fibreglass, as these are widely used in the construction of large wind turbine blades. Multiple vendors were approached both domestically and internationally, with the resulting quotation and unit costing being prohibitively expensive. CNC milling of hoop pine timber was also investigated due to its easy machining and fatigue resistance properties. While the initial stock was cheaper, the material wastage and mill cycle time are high. Milling fine geometric details such as the blade tip and trailing edge would also be challenging and prone to suffering from the effects of tool chatter.

The use of injection moulding was decided upon for the production of the blades as it satisfied the stiffness to mass ratio requirements, while being cost-effective even at lower volume production runs. A mould was made in two halves by CNC milling incorporating the design of the sprue, runner, and ejector pins. The blade was constructed from a proprietary nylon/carbon fibre fill material, which is injectable for our dimensions in question. Post-injection, only a small amount of finishing work is required to remove the sprue and flashing from the parting lines (Figure 14.2). Due to the homogeneous nature of the material and tight geometric tolerance derived from this method, the blades require little to no effort for balancing.

The use of reinforced composite polymers for the diffuser was quickly abandoned due to the high cost. It is important that the cost of the diffuser does not offset the monetary gains of the increased efficiency over a bare wind turbine. One factor that worked in our favour is that the diffuser was never designed nor intended to be a structural component (unlike the blades). These structural loads due to wind actions are supported by the cross members and centrebody of the wind turbine, they are then transmitted through the central yaw bearing and into the tower. After much investigation, the only viable manufacturing technique was rotational moulding where an HDPE blend (in solid granular form) is placed into a cavity mould. This is then heated in an oven to melt the polymer stock and rotated on two axes to ensure an even and regular coating. Upon cooling, this is then released from the mould. Final finishing is then undertaken to remove any excess flashing (Figure 14.3).

Compared to the blades and the diffuser, the other mechanical components such as the centre body, cross supports, and yaw mechanism are trivial to design

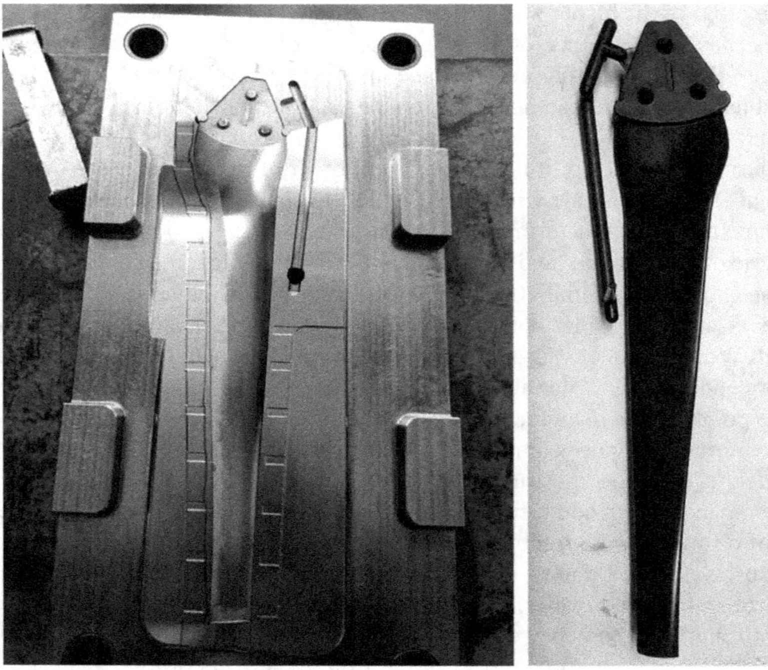

Figure 14.2 The completed mould (left), and resulting blade (right)

and manufacture. These were built at a local steel fabricators workshop from readily available stainless steel and aluminium stock. As such, these details will not be presented here and in any matter are likely to be of little interest to the reader.

14.3.3 Electrical and controller design

Now that the wind turbine mechanical design has been specified, attention can be turned towards the generator and electrical control system. Creating an efficient electrical system complimentary to the turbine output is a difficult task that has plagued small wind manufacturers. Poor historical generator electrical controller design has had a negative impact on small wind [6]. The following design criteria for a generator and electrical system were established to avoid these foreseeable problems include the following:

1. Minimal moving parts so as to reduce the need for system maintenance. For remote or off-grid locations, reducing or eliminating maintenance is an important aspect. For some customers, this is a compulsory requirement.
2. Use of an 'off the shelf' generator to reduce capital costs.
3. Utilisation of a maximum power point tracking (MPPT) controller scheme.
4. Incorporation of overspeed control to avoid the generation of high voltage/current which could damage electrical systems not only in the turbine, but downstream components (batteries).

Figure 14.3 The diffuser mould (top), and the unfinished diffuser part (bottom)

5. Output of a 24/48 VDC source as to interface with microgrid battery systems.
6. Ability to interface with a wide range of electrical system architecture, batteries, battery management systems (BMSs), solar PV modules, and diesel generator set.

To address the first point, the obvious criterion is to select a permanent magnet generator (PMG). This generator does not use slip rings (compared to a DC generator) meaning that carbon brushes that are a source of wear and component failure will be avoided altogether. Furthermore, a design was chosen such that the stator was central to the rotating magnets (as opposed to the rare earth magnet armature spinning internally). This was done in order to avoid the situation where at high temperatures, the glue could degrade over time and cause the magnet attached to an armature to break free, causing failure of the generator. This has been anecdotally reported by customers who have previously used white-label small wind turbines purchased at low costs from online retailers.

The ability to manufacture an in-house generator was considered in view of point two. A cursory analysis of the costs involved in designing and manufacturing a generator from scratch was undertaken. The capital expenditure involved with designing, 'tooling up', and undertaking a small batch (<100) unit run would be prohibitive. Costs would only become more competitive once exceeding 1,000 units were reached – a change in production magnitude. Costs aside, the time investment would also be extensive and would require specialised expertise that existed outside that of the current authors' skillsets. PMGs are commonplace among most small-scale wind systems, due to recent reductions in cost and physical size but off-the-shelf components often do not match turbine performance. An extensive international search was undertaken to find the most appropriate supplier of PMGs.

Design criteria three focused on the requirements for an MPPT control algorithm whereby the optimal generator rotor speed (and hence voltage) is matched for the aerodynamically optimal rotor speed. This ensures that the rotor is performing at, or close to, its designed tip speed ratio as variations in the inlet wind speed occur. MPPT control methodology as it applies to small wind turbines is detailed in many references such as [22] and will not be repeated here in detail. For controlling the power output, in many cases, a simple solar PV module MPPT controller is used, instead of a wind-turbine-specific controller. Some solar PV MPPT controllers are even marketed as a small wind turbine MPPT controller, irrespective of system design. While this theoretically could control a small wind turbine, it is highly undesirable and is likely to result in a poor performance, potentially contributing to the poor reputation of small wind turbines.

Small wind turbines typically do not have the means to measure the incoming wind speed for input into an MPPT control algorithm, such as via ultrasonic anemometers or lidar systems. This would significantly increase the overall system cost and adds an additional challenge to the control scheme. For the end user, this would mean an additional component that would need to be installed, maintained, and calibrated regularly (an undesirable trait for remote sites).

Many customers raise the question of overspeed protection to address point four. While this is an important design point of the wind turbine manufacture, safety in high wind events could be the critical path in a business case for the deployment of new technologies (especially in cyclonic regions). Compared to say a solar PV module, the turbine cannot simply be open circuited to prevent power generation once the batteries of a system are at full capacity, thereby preventing overcharging and damaging the batteries. This open-circuit condition would mean that the rotor is unloaded and would quickly accelerate to high rotor speeds. This would create substantial noise, drag loading, and potentially catastrophic rotor failure. Undue cycling of the PMG bearings will occur if operated unloaded beyond at high rpm causing unnecessary wear. Mechanical brakes are rarely seen on small-scale wind turbines, likely due to cost, mechanical complexity, and maintenance requirements.

To mitigate overspeed events, the authors have executed a threefold approach, (1) the rotor and diffuser aerodynamics are designed to operate suboptimally at high-speed events, reducing the turbine output, (2) a dynamic brake is applied by the control system as a back torque to slow the rotor, with energy dissipated into the dump resistor as waste heat, and (3) in extreme cases the generator windings are shorted, rapidly reducing the rotor to very low rpm causing complete stall. This system has been designed and engineered to safely shut down in a wind up to 180 km/h. Structural and electrical components have been individually tested to these levels. These overspeed events are difficult to test in laboratory conditions and rely on the field measurements, both for internal verification and customer proof points. This has been tested in the field at 108 km/h. This wind event consisted of many subsequent gusts and wind speed fluctuation. The turbine safely shut down with no reported damage or negative impacts to the system. Furthermore, some customer tower systems in the communications market have been designed to be lowered (either via a gin pole or hinge mechanism) for stowage during forecast high-speed events such as what can occur in cyclone regions.

Point five. Despite the fact that an aerodynamically ideal rotor may be designed, the need for a suitably idealised generator and controller system is necessary to translate the aerodynamic efficiency gains into electrical power for the end user. The microgrid systems are designed to operate at 12–48 VDC and run from a stand-alone battery system. They are, therefore, not designed to be grid tied meaning the small wind turbine controller will not need an inverter to produce an alternating current source at grid voltage (240 V, 50 Hz Australian grid frequency). Recall that the generator selected for use is a permanent magnet configuration which by nature will produce an alternating voltage/current output. As the wind turbine can operate across a number of inlet wind speeds, the amplitude of the power output can fluctuate substantially as per Figure 14.4.

The design challenge becomes the fact that the variable voltage/current output must be converted into a constant voltage given by the system bus voltage. This set point is usually controlled by an existing BMS that must be followed by the wind turbine controller so as not to compete with the BMS. In the first instance, this means that the alternating power at the PMG must be rectified into a direct current source by means of an onboard rectifier. This rectified current is then sent to the boost converter; a DC/DC converter that steps up the voltage to match the voltage of the microgrid DC bus, illustrated in Figure 14.5. The controller must then match the output current to the power curve given that the voltage is proportional to the rotor rpm. Small wind turbine rotors have a very low inertia so voltage fluctuations can be very fast, in the order of <1 s; the MPPT loop speed must be fast enough to match the SWT voltage. Not doing so results in typical problems of small wind turbines; noise (due to overspeed) and low power output performance.

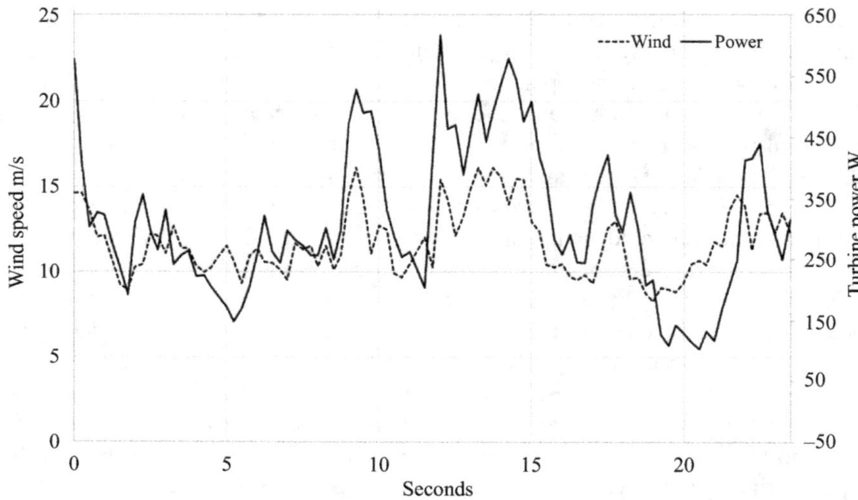

Figure 14.4 A 10-s operating window of wind turbine power output and wind speed

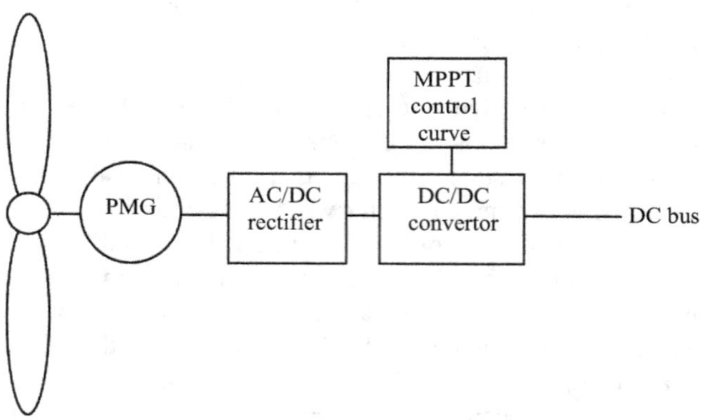

Figure 14.5 Control logic diagram of the wind turbine MPPT controller

An additional challenge to note during the design of small wind turbine control systems is that small wind turbines often operate at lower voltages with comparatively higher currents which must be carefully considered in the electrical system. This can cause difficulty in the selection of components when populating the controller board.

As a final point of consideration, the end user can operate a wide array of power systems, including components such as solar PV modules, diesel generators, batteries (of varying chemistries), and BMSs. It is incumbent upon the small wind turbine manufacturer to ensure that the system can readily integrate electrically in a manner that is 'plug and play'. This often overlooked step is imperative for customer adoption and the relevance of small wind.

14.3.4 System commercials

Currently small wind turbines face challenges in being cost competitive with incumbent sources of centralised energy generation (i.e. coal, gas, oil, and utility-scale renewables). Designing a grid-connected small wind turbine provides extra challenges both in the electrical control system and certification. While rotor optimisation is the topic of interest in most studies, designing and manufacturing a complete small wind system include a number of components beyond the rotor design. Due engineering effort has to be undertaken in the design and manufacture of the tower, nacelle and yaw system, generator selection, electrical control, and grid connection. These subcomponents can be expensive, for example the breakdown of a 5-kW HAWT and 200-W DAWT is given in Table 14.1.

It can be seen from Table 14.1 that the two most expensive components of the 5-kW small wind system are the tower installation, and grid connection. When considering the 200-W DAWT, the diffuser accounts for only 6% of the installed system cost. The nacelle platform and yaw mechanism is the highest percentage of mechanical system cost and is likely due to the high cost of 'one-off' machining. Higher volumes would see this reduce in absolute terms. The lack of a tail fin also represents a cost saving.

The authors have taken a lean approach to hardware development, whereby superfluous system components are excluded from the design and manufacture phase. We have, therefore, solely targeted customers and applications that are not grid-connected, and who already have an existing tower structure where the turbine could be mounted.

14.3.5 Market applications in the Australian context

The commercialisation efforts detailed in this chapter are undertaken within the Australian region. It is, therefore, necessary to detail market applications within this geographical context.

Table 14.1 Typical costs of two small wind turbines

Component	5-kW HAWT [21]	200-W DAWT
Blades	7%	3%
Diffuser	N/A	6%
Platform, tail fin	5%	21%
Gearbox, generator, break	6%	12%
Nose cone and cover	3%	13%
Controller/inverter	18%	20%
Tower	32%	N/A
Installation and grid connection	29%	25%

Despite the significant uptake of domestic and residential roof-mounted solar PV systems, there is little market for small-scale wind systems, with only one known small wind turbine manufacturer. In the Australian context FITs have existed for solar PV systems; as has much of the political and financial interest has focused on solar PV systems [23]. Roof-mounted wind turbine systems also bring into play other issues not relevant to remote sites, including local council approval, noise, and visual amenity complaints from the public. For comparison, the LCOE of rooftop solar† is in the order of 12 c/kWh AUD compared to small wind turbines‡ which is in the order of 26 c/kWh AUD.

Two major driving factors in the Australian off-grid microgrid market are increased renewable energy penetration and energy resilience for the mitigation of natural disasters. This has created a market need for an additional energy resource for remote users. For example, the extreme Australian bushfire season in 2019/2020 resulted in 1,390 reported impacts to the telecommunications network, with power outages accounting for 88% respective telecom sites. Few sites were directly impacted by fire, rather upstream damages to damage to feeders and single-wire earth return line infrastructure, the inability to access sites due to damage at off-road four-wheel drive access points, and the declaration of a national emergency whereby the RAPS site is under control of emergency authorities [24]. For context, these sites historically required 8 h of a backup energy supply. This has since been extended to 12 h – a figure that is insufficient by previous events. Many operators are now self-mandating extended system battery autonomy between 3 and 5 days.

Smoke generated from the 2019/2020 fires also reduced the output of solar PV modules during this season. Air quality throughout Australia was declared hazardous for 47 days during this summer period. Results from a study by the Australian Energy Market Operator estimated that smoke haze results in a reduction in utility-scale solar generation by 6%–13% throughout the 2019/2020 summer period. To the best of the author's knowledge, no reportable data exist for these off-grid energy systems, but the findings from the report in [25], when generalised to off-grid systems, indicate reduced energy generation. Wind, therefore, provides an opportunity to mitigate against smoky conditions.

The future application of small wind turbine systems is not where the LCOE is cheapest due to an abundance of energy supply (i.e. cities), but in remote or off-grid locations. For example, the ARENA released a report that showed some 2% of the Australian population is considered off-grid but yet accounts for 6% of the national energy consumption. This means that the greatest need and demand is in regional areas further away from the population centres. These customers are usually serviced by microgrids independent of the National Energy Market [26].

Market applications of small wind turbines in this instance would include remote sites, monitoring stations, telecommunications, mining, agriculture, aquaculture, yachting, and off-grid dwellings. Several installations are shown in Figure 14.6.

†https://www.energycouncil.com.au/media/15358/australian-energy-council-solar-report_-january-2019.pdf

‡https://distributedwind.org/wp-content/uploads/2016/05/SMART-Wind-Roadmap.pdf

Figure 14.6 Site installations across a number of markets, including telecommunications, mining, agriculture, and domestic

14.3.6 Customer channels

The sale of utility-scale wind turbines is classed as a complex or enterprise sale whereby many units are sold in a single high-value sales contract. These contracts may be negotiated over many months or years, as turbines of this scale are not sold by simple transactional sales. A small wind company faces the challenge of selling a product that is not usually purchased in bulk (i.e. a single unit is often sold), nor is it a simple transaction sale like smaller perishable or single use commodity items. Sales are often drawn-out if they involve wind resource siting. Most small wind sites are typically found by the 'suck it and see' method in lieu of a detailed (and costly) wind resource assessment [27]. Small wind turbine purchasers may enter detailed discussion with the turbine supplier and a comparison can be made with competing small wind units by detailed system modelling, such as that offered by Homer Pro. This is indeed a time-consuming and costly sales process, one which small wind manufacturers must take into due consideration.

Ideally, a small wind turbine company would target higher value enterprise sales, whereby an order of many units could provide the capital to trigger a higher volume manufacturing order. Business to business sales could eliminate the challenge of dealing directly with the customer, as retail customers often undertake extensive research on the purchase of even a single product. Nevertheless, it is prudent for any small wind developer to have a well identified customer prior to the commencement of vendor selection.

14.3.7 Field installation

To field test our wind turbine in a commercial setting, we have undertaken an installation with an Australia-based remote telecommunications provider. Compounding the difficulties in the previously discussed controller and electrical system design, most telecommunication operators use 48 VDC systems for their towers. It is very difficult for a small wind generator to achieve the required voltages to charge the batteries in the system and a boost converter is therefore required. It has been challenging to source a suitable boost converter for this system and may require a very significant further investment of time and money to develop one.

The authors note that due to the challenges of accessing remote locations, the control system should be as simple to install as possible. As access to the site is limited, installation should take place ideally as one body of work, without the need for additional visits. This field campaign is still ongoing, with future work to report a validated power curve.

An additional benefit of the diffuser is for safety and installation considerations – the diffuser provided a good lifting point for installation and provided an extra level of safety for the technicians who would have otherwise been working at heights in close proximity to exposed wind turbine blades (Figure 14.7). Conversely, the very rugged diffuser also protected the turbine blades from damage during the lifting and installation process.

14.4 Conclusions

There are several barriers to the adoption of small wind generation, including poor historical performance, high LCOE, and safety and noise concerns. DAWTs hold potential in increasing power output and reducing LCOE. Several challenges have been identified in commercialising this technology:

- Small wind turbines are a complex multidisciplinary system, involving aerodynamics, dynamics, structural analysis, electrical control, material science, manufacturing supply chains, and business operation knowledge.

Figure 14.7 Installation at a remote site

- Manufacturing supply chain issues due to high cost for low volume, exotic materials, and challenging manufacturing techniques.
- Channels to market, including the balance between enterprise vs transactional sales and the time-cost associated in making sales.

This chapter details the commercialisation efforts of novel DAWT technology. A 200-W unit was designed, manufactured, installed, and piloted at a customer's remote communications site. Other niche market applications for small wind were identified in the following markets: remote monitoring sites, telecommunications, mining, and agriculture.

14.5 Recommendations and future work

The author's detail the following lessons learnt and recommendations:

- The small wind turbine design standard, IEC 61400.2, is only applicable to open-bladed HAWTs, with revision to include other small wind turbine configurations recommended, i.e. DAWTs and vertical axis turbines. While much of the standard is applicable (i.e. HAWT rotor), specific guidance with respect to the design of other components such as the diffuser shroud and support struts should be given. Attention to the interfacing of a small wind turbine electrical system to off-grid batteries should also be expanded further. This is important due to the rapid cost reduction of renewable energy storage, which is an enabler of the benefits of wind – especially since the output is more stochastic and uncorrelated to the solar PV output, for which a system is usually designed for.

- System integration is key for customer adoption. Small wind turbine manufacturers should ensure that and resulting electrical systems or output of MPPT controller is plug-and-play. Retrofitting small wind turbines to existing structures can lower, or essentially eliminate, the costs associated with tower installation (including transportation to site, foundation work and erection).
- Customer education is important and often overlooked. While many users are aware of the benefits that solar PV and battery systems can bring, less can be said about small-scale wind turbine systems, for example as a hedge from high penetration solar PV, or during smoky bushfire conditions.
- Hardware designs that require minimal maintenance or are maintenance free. For remote sites, it is highly desirable to visit them as infrequently as possible. Small wind systems should be designed with few moving parts. Ideally, the electrical controller should be separated from the centre body or hub to allow for troubleshooting, repair, or replacement without needing to swap out the physical hardware. This factor has been highlighted during the COVID-19 pandemic where workforce could not be mobilised interstate within Australia.

Future work will be centred on performance assessment of the 200-W turbine in field conditions. Data logging is being undertaken on the site wind resource, turbine output power, and rotor rpm. These measurements are expected to form an experimentally validated power curve. The authors note that this work will be undertaken in conjunction with the ARENA. Results of this project will be released in the public domain in due course.

Ascertaining at what physical scale a DAWT is no longer commercially viable will provide future challenges. It is currently unknown at what rotor diameter (and hence power output) the physical structure of the diffuser would become untenable for field operation. The commercial failure of Vortec7 indicates that a 7-m DAWT is beyond the upper range of commercial viability. There is a trade-off between increased energy production and cost of materials, with this having an unknown impact on LCOE at larger scales.

Conflict of interest

Work in this chapter was part funded by Diffuse Energy Pty Ltd, in which the authors have a business and/or financial interest. The authors have disclosed those interests fully to the editors and have in place a plan for managing any potential conflicts arising from the publication of this chapter.

References

[1] International Electrotechnical Commission IEC61400.2-2013. Wind Turbines – Part 2: Small wind turbines; 2013.

[2] Bergey Windpower Co. Trials of building mounted wind turbines; 2019. Available from: http://bergey.com/technical/warwick-trials-of-building-mounted-windturbines [Accessed: 14-April-2019].

[3] Clausen PD and Wood DH. Research and development issues for small wind turbines. *Renewable Energy*. 1999;16(1–4):922–7.

[4] Medd A and Wood DH. Small wind turbine performance assessment for Canada. In *Research and Innovation on Wind Energy on Exploitation in Urban Environment Colloquium* 2017 (pp. 155–163). Springer, Cham.

[5] Ross SJ, McHenry MP, and Whale J. The impact of state feed-in tariffs and federal tradable quota support policies on grid-connected small wind turbine installed capacity in Australia. *Renewable Energy*. 2012;46:141–7.

[6] Wood D. Small wind turbines for remote power and distributed generation. *Wind Engineering*. 2010;34(3):241–54.

[7] Manwell JF, McGowan JG, and Rogers AL. *Wind Energy Explained: Theory, Design and Application*. 2nd ed. Chichester: John Wiley & Sons; 2009.

[8] Phillips DG. An Investigation on Diffuser Augmented Wind Turbine Design. Doctoral dissertation, University of Auckland, 2003.

[9] Gilbert BL and Foreman KM. Experiments with a diffuser-augmented model wind turbine. *Journal of Energy Resources Technology*. 1983;105(1):46–53.

[10] Venters R, Helenbrook BT, and Visser KD. Ducted wind turbine optimization. *Journal of Solar Energy Engineering*. 2018;140(1):011005.

[11] Tang J, Avallone F, Bontempo R, van Bussel GJ, and Manna M. Experimental investigation on the effect of the duct geometrical parameters on the performance of a ducted wind turbine. *Journal of Physics: Conference Series*. 2018;1037(2):022034.

[12] Kanya B and Visser KD. Experimental validation of a ducted wind turbine design strategy. *Wind Energy Science*. 2018;3(2):919–28.

[13] https://distributedwind.org/wp-content/uploads/2016/05/SMART-Wind-Roadmap.pdf.

[14] Assessing the future of distributed wind: Opportunities for behind-the-meter projects. https://www.nrel.gov/docs/fy17osti/67337.pdf.

[15] Kesby JE, Bradney DR, and Clausen PD. Determining the performance of a Diffuser Augmented Wind Turbine using a combined CFD/BEM method. *Renewable Energy and Environmental Sustainability*. Les Ulis Cedex: EDP Sciences. 2017;2:22.

[16] Peterson P and Clausen PD. Timber for high efficiency small wind turbine blades. *Wind Engineering*. 2004;28(1):87–96.

[17] Clausen PD, Reynal F, and Wood DH. Design, manufacture and testing of small wind turbine blades. In *Advances in Wind Turbine Blade Design and Materials* 2013 (pp. 413–431). Woodhead Publishing.

[18] Anup KC, Whale J, and Urmee T. Urban wind conditions and small wind turbines in the built environment: A review. *Renewable Energy*. 2019;131:268–83.

[19] Evans SP, Bradney DR, and Clausen PD. Assessing the IEC simplified fatigue load equations for small wind turbine blades: How simple is too simple? *Renewable Energy*. 2018;127:24–31.

[20] Goodfield D, Evans SP, KC A, *et al.* The suitability of the IEC 61400-2 wind model for small wind turbines operating in the built environment. *Renewable Energy and Environmental Sustainability*. 2017;2:31.

[21] Wood D. Small wind turbines analysis, design, and application. 2011. Berlin: Springer.

[22] Batt N and Coates C. Maximising output power of self-excited induction generators for small wind turbines. In *2012 XXth International Conference on Electrical Machines* 2012 (pp. 2105–2111). Marseille, France: IEEE.

[23] Warren M. *Blackout*. 1st ed. Melbourne: Affirm Press; 2019 [Chapter 7].

[24] Australian Communications and Media Authority. Impacts of the 2019–20 bushfires on the telecommunications network; 2020. Available from: https://acma.gov.au/publications/2020-04/report/impacts-2019-20-bushfires-telecommunications-network [Accessed: 29-July-2020].

[25] Australian Energy Market Operator. 2019–20 NEM summer operations review report; 2020. Available from: https://aemo.com.au/-/media/files/electricity/nem/system-operations/summer-operations/2019-20/summer-2019-20-nem-operations-review.pdf?la%BCen [Accessed: 04-August-2020].

[26] Australian Renewable Energy Agency. Off grid; 2020. Available from: https://arena.gov.au/renewable-energy/off-grid/ [Accessed: 04-August-2020].

[27] Environment NSW. Small wind turbine consumer guide; 2010. Available from: https://environment.nsw.gov.au/resources/households/NSWSmallWindTurbineConsumerGuide.pdf [Accessed: 15-August-2019].

Chapter 15

A tide like no other: harnessing the power of the Bay of Fundy

Tony Wright[1]

This chapter provides an overview of efforts to demonstrate tidal stream technologies at the Fundy Ocean Research Centre for Energy (FORCE) in Minas Passage, Bay of Fundy. The power extraction potential in the Minas Passage is estimated at nearly 8,000 megawatts, roughly equivalent to the power needs of 3 million homes. Established in 2009, FORCE was created to explore the potential for tidal stream energy to contribute to Canada's future clean energy supply. While still an emerging technology, tidal stream technology has the potential to provide a predictable source of renewable energy and thereby reduce GHG emissions from electricity generation and help respond to climate change. Over the past decade, supportive government policy, combined with investment in technology demonstration infrastructure and technical and environmental research, has spurred the participation of over 500 companies in Atlantic Canada. Challenges remain: the sector has not yet converged on a single design solution for energy capture or mooring, working with new technologies in marine environments is expensive, and more research is required to understand any potential environmental effects. Interest remains high, however, as 2021 witnessed the construction of two new technologies in Nova Scotia, in anticipation of later deployment at FORCE.

15.1 Introduction

Marine renewable energy (MRE) has been harnessed for centuries – powering mills, transporting nutrient-rich sediment, moving vessels, supporting marine life migration, even terrifying surfers – but only recently has it been understood as a vast untapped reserve of power. Seventy-one per cent of the Earth's surface is composed of moving water, all containing energy that can be converted to electricity.

Theoretical estimates for global MRE potential indicate resources exceeding 100,000 terawatt hours (TWh) of electricity, equal to the power needs of over

[1]Fundy Ocean Research Center for Energy, Parrsboro, Nova Scotia, Canada

8 billion Canadian households* – more than the current power demands of the entire planet.

Canada continues to deepen its exploration of MRE as a potential solution to clean energy, greenhouse gas emission (GHG) reduction, and economic growth targets. Given the country's natural resource assets as well as existing expertise in the marine sector, tidal energy has the potential to Canada's clean energy bottom line. Canada has an estimated 35,700 megawatts (MW) of tidal energy potential, enough clean power to displace over 113 million tonnes of CO_2 (equal to removing over 24 million cars off the road). No single site has more power potential than the Bay of Fundy.

Home to the highest tides in the world, the Bay of Fundy stretches 270 km along the US and Canadian coastlines, pushing roughly 160 billion tonnes of water – more than all the freshwater rivers and streams in the world combined. Often referred to as the 'Everest of Tidal Power', Fundy's tidal flow speeds up as it pinches through a narrow section called the Minas Passage, resulting in water speeds that can exceed 20 km/h. It is a lot of water, moving very fast.

The energy potential is staggering: roughly 8,000 MW, enough to power 3 million homes; it is estimated approximately 2,500 MW may be extracted without significantly changing the tide height [1,2].

That equates to an annual reduction of 5,000,000 t of GHG emissions based on a Nova Scotia fossil electricity intensity factor of 0.855-kg CO_2 eq/kWh. The opportunity includes both large-scale transmission projects and small, distributed community generation. More than 500 companies have already found work in the Bay of Fundy's emerging tidal stream sector.

Fundy's unparalleled tides have been recognized as an energy source for centuries. There is evidence of up to seven tidal grain mills along the Fundy coastline in the 1600s, built by the Acadians. By the early 1900s, engineers in both Canada and the United States began to consider the potential to generate electricity by building dams and holding tanks. In 1963, President John F. Kennedy even spoke about that potential from the White House lawn, saying: 'This can be one of the most astonishing and beneficial joint enterprises that the people of the United States have ever undertaken'. But nothing happened until soaring oil prices in the 1970s rekindled interest. And in 1984, Nova Scotia Power built a 20-MW tidal barrage along the Annapolis River. One of only a handful of tidal barrages anywhere the world, the Annapolis Tidal Station ceased operation in January 2019 and the project was cancelled after clearly showing the Bay of Fundy can generate electricity. But because Annapolis is effectively a dam, it also had many unwanted impacts – both on sediment and marine life.

*Electricity Measurement and Conversion: Electrical capacity is denoted in watts (W), kilowatts (kW), megawatts (MW), and terawatts (TW), and electricity usage is most often measured in kilowatt hours (kWh), megawatt hours (MWh), and terawatt hours (TWh). The watt (W) is a measure of electric power. Power is the rate of doing work or producing or expending energy. A 1,000-W of power produced or used for 1 h equals 1 kilowatt hour (kWh). On average, a typical household in Canada uses about 1,000 kWh per month resulting in annual household consumption of 12,000 kWh of electricity per year (1 kWh = 0.001 MWh; 1,000,000 MWh = 1 TWh).

Tidal stream technology – which operates like a wind turbine under the water – has the potential to generate power with a relatively small environmental footprint. Unlike a dam, tidal stream devices do not force marine life to pass through them and have smaller effects on both water flow and sedimentation. Tidal stream technology also offers a significant strategic advantage over wind and solar: it is predictable. Tidal energy is not weather dependent; its power is derived by the movement of the earth combined with the gravitational influence of the sun and moon. This offers utilities and ability to reliably predict tidal resource levels years in advance.

The Province of Nova Scotia and the Government of Canada invested in a technology demonstration facility in the Bay of Fundy; the aim was to demonstrate a number of different tidal stream devices from this emerging sector. Scientists, studying both the bathymetry and hydrodynamics in the Bay of Fundy, identified a unique test site in the Minas Passage, able to accommodate a number of turbines.

The Fundy Ocean Research Centre for Energy (FORCE), established in 2009, was created to explore the potential for tidal stream energy to contribute to Canada's future clean energy supply. While still an emerging technology, tidal stream technology has the potential to

- provide a predictable source of renewable energy;
- reduce GHG from electricity generation;
- contribute to action on climate change;
- spur industrial and local economic growth by capitalizing on skills and assets already present in other sectors;
- meet up to 100% of Nova Scotia's energy needs during flow and ebb tides.

Over the past decade, the Governments of Nova Scotia and Canada have established supportive policy, made key investments in technology demonstration infrastructure and technical and environmental research, and spurred activity by connecting a steadily growing supply chain of Atlantic Canadian businesses – over 500 companies have participated in the sector to date. As a result, Nova Scotia is often recognized for its strategic approach, slowly building the experience, knowledge, and innovation necessary to advance tidal sector activity.

Challenges remain as follows: the sector has not yet converged on a single design solution for energy capture or mooring, working with new technologies in marine environments is expensive, and more research is required to understand any potential environmental effects.

Nova Scotia's tidal stream sector faces many of the same challenges the global sector faces: challenges attracting finance, knowledge and technology gaps, and depending on location, insufficient infrastructure. FORCE was created to respond to these challenges.

FORCE has two major roles: host and steward to tidal stream energy activity.

15.2 FORCE as host

As host to tidal stream developers, FORCE provides a shared observation facility, subsea power cables, and grid connection. Since securing its original Environmental Assessment (EA) approval in 2009, FORCE has the following:

- Built and installed an 11-km network of 34.5-kV subsea power cables (see Image 15.1).
- Built and activated a 30-MW substation and associated electrical equipment.
- Built and maintained a public observation facility with over 30,000 guests since opening.
- Hosted three tidal stream device deployments.
- Conducted over 10,000 h of site characterization work.

There are currently five sites permitted for a total of 22 MW at FORCE, allotted in Table 15.1, and located per Figure 15.1.

OpenHydro, now defunct, deployed three tidal stream devices at FORCE; while the company ultimately failed, they reached critical milestones in the demonstration process: a tidal stream device in the Minas Passage can be deployed, retrieved, connected to subsea cabling, generate power, and supply electricity to the provincial grid.

As host, FORCE is expecting activity in 2021: Sustainable Marine announced successful energy generation from its 280-kW floating platform PLAT-I in Grand Passage in 2019 (see Image 15.2) and has been granted a license for a 1.26-MW

Image 15.1 Power cables come ashore at FORCE

Table 15.1 Present FORCE berth allotments

Berth	Entity
A	Minas Tidal Limited Partnership (MTLP)
B	Rio Fundo Operations Canada Ltd (RFOCL) (DP Energy-owned)
C	Sustainable Marine Energy Canada (Sustainable Marine)
D	Big Moon Power Canada (BMP) (former Cape Sharp Tidal Venture berth)
E	Halagonia Tidal Energy Ltd (HTEL) (DP Energy-owned)

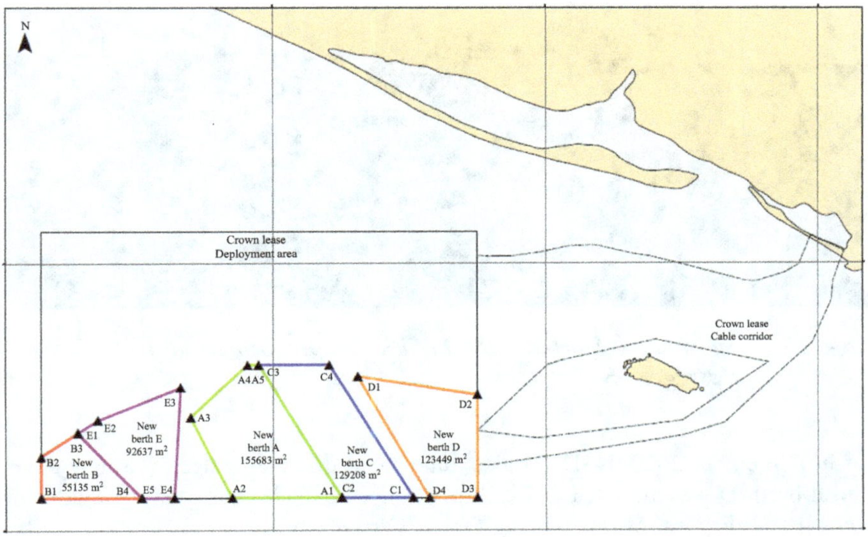

Figure 15.1 Locations of berths at FORCE

tidal array at FORCE in 2021 from Nova Scotia's Department of Energy and Mines. In early 2021, Sustainable Marine launched its next generation 420-kW PLAT-I 6.40, the world's first floating tidal energy array in Nova Scotia. Constructed by Meteghan, Nova Scotia's A.F. Theriault & Son Ltd, the tidal energy platform will first undergo commissioning and testing in Grand Passage. It will then be moved to the FORCE site as part of the first phase of the project. Sustainable Marine is supported by the Government of Canada with $28.5 million in funding. When complete, the project will deliver up to 9 MW of electricity in total to the Nova Scotia grid, enough to supply roughly 3,000 homes, and reduce GHGs by 17,000 t.

DP Energy announced the successful deployment of Acoustic Doppler Current Profilers (ADCPs) at berth E with support from Nova Scotia companies Seaforth Geosurveys and Huntley's Sub Aqua Construction in 2019, in preparation for the deployment of six Andritz Hydro Mk1 1.5-MW seabed mounted tidal turbines.

Image 15.2 Sustainable Marine's 280-kW PLAT-I platform installed in Grand Passage, NS

In September 2020, NSDEM also announced that the tender for a new developer at berth D was awarded to Big Moon Power Canada; this tender includes the removal and disposal of the existing OpenHydro turbine.

15.3 FORCE as steward

FORCE acts as a steward to the site, in accordance with the conditions of its EA approval and fundamental to FORCE's mandate: to better understand the effects turbines have on the environment and report these findings to the public.

FORCE has slowly built up its science capacity, enabling in-house study design as FORCE continues to adaptively manage its environmental effects monitoring programmes (EEMPs). Staff are now primarily science based, including a core group of research scientists and ocean technologists, who are working to better understand the physical and biological conditions of this site. Since the start of the project in 2009, FORCE has conducted baseline environmental studies, applied research, and environmental monitoring work at its site in the Minas Passage. FORCE receives ongoing monitoring advice through an independent Environmental Monitoring Advisory Committee (EMAC), with membership from the scientific, fishing, and First Nations communities.

To date, international research studies of tidal stream devices have not observed collisions with fish [3]. But this must be monitored in the Minas Passage. This remains difficult: the intense turbulence characteristic of the FORCE site makes it challenging for sensors to capture optical and acoustic data. This means not only environmental monitoring but also research and data modelling programmes are all critical to retiring risk and uncertainty about potential impacts.

Much of FORCE's science work is conducted with partners: these include the University of Maine, the Sea Mammal Research Unit Consulting (Canada), Envirosphere Consultants, Acadia University, TriNav Fisheries Consultants, JASCO Applied Scientists, Ocean Sonics, Dalhousie University, Nexus Coastal Resource Management, and GeoSpectrum. These studies are examined in further detail later in this chapter.

Monitoring: FORCE's EEMP is designed to better understand the natural environment of the Minas Passage and the potential effects of turbines as related to fish, seabirds, marine mammals, lobster, marine noise, benthic habitat, and other variables. All documents are available online: fundyforce.ca/document-collection.

Since 2016, this work now represents more than 5,000 'C-POD' marine mammal monitoring days (see Image 15.3), more than 500 h of hydroacoustic fish surveys, biweekly shoreline observations, 49 observational seabird surveys, as well as lobster surveys, drifting marine sound surveys and additional sound monitoring.

Research: FORCE also advances research through its growing team of science personnel and onshore and offshore assets. Dr Dan Hasselman and Dr Joel Culina, two leading experts in fish monitoring and hydrodynamics, together with FORCE's team of ocean field technologists, are able to deliver a sector-leading science programme in partnership with universities and other research entities. FORCE's onshore assets include a meteorological station, video cameras, an X-band radar system, and tide gauge. Offshore assets include modular subsea platforms for both autonomous and cabled data collection and a suite of instrumentation for a variety

Image 15.3 C-PODs: Ocean technology staff haul a C-POD SUBS package aboard during recovery

of research purposes. Research priorities are based on consultation with regulators, other Fundy users, academics, and industry; each project is aimed at reducing risk.

15.4 Turbines in the water

On 12 November 2009, Nova Scotia Power began testing a 1-MW OpenHydro turbine at FORCE (see Image 15.4). The 10-m diameter open-centre in-stream tidal turbine was the first commercial-scale tidal turbine deployed in North America.

Once on-site, the turbine and its 400-t subsea gravity base were successfully deployed. It was later discovered that an acoustic modem intended to allow data to be recovered from the turbine had not been functioning. Alternate measures were used to monitor the turbine and ensure that it remained in position. In May 2010, OpenHydro was able to capture limited video footage of the turbine that showed some damage to the rotor and led NSP and OpenHydro to remove the unit from the Basin.

Image 15.4 2009 Nova Scotia Power/OpenHydro deployment

The turbine and gravity base were successfully recovered on 17 December 2010, from the Minas Passage, a world first. OpenHydro confirmed that the structure appeared to be in good condition overall and, consistent with previous images, the blades were missing from the centre of the unit. FORCE representatives were on site for the retrieval and confirmed the condition of the unit.

OpenHydro reported that blade damage appeared to be the result of peak velocities of up to 2.5 times stronger than originally predicted. They also confirmed that the unit had stopped functioning from 4 December, on the second spring tide after deployment.

The FORCE monitoring programme observed no environmental effects from the OpenHydro unit. The unit itself was deployed and recovered without environmental damage, biofouling, or damage to the structure itself. No significant changes were observed at the turbine site, except

- some unidentified debris on the seabed thought to be part of the damaged turbine, and
- two small pits in the bedrock surface from the support legs (the unit weighted approximately 450 t).

In 2009, OpenHydro faced a dilemma: deploy its turbine in a location where the current was known – or at least thought to be known – but the bottom unknown due to its covering of boulders and cobbles; or in a location where the current was unknown but the bottom was known. It chose the known bottom. As a result, the base was stable for the 13 months of the deployment, but the power was more than twice the intended operating range of the turbine. The blades were blown out within 3 weeks of deployment during the convergence of the second highest spring tide of the year and a major storm.

The Nova Scotia tidal sector learned that high-energy sites are more complex than previously thought. And a much more sophisticated model of the site was needed. Characterization of the FORCE site has required the development of novel techniques for the deployment of acoustic and current sensors and has spurred the development of the next generation of sensors. This work continues today.

In November 2016, Cape Sharp Tidal Venture (CSTV) deployed the first OpenHydro open-centre turbine at the FORCE site (see Image 15.5). The operation was completed in one tidal cycle, while holding station over the sea floor. Atlantic Towing's Kingfisher towed the Scotia Tide deployment barge from West Bay to the site; once in position, the turbine was lowered to the sea floor in a 4-h operation on an ebb tide. More than 25 local crews participated, including Atlantic Towing, Seaforth, Sand and Sea Marine, and RMI.

CSTV's marine operations team then connected the turbine cable tail with a 300-m interconnection cable (installed the Minas Passage in the previous winter). This connected the turbine's power and data systems to FORCE's existing 16-MW subsea cable and on to the onshore substation (see Image 15.6).

Following cable connection, CSTV announced successful power generation and entered a period of demonstration, commissioning and monitoring.

In June 2017, Cape Sharp retrieved the turbine for work to its turbine control centre (TCC) in Saint John, New Brunswick (see Image 15.7). The TCC transforms

Image 15.5 OpenHydro turbine is lowered to sea floor, 2016

Image 15.6 Cape Sharp Tidal Ventures (CSTV) subsea cable connection, 2016

the raw power from the generator into grid-compatible AC power and sends operational and environmental sensor data to shore in real-time via the power cable.

In July 2018, deployment operations began for the second, identical 2-MW turbine in the Minas Passage. The turbine was deployed on the afternoon of

Image 15.7 CSTV OpenHydro turbine I in Port of Saint John for upgrade work, 2017

Sunday, 22 July and grid-connected on Tuesday, 24 July. After undergoing initial commissioning, including both operational and environmental monitoring device testing by OpenHydro personnel, Naval Energies announced an end to all its investment in tidal stream; OpenHydro was placed in immediate liquidation. OpenHydro staff disconnected the turbine from the substation. In September, an OpenHydro inspection of the turbine found the rotor had stopped; Cape Sharp announced reconnection of near-field monitoring equipment.

In 2019, following a period of commissioning and testing environmental monitoring equipment, Sustainable Marine Energy Canada began operating tidal turbines developed by SCHOTTEL on the PLAT-I platform installed in Grand Passage, Nova Scotia. On 23 February 2019, the system generated first power.

In 2020, the Government of Canada announced $28.5 million to Sustainable Marine to deliver Canada's first floating tidal energy array. In February 2021, Sustainable Marine launched its first 480-kW device assembled in Meteghan, Nova Scotia (see Image 15.8) for testing before being relocated to FORCE. The objective of the Sustainable Marine project is to provide up to 9 MW of electricity to Nova Scotia's power grid, thereby reducing GHG s by 17,000 t of carbon dioxide annually.

Video of the Sustainable Marine launch is available on their YouTube channel: https:youtube.com/channel/UCsIzDumme7L2RwpDs3vxEYQ/videos.

15.5 Operations and infrastructure

15.5.1 Site characterization

Developers need to know as much about their marine site as wind developers know about their sites.

Image 15.8 Sustainable Marine's 420-kW PLAT-I 6.40 floating tidal energy platform, completed at A.F. Theriault & Son Ltd, launches from Meteghan, Nova Scotia in February 2021

Marine high flow site characterization is a challenge for experts, a necessity for tidal power extraction, and an economic opportunity for Canada. It requires sophisticated data on tidal ranges, turbidity, salinity, temperature, acoustics, weather, current profiles, and waves. It governs energy assessment, engineering, and marine operations. And no one has yet mastered it.

Working at high flow sites such as FORCE is expensive and difficult. Acoustic Doppler current profiling, used to determine currents, is a good example because FORCE and its contractor Oceans Ltd of St. John's and Halifax now have more than 10-year experience at the site. At this early stage, each deployment of an ADCP and analysis of the data was extremely expensive, costing FORCE about \$50,000 (the largest part of which was boat time) and in 2011 two ADCPs were lost, one snagged by a fisher and the other buried in a gravel wave. An ADCP unit costs about \$50,000. So \$50,000 for a success; \$100,000 for failure (necessitating a second unit) – and there is no way to know the difference until recovery is attempted. (Since 2011, costs have decreased significantly; FORCE handles ADCP deployment and data analysis in-house.)

Furthermore, OpenHydro's experience with its first deployment showed that conventional ADCP sample rates tended to understate velocity and turbulence at high flow sites – in this case, by over 200%.

Experienced operators characterize it thus: 'you should never put anything in the ocean you really need to get back'. A list of things that have gone wrong in the ADCP programme – some of which have resulted in the loss of equipment and data – despite FORCE's contractors' years of experience in ocean research and marine operations includes:

- Missing moorings – snagged, dragged, buried, or broken.
- The acoustic release (which dispatches a recoverable buoy on a line to the surface from a sound signal) did not answer or did not release.
- The instruments fail through being struck by debris, vibration/chafe, or being dragged during recovery.
- Cable corroded and broke under load while another length of cable from the same reel was recovered in pristine condition.
- Navigation errors, computing errors and vessel problems, seasickness, fatigue, and human error.
- Storms and weather.

It was necessary to develop new approaches. An entirely new bottom stand configuration was developed to deal with gravel waves[†] and FORCE created the unique practice of deploying two adjacent ADCPs simultaneously for a period of a month, one sampling continually at a narrow height[‡] band and the other sampling two of every 5 min throughout the water column.[§] The first gives extremely detailed data on turbulence but only at one spot and in a very large data set. The second gives more conventional ADCP data, which the machine itself 'smooths' by disregarding outlying 'pings'.

However, there is debate among experts as to the significance of the outlying pings and traditional quality control methods are not applicable to the high-frequency data. As a result, at FORCE's suggestion, Dr Joel Culina received an Industrial Research & Development Fellowship from NSERC[¶] to work with Oceans Ltd to examine the FORCE data sets in greater detail, conduct further research on currents and turbulence at our site, and generally advise FORCE on such matters.

While it is clear that new sensing technologies are required and being developed under FORCE's stable of sensor platforms, nevertheless, by adapting practices and equipment intended for use in more conventional marine environments, FORCE has made significant progress in the following fields:

[†]A great deal of innovation has taken place in deployment, mooring, and recovering equipment. As a result since 2012, we have recovered all the ADCPs deployed. The work of improving all aspects of mooring design, construction, deployment, and recovery is now part of the FAST programme and will be recounted in that project's completion report.

[‡]ADCPs sit on the bottom and look up so we speak of height in the water column rather than depth.

[§]This approach was prompted by the OpenHydro experience (where twice as much power was encountered as 2008–9 data techniques predicted) and results on Site B, where Alstom had asked for continuous readings that did in fact show much higher currents and greater turbulence than previously revealed by conventional techniques.

[¶]FORCE had applied to NSERC for this grant but an initial favourable ruling by NSERC's regional office was reversed in Ottawa and FORCE was declared ineligible for NSERC funding.

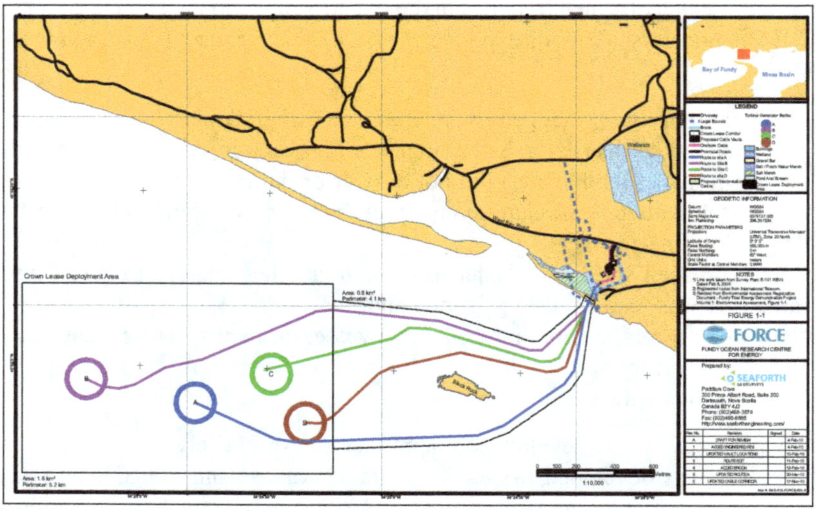

Figure 15.2 Initial berth sites at FORCE

- Characterizing the berths (Figure 15.2) and cable routes in respect of key factors of current, turbulence, and tidal range.
- Techniques for the deployment and recovery of equipment in high-energy sites.
- Extending the operational parameters of existing equipment.
- Data analysis.

15.5.2 Hydrodynamic loading study

The Halifax office of Dynamic Systems Analysis Ltd (DSA)[‖] was contracted to provide advice on cable deployment. DSA used its dynamic analysis software ProteusDS to simulate cable lay operations for the Site D cable in the full range of current conditions.

The simulations showed that there was little risk of cable damage due to excessive cable bending in the touchdown area as the minimum-bending radius seen in the simulations was 115 m and the minimum allowable bending radius for the cable is 2.2 kN while under tension. However, the forces on the barge due to the cable increased dramatically, indicating that deploying cable in these currents poses a risk to barge navigation – which of course is central to cable seabed touchdown.[**]

[‖]DSA is a small Canadian company that provides numerical modelling services and software to the defence, marine, offshore, and subsea industries. DSA creates engineering analysis software that tackles tough ocean engineering problems. DSA has offices in Victoria, BC and Halifax, NS. DSA's team of six engineers and software experts work with partners around the globe. Source: http://dsa-ltd.ca/about/dsa/.

[**]The maximum loads on the barge as a result of the cable in these currents were 132 kN in the surge direction and 207 kN in the sway direction. Hydrodynamic loading of the cable on the barge only has a significant effect in extreme current conditions. If these conditions were to arise during the cable lay, hydrodynamic loading on the cable may cause the barge to become unstable.

A vortex induced vibration (VIV) study was performed to determine whether or not the cable would undergo any vortex-induced motion, and what effect this might have on the cable lay. The study determined that some VIV is expected to occur for the cable at some select current speeds, depths, and cable lay tensions. The motion may increase the drag coefficient and hydrodynamic loading on the cable. However, in expected current conditions, it was concluded the motion should not have a substantial effect on the cable lay operations.

A sensitivity study modelling seabed/cable interaction showed that under the maximum flood tide current conditions, the cable was unlikely to move once laid.

15.5.3 Continuous site monitoring

FORCE has operated an X-band radar overlooking its turbine demonstration site since 2015 that, with the assistance of Acadia University computer programming, reports and generates maps of surface currents and the wave field of the FORCE area. In collaboration with the National Oceanography Centre, United Kingdom, their software was adapted for high-resolution surface velocity mapping of FORCE region waters. Fixed-position, land-based X-band marine radar is unique among sensors in its ability to provide continuous horizontal measurements, of multiple variables of interest to turbine developers and regulators. These variables can be grouped into two classes: hydrographic and trajectory-following variables, at and above the surface of the ocean.

A digital tide gauge was installed August 2013. This allowed FORCE (and the public, through Ocean Networks Canada) to monitor and evaluate actual tidal conditions in the Minas Passage in real time, something that previously could only be inferred from the single tide gauge in the Bay of Fundy in Saint John harbour.

15.5.4 The subsea power cables: plug and play

By providing common infrastructure and an approved, well-characterized site, FORCE allows its berth holders and their suppliers to focus on innovation and development.

FORCE is the world's only site that supports the development of turbine arrays. The European Marine Energy Centre (EMEC) has more berths and has been operating longer but is limited by its 11-kV cables and remoteness from the national grid. Central to Canada's advantage in the next stage of tidal stream development is FORCE's selection of 34.5-kV submarine cables. Cable deployment is costly and risky and should only be undertaken once, so adaptable cables are well worth the extra expense.

FORCE was initially envisioned as a Canadian version of EMEC with capacity for one tidal device per cable. In August 2008, FORCE received quotations for cable supply and learned our capital was inadequate to proceed. The cost of cable was much higher than FORCE's engineering team had been led to expect.

But this setback caused FORCE to consider its natural advantages:

- FORCE's location near the Eastern North American grid makes transition from testing to commercial development possible.

- Tidal energy is an evolving suite of technologies so adaptability is critical.
- The Minas Passage is large, with abundant kinetic energy, so if tidal stream were to move to larger scale, it would happen here.

Through a competitive process conducted in 2009, FORCE identified a preferred supplier: IT International Telecom (IT), which submitted a turnkey bid offering to obtain and install submarine cable to FORCE's specifications of either 13.8- or 34.5-kV power cores. Analysis of the bids conducted in November 2009 indicated the incremental cost of the 34.5-kV cable over the 13.6 kV was 20% in supply and no effect on installation costs; or, in other words, an overall incremental cost of 10%.

Due to the greater utility FORCE decided to order four (three + a 500-m spare) 34.5-kV cables.[††] Each cable is a three-conductor 34.5 kV, EPR insulated (133%) and double-wire armoured submarine composite cable.

The same approach, of building for flexibility where practicable, FORCE adopted for the substation and transmission line. FORCE considered five options for the substation and three for the power line – one distribution hookup and two transmission line options. The substation options ranged from a 5-MW substation which offered the lowest initial cost but least utility, to a full-fledged facility capable of serving a commercial field but which was estimated to cost more than $13M. FORCE took the middle road, a $7M substation on an oversized pad with some 'overbuilt' components operated at 25 kV.

When it was found that upgrading the local distribution system would cost nearly as much as a transmission line directly to the NSPI Parrsboro substation, FORCE opted for a transmission line built to 137-kV standards operated at 25 kV. As a result of this foresight, the interconnection can be upgraded to with minimal stranded assets.[‡‡]

15.5.4.1 First cable rehearsals

IT, then holders of a supply/install cable contract with FORCE, conducted deployment trials at FORCE on 19 and 20 August 2012 (see Image 15.9). IT's subcontractor, Atlantic Towing Ltd (ATL), mobilized two 4000 HP Z-drive tugs, and a barge to demonstrate station keeping and the ability to drive cable route D. This was intended as a rehearsal of the actual cable deployment IT planned in late September.

Despite trying several vessel configurations, these trials showed the following:

- The tugs and barge could not maintain station.
- The tugs could not maintain adequate control of the barge along the cable route.
- Aspects of shore end operations were not tested nor demonstrated adequately.
- Additional analysis of site data and modelling is required.

[††]The cable contract with IT was executed by FORCE in September 2010.

[‡‡]In March 2014 Province awarded FORCE a $4.4M grant for a phase 2 upgrade to FORCE's electrical facilities that will enable delivery of 20 MW (and possibly as much as 30 MW) of power to the grid.

Image 15.9 First cable deployment trials at FORCE

IT elected to conduct additional trials in the Saint John Harbour. This eliminated any chance of a 2012 cable deployment.

The second set of trials in Saint John was conducted 10–11 October 2012, with ATL and IT using four Z-Drive tugs. The trials demonstrated that four tugs were needed to provide sufficient power to manoeuvre and control the barge. However, the Hydraulic Power Unit provided by Prysmian to unspool the cable did not turn fast enough to overboard the cable at the minimum speed required to maintain control. It was also apparent that beaching the barge is essential because, even with enough power, the crews cannot hold station long enough to offload the cable end to the beach.

As 2012 ended, there were more questions than answers; the answers there were, were negative. The trials had demonstrated that two Z-Drive tugs and a barge (the method IT's bid was predicated on) would not work and only that the vastly more expensive four tug solution *might* lay the cable – at a pace perhaps four times the maximum the equipment could unreel the cable – but could not land it. Concerns also persisted regarding cable end termination and bottom conditions.

Z-drive tugs, at the time, cost about $10,000/day each plus fuel ($5,000/day approx.) and FORCE's site is a full 2 days of steaming from their homeport of Saint John. They are also deep draft vessels, which means that additional, shallow draft tugs would be required to beach the barge, and the tugs themselves are subject to high current loading. In other words, the tugs themselves added to the navigational difficulties and in effect become part of the problem.

As a result of the sea trials, and reconsideration of site data, a number of significant decisions were made.

IT had initially planned to lay from shore to sea to avoid the risk of coming up short. FORCE had urged a sea to shore lay because site data indicated navigation was less difficult in that direction and shortfall risk was minimal because had prudently ordered extra cable on each reel. Following the sea trials IT agreed with FORCE.

Cable routes were initially selected to maximize separation of cables in order to facilitate cable repair operations. However, after experience with inshore ADCP deployments, and detailed analysis of the bathymetry, it was determined that is impractical and maybe impossible to repair a broken power cable unless the break occurs near one end or the other of the cable.

Tug and barge trials and computer simulations of cable catenaries demonstrated the need to simplify routes and reduce cross current segments. In short, precision was neither advantageous nor possible.

There were no wet-mate connectors (equipment that allows underwater electrical connection) available for the FORCE cable and wet-mate connectors require ROVs in any case (and ROVs are extremely challenging to operate in tidal flows like those of the Minas Passage). Dry-mate connectors on the other hand require very precise station keeping. Experience at EMEC suggested at one time splicing would take not less than 40 h. This has been reduced since but it was the situation at the time. Although ATL had said it could maintain station for 7–10 days FORCE doubted this – correctly as the sea trials proved. EMEC has longer slack tides and slower currents but uses a powerful dynamic positioning vessel. EMEC has the advantage of proximity to North Sea petroleum industry assets. The cost of such vessels at FORCE is prohibitive. FORCE elected to have its own dry-mate cable termination (CAT) designed and manufactured.

Seaforth Geosurveys Inc. (Seaforth) and Atlantic Marine Geological Consulting Ltd (AMGC) were contracted October 2012 to provide a detailed quantitative and qualitative description and the assessment of the four cable routes from existing and new data and as a result portions of the cable routes were realigned.

FORCE felt that the sea trials demonstrated that a radically different approach was required and began considering using smaller, shallow draft tugs, and barges. FORCE assumed the management of the data cable deployment utilizing small conventional tugs and a small barge. IT provided the cable, and the deck spread and cable-handling personnel. The data cable was successfully laid in December 2013.

FORCE utilized the same senior team for the power cable operations as the data cable deployment under the leadership of Tony Wright P. Eng. who had become the general manager of FORCE effective from 1 January 2014.

Cable vaults were installed on the beach to reduce the amount of submarine cable to be hauled off the barge, reducing the amount of time the barge is grounded; and to enable the shore-side cables to be installed during the unusually dry summer of 2014, minimizing impact on the wetlands in which they are buried.

FORCE decided to use a heavy lift ship (HLS) and much smaller vessels than planned by IT. The HLS could self-load all four cable reels in Saint John, bring them to the FORCE site in one trip, and then load each one on the lay barge for each deployment. This meant a mid-sized barge and conventional tugs could be used and, weather permitting, cables could be laid on consecutive days. The HLS would have reduced transit time, saved money, and reduced operational risk. Placing 150t cable reels at very precise locations on the deployment barge is technically very challenging and must be performed using the proper equipment and experienced crew.

Unfortunately, there was no suitable Canadian vessel available when required. An alternative approach was quickly initiated – the deployment barge would do double duty as a transit vessel. A large crane was secured to load the reels on the barge in Saint John for transport to FORCE. Unlike the HLS the deployment barge could only carry one reel at a time. The three extra transits increased the cost and the duration of the project (the round trip was 3 days per cable) and risk of weather delays. The crawler which moved the reels from the warehouse was required for a month instead of a few days and, of course, the heavy crane was needed for the same period.

The Port of Saint John required FORCE to obtain a structural analysis of the wharf because the forces associated with moving the cables and reels, which would have been borne by the HLS, would now be borne by the crane on the wharf. The analysis was completed on 3 October and the Port authorized the work to proceed that evening.

As a result, the cable was loaded on 6 October 2014 and transited to FORCE on 7 October to be ready for sea trials. Unfortunately, high winds out of the south and west created sea states that hampered marine operations and made deployment impossible; sea trials were conducted on 8, 9, 11, 12, and 14 October. Cable C was successfully deployed on 16 October. Cable B was loaded on 18 October and deployed on 19 October. Cable A was installed on 26 October; Cable D was installed on 28 October. The cables were surveyed and buried over the next 2 days. The cables were fully connected and tested by 20 November.

Central to FORCE's role as host to technology is the ability to transmit power to the electricity grid. The installed four underwater power cables have a combined length of 11 km and a total capacity of 64 MW, equivalent to the power needs of 20,000 homes at peak tidal flows. Each 34.5-kV cable, together with its reel, weighed over 100 t during installation. Video is available here: https://vimeo.com/110095612.

15.6 Research and monitoring

Tidal stream devices have been installed in a number of locations around the world, including the United Kingdom, France, and the Unites States. To date, international research studies indicate that:

- *'Small numbers of operational marine energy devices are unlikely to cause harm to marine animals, including marine mammals, fish, diving seabirds, and benthic animals; change habitats on the seafloor or in the water significantly; or change the natural flow of ocean waters or waves'.*
- *'Despite our findings, we still need more data about what might, or might not, happen to animals swimming close to operating turbines underwater'.* [4]

Figure 15.3 The scale of a single turbine (based on the dimensions of the OpenHydro turbine deployed by CSTV, indicated by the red dot and above the blue arrow) in relation to the cross-sectional area of the Minas Passage. The Passage reaches a width of ~5.4 km and a depth of 130 m

Since 2009, as a condition of its provincial EA approval, FORCE has been conducting an EEMP to better understand the natural environment of the Minas Passage and the potential effects of turbines. All documents are available to the public at https://fundyforce.ca/document-collection.

FORCE's monitoring programme is developed by the FORCE science director Dr Daniel Hasselman with input and review by national and international experts and scientists, provincial and federal regulators, and FORCE's EMAC, which includes representatives from scientific, Mi'kmaw, and fishing communities. The EEMP is designed to

- monitor the environmental effects of operating turbines;
- focus on five subject areas: lobsters, fish, marine mammals, marine seabirds, and marine noise;
- be adaptive, based on monitoring results, lessons learned and input from regulators and EMAC.

Since 2009, more than 100 related research studies have been completed or are underway considering physical, biological, socioeconomic, and other research areas related to tidal energy in the Minas Passage. But many questions remain.

Overall, the risks associated with single device or small array projects are anticipated to be low given the relative size/scale of devices. For example, at the FORCE site the 2-MW OpenHydro turbine occupies ~1/1,000th of the cross-sectional area in the Minas Passage (see Figure 15.3).

A full evaluation of the risks of tidal stream energy devices, however, will not be possible until more are tested over a longer term period with monitoring that documents local impacts, considers far-field and cumulative effects, and adds to the growing global knowledge base.

As well, a full understanding of turbine–marine life interactions will not be possible until turbine-specific and site-level monitoring efforts are integrated, and additional data are collected in relation to operating turbines. Further, multi-year data collection will be required to consider seasonal variability at the FORCE test site and appropriate statistical analyses of these data will help to obtain a more complete understanding of marine life–turbine interactions.

The most recent monitoring programmes at the FORCE site were initiated in the anticipation of OpenHydro turbine deployments in 2016. These efforts are divided into two components: mid-field monitoring activities led by FORCE ($>$ 100 m from a turbine), and near-field or 'turbine-specific' monitoring led by project developers ($\leq$ 100 m from a turbine) at the FORCE site.

Table 15.2 The objectives of each of the 'mid-field' environmental effects monitoring activity, which consider various valued ecosystem components (VECs), led by FORCE

Mid-field environmental effects monitoring VEC	Objectives
Lobster	• To determine if the presence of a tidal stream energy turbine affects commercial lobster catches
Fish	• To test for indirect effects of tidal stream energy turbines on water column fish density and fish vertical distribution • To estimate probability of fish encountering a device based on fish density proportions in the water column relative to turbine depth in the water column
Marine mammals	• To determine if there is permanent avoidance of the mid-field study area during turbine operations • To determine if there is a change in the distribution of a portion of the population across the mid-field study area
Marine sound (acoustics)	• To conduct ambient sound measurements to characterize the soundscape prior to and following deployment of the in-stream turbines
Seabirds	• To understand the occurrence and movement of bird species in the vicinity of tidal stream energy turbines • To confirm FORCE's environmental assessment predictions relating to the avoidance and/or attraction of birds to tidal stream energy turbines

Mid-field monitoring at the FORCE site presently consists of monitoring for fish, marine mammals, seabirds, lobster, and marine sound (see Table 15.2). Since the start of this latest monitoring effort in 2016, FORCE has completed the following:

- ~564 h of hydroacoustic fish surveys;
- more than 5,083 'C-POD' marine mammal monitoring days;
- biweekly shoreline observations;
- 49 observational seabird surveys;
- four drifting marine sound surveys and additional sound monitoring; and
- 11 days of lobster surveys.

Table 15.2 outlines the objectives of the respective mid-field monitoring activities conducted at the FORCE demonstration site. Near-field monitoring summaries will be updated as turbines are scheduled for deployment at FORCE. At this time, and considering the scale of turbine deployments in the near-term at FORCE, it is unlikely that significant effects in the far-field will be measurable.

Far-field studies such as sediment dynamics will be deferred until such time they are required. However, recent discussions with scientists serving on FORCE's EMAC suggest that the natural variability inherent to the upper Bay of Fundy ecosystem far exceeds what could be measured by far-field monitoring efforts. Moreover, the scale of tidal power development would need to surpass what is possible at the FORCE tidal demonstration site to extract sufficient energy from the system to have any measurable effects. In short, far-field monitoring would be futile unless tidal power development transitions from demonstration scale to commercial arrays.

As more devices are scheduled for deployment at the FORCE site and as monitoring techniques are improved, monitoring protocols will be revised in keeping with the adaptive management approach. These studies will be developed in consultation with EMAC, regulators, and key stakeholders.

15.6.1 Lobster

FORCE conducted a baseline lobster catchability survey in fall 2017 (NEXUS Coastal Resource Management Ltd, 2017). This catch-and-release survey design was conducted over 11 days and consisted of commercial traps deployed at varying distances around the future location of the CSTV turbine deployment planned for 2018. Captured lobsters were measured (carapace length), had their sex and reproductive stage determined (male, female, and berried female), and shell condition evaluated. This baseline survey captured 351 lobsters and reported a high catchability rate (> 2.7 kg/trap).[§§] Preliminary qualitative analyses indicated that catch rates declined during the survey and were associated with increasing tidal velocities; a statistically significant negative relationship was detected between catch rates and maximum tidal range. No significant difference in catch rates was detected across separate locations from the proposed turbine location. Cumulatively, these results suggested that the impact of turbines may be higher on lobster catchability than anticipated in the EA, but a repeat of the study in the presence of an operational turbine is required to verify this prediction.

15.6.2 Fish

FORCE has been conducting mobile fish surveys since May 2016 to test the EA prediction that tidal stream turbines are unlikely to cause substantial impacts to fishes at the test site. To that end, the surveys are designed to

- test for indirect effects of tidal stream energy turbines on water column fish density and fish vertical distribution; and
- estimate the probability of fish encountering a device based on any 'co-occurrence' relative to turbine depth in the water column.

Moreover, these surveys follow a 'BACI' (before/after, control/impact) design to permit a comparison of data collected before a turbine is installed with data collected while a turbine is operational at the FORCE site, and in relation to a reference site along the south side of the Minas Passage. These 24-h mobile surveys encompass two tidal cycles and day/night periods using a scientific echosounder, the Simrad EK80, mounted on a vessel, the Nova Endeavor (Huntley's Sub-Aqua Construction, Wolfville, NS). This instrument is an active acoustic monitoring device and uses sonar technology to detect fish by recording reflections of a fish's swim bladder.

[§§]This is classified as 'high' according to DFO's Catch Per Unit Effort (CPUE) index [7].

Analyses of hydroacoustic fish surveys completed during baseline studies in 2011 and 2012 [5] and surveys during May 2016 to August 2017 [6] evaluated changes in fish densities in association with diel stage (day/night), tidal stage (ebb/flood), and turbine presence or absence (an OpenHydro turbine was present November 2016 to June 2017). While preliminary results support the EA prediction that tidal stream devices have minimal impact on marine fishes, additional surveys in relation to an operating turbine are required to fully test this prediction.

15.6.3 Marine mammals

FORCE conducts two main activities to test the EA prediction that project activities are not likely to cause significant adverse residual effects on marine mammals within the FORCE test site. These activities have been ongoing on a regular basis since 2016. Specifically, FORCE in continuing to

- conduct passive acoustic monitoring (PAM) using ‘click recorders’ known as C-PODs and
- implement an observation programme that includes shoreline, stationary, and vessel-based observations.

15.6.3.1 Passive acoustic monitoring

The first component of FORCE’s marine mammal monitoring programme involves the use of PAM instruments known as C-PODs which record the vocalizations of toothed whales, porpoises, and dolphins.[¶¶] The programme focuses mainly on harbour porpoise – the key marine mammal species in the Minas Passage that is known to have a small population that inhabits the inner Bay of Fundy. The goal of this programme is to understand if there is a change in marine mammal presence in proximity to a deployed tidal stream energy device and builds upon baseline C-POD data collection within the Minas Passage since 2011.

From 2011 to early 2018, more than 4,845 ‘C-POD days’[‖‖] of data were collected in the Minas Passage. Over the study period, it was found that harbour porpoise use and movement varies over long (i.e., seasonal peaks and lunar cycles) and short (i.e., nocturnal preference and tide stage) timescales. This analysis, completed by Sea Mammal Research Unit (Canada) (Vancouver, BC), showed some evidence to suggest marine mammal exclusion within the near-field of CSTV turbine when it was operational (November 2016 to June 2017). This analysis revealed that the C-PODs in closest proximity to the turbine (230- and 210-m distance) had reduced the frequency of detections, but no evidence of mid-field avoidance with a turbine present and operating. The latest findings also revealed a decrease in detections during turbine installation activities; consistent with previous findings, but requiring additional data during an operational turbine to permit a full assessment of the EA predictions.

[¶¶]The C-PODs, purchased from Chelonia Limited, are designed to passively detect marine mammal ‘clicks’ from toothed whales, dolphins, and porpoises.

[‖‖]A ‘C-POD day’ refers to the number of total days each C-POD was deployed times the number of C-PODs deployed.

SMRU provided their third year report of harbour porpoise monitoring using C-PODs at the FORCE test site. The report describes the results of C-POD deployments 7–10 (i.e., 416 days from 5 May 2018 to 14 August 2019) and places the results in the broader context of the overall marine mammal monitoring programme at FORCE. This ongoing monitoring programme continues to show the prevalence of harbour porpoise at FORCE, with the species being detected on 98.8% of the 1,778 calendar days since monitoring with C-PODs commenced in 2011. Harbour porpoise detections at FORCE vary seasonally, with peak activity occurring during May–August, and lowest detections during December–March.

SMRU provided an interim report covering C-POD deployment 11 (August 2019 to January 2020); the report supports the findings of previous monitoring activities that harbour porpoise are prevalent at the FORCE test site. The report suggests that enough baseline data have been collected to meet the goals of the EEMP.

15.6.3.2 Observation programme

FORCE's marine mammal observation programme includes observations made during biweekly shoreline surveys, stationary observations at the FORCE Visitor Centre, and marine-based observations during marine operations. All observations and sightings are recorded, along with weather data, tide state, and other environmental data. Any marine mammal observations are shared with SMRU Consulting to support validation efforts of PAM activities.

FORCE has started using an Unmanned Aerial Vehicle (UAV) for collecting observational data along the shoreline and over the FORCE site using transects by programming GPS waypoints in the UAV to standardize flight paths. Several FORCE staff received training to operate FORCE's UAV and have acquired UAV pilot certification by successfully passing the 2019 Canadian Drone Pilot Basic Operations Examination, administered by Transport Canada. These staff are now licensed to safely operate the UAV at the FORCE site. FORCE also hosts a public reporting tool that allows members of the public to report observations of marine life: mmo.fundyforce.ca.

15.6.4 Marine sound

Marine sound – often referred to as 'acoustics' or 'noise' – monitoring efforts are designed to characterize the soundscape of the FORCE test site. Data collected from these monitoring efforts will be used to test the EA predictions that operational sounds produced from functioning tidal stream turbines are unlikely to cause mortality, physical injury, or hearing impairment to marine animals.

Results from previous acoustic analyses completed at the FORCE site indicate that the CSTV turbine was audible to marine life at varying distances from the turbine but only exceeded the threshold for behavioural disturbance at very short ranges and during particular tide conditions. This is consistent with findings at the Paimpol-Bréhat site in France where an OpenHydro turbine was also deployed – data suggest that physiological trauma associated with a tidal turbine is improbable, but that behavioural disturbance may occur within 400 m of a turbine for marine mammals and at closer distances for some fish species [8].

15.6.5 Seabirds

FORCE's seabird monitoring programme is designed to test the EA prediction that project activities are not likely to cause adverse residual effects on marine birds within the FORCE test area. However, there has been limited opportunity to determine potential effects of an operational turbine on seabirds at the FORCE test site and to test the EA predictions.

Since 2011, FORCE and Envirosphere Consultants Ltd (Windsor, NS) have collected observational data from the deck of the FORCE Visitor Centre, documenting seabird species presence, distribution, behaviour, and seasonality throughout the FORCE site [9]. FORCE recently commissioned Envirosphere Consultants Ltd and Dr Phil Taylor (Acadia University) to synthesize the results of its observational seabird surveys (2011–18) at the FORCE test site, and to evaluate advanced statistical techniques for analysing seabird count data in relation to environmental predictor variables. The seabird count data were examined using Generalized Additive Models (GAMs) to characterize seabird abundance and to better understand the potential impacts of tidal turbines on seabirds at the FORCE test site.

The results of the analyses revealed that overall model fit is suitable to characterize count data for some species, and that there are clear patterns of effects of time of year, wind speed and direction, tide height, and time of day on the number of seabirds observed. However, the analyses also revealed that not all species reported at FORCE have been observed frequently enough to be modelled effectively using the GAM approach. This is due in part to the variability in count data that are particularly relevant for modelling the abundance of migratory species that are only present at the FORCE site for brief periods during annual migrations. This is consistent with observational data collected over the course of these surveys that have demonstrated that the FORCE site has a lower abundance of seabirds in relation to other areas of the Bay of Fundy, and even other regions of Atlantic Canada.

Given these results, the report recommends that future monitoring and analyses focus on locally resident species (i.e., great black-backed gull, herring gull, black guillemot, and common eider) so that the EA predictions can be tested most effectively. This work contributes to the development of appropriate analytical methods for assessing the impacts of tidal power development in the Minas Passage on relevant seabird populations and supports the continued responsible development of tidal energy at FORCE.

15.6.6 The Pathway Program

The Pathway Program is a collaborative effort between FORCE and OERA to identify an effective and regulator approved monitoring solution for the tidal energy industry in Nova Scotia. The Pathway Program involves several phases, including (i) Global Capability Assessment, (ii) Advancing Data Processing and Analytics, and (iii) Technology Validation.

The first phase of this programme, a Global Capability Assessment, involved a comprehensive literature review about the use of different classes of environmental monitoring technologies (i.e., PAM, imaging sonars, echosounders) for monitoring tidal energy devices around the world. Subject matter experts were commissioned to provide reports on these instrument classes, and these reports are publicly available.***

Phase II of the programme ('Advancing Data Processing and Analytics') involved the development of automation tools for expediting post-processing and reporting of monitoring data. To that end, the Pathway Program partnered with DeepSense (Dr Scott Lowe) in the Computer Science Department at Dalhousie University and used machine learning methods to develop a new tool (i.e., 'EchoFilter') for post-processing raw hydroacoustic data. This new tool accurately (Jaccard index > 94%) identifies the boundary between turbulence and biological targets on raw echograms and reduces the time required for manual post-processing by 45%–59%. This tool is now being paired with a streamlined analysis pipeline developed by Dr Louise McGarry that will generate standardized figures and tables for inclusion in quarterly and annual reports for regulators. Cumulatively, this work reduces the time between data collection and reporting and allows regulators to see the results of monitoring activities more quickly.

The Pathway Program also partnered with Dr Dave Mellinger at Oregon State University to develop a harbour porpoise click detector and classifier algorithm that would be suitable for use in Minas Passage. That effort has generated a new python-based tool, which recently underwent beta-testing and proved to be a valuable method for detecting harbour porpoise echolocation clicks in tidal channels dominated by noise contamination from flow and ambient noise. This new tool is also being paired with a streamlined analysis pipeline to generate standardized figures and tables for inclusion in quarterly and annual reports for regulators.

FORCE is collaborating with Sustainable Marine and using the floating tidal energy platform (PLAT-I) deployed in Grand Passage, NS, to conduct four projects; three of these projects focus on evaluating the utility of echosounders for quantifying biological targets in high flow environments. These projects evaluate the performance of echosounders in bottom and surface deployments and using a suite of complementary technologies (optical cameras and imaging sonars) to investigate target detections. The fourth project involves an assessment of the relative performance of PAM instruments for detecting synthetic harbour porpoise clicks in high flow environments using similar bottom and surface deployments (see Image 15.10).

15.6.7 Risk Assessment Program

The Risk Assessment Program (RAP) for in-stream tidal energy is a collaborative effort between FORCE, Dalhousie and Acadia Universities, Mi'kmaw Conservation Group, and industry to advance our understanding of the environmental risks of tidal stream development in Minas Passage. The greatest potential risk of tidal turbine operations continues to be perceived by regulators and

***These are available online at https://oera.ca/research/pathway-program-towards-regulatory-certainty-instream-tidal-energy-projects.

Image 15.10 Deployment of the FAST platform in Grand Passage for an assessment of PAM instrument performance

stakeholders as collisions between marine animals and turbines blades [10]. However, these types of interactions are difficult to observe directly due to the environmental conditions under which they would occur (i.e., fast flowing, turbid waters) and using the suite of environmental monitoring instrumentation currently available (i.e., standard oceanographic and remote sensing instruments intended for use in more benign marine conditions) [11] but can be modelled using appropriate baseline data. The objective of the RAP is to develop statistically robust encounter rate models for migratory and resident fishes with tidal turbines in the Bay of Fundy using a combination of physical oceanographic data related to flow and turbulence in the Minas Passage and hydroacoustic tagging data for various fish species curated by the Ocean Tracking Network (OTN) at Dalhousie University.

Recent research has revealed how hydrodynamics (flow and turbulence-related features) in tidal stream environments can influence the distribution of marine

Image 15.11 One of two high-resolution radars constructed near the FORCE site to be used for the RAP

animals, including fish. The Minas Passage is characterized by a series of turbulent hydrodynamics features (i.e., vortices, eddies, whirlpools, wakes, and shear currents) that could impact the spatiotemporal distribution of various fishes. The RAP will use a series of mobile ADCP transects combined with a high-resolution radar network (see Image 15.11) to create the first spatiotemporal flow atlas of the Minas Passage to understand these hydrodynamic features. Concurrently, hydroacoustic data for various migratory and resident fish species in the Bay of Fundy that are curated by OTN will be compiled and analysed to understand their spatiotemporal distributions. The hydrodynamic and hydroacoustic data will then be combined with information about turbine-specific parameters (e.g., turbine blade length, swept area, turbine height off the seabed) to develop encounter rate models for various fish species. These models will then be refined and validated through a series of hydroacoustic tagging efforts, ultimately leading to the development of a user-friendly Graphical User Interface (GUI) similar to what is available for the offshore wind energy industry in the United Kingdom. Ultimately, the RAP will contribute towards improving our understanding of the risks of in-stream tidal power development for fishes of commercial, cultural, and conservation importance in the Bay of Fundy and will assist in the development of future EEMPs.

Since the programme commenced in April 2020, OTN has acquired acoustic tag data from 22 contributors, covering nine species of fish in the Bay of Fundy (i.e., alewife, American shad, American eel, Atlantic salmon, Atlantic sturgeon, Atlantic tomcod, spiny dogfish, striped bass, and white shark). FORCE has also established a high-resolution radar network in Minas Passage to begin quantifying hydrodynamic features (turbulence, flow, etc.) of Minas passage. The integration of physical habitat variables with acoustic tag data has commenced, and a process for combining these data sets to develop encounter rate model is emerging and appears promising. Discussions are also underway with project partners about where to deploy acoustic receivers and where to tag target species to support model validation efforts.

15.7 Social license

Social license for tidal stream technology depends on public and regulator confidence that the effects of tidal devices on marine life and the environment are understood and acceptable.

Over the span of the project to date, FORCE has significantly enhanced its engagement capacity. FORCE created and hired a full-time engagement coordinator position, expanded outreach efforts in Cumberland County via FORCE's full-time facility manager, and shifted work priorities to be able to meet with over 100 different groups to understand issues, interests, and concerns. Importantly, FORCE's entire staff composition has changed. While FORCE was primarily an electrical, civil, and marine engineering construction project from 2009 to 2014 as the onshore and offshore facilities were built, FORCE has transitioned into a primarily science-focused research centre. After the substation was complete, FORCE eliminated its civil engineering position and created five science positions: a director of science, a research assistant, and three ocean technologists. This capacity has enabled FORCE to meet directly with organizations on a semi-regular basis to keep them informed on research activity and monitoring outcomes. This remains a large staff demand.

As part of the Terms and Conditions of the 2009 EA Approval, FORCE established the Community Liaison Committee (CLC) and the EMAC. The CLC is a community-based committee, including members from the general public, fishers, and First Nations and is intended to keep the community updated on FORCE project activities. In addition, it provides an opportunity for questions and feedback from the community regarding the project. EMAC, described in Section 15.3, is also an ongoing conduit for participation and advice to FORCE on the design and implementation of its monitoring programme.

Polling of Nova Scotians in 2016 (and again in 2018) indicates generally strong support for tidal energy testing (approximately 90% of residents in support, including over-sampling in communities along the Bay of Fundy). Tidal energy also has its critics. As Cape Sharp Tidal prepared to deploy its first 2-MW turbine in the Minas Passage, representatives from fishers' associations, Mi'kmaw communities, and some members of the public expressed concerns about potential impacts. Some of their concerns relate to the following:

- *Financial risk*: e.g., some lobster fishers carry high debt load for license, boat, and equipment, and any change to their catch can have significant implications to their livelihood.
- *Displacement*: potential loss of marine space with increased footprint of tidal energy areas.
- *Species*: species at risk, and/or species important to food, social, and ceremonial fisheries such as American Eel, Striped Bass, Atlantic Salmon.
- *Rights*: the inherent, treaty, and legal right the Mi'kmaq have to fish the Fundy watershed.
- *Fish migration*: even if collision risk proves to be low, avoidance behaviour could change migratory routes and therefore impact traditional spots.
- *Engagement*: a need to broaden outreach to more groups and wider geography.

As mentioned, both EMAC and CLC have representation from the Nova Scotia Mi'kmaq and fishing communities, and FORCE has had ongoing involvement with lobster fishers who work near the test area. However, there is an ongoing need to engage with groups from far outside the direct project site. FORCE expects that need will rise each time a developer's device approaches deployment, as media and therefore public attention increase.

In addition, there remains an ongoing need to share information about local content with project developers and opportunities available to Nova Scotia's diverse ocean SME community. In light of OpenHydro's insolvency, there is a need to ensure confidence in the sector's viability and to encourage some SUSTAINABLE MARINEs to participate in the sector.

Ultimately, tidal energy needs to coexist with other users of the marine environment. Tidal stream technology also has the potential to connect maritime communities by protecting marine life from the impacts of climate change – including ocean acidification, erosion, storm surges, and population level shifts in the marine ecosystem.

15.8 Visitor centre

The FORCE visitor centre (see Image 15.12) is the most vital part of FORCE's engagement efforts with the public. The centre includes interpretive panels, videos, interactive displays, and a direct view of the ocean test site. More than text and image content, however, is face-to-face interaction. In delivering public programmes, each summer FORCE hires two student interpretive assistants and (for

Image 15.12 The FORCE visitor centre overlooks the ocean test site in the Minas Passage

the last 4 years) an ocean technologist intern from NSCC's ocean technology programme. Four ocean technologists have found full-time work with FORCE after the summer as part of FORCE's science programme.

As of the end of 2020, FORCE has welcomed over 30,000 people to the site. Roughly 60% of visitors are from Atlantic Canada, 25% from Central/Western Canada, 10% from the United States (in particular NE states), and roughly 5% from Europe (in particular the United Kingdom, Germany, and the Netherlands).

During COVID, the project's online presence was a critical part of FORCE's ability to reach audiences and share information about tidal stream energy activity and research. One of the strategic initiatives was to place a greater emphasis on FORCE's research mandate, which included a new website, document library, and user interface for live data (www.fundyforcelive.ca). Social media activity has increased over the last 5 years: on Facebook, reach has increased by 240% (935–2,240 likes); on twitter, an increase of 250% (765–1,945); on Instagram, launched in 2017, to 336. Total reach continues to grow, with social media posts receiving over 100,000 views per year.

15.9 Emission displacement

Table 15.3 predicts the emissions avoidance as tidal devices are added to the grid. The maximum projected potential for the TISEC technology (best case scenario), with 2,500 MW of installed capacity, in continual operation over a year, at 50% capacity factor, offers a potential yield up to 10,950 gigawatt hours (GWh) of energy, or 9.1 million tonnes of displaced CO_2.

Table 15.3 Project emissions displacement

Estimated year	Installed MW	Potential annual generation[a] (GWh)	Share of total Nova Scotia generation[b] (%)	Potential annual CO_2 displacement[c]
2020	Baseline (no tidal)	0	0.00	0
2021	Single device (1 MW/grid connect)	3.33	0.03	2,774 t
2022	First array (5 MW)	16.6	0.14	13,680
2025	Multiple arrays (20 MW)	66.58	0.55	54,720
2030	Full cable arrays (64 MW)	212.8	1.77	177,080
2050	NS strategic goal (300 MW)	998.64	8.32	0.8 Mt
2100	Estimated potential (2,500 MW)	8,322	69.35	6.9 Mt

[a]Assumes ongoing annual operation at 38% capacity factor [(MW)×24 (h)×365 (days)×0.38 (capacity factor based on NS UARB Developmental FIT)/1,000 (gigawatt conversion)].

[b]Assumes constant of 12,000-GWh total Nova Scotia generation.

[c]Assumes constant of 0 emissions from tidal; assumes constant of 10-Mt total NS CO_2 emissions from electricity (actual emissions projected to decline due to GHG/renewable regulations, efficiency programmes).

15.10 Final thoughts

The global climate is warming. This warming is mostly the result of human activity – predominantly, the burning of fossil fuels. In addition, about 6.5 million deaths worldwide are linked to air pollution each year, a number that could grow in coming decades unless the global energy sector responds.

Nova Scotia is already a leader here, achieving some of North America's most significant per capita gains in renewable energy and energy efficiency. But there is still work to do: most of Nova Scotia's electricity is still generated by burning imported fossil fuels (mostly coal). These emissions contribute to climate change impacts, including sea level rise, storm surge, extreme weather events, ocean acidification, and population level shifts in the marine ecosystem on which Atlantic Canada is particularly dependent.

These realities underline the pressing need to continue to explore, advance, and adopt alternative approaches to electricity generation – in this case, an exploration of the potential to use tidal stream energy technology to generate electricity from the tides in the Minas Passage.

Not only does tidal technology have the potential to reduce dependence on fossil fuels and meet electricity demand, it also has significant economic potential for companies offering technology and expertise. Estimates suggest the following:

- Industry could be worth $2B to Canada in exports by 2030 (*MRE Technology Roadmap*).
- Potential for 22,000 full-time positions in Canada over the next 25 years (*Ocean Energy Research Association*).

- Potential capacity of 210 GW by 2050, representing more than $60 billion in annual investment (*International Energy Agency*).
- To explore these opportunities, the industry must continue to refine tidal stream technology, lowering technical and financial risk; continue to monitor devices, helping to understand and mitigate any potential effects; communicate with rights holders, stakeholders, and taxpayers about project activity – both successes and failures.

FORCE is designed to demonstrate in-stream tidal technology; principally, to understand if the technology is both environmentally sustainable and viable. FORCE has attracted international investment based on a number of key assets:

- The Minas Passage's unparalleled resource, estimated at 8,000 MW.
- Completed onshore and offshore electrical infrastructure, allowing access to the transmission grid.
- A feed-in tariff (FIT), providing an incentive for developers.
- A growing knowledge base of monitoring and site data.

Looking ahead, the next wave of companies bringing their technology to FORCE – with Sustainable Marine already having built and now testing their device – is estimated to generate up to an additional $200M in Canadian development over the next several years. Activity in the Bay of Fundy has increased, including the following:

- Sustainable Marine generating energy from its 280-kW floating tidal platform in the outer Bay of Fundy.
- DP Energy announced the successful deployment of tidal current sensors on their FORCE site.
- Jupiter Hydro and Big Moon Power received permits to begin new tidal device testing.
- Nova Innovation has been approved for a 4-MW tidal project near the FORCE site.
- Big Moon Power has won a new 4-MW berth site at FORCE.
- Sustainable Marine launched the 480-kW next iteration of their turbine design in February 2021.

There is still work to do, both in terms of technology maturation and understanding risks. That means a commitment to rigorous science and monitoring. The tidal energy sector depends on public and regulator confidence that any potential effects on marine life and the environment are understood and acceptable. Ultimately, tidal energy needs to coexist with other users of the marine environment. Members of Mi'kmaw and non-Mi'kmaw fishing communities, while supporting tidal energy in principal, have expressed concerns about potential impacts – and answering those concerns will take time, and science. There is also a sense of urgency, as the impacts of climate change loom larger, and Nova Scotia remains highly reliant on fossil fuels. The race to build a turbine equal to Fundy's massive tides, and a benign resident to its ecosystem, continues.

References

[1] R. Karsten, J. McMillan, M. Lickley, and R.D. Haynes. Assessment of tidal current energy for Minas Passage, Bay of Fundy. *Proceedings of the Institution of Mechanical Engineers, Part A: Journal of Power and Energy*, Vol. 222, pp. 493–507, 2008 (http://pia.sagepub.com/content/222/5/493.short).

[2] R. Karsten, A. Swan, and J. Culina. Assessment of arrays of in-stream tidal turbines in the Bay of Fundy. *Philosophical Transactions of the Royal Society A*, Vol. 371, pp. 20120189, 2013 (http://dx.doi.org/10.1098/rsta.2012.0189).

[3] C. Sparling, A. Seitz, E. Masden, and K. Smith. 2020 State of the Science Report – Chapter 3: Collision Risk for Animals around Turbines, September 30, 2020, pp. 29–65.

[4] A. Copping and L. Hemery. 2020 State of the Science Report: Environmental Effects of Marine Renewable Energy Development Around the World, 2020.

[5] G.D. Melvin and N.A. Cochrane. Investigation of the vertical distribution, movement, and abundance of fish in the vicinity of proposed tidal power energy conversion devices. *Final Report for Offshore Energy Research Association, Research Project 300-170-09-12*, 2014.

[6] A. Daroux and G.B. Zydlewski. Final Report 2016-2017 Mobile Acoustic Fish Monitoring of the Fundy Ocean Research Center Crown Lease Area. *Report submitted to the Fundy Ocean Research Center*, 2017.

[7] A. Serdynska and S. Coffen-Smout. Mapping inshore lobster landings and fishing effort on a Maritimes region statistical Grid (2012–2014). *Fisheries and Oceans Canada*, 2017.

[8] J. Lossent, M. Lejart, T. Folegot, D. Clorennec, L. Di Iorio, and C. Gervaise. Underwater operational noise level emitted by a tidal current turbine and its potential impact on marine fauna. *Marine Pollution Bulletin*, Vol. 131, pp. 323–334, 2018.

[9] P.L. Stewart, F.L. Lavender, and H.A. Levy. Marine Mammal and Seabird Surveys, 2010. (https://fundyforce.ca/resources/f1c21770b591141114866df4d491471a4/Appendix2-2012-Seabird-and-Marine-Mammal-Surveys-Envirosphere.pdf).

[10] A.E. Copping, L.G. Hemery, D.M. Overhus, *et al.* Potential environmental effects of marine renewable energy development – the state of the science. *Journal of Marine Science and Engineering*, Vol. 8, No. 11, p. 879, 2020.

[11] D.J. Hasselman, D.R. Barclay, R. Cavagnaro, *et al.* 2020 State of the Science Report, Chapter 10: Environmental Monitoring Technologies and Techniques for Detecting Interactions of Marine Animals with Turbines (No. PNNL-29976CHPT10). Pacific Northwest National Lab. (PNNL), Richland, WA (United States), 2020.

Chapter 16
Sustainable materials for small blades

Igor dos Santos Gomes[1], *Roberto Tetsuo Fujiyama*[1,2], *Jerson Rogério Pinheiro Vaz*[1,2] *and David Wood*[3]

Increasing the power of hydrokinetic turbines (HKTs) and wind turbines (WTs) using blades with more efficient geometries is important. However, there is another important aspect: the materials that can be used in these applications. The requirements may include, for example, a combination of low density, high mechanical strength, high modulus of elasticity, high capacity to absorb energy on impact and high chemical and environmental resistance. In achieving these goals, synthetic-fiber-reinforced composites have advantages, such as ease of mixing with matrices, adaptation to various manufacturing techniques, very high specific resistance and good fiber–matrix interface, due to chemically reacting with the matrix that increases the composite strength. Regardless of the manufacturing techniques, it is necessary to have the conformity of the manufactured to the design shape. Small blades, therefore, need high tolerances in the tip region where the chord is smallest. Furthermore, the synthetic fibers, when combined with thermoset resins, for example, are energy-intensive and expensive due to the heating needed to cure them and are also very difficult to recycle, which it makes the blades of any size the least recyclable component of a WT. This creates a great opportunity for developing new materials, new methodologies and effective processes to produce composites reinforced by sustainable materials. This chapter draws a parallel between conventional materials with manufacturing methods and the newest materials with alternative methodologies for manufacturing small HKT and WT blades, and how the properties of sustainable materials are suitable these types of applications. We describe a range of natural reinforcement materials from the Amazon region in Brazil and a possible foam core replacement from a Brazilian palm tree.

[1]Graduate Program in Natural Resources Engineering, Institute of Technology, Federal University of Pará, Belém, Brazil
[2]Graduate Program in Mechanical Engineering, Institute of Technology, Federal University of Pará, Belém, Brazil
[3]Department of Mechanical and Manufacturing Engineering, Schulich School of Engineering, University of Calgary, Calgary, Canada

16.1 Introduction

In HKTs and WTs, the main components are the rotor, shaft, blades, couplings, gearbox, output shaft, brake and generator. The blades are responsible for converting the kinetic energy of water or wind into mechanical power as described in Chapters 4 and 7. However, there is something more to be considered: the most appropriate material for the blades.

Currently, rotor blades are manufactured from a combination of glass and/or carbon fiber composite materials with a thermoset resin such as epoxy [1]. A number of factors such as increased competition, the finitude of the materials used, the reduction in costs and issues of recyclability have led manufacturers to consider new materials. A similar situation has arisen in the aeronautical industry, which has increasingly used composite materials and renewed its manufacturing processes [2]. Modern structural designs often require combinations of material properties that are difficult to obtain with conventional materials. These requirements may include, for example, a combination of low density, high-directional mechanical strength, high modulus of elasticity, high capacity to absorb energy on impact and high chemical and environmental resistance [3].

Although composites reinforced by carbon fiber, boron, glass and Kevlar have been widely accepted for structural and nonstructural uses, including turbine blades [4], environmental concerns are increasing the demand for replacing existing synthetic fibers with biodegradable, renewable and low-cost vegetable fibers for the manufacture of composite material. In comparison with traditional reinforcement materials, plant fibers, also called lignocellulosic materials [5], have acceptable specific strength properties, low density, low abrasion multifunctionality, good thermal properties, improved energy recovery and cause less skin irritation and respiratory [6–8].

In the material selection, rigidity and resistance are properties that must be considered to improve the elastic qualities [9]. Then, composite materials allow designers to adapt the structure to meet the objectives established in the project. Different types of laminates are generally used for manufacturing different structural elements of turbine blades. The configurations can include unidirectional material on the spar that runs through the center of the large blade shown in Figure 16.1, and several forms of multidirectional reinforcement, as biaxial and triaxial fabrics, both internally and externally. The laminate skin on the small blade also requires biaxial and triaxial fabrics. The configuration significantly influences the strength of the blade and its fatigue life, depending on the type of fibers and resin used, the shape and direction of loading [12,13]. WT blades, for example, are subjected to a high number of fatigue cycles so it is essential that the material of the blades has an extremely long fatigue life [14]. Therefore, the design and material selection of the blades are crucial for obtaining low-cost blades, considering their influence on the total cost, efficiency and robustness of the turbine. Several authors have suggested a wide range of materials for the construction of the blades, such as E-glass fiber, carbon fiber, reinforced plastics and others [15–17].

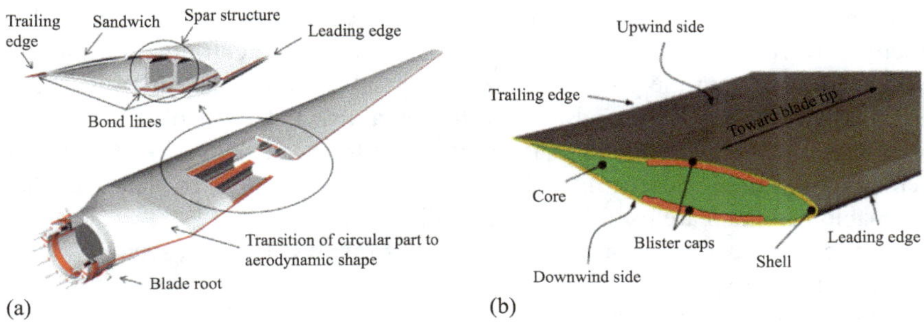

Figure 16.1 Example of structural details in a (a) large [10] and (b) small [11] wind turbine blade that may be represented in subcomponents

Whatever technique is used to make turbine blades, good operational performance requires the conformity of the built shape to the design shape. For small blades, it is necessary high tolerances in the tip region where the chord is smallest [18–20]. The synthetic-fiber-reinforced composites have advantages such as ease of mixing with the matrix, suitability for several manufacturing techniques [21], high stiffness and strength to weight ratio [22] and good fiber–matrix interface, besides chemically reacting with the matrix that improves the composite strength [23]. However, these materials, as reinforcements of thermoset–matrix composites, are energy-intensive and expensive because of the heating process required to cure [1]. The blades manufactured with this type of matrix require adhesives to bond the two blade skins, which are normally made separately. This is a time-consuming process that results in stress concentrations and areas of weakness and can decrease the reliability and life span of blades [24]. Besides these issues, not all of these materials are available to small communities or make economic sense for small-scale power generation. Beside this, reliability is important, but the life cycle energy usage needs emphasis, considering that only a small percentage of turbine blades are recycled during the decommissioning stage because of the complexity of their composition and materials used [25] and the cross-linked nature of thermoset materials [26].

Therefore, reducing the environmental impact of HKT and WT blades represents a great opportunity for developing new materials, new methodologies and effective processes on composites reinforced by sustainable materials. Among their several benefits are the reduction of the weight of the blades, an increase in the specific resistance and a reduction in their starting time, considering that the great majority of vegetable fibers have low density. However, the development of low-cost, aligned, vegetable fiber semi-products is a limiting factor in structural composites. For this reason, more studies are needed to understand the behavior of these materials when employed in specific applications/structures, instead of just making analysis of materials to data extracted from static coupon tests [11]. These tests have to be a first step to develop appropriate "material allowables" for design applications as well as to qualify final structural components and full-scale structures for their intended ultimate use [27].

Many studies use different processes, equipment, tools and materials, which directly influence on the final product. Effective manufacturing processes are necessary to maintain the power-extraction capability of the blades, on which the output power of a WT depends. This chapter aims to draw a parallel between conventional materials with manufacturing methods and the newest materials with alternative methodologies for manufacturing small HKTs and WTs blades and how sustainable materials are useful in these applications.

The remainder of this chapter is organized as follows. The next section presents the generalities of small HKTs and WTs blades, example of their designs, with the functions of some components and subcomponents, how and where are applied the conventional materials used for manufacturing them, opening the discussion about using sustainable materials rather than synthetic ones. In Section 16.3, we consider sustainable reinforcement materials for small HKT and WT blades and the natural fibers from the biodiverse Amazon and other regions of Brazil. They are described in detail, to show their interesting and potential properties for these types of application, such as petiole of miriti, which can be useful in the core of small blades. Further, some methodologies for the manufacture of blades demonstrate the uses of different processes, equipment, tools and sustainable materials. Section 16.4 summarizes the final considerations, specially concerning the importance of coupon tests as a first stage for understanding the behavior of the composite reinforced by natural fibers to determine the stiffness and strength of the material in static and dynamic states, and ahead the subcomponents and full-scale blades tests, to make possible the development of numerical models capable of predicting their performance in different operational conditions. Since we have not researched bioresins or thermoplastic resins, these important topics are not considered.

16.2 Small hydrokinetic and wind turbine blades

Optimization of the use of composite material is becoming increasingly necessary, especially for the inlet fans of large aeroengines, considering that the weight of the blade increases significantly as its power generation capacity increases [28]. Although it is known that very long blades are difficult to make, without a solid knowledge of the material behavior under the dominant loads including the weight of the blade [29], for small blades this knowledge has also a great importance, to provide low inertia for starting, minimizing gyroscopic moments during yaw, and fatigue life prediction [18]. Further, these components are subjected to an unusual loading environment, characterized by a variety of external environmental conditions, severe loads and complex stress states in their internal and external structures over a long lifetime.

Blades can be manufactured in many ways [30]. The large blade shown in Figure 16.1(a) has subcomponents, such as the blade root, trailing and leading edge, transition of circular to aerodynamic shape, sandwiches, flanges and shear web, spar structure (typically one or two I-beams, or box beam), bond lines and joints throughout the blade [10,31].

The main structural function is performed by an internal spar with spar caps at the location of maximum thickness, to resist flapwise bending and one or two shear webs, to resist torsional and flexural loads [11,32]. The spar caps can be manufactured in angle ply and unidirectional fabrics, while the leading and trailing edges can be made from thick layers of angle plies, with a core in a sandwich structure or filling the empty space inside a small blade [32]. As indicated in Figure 16.1(b), small blades, including those described in the remainder of this chapter, typically do not have a spar.

Furthermore, it is possible to add a lightweight core layer to the spar cap, to improved buckling capacity at the expense of increased global deflection [30,33]. The core separates and fixes the skin, carries the transverse shear load and provides other structural or functional duties, etc. [34]. The blades of HKTs can be manufactured in a single material, such as aluminum or stainless steel, without divisions between cores in shells.

Figure 16.1(b) shows a small blade that consists in a core and fiber-reinforced composite structural blister caps and constant-thickness outer shell. The core provides resistance against buckling, the unidirectional fiber-reinforced blister caps provide maximum axial (tensile) and bending (flexural) stiffness and strength, and the multi-axial fiber-reinforced outer skin provides resistance against torsional loads [11].

Blade cores are made from foams with high stiffness and strength-to-weight ratio, shear properties, fatigue performance, high impact strength, very good formability, such as polyvinyl chloride, polyethylene terephthalate, polyurethane (PUR), balsa wood, expandable polystyrene [35–39]. In these types of application, the miriti petiole, for example, can be a potential sustainable alternative, considering its low density, easy for manufacturing and to adapt to different blades' geometries.

16.2.1 Conventional materials used for manufacturing

The dimensioning of HKT blades must take into account the hydrodynamic, structural and dynamic aspects. The selection of materials to be used should consider, mainly, the following factors [40]:

1. corrosion protection;
2. impermeability to water;
3. buoyancy;
4. mechanical stiffness;
5. resistance to fatigue and impact;
6. resistance to cavitation.

The last three apply to WT blades as well. The materials and methods of manufacturing WT blades can be used in underwater applications, as long as the first three factors are taken into account [40]. If the blade has a foam core, this has to be protected from water ingress through the bolt holes [19]. It is not shown in Figure 16.1, but small blades usually have a rectangular or similar-shaped attachment section that is bolted to a blade holder with the bolts horizontal, rather than radial as on large blades.

Considering the high specific strength and stiffness, excellent resistance to corrosion and fatigue, good fracture toughness, ease of molding to complex surfaces, the ability to develop a product compatible with loading, pre-impregnated carbon/epoxy sheets of the AS4/3501-6 type are often used for HKT blades [40]. Because they are lighter and have a relatively better corrosion resistance, excellent durability, weldability and deformability aluminum alloys are good candidates for this type of application, where the components are subjected to a dynamic loading [41], especially in turbines that do not have cavitation in service [42]. As an example, the AA6061-T9 and AA6061 alloys are excellent candidates. In particular, the Al–Mg–Si alloy (AA6061-T6) has a high yield strength and a stiffness/weight ratio comparable to that of structural steels. It has about three times less stiffness than steel, but it is also about three times lighter.

Another example is the 6082T6 aluminum alloy, used for the manufacture of HKT blades [43], through computer numerically controlled (CNC) machining, as shown in Figure 16.2. According to [43], the aim was to manufacture blades in stainless steel 304 to maximize rigidity. However, due to time constraints, the initial set of blades was made from aluminum alloy, which is much faster than machine stainless steel 304. In addition, the blades have been anodized to improve their corrosion resistance.

On the other hand, carbon/epoxy composites, for example, are immune to corrosion, have a higher mechanical resistance compared to the AA6061-T6 alloy and, in addition, have a higher stiffness/weight ratio [44]. Furthermore, most blades that are manufactured using them are solid (without empty spaces inside) and, depending on the length of the blade, it makes the application expensive, because aluminum alloys are very expensive [3].

What about making a hollow aluminum blade, to reduce the use of this material and increase the cost-effectiveness? In this case, it is important to note that the smallest blades have higher angular velocity and their free-yaw behavior leads to high gyroscopic moments [19] and the fatigue issues. Small blades can experience up to two orders of magnitude more fatigue cycles than larger ones over a nominal 20-year lifetime [18].

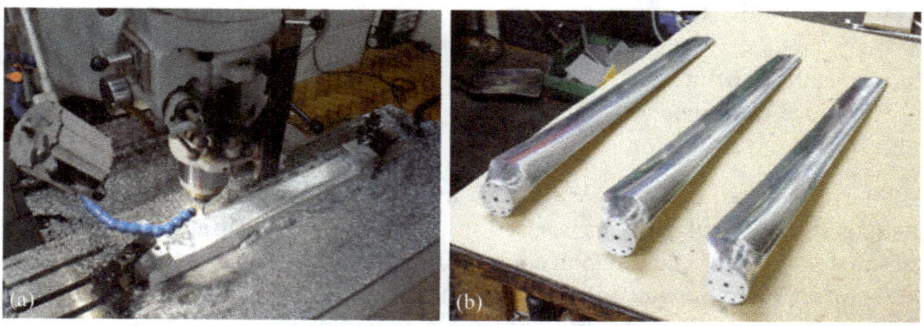

Figure 16.2 (a) Manufacture of aluminum blades in CNC machining and (b) the manufactured 0.594 m blades [43]

Aluminum alloys offer lower fatigue strength [45–47] and the AA6061-T6 alloy has the shortest total fatigue lives under constant amplitude loading, followed by the high-to-low and low-to-high loading sequences for both room and elevated temperatures [47]. For this reason, metal blades, frequently rolled from galvanized steel sheet, may be more appropriate for turbines with slow rotation, which means that typical blade stresses are low. One example that can be cited for this case is the water-pumping windmills, which are described in Chapter 6.

16.2.2 Synthetic materials used for manufacturing

In most composite processing, material costs make up a significant proportion of the overall component cost [48]. There is a huge number of small blade manufacturing techniques using composites, such as hand layup, vacuum infusion and resin transfer molding (RTM) [49,50]. The laminate, for example, is made by incorporating reinforcement material, such as fiberglass or carbon fiber into a resin. Due to the high centrifugal forces, considerable reinforcement is required in the radial direction, often achieved using unidirectional E-glass, laid in the radial direction [51].

There is also a considerable choice in resins that can be used. The most common are vinyl ester, polyester and epoxy resins. Epoxy resins tend to be more expensive, but in terms of total material costs for a blade, the difference is likely to be small. They also tend to have a long service life and excellent fatigue properties. Polyester resins are generally easier to handle but have greater shrinkage and higher curing temperature [18]. Beside this, the properties of the resin which are the key for manufacturing turbine blades are [18] as follows:

1. The viscosity, which regulates how well it flows and wets out a mold.
2. Pot life that determines the time during which a blade can be manufactured.
3. The glass transition temperature, which governs the curing temperature of the resin.

The glass transition temperature required by Germanischer Lloyd guidelines [52] is at least 65°C and higher than any operational temperature [52]. It also specifies that the material safety factors for composites must take into account the environmental degradation and operating temperatures. Furthermore, a wide variety of gel coats can be used for shaping blades to provide a good surface. Some resins are more suitable than others for gel coat: epoxy resins, for example, normally require a gel-coat-bonding primer [18].

Most of the blades manufactured today have polyester as their matrix, but epoxy is also widely used. The density of them is similar, but the fatigue performance of epoxy matrix composites is better; they offer easier fabrication and have no toxic emissions during manufacture [53]. The fibers immersed in polymeric matrix result in some resistance to compression, but mainly in the improvement and maintenance of the fibers [51].

Some manufacturers have exchanged fiberglass for carbon fiber, because it is considerably stronger and is used in the aeronautical industry [53]. Possibly the main difference between the applications of carbon and glass fibers would be the way the latter are arranged in a relatively coarser polymeric matrix.

16.2.3 Natural materials used for blades

The choice of manufacturing methods depends on the specified materials, blade size and airfoil, as well as economic considerations. Timber can be used for the manufacturing of small WT blades. This alternative may be interesting, as wood has a long fatigue life and its cost as a raw material for manufacture is acceptable [18]. However, wood is not the most convenient material for making blades of more complex shapes, which are necessary to achieve high aerodynamic efficiency. This effect is seen when it is necessary to introduce concavities in the lower surface, that are quite common in the sections of many high-performance blades, making it difficult to reproduce them in wood. In addition, each blade must be machined separately to maintain dimensional tolerance [54].

Timber is potentially interesting because of its low specific gravity, but its relatively low stiffness makes it difficult to use due to the deflection limit for large WTs blades. It means that timber is an ideal material for short blades when it is possible to obtain clear blanks, although soft and porous timbers suitable for small blades are susceptible to degradation due to exposure to the sun and erosion from dust particles in the air [19]. However, a range of blade was tested coating materials primarily for use in Nepal and found that enamel paint with a primer undercoat offered excellent surface protection [55].

Nevertheless, the limitation on the length of solid timber blades is the cost of knot-free blanks [54,55] and it can also be difficult to reproduce accurately the three-dimensional shape of a well-designed blade [19], such as the trailing edge. This can be a problem for small blades because of the conflicting requirements of finite thickness for safety and integrity and minimal thickness to reproduce the design shape and suppress trailing edge noise [18].

The tensile strength and fatigue resistance of *Douglas fir*, a North American timber, and *Sitka spruce*, a species widely cultivated in Europe were measured in [56]. Analyzing the woods experimentally, the study recommended for using these firs on small WT blades. In order to supply energy to remote areas of Nepal, local wood was applied to WT blades and towers [55,57]. Of the woods studied, Lakuri was selected on the basis of a variety of criteria, including mechanical properties, weathering effect on coatings, price, growth and availability. In this same perspective, a comparative study was carried out between four different types of Iranian woods, alder, ash, beech and hornbeam, based on the performance of the turbine, considering not only the output power, but also the starting performance which is influenced by the blade density [58].

The easiest way to reduce the inertia of any blade without a significant impact on its stiffness is to make it hollow [58]. Thus, a hollow wooden shovel not only represents a great challenge but also an excellent opportunity, because although they may take longer to make, they can provide a range of wood block utilization, from which the blades can be cut. The blocks must be free of defects, which becomes more difficult to achieve as their size increases.

As can be seen, there are many promising natural materials to use in turbine blades. One is balsa wood, which has low density and specific resistance. LM Wind Power blades, for example, are made from fiberglass and balsa wood, joined with resin and gel coats [59]. Considering the fact that propellers made of wood have had successful applications in small and large aircraft in the past and considering that this material has good properties for underwater application [60], a wood called Belian has been proposed for HKT blades, which is abundant in remote regions of Sarawak, Malaysia [61].

Belian wood is known for its resistance to insects, for having super durability, even when submerged in water. Sarawak residents use it on the pillars of houses and electric utilities on electric poles [62]. The use of this wood in the construction of an HKT blade can be accomplished using common tools and materials through a simple method, which allows it to be applied in remote locations [63]. This can significantly reduce the total cost of a turbine system because, by producing turbine components locally, the expensive exchange and import costs can be avoided. In addition, local construction creates job opportunities that can help people financially.

Wood has cellulosic fibers that can easily be aligned in the radial direction—the direction of the greatest load on the rotor. CNC wood milling machine is simple, but probably expensive for the large volume of production [64]. Copying machines are much cheaper, but a commercial unit designed for gun stocks and related items were found inadequate for blades where the tip chord is of the order of 5 cm [56]. CNC machining, for example, was used to manufacture HKT blades [63], as shown in Figure 16.3, using *Pinus radiata*. According to the authors, this timber is soft, of low cost and fast and easy to work.

Mechanical and corrosion resistance is improved by coating the wood core with epoxy resin layers reinforced by fiberglass mats. The authors also determined that this combination of materials is the least expensive, given the costs of raw

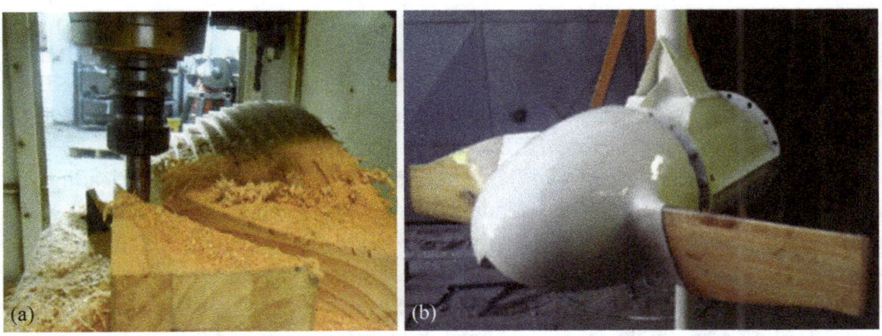

Figure 16.3 (a) CNC machining of HKT blades and (b) blades coupled to the hub [63]

materials in Chile [63]. An important problem with timber is the high uncertainty of mechanical properties and variations in density [65–67].

Balsa wood, for example, is one of the widely used structural core materials in sandwich structures owing to their particularly high strength to weight ratio [68,69]. In addition, balsa wood also exhibits excellent shear strength, high fatigue properties, ease of assembly and high absorption energy [70].

Balsa wood is used in the manufacture as core in sandwich panels in the shells of large WT blades, which carry high bending loads [71,72]. In some cases, the balsa core can crack, but without failing due to fatigue, because the cracks are very local [72]. In other words, the stresses in the core are rarely high enough to contribute to a significant probability of failure [69].

Fatigue tests on a 1-m solid blade for a 600-W rated power turbine using two types of wood commonly available in Australia [54] showed that both *P. radiata* and hoop pine were suitable, although there was a difference in strength between the two timbers. In addition, it was anticipated that these timbers should be suitable for small turbine blades, up to 5 kW, with blades up to 2.5 m [54].

In the past, one of the greatest difficulties in using composites was the high cost for manufacturing the molds with high-dimensional integrity, which requires them to be CNC machined. Another difficulty that the use of molds for composite blade faced, besides the costs, was the manufactured of the mold itself, due to the complicacy of reproducing precisely the shape of the blade [19].

Making blade molds is becoming an increasingly easy task due to the progress with functionality and ease of use of computer-aided manufacturing [73]. This is associated with the reduction in the cost of high performance computers and has encouraged small companies to build high quality and efficient blades for WTs [74].

The performance of HKTs and WTs depends on manufacturing highly accurate geometries [20]. Accuracy problems are exacerbated by the dominance of the tip region in producing power [19]. This is, therefore, the region where not only the greatest accuracy is needed, but it also has the smallest chords on the blade. In some situations, a linear taper can make material distribution and manufacturing processes much simpler [75].

Often the same profile is used for the whole blade on small turbines, but a few geometric liberties are still possible, mainly in the attachment section. To have good starting performance, the blade chord must be large in the root section but then it normally decreases toward the tip. Small blades are usually mounted in a blade holder so the attachment is a rectangular prism rather than a circular cylinder as used on large blades, Figure 16.1.

If the blades are made in two halves, joining is usually done at the leading and trailing edges. This may result in a blade that is too thin or thick, in comparison to the design, probably as a result of the way the two halves were joined, which makes, for example, the trailing edge a problem [18], insofar as trailing edge noise arises from vortex shedding behind a trailing edge of finite thickness and usually depends on the ratio of trailing edge thickness to the chord. It is difficult to keep this parameter small for small blades [18,19]. Providentially, noise calculations suggest that trailing edge thicknesses of 1–3 mm are tolerable for most small blades [76].

16.3 Sustainable reinforcement materials for small hydrokinetic and wind turbine blades

Lignocellulosic materials, due to their relative abundance, low cost, low density, high specific properties and positive environmental profile [77], are potential substitutes for composite reinforcements, such as those reinforced by E-glass fibers. Since 87% of the global fiber-reinforced plastic (FRP) market, which represents 8.7 million tons, is based on E-glass composites, natural fibers and their composites have a great opportunity for development and market capture [78].

WT blades have been made for comparing the use of polyester matrix reinforced by flax fiber fabric and polyester matrix reinforced by E-glass fibers fabric [11]. These blades have a core, for resistance in flexing and buckling. The core was manufactured in CNC machine, the material used in it, however, is not detailed. The shells of these blades consisted in laminates made of fabrics with unidirectional fibers, to provide maximum rigidity and resistance in tensile and flexion, and fabric with fiberglass multiaxial fibers, to provide resistance against shear loads related to torsion. These shells were made using RTM [11].

It was observed that the lower density of vegetable fibers provides weight savings [11]. Thus, the flax/polyester blade was 10% lighter than the E-glass/polyester blade (that represented 45% fiber mass savings). The static tests on the blades, according to the British Standard-European Norm, BS-EN 61400-23:2002 [79] and BS-EN 61400-2:2006 [80], confirmed that, as the E-glass/polyester blade, the flax/polyester blade satisfied the design and structural integrity requirements for a 11-kW WT under normal operation and ultimate load [11].

In Brazil, vegetable fibers, such as jute [81], mallow [82], piassava [83], curauá [84], miriti [85], palha da costa [81], sisal [86], açaí [87], banana [8], tururi [88] and petiole of miriti [89], have been studied and characterized, presenting interesting properties and that can be potential alternatives for use in blades of HKTs and WTs.

For example, jute fiber has been used to manufacture, through the process of manual lamination, a small WT blade, with a PUR core [90]. The blade supported wind loads equivalent to a maximum force of 395 N, with no sign of structural failure. Furthermore, a methodology for manufacturing HKT and WT blades was developed using the vacuum lamination method [91]. The blades were manufactured with bidirectional E-glass fiber–reinforced polyester resin shells and a petiole of miriti as core, which gave the blade a low density and rotational moment of inertia, contributing to the starting performance of a small WT. For HKTs, the starting of turbines is really important, mainly when it occurs at low stream.

16.3.1 Vegetable fibers from Brazil

Vegetable fibers are themselves natural composite materials. They have a complex structure, as shown in Figure 16.4, consisting of a thin primary and secondary wall formed by three layers, with the medium thickness layer determining the mechanical properties of the fiber (some authors cite as primary, secondary and tertiary walls) [93]. Vegetable fibers are like microscopic tubes.

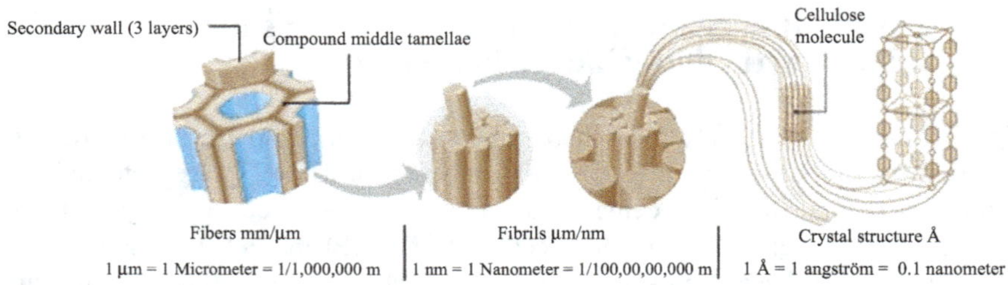

Figure 16.4 Structure of a vegetable fiber [92]

Cell walls are made up of microfibrils joined by a matrix of lignin and hemicellulose and have different orientations in each cell wall. Microfibrils have a diameter of about 10–30 nm and are made up of 30–100 cellulose molecules [93–95]. The microfibrillar angle, or medium angle, formed between the microfibrils and the fiber axis varies from one fiber to another and may be responsible for the mechanical properties of the fibers. Those with small angles generally have greater strength and stiffness, while others with greater angles provide ductile behavior [96].

Lignin is responsible for connecting the filaments that form a bundle, and pectin connects that bundle to the stem. As lignin and pectin are weaker polymers than cellulose, they must be removed before the fibers can be used as reinforcement in composites. Much of the pectin is removed when the bundles are separated from the stem through the process of maceration (immersion) and separation [95]. Although the natural fibers have the same cell structure (cell walls, lamella media and lumens), they have differences in terms of total cross-sectional area, cell wall thickness, quantity and area of lumens. These factors affect the fiber diameter and their mechanical strength [95].

Another factor that can also influence mechanical resistance is the chemical composition. Vegetal fibers mainly consist of three chemical components: cellulose, hemicellulose and lignin, which are responsible for the mechanical and structural behavior of the plant, giving it good strength, rigidity and durability [5]. Cellulose is mainly responsible for resistance. Sisal fiber, for example, is made up of 73% cellulose, 10% hemicellulose and 7.6% lignin [97], while jute fiber has about 65% cellulose, 17% hemicellulose and 14% lignin [98].

The shape, size and mechanical strength of natural fibers can vary widely, and, the quality of the fiber depends on factors such as plant growth, which depends on the species, the type of cultivation, the way the fiber is grown in the plant and the location [83,99]. Quality also depends on the harvest, as the maturation of the fiber in the plant is related to the thickness of the cell wall and the fiber's roughness. There is a further dependence on the methods of fiber extraction, decortication, and maceration processes, and on the supply of this raw material; the transport, storage conditions and age of the fiber [99].

Natural fibers are obtained from different parts of a given plant, such as the bast, leaf, fruit or seeds. Among the fibers widely grown in Brazil, jute, mallow and piassava, shown in Figure 16.5, are fibers from the bast. On the other hand, the curauá, miriti, palha da costa and sisal, shown in Figure 16.6, produce fibers from the leaf. The açaí fiber, shown in Figure 16.7, comes from the seed. The banana tree, the miriti and tururi palms, shown in Figure 16.8, yield fibers from other parts of the plant.

The vast biodiversity in the Amazon Region offers a high variety of plant fiber species, some from cultivation (either from the fiber itself or from the fruit) and others are native. Cultivated fibers include jute, mallow, curauá and açaí. Among the native ones are miriti, palha da costa, piassava and tururi.

Jute (*Corchorus capsularis*) is originally from India, comes from the family of tilias, a woody herb, can reach a height of 3–4 m and its bast is approximately 20-mm thick, growing in humid and tropical climates [110]. This fiber was introduced in Brazil by the Japanese and became one of the main economic activities of the riverside populations of the Amazon region, being a fundamental factor in the establishment of more than 15,000 families [111].

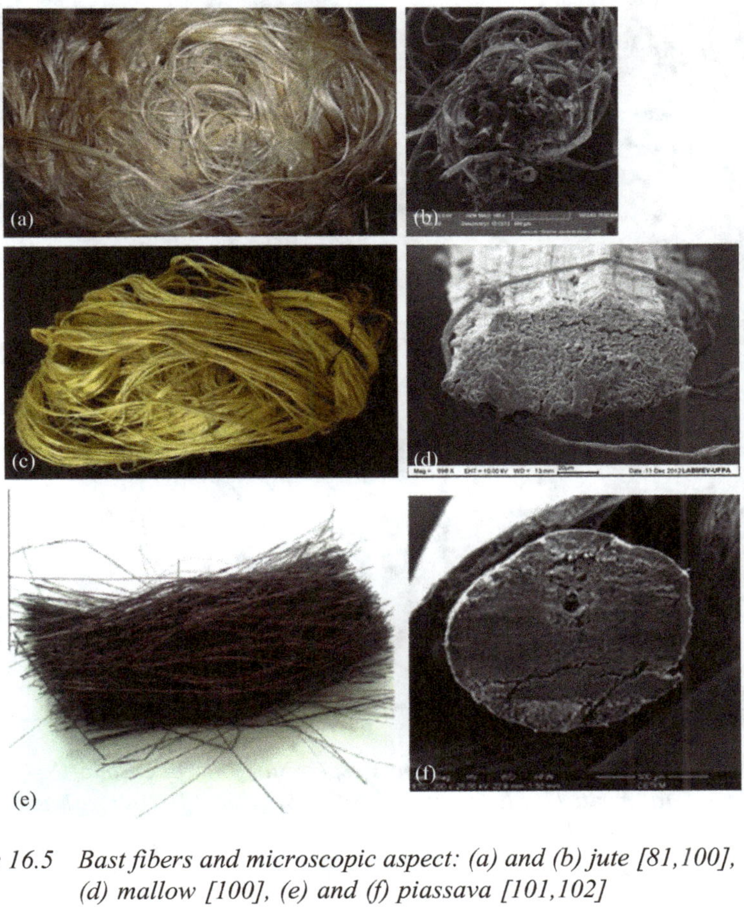

Figure 16.5 Bast fibers and microscopic aspect: (a) and (b) jute [81,100], (c) and (d) mallow [100], (e) and (f) piassava [101,102]

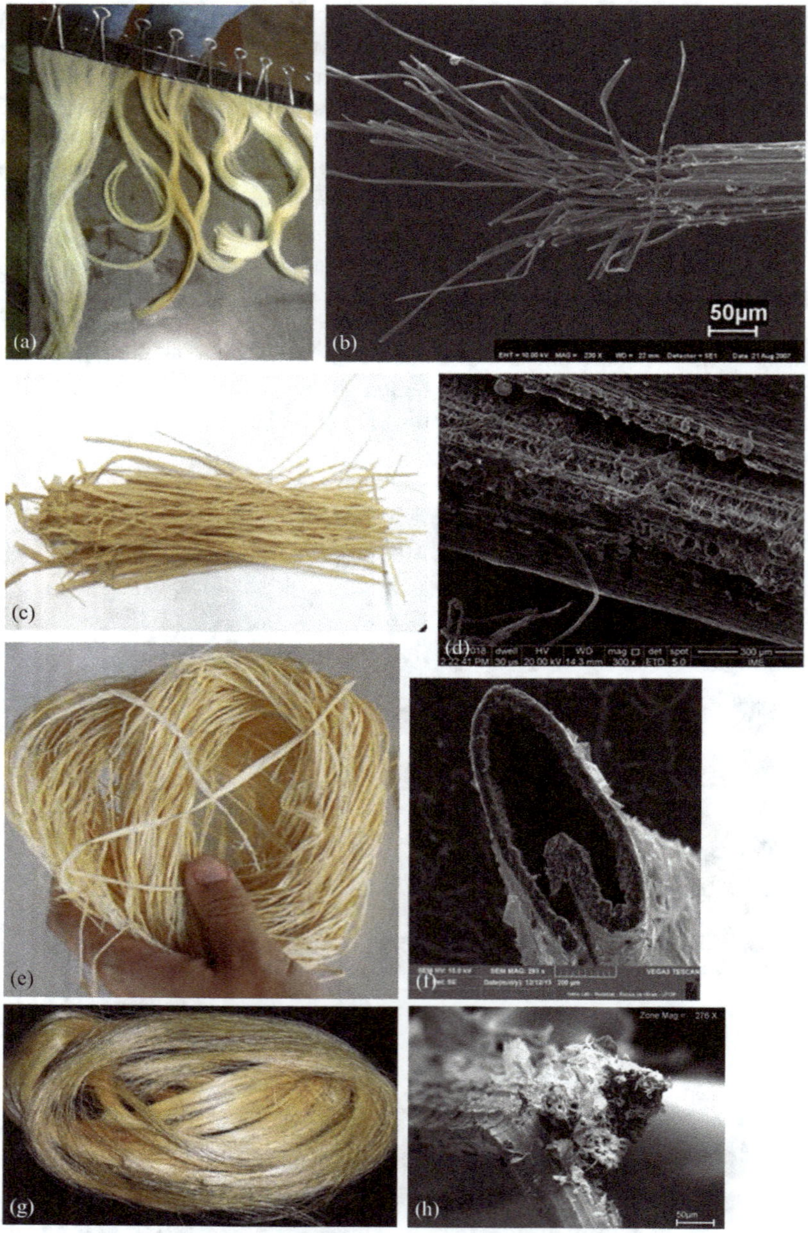

Figure 16.6 Leaf fibers and microscopic aspect: (a) and (b) curauá [81], (c) and (d) miriti [103], (e) and (f) palha da costa [81,104], (g) and (h) sisal [100]

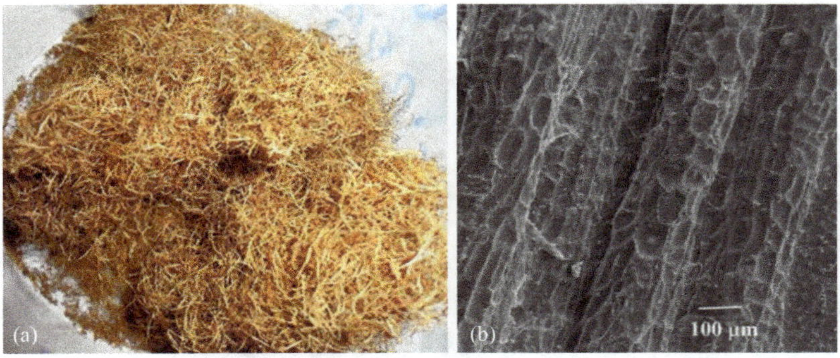

Figure 16.7 Seed fiber and microscopic aspect: (a) and (b) açai [105,106]

The annual world production of jute exceeds 3 million tons, led by India. Starting from a modest position in recent years, Brazilian production surpassed 100,000 t/year in the early 1980s and current forecasts of growth are 15%–20%/year due to the growing consumption by products with ecological appeal [112].

The native mallow (*Urena lobata*), belonging to the family *Malvaceae*, has been well adapted to the conditions of lowland cultivation in the states of Amazonas and Pará, since the 1930s [113]. This species provides a fiber with greater resistance than jute, but less silky and shiny. It is a plant that has high productivity in a short period of time. In 2011, the production was 11,560 t [114].

The natural habitat of Curauá (*Ananas erectifolius*) is the highest areas of the banks of the rivers of the Amazon Region. It produces a high-strength fiber that can produce a reinforced polymer with a lower density, for use in many applications [115]. Curauá fibers have a tensile strength five to nine times greater than that of sisal and jute. Thus, the demand for this fiber is increasing [84]. In Santarém, Pará, production is around 98 t/year [116].

The açaí fruit comes from a palm tree *Euterpe oleracea* Mart., typical of the Amazon Region [117,118]. Although the açaí tree is also found in other states, Pará is the largest national producer, producing more than 90% of what is consumed in the country [119]. This high production of açaí results in the large amount of seeds. Açaí fibers are fixed around the seed of the fruit, and after processing the juice, they are located in the mesocarp of the fruit, exactly where the pulp is located. The fibers are, therefore, a by-product of the extraction of juice or pulp from açaí. Once dehydrated naturally, they can easily be removed using the hands. They have an average length of around 12 mm [119]. Research has been carried out in order to dispose of fibrous residues from açaí. Among these are the use of materials and products for energy generation, fertilizer production, paper manufacture and handicrafts.

Another palm tree found in the Amazon estuary is the miriti (*Mauritia flexuosa L. f.*), which is prominent in regional culture, since it is specifically used for food, building houses and making work utensils and handcrafts [120]. The fibers are

Figure 16.8 Other types of fibers and microscopic aspects: (a) and (b) banana [107,108], (c) and (d) petiole of miriti [89,91], (e) and (f) tururi [109]

extracted from the leaves and the long sticks that support the leaves, the petioles. Both are widely used in local handicrafts [121], in the production of basketry and toys, which are admired for their beauty and originality [121,122]. Toys, for example, are commonly sold during the Círio de Nazaré festival in Belém do Pará.

The palha da costa (*Raphia vinífera*), known as jupatí, is also a common palm in the Amazon Region, on the banks of rivers and streams. It has several uses: the fruit for food and medicines, fibers of the leaves and bark of the petiole for confectionaries and manufacturing handcrafted objects and leaves used in the roof of different types of constructions [104]. National production of palha da costa is not large. However, the fibers are exported and used as garden accessories or as natural twine in many countries [123].

In addition to fibers from the stem, leaf and seed, tururi, a natural fabric from the ubuçú palm (*Manicaria saccifera*), is common in lowland forests in the

Amazon Region. This fabric consists of a kind of envelope (bract) that protects the maturing fruit [124]. Tururi is a fibrous, flexible, resistant material, dark brown in color, which can be bleached and dyed in lighter colors, used in manufacture of hats, backpacks, bags, wallets, bags, slippers, clothing, decoration in general [125].

Piassava, *Leopoldinia piassaba Wallace*, is concentrated in the state of Para but is also cultivated in Bahia, northeast Brazil, where it is the name of *Attalea funifera Mart* [126]. Piassava of Pará has a much softer texture and greater commercial value. Its fiber is flexible and more elastic, whereas piassava from Bahia has the advantages of being impermeable, retaining its elasticity when moistened and of forming longer fibers [127]. Among these, that of Bahia has the highest production in the country, corresponding to 95.2% of national total [128].

The Northeast Region of Brazil has great potential for growing vegetal fibers. As an example of this, in addition to the aforementioned piassava, sisal and banana fibers can be mentioned. Sisal (*Agave sisalana*) is a tropical vegetable and that is why there are so few commercial plantations with fibrous plants outside that environment [129]. However, production in Brazil, in the first half of 2018, was around 10,300 t [114].

The leaves of this plant have sandwich-type structures and from each one, 700–1,400 fibers can be extracted, ranging from 0.5 to 1 m in length [130]. Each sisal fiber is made up of a 100 elementary fibers connected together. It has a high content of cellulose, excellent properties of resistance to rupture and elongation and good resistance to salt water [131].

There are many species of bananas. Of the 7 million tons produced in Brazil each year, the largest producers are the states of Bahia, São Paulo, Minas Gerais, Pará and Santa Catarina, with a cultivated area of 240,000 ha [132]. The fibers of the *Musa sapientum* sp., for example, are extracted from the pseudobast [108]. Depending on the layer of the pseudobast from which it is extracted, the texture can change [133], from pelisse or leather (inner layer), lace (intermediate layer) and straw or cover (outer layer) [134].

Table 16.1 presents some of the main physical and mechanical properties of vegetable fibers from Brazil, which make these fibers interesting for application in HKT and WT blades, considering their low density when compared to synthetic fibers, E-glass and carbon, commonly used in blades.

Other important factors are the mechanical strength of the fiber, its strain capacity and the modulus of elasticity. For understanding the behavior of composite materials reinforced by vegetable fibers, a good start is understanding the behavior of the reinforcement and the matrix, separately. The mechanical properties of fibers, for example, tend to change with their dimensions; both length and diameter can influence the tensile strength, elasticity modulus and rupture strain. It is also important to mention that the properties are better used if the fiber is easy to work with, because this allows them to be organized in the form of unidirectional, bidirectional fabrics and in the form of mats.

Bast fibers, such as jute and mallow, can be used in the form of fabrics and mats, which can be shaped in different geometries, without difficulty. Leaf fibers, such as curauá, miriti, palha da costa and sisal, banana and tururi, also have this

Table 16.1 *Some of the main physical and mechanical properties of vegetable fibers from Brazil*

Type of fiber	Fiber	Diameter (μm)	Density (g/cm^3)	Moisture content (%)	Failure strain (%)	Tensile strength (MPa)	Tensile modulus (GPa)	Cellulose (%)	Hemicellulose (%)	Lignin (%)	Source
Bast	Jute	–	–	–	–	–	–	65	17	14	[98]
		–	1.35–1.45	–	1.4–3.1	393–773	25–55	–	–	–	[135]
		–	1.3	–	–	393–773	–	61–71	14–20	12–13	[136]
		62±14	–	–	–	337	–	–	–	–	[137]
		617±16[a]	1.308	9.91±0.75	2	136.53	9.38	–	–	–	[81]
	Mallow	42.6	1.4	–	5	160	–	76	–	10	[138]
		–	–	–	–	–	–	69.07	11.64	26.63	[139]
		42	1.3	–	4	180	–	–	–	–	[82]
		45±15	1.48±0.01	11.0±0.5	5±2.03	396±89	–	–	–		[100]
	Piassava	–	–	–	11.9±4.3	131.1±27.1	2.58±0.39	–	–	–	[83][b]
		–	1.10–1.45	–	–	109–1,750	5–6				[140][b]
		–	–	–	–	–	–	25.6	12.6	55.5	[112][b]
		–	1.04±0.10	15.22±0.59	–	–	–	–	–	–	[126][c]
Leaf	Curaua	–	–	–	4.49–4.57	131–310	48.7	–	–	–	[84]
		–	0.57–0.92	–	–	117–3,000	27–80	–	–	–	[140]
		–	–	–	1.0±0.2	543±260	63.7±32.5	59.4	–	–	[112]
		64±6	1.221	9.48±0.142	2.4	249.64	20.54	–	–	–	[81]
	Miriti	0.172[d]	–	–	–	–	–	69.41	16.10	19.94	[141]
		–	–	–	–	684	36.26	–	–	–	[85]
		–	0.63–1.12	–	–	129–254		–	–	–	[140]
		325–387	0.886	–	2.60	492±102.29	24.87±5.37	–	–	–	[89]
	Palha da costa	597±27	0.965	9.47±0.051	1.9	103.66	7.47	–	–	–	[81]
	Sisal	–	1.40–1.45	–	2.0–2.9	347–700	9–28	–	–	–	[86]
		–	–	–	–	–	–	73	10	7.6	[97]
		260±49	1.48	–	5.80±1.97	304±116	–	–	–	–	[142]
		250±50	1.40±0.01	12±0.5	6.5±3.21	432.48±106.19	–	–	–	–	[143]

Seed	Açai	–	0.88	–	–	–	–	–	–	–	[144]
		110±30	1.362±0.007	8±2	–	16.48±7.08	–	46.51	18.65	30.35	[119]
		104	0.74	6.4±0.3				27.5±1.4	13.9±1.6	39.1±1.2	[87]
		120±30	1.374±0.05	7.50±1.4	–	17.80±2.6	15.70±2.9	46.43	17.21	31.12	[145]
Others	Banana	–	–	–	–	–	–	60–65	6–8	5–10	[8]
		–	0.67–1.50			700–800	27–32	–	–	–	[140]
		80–250	1.35		10.35						[146]
				10–12				63–64	19	5	[147]
	Tururi without stretch open	–	1.18	–	–	29.95	–	–	–	–	[88]
	Tururi 100% stretch open	–	1.18	–	–	9.38	–	–	–	–	
	Tururi bract[e]		0.99±0.08	9.67±0.91	5.24±0.98	75.58±15.71	2.29±0.36	74.1	12.0	31.1	[109]
	Petiole of miriti	7.95[f]	0.06±0.01	–	17.52±4	4.26±1	0.050±0.023	–	–	–	[89]
		–	0.847±0.134	–	–	253.4±71.79	18.55±2.91	51.29	18.80	16.37	[148][g]
		350–950	1.1–1.15			133–260	9.0–15	67.8	24.0		[103]
		–	0.1	–	–	–	–	–	–	–	[149,150]
	Balsa[h]	–	0.1–0.25	–	–	–	–	40–45	–	25	[151,152]
		–	0.06–0.38	–	–	–	–				
Sintetic	E–glass[h]	–	2.50–2.58	–	–	2,000–3,450	70–73	–	–	–	[153,154]
	Carbon[h]	–	1.78–1.81	–	–	2,500–6,350	230–400	–	–	–	

[a]This value is so high because the author performed the characterization of the jute yarns, which are formed by microfibrils, while the other authors characterize the jute fibers.

[b]Piassava of Bahia.

[c]Piassava of Pará.

[d]Diameter of leaf fibrils.

[e]Density, moisture content, failure strain, tensile strength, tensile modulus are taken from the region of the middle of the tururi bract and the percentage of cellulose, hemicellulose and lignin and take from the tururi bract as a whole.

[f]Diameter of petiole fibers.

[g]Properties of petiole fibers.

[h]These materials are here to compare its properties, special density, to the others ones.

characteristic. Piassava, being a rigid fiber, is difficult to form into mats. Its best use is in the form of short fibers, comminuted or in the form of fillers, such as açaí fiber. Another material, workmanship of which is excellent, is the petiole of the miriti, which is a porous, spongy material. From this, it is possible to make from sticks, billets and thin sheets, all because there is no difficulty in making cuts to get several geometries. This material is similar to balsa wood, but softer with a density at the low end of the range of densities for balsa.

The porosity of the vegetal fibers contributes to an efficient adhesion between it and the matrix it is reinforcing, since the matrix entering the pores provides its anchorage in the fiber and, therefore, an effective transfer of efforts. However, the fibers may be hydrophilic and moisture-containing, with water possibly occupying their pores and recesses, which can reduce adhesion. This difficulty can be overcome by drying the fibers to remove moisture. If there are components other than water in the pores and recesses, chemical treatment with NaOH, for example, is needed to clean the fiber.

Depending on the characteristics presented, these fibers, from the bast, leaf and others, have the potential to be used in different parts of turbine blades, like shells, and the petiole of miriti is a potential sustainable material to the core to support the spar caps, bear shear load or to form the external geometry shape of small and large blades, as previously suggested in Section 16.2.

The petiole of a palm tree is flexible and strong and can both support the weight of the leaf and withstand wind loads [155]. The petiole of the miriti palm supports each leaf and is formed by a rigid outer surface involving a foam material, which has a density of around 0.06 and 0.1 g/cm^3.

A comparison can be made to balsa wood, widely used in turbine blades as core and laminate component, which has a density between 0.1 and 0.25 g/cm^3, although it can vary as much as from 0.06 to 0.38 g/cm^3. The low density is extremely valuable in applications that require lightweight materials with good mechanical performance and the petiole of miriti can be a good choice.

16.3.2 Composite materials reinforced by vegetable fibers

In the last few decades, composites reinforced by vegetable fibers have gained great importance, because of the unique combination of high performance, great versatility and processing advantages at favorable costs by permutation and combination of different fibers and matrices [154] and also because these materials have valuable properties such as high specific strength and stiffness, good fatigue performance and damage tolerance, low thermal expansion, nonmagnetic properties, corrosion resistance and low energy consumption during manufacture [156]. These properties are associated with the use of the plant fibers, also called lignocellulosic materials [5], that, in comparison with traditional reinforcement materials, have acceptable specific strength properties, low density, low abrasion multifunctionality, good thermal properties, improved energy recovery and cause less skin and respiratory irritation [7,8].

Among the multiple applications of plant fibers as a building material is its use as a reinforcement for polymeric resins [157–163]. These studies include the effect of chemical treatment, the influence of fiber volumetric fractions on the composite and the physical and mechanical properties of the fibers, and the behavior of the fiber/matrix interface has been found important for good mechanical performance [51]. Table 16.2 [164] shows some of the several advantages of vegetable fibers over synthetic ones, with regard to economics, technological properties and ecological issues of these materials.

As 87% of the global FRP market of 8.7 million tonnes is based on E-glass compounds, plant fibers and their compounds have a great opportunity to capture the market [164]. However, the main disadvantage of these natural fiber polymeric composite materials appears to be the compatibility between the natural hydrophilic fibers and the hydrophobic matrix which makes it necessary to use compatibilizers or coupling agents in order to improve the adhesion between fiber and matrix [165]. Besides modifying the surfaces of the fibers through chemical treatment to improve the anchoring of the fiber in the matrix, it is necessary to dry the fibers properly, to remove moisture.

In addition to the disadvantages, it is extremely important to know the factors that influence the properties of composite materials reinforced by plant fibers, in

Table 16.2 Comparison between plant and synthetic fibers [164]

Properties	Plant fibers	Glass fibers	Carbon fibers
Economy			
Annual global production of fibers (t)	31,000,000	4,000,000	55,000
Distribution of fibers for FRPs in EU (t)	Moderate (~60,000)	Wide (600,000)	Low (15,000)
Cost of raw fiber (£/kg)	Low (~0.5–1.5)	Low (~1.3–20.0)	High (>12.0)
Technical			
Density (g/cm^3)	Low (~1.35–1.55)	High (2.50–2.70)	Low (1.70–2.20)
Tensile stiffness (GPa)	Moderate (~30–80)	Moderate (70–85)	High (150–500)
Tensile strength (GPa)	Low (~0.4–1.5)	Moderate (2.0–3.7)	High (1.3–6.3)
Tensile failure strain (%)	Low (~1.4–3.2)	High (2.5–5.3)	Low (0.3–2.2)
Specific tensile stiffness (GPa/g cm^3)	Moderate (~20–60)	Low (27–34)	High (68–290)
Specific tensile strength (GPa/g cm^3)	Moderate (~0.3–1.1)	Moderate (0.7–1.5)	High (0.6–3.7)
Abrasive to machines	No	Yes	Yes
Ecological			
Energy consumption/kg of raw fiber (MJ)	Low (4–15)	Moderate (30–50)	High (>130)
Renewable source	Yes	No	No
Recyclable	Yes	Partly	Partly
Biodegradable	Yes	No	No
Hazardous/toxic (upon inhalation)	No	Yes	Yes

order to prevent failures and better dimension the components that use these materials.

Further, the vegetable fibers that reinforce these materials in the matrix are heterogeneous materials. While the fibers provide strength and stiffness to the composite, the matrix transmits externally applied loads, through shear stresses at the interface, to the reinforcement fibers and protects the fibers from external damage. Thus, the advantage of coupling the two distinct constituents is that the fibers remain highly resistant and stiff, which in practical situations would be difficult to achieve.

One of the many advantages of composite materials, in general, is the possibility of adapting the material's properties to meet different requirements. It is well known that the macromechanical behavior of composites that have plant fibers as reinforcements depends on the following parameters [165]:

1. the fiber properties;
2. the volumetric or mass fraction of the fiber in the composite;
3. the geometry of the fibers and the properties of the fiber/matrix interface;
4. the arrangement, orientation and stacking sequence of the fibers in the composite;
5. the properties of the matrix.

The effect of all these parameters can be demonstrated by the fundamental equations for composite strength: the generalized rule of mixtures for the tensile and strength module of staple fiber composites and its modifications [51].

We will not discuss the analytical determination of strength further, but we note some properties of composites reinforced by vegetable fibers in Table 16.3. These are tensile strength, tensile modulus, flexural strength, flexural modulus and impact absorbed energy, which can be important for HKT and WT applications. Such properties are of the composites reinforced by the fibers from Brazil, presented in Section 16.3.1.

These properties were determined through standardized experiments using coupons. This, by the way, is considered the first important step in the pyramid for the analysis of turbine blades [10]. For reducing the uncertainties due to the use of composite materials and the overall structural behavior of the blades, the WT standards of the International Electrotechnical Commission, IEC 61400-1:2005 [194] and IEC 61400-23:2001 [195] prescribed that experimental tests be conducted in three levels. These are level 1 (coupons of materials constituent, ply, laminate); level 2 (subcomponents, as beam, flanges, webs, connections); and full-scale level 3 (the blades). Thus, the main purpose of the coupon-level tests is to determine the stiffness and strength of the material in both ultimate and fatigue limit state. Generally, these tests are performed on small test specimens with basic or bulk material, called material testing. On the other hand, full-scale tests are conducted to verify the strength of the blades in both ultimate and fatigue limit state.

These tests also provide data that serves as the input to design and analysis software to calculate the ultimate performance of the final structure [28]. It is

Table 16.3 *Literature survey of mechanical properties of composite reinforced by different types of vegetable fibers from Brazil*

Manufacturing technique	Fiber type	Reinforcement form	Composite	Fiber content (%)[a]	Tensile strength (MPa)	Tensile modulus (GPa)	Failure strain (%)	Flexural strength (MPa)	Flexural modulus (GPa)	Deflection (mm)	Impact energy	Standard used[b]	Source
Hand lay-up, molding with compression	Bast	4 layers of fabrics	Jute/vinyl ester	40 *m*	41	6	0.8	–	–	–	–	ASTM D3039M-95	[166]
Hand lay-up, molding with compression		3 layers of jute yarns aligned at 0°	Jute/terephthalic polyester	16.43 *m*	57.28±5.58	1.73±0.021	3.39±0.41	–	–	–	–	ASTM D3039/D3039M-14	[167]
Infusion at 53.3 kPa vacuum		2 layers of fabrics		19 *v*	22.66±2.04	0.825±0.023	2.8±0.1	–	–	–	2.13±0.39 (kJ/m^2)	ASTM D3039; ASTM D5942	[81]
Molding without compression		15-mm fibers randomly distributed		1.34	29.3±2.85	–	–	57.58±5.88	–	2.6	–	ASTM D638-10; ASTM D790-02	[168]
				7.77 *m*	30.57±3.49	–	–	–	–	–	–	ASTM D638M-89	[169]
Molding without compression		15-mm fibers randomly distributed	Mallow/terephthalic polyester	5 *m*	25.48±1.17	–	–	–	–	–	–	ASTM D638M-89	[170]
Molding with compression		1 layer of fabrics	Mallow/orthophthalic polyester	40 *v*	110±2	11±1	–	–	–	–	–	ASTM D638	[171]
		Continuous fibers aligned at 0°		30 *v*	–	–	–	1,058.3±267.4	–	20.79±13.2	716.2±134.8 (J/m)	ASTM D790-17, ASTM D256-10	[82]
			Mallow/epoxy	30 *v*	–	–	–	–	–	–	905.49±94.56 (J/m)	ASTM D6110-10	[172]
				30 *v*	177.49 ±17.80	20.50±0.10	–	191.27±22.68	10.9±0.58	–	498.86±34.67 (J/m)	ASTM D638-08; ASTM D790-07; ASTM	[173]

(Continues)

Table 16.3 (Continued)

Manufacturing technique	Fiber type	Reinforcement form	Composite	Fiber content (%)[a]	Tensile strength (MPa)	Tensile modulus (GPa)	Failure strain (%)	Flexural strength (MPa)	Flexural modulus (GPa)	Deflection (mm)	Impact energy	Standard used[b]	Source
												D256-06	
Molding without compression		15-mm fibers randomly distributed	Piassava[c]/terephthalic polyester	16.30 *m*	19.56±1.32	–	–	–	–	–	–	ASTM D638M-89	[174]
				14.82 *m*	13.65±2.75	0.636	2.16	–	–	–	–	ASTM D638-89	[175]
Molding with compression		Continuous fibers aligned at 0°	Piassava[d]/orthophthalic polyester	40 *m*	–	–	–	–	–	–	94.7±20.6 (J/m)	ASTM D256-06	[176]
		2 layers of fiber fabrics aligned unidirectionally at 0°		[e]	43.00±6.12	4.07±0.99	6.00±2.58	43.95±6.41	0.051±0.007	1.65±0.06		ASTM D3039-12; ASTM D790-13	[177]
Molding without compression		Continuous fibers aligned at 0°	Piassava[d]/epoxy	40 *v*	89.33±18.51	2.87±0.57	–	78.41±9.01	3.46±0.69	–	210±20.6 (J/m)	ASTM D638; ASTM D790-03; ASTM D256	[101]
Hand lay-up	Leaf	Continuous fibers aligned at 0°	Curaua/orthophthalic polyester	20 *m*	–	–	–	96.6±13.0	–	–	–	ASTM D790	[178]
Molding with compression			Curaua/epoxy	20 *v*	78.98	44.54	3.33	106.75	19.2	–	–	ASTM D3039/D3039M-00; ASTM D790-03	[179]
Infusion at 101.3 kPa vacuum		2 layers of continuous fibers aligned at 0°	Curaua/orthophthalic polyester	38 *v*	86.71±2.09	3.883±0.233	2.8±0.1	–	–	–	14.49±1.75 (kJ/m^2)	ASTM D3039; ASTM D5942	[180]

	2 layers of bidirectional fabric			30.10±3.88	1.113±0.099	2.5±0.2	–	–	–	–	ASTM D3039	
			39 *v*	–	–	–	–	–	–	6.14±0.57 (kJ/m^2)	ASTM D5942	
Molding with compression	3 layers of long fibers aligned at 0°	Miriti/terephthalic polyester	10 *v*	40.83±7.60	2.73±0.12	–	51.30±5.18	4.12±0.41	–	–	ASTM D638-08, ASTM D790-03	[181]
	3 layers long fibers aligned at 0°, 90° and 0°		10 *v*	17.63±4.19	0.48±0.096	–	58.59±7.43	4.36±0.21	–	–		
	10 layers of bidirectional fabric	Miriti/epoxy	30 *v*	–	–	–	–	–	–	305.51±28.12 (J/m)	ASTM D6110-97	[182]
Molding without compression	15-mm fibers randomly distributed	Palha da costa/terephthalic polyester	9.98 *m*	16.48±1.31	–	–	–	–	–	–	ASTM D638-89	[104]
Hand lay-up, molding with compression	25-mm fibers randomly distributed		7.33 *m*	8±2.02	0.89±0.19	0.9±0.06					ASTM D3039/D3039M-08	[183]
Infusion at 101.3 kPa vacuum	2 layers of fibers aligned unidirectionally at 0°	Palha da costa/orthophthalic polyester	40 *v*	22.97±1.58	1.303±0.090	2.2±0.1				4.70±0.93 (kJ/m^2)	ASTM D3039; ASTM D5942	[180]
	2 layers of bidirectional fabric		43 *v*	20.27±1.88	0.913±0.124	2.9±0.4				2.36±0.48 (kJ/m^2)		
Molding with compression	35-mm fibers distributed longitudinally in relation to the effort	Sisal/terephthalic polyester	30 *m*	115.82±3.01	3.07±0.09	6.28±0.27	–	–	–	–	ASTM D3039	[184]
Hand lay-up, molding with compression	1 layer of long fibers aligned at 0°		41.26 *v*	129.49±17.89	4.2±0.44	–	–	–	–	–	ASTM D3039M-00	[185]

(Continues)

Table 16.3 (Continued)

Manufacturing technique	Fiber type	Reinforcement form	Composite	Fiber content (%)[a]	Tensile strength (MPa)	Tensile modulus (GPa)	Failure strain (%)	Flexural strength (MPa)	Flexural modulus (GPa)	Deflection (mm)	Impact energy	Standard used[b]	Source
Molding without compression		15-mm fibers randomly distributed		3.90 *m*	26.16±2.15	–	–	–	–	–	22.36±4.72 (kJ/m^2)	ASTM D638M-89; ASTM D5942-96	[142]
				1.07	35.61±1.24	–	6.25±0.05	60.53±8.86	2.6±0.01	–	–	ASTM D638-10; ASTM D790-02	[168]
	Seed	5- to 15-mm fibers randomly distributed	Açai/terephthalic polyester	3.14 *m*	23.38	–	–	–	–	–	–	ASTM D638-89	[105]
Molding with compression		20-mm fiber mat randomly distributed	Açai/orthophthalic polyester	6.5 *v*	13±4	–	9±1.1	15±2	0.837±0.236	–	–	ASTM D638; ISO 178	[87]
Molding without compression		Short fibers randomly distributed	Açaí/isophthalic polyester	20 *v*	–	–	–	69.5	–	–	0.8 (J)	ASTM D790-02; ASTM D5942	[186]
			Açaí/terephthalic polyester	5 *v*	35.94	–	–	–	–	–	–	ASTM D638	[187]
	Other	Inner layer 5-mm fibers randomly distributed	Banana/terephthalic polyester	5.41 *m*	35.39±1.21	–	–	–	–	–	–	ASTM D638M-89	[188]
Hand lay-up, molding with compression		Short fibers randomly distributed	Banana/epoxy	28 *m*	83.55	0.298	–	46.04	–	–	2.25 (J)	ASTM D3039; ASTM D790; ASTM D4812	[189]
		6-cm fibers randomly distrib-		30 *v*	85	3.4	–	48	1.5	–	–	ASTM D638; ASTM	[190]

	uted										D790	
Molding with compression	Long fibers aligned unidirectionally at 0°	Petiole of miriti/ orthophthalic polyester	40 v	100.15±10.72	1.66±0.27	5.4±0.5	–	–	–	–	ASTM D638	[191]
		Pecíolo/ epoxy	30 v	76.07±4.73	1.52±0.11	0.05±0.01	154±14.27	65±5	4.8	114.17±14.37 (J/m)	ASTM D638; ASTM D790-03; ASTM D256	[148,192]
		Petiole of miriti/ orthophthalic polyester		126.33±12.55	1.53±0.22	0.06±0.03	163.7±22.2	22.1±3.2	2.4	121.00±53.00 (J/m)		[148]
Vacuum infusion	2 layers of fabric oriented at 0°/ 90°, without stretch open	Tururi/epoxy	29 v	5.8[f]	2.12	3	–	–	–	–	ASTM D3039	[193]
	2 layers of fabric oriented at 0°/ 90°, with 100% stretch open		23 v	4.2[f]	2.4	1.6	–	–	–	–		
Hand lay-up, molding with compression	3 layers of fabric oriented at 0°, without stretch open	Tururi/terephthalic polyester	14.39 m	35.76	–	–	–	–	–	–	ASTM D3039-05	[88]
	3 layers of fabric oriented at 0°, with 100% stretch open		7.19 m	16.8	–	–	–	–	–	–		

Note: The values presented in this table are average values and defined in the maximum load condition.

[a] *m* Refers to the mass fraction and *v* to the volumetric fraction of the fibers in the composite.

[b] Standards for tensile tests: ASTM D3039/D3039M and ASTM D638/D638M; Standards for Flexural Tests: ASTM D790 and ISO 178; Standards for Impact Test: ASTM D5942, ASTM D256 and ASTM D4812, where standard ASTM is from American Society for Testing and Materials and ISO from International Organization for Standardization.

[c] Piassava of Pará.

[d] Piassava of Bahia.

[e] The author did not provide the fraction of fiber used in the manufacture of the composite, only the density of the composite (0.9958 g/cm^3) and that of the resin used (1.12 g/cm^3).

[f] Due to irregularities in the thickness of the coupons tested, the author reported that these values correspond to the maximum load in N, normalized by the area density (g/m^2) of the fabric without and with 100% stretch open, with the maximum load being 1163 N and 665 N, respectively.

essential to develop numerical models capable of predicting structural deformation of a blade when it is subjected to different loads, in different dynamic conditions, and also properties of the blade and its components, including adhesive joints, ply drops, even in the presence of material defects [196,197].

Mechanical properties of composites reinforced with vegetable fibers are presented in Table 16.3 and change considerably, depending on the following:

1. Manufacturing method: This aims to achieve good structural quality, excellent distribution and homogenization of the matrix on the fibers, the absence of voids, air bubbles, which are concentrated tension mechanisms, excellent proportions between fiber and matrix, in order to reduce the mass of the material and increase its strength, processing savings from the use of low-cost mechanisms, materials and tools, and easy workability, making the process fast.
2. Fiber type: Generally, greater mechanical performance of the fiber is achieved with higher cellulose content and with cellulose microfibrils aligned in the direction of the fiber. The composite properties will also depend on the type of fiber, orientation and layout in the matrix, as well as the adherence existing in the fiber/matrix interface, i.e., their bonding that transfers load from the matrix to the fibers.
3. Type of matrix: The matrix is an essential part of a fiber-reinforced composite, as it provides a barrier against harmful environments to the fiber, protects the fiber surface from mechanical abrasion and transfers load to the fibers. Although the hydrophilic and hydrophobic nature of fibers and matrices, respectively, it is necessary a good interface fibre/matrix to provide the effective resistance for effords.

Beside these three topics, it is also important to take into account standardized tests to allow meaningful comparisons with the properties showed in Table 16.3. The deflection of a sample, for example, depends on its size, so it can be difficult to make sense to deflection values for different-sized samples. Because of this, the comparisons must occur in tests that used the same standards. Thus, these considerations of the properties of the material, together with the operating environment of the turbine and the production costs, form the basis for selecting the material suitable for the types of efforts existing around the blades and normal operating conditions. Hence, there is a need for studies that makes it possible to understand the properties of fiber-reinforced composites not only in static but also in dynamic conditions.

16.3.3 Manufacturing of small turbine blades using sustainable materials

Sustainable materials have been studied as potential alternatives for application in HKT and WT blades. Design and manufacturing guidelines have been developed for low-cost HKT blades using *P. radiata* timber laminated with epoxy resin reinforced by E-glass fiber on the outside [63], and the design and techniques of local construction of HKT blades using belian timber [61].

Structural analysis of small WT blades manufactured using a PUR core coated with ester vinyl resin reinforced by jute fiber fabric was done in [90], and for flax fiber replacing E-glass fiber in [11]. For both HKTs and WTs, we have the methodology of manufacturing blades using the petiole of the miriti palm as a core coated with E-glass fiber reinforced polyester resin [91]. This methodology generated two patent applications: BR 10 2019 014434-3, for HKT blades, and BR 10 2019 017638-5, for WT blades. Figure 16.9 shows some of the blades developed and manufactured in the researches cited.

The blade manufactured using *P. radiata* as core, coated with epoxy resin reinforced by E-glass fibers [63], was developed to reduce the investment cost, mainly in relation to the design, materials and manufacturing itself, in order to obtain an economical small HKT. The wooden cores of this blade were made in a CNC machine based on geometric data produced by the Turbem software [63,198,199]. An epoxy resin reinforced by layer of 2 mm of E-glass fibers was then applied. The volumetric proportion of fiber/matrix was 60/40. Two blades were manufactured and Table 16.4 shows the characteristics of the materials used.

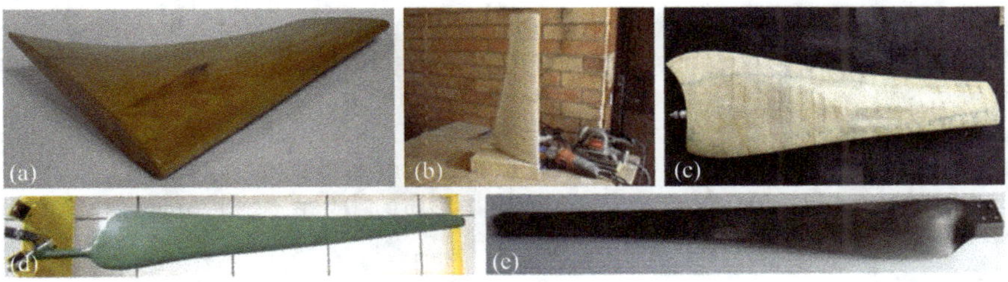

Figure 16.9 (a) 0.75-m blade, using Pinus radiata timber as a core coated with epoxy and E-glass fibers [63]; (b) proposed 0.8-m blade, using belian wood [62]; (c) 0.598-m blade, manufactured using petiole of miriti as core coated with polyester and E-glass fiber [91]; (d) 1.07-m blade, made using polyurethane as core coated with vinyl ester and jute fiber fabric [90]; and (e) 3.5-m blade, made using polyester and flax fiber [11]

Table 16.4 Characteristics of the materials used in the manufacturing of a 0.75-m blade [63]

Material	Mass per rotor (kg)	Elasticity modulus (GPa)	Failure stress (MPa)
Pinus radiata (dry)	12	10.07	49
E-glass fiber	6.4	72.3	3,445
Epoxy resin	4.6	10.5	85
Composite reinforced by E-glass fibers	–	25.21	537.8

The blades were tested in an HKT under normal operating conditions. The maximum stresses, estimated during operation at a given flow speed, were computed, allowing determination of the maximum flow speed at which the rotor can be safely operated. As a result, it was found that a blade made entirely of resin and E-glass fibers has about ten times the load capacity of one made of timber. The timber blade had a maximum deflection [63], 32.2 mm at the tip, while the blade made of composite (blade with wood as core coated with epoxy resin reinforced by E-glass fibers) deflected 11.8 mm, under maximum load conditions as predicted by Turbem. However, timber blades generally have excellent fatigue behavior, as further described in [18], and the epoxy coating protects the blade from corrosion and impact damage.

A copying router was developed to make approximately 1-m long blades from *Douglas fir* and *Sitka spruce* timbers blades [56]. The blade profiles were compared to the master blade using a coordinate measuring machine. The results show that, even so the trailing edge region was not accurately reproduced, the copying router produces blade profiles of similar shapes and with correct angles of twist. As far as we are aware, this is the only measured comparison of actual to design shape.

According to the sequence shown in Figure 16.10, taken from [62], it is possible to produce a model blade using the templates of the profiles for each section of the blade, which were printed in scale, according to Figure 16.10(a), cut and placed on a piece of soft wood (pine was used for this) according to Figure 16.10(b).

As detailed by [62], the contours in the wood were then carefully cut using a saw and smoothed with a suitable file. Subsequently, the profiles of the sections were organized in order, as seen in Figure 16.10(c). Then, sections were made to

Figure 16.10 (a) Arrangement of the hydrokinetic profiles for (b) the cuts in wood, (c) distribution of the profiles for making (d) of the sections of the blade, (e) bonding the sections and (f) manufactured model blade [62]

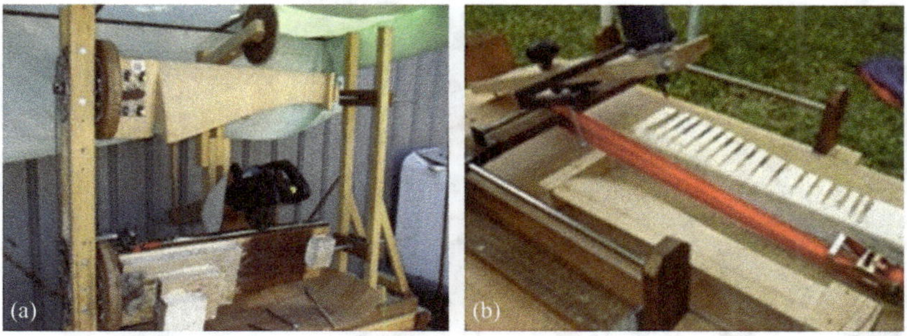

Figure 16.11 Manufacturing techniques for (a) circular saw jig and (b) router jig [62]

fill the spaces between the profiles and then they were organized according to the corresponding twist angle, which can be seen in Figure 16.10(d) and (e). Finally, after drying the glue, the blade was carefully polished. The result, in Figure 16.10 (f), was a 0.8-m model blade, used to make three copies, using two different techniques of copy jigs: circular-saw-based and router-saw-based, shown in Figure 16.11(a) and (b), respectively.

Comparing circular and router saw based, according to [62], the advantages of the first are the possibility of making much faster and deeper cuts in a single pass. The disadvantages are related to the irregular depth of cuts due to the flexing caused by the weight of the circular saw, the difficulty of controlling the operation. In addition, the process can be dangerous because the saw is exposed and close to the operator's body. It also may take longer to rough, repair and sand the cuts due to the fact that the saw is fine and the copies are produced without precision.

The second method, on the other hand, as stated by [62], has the advantage of being fast, easy to operate and control the router saw, with less danger because the router bit is short and there is no bending and thus the blades are produced with greater precision. There is also a small time spent on straightening and finishing, due to the uniform cut, although very wavy. However, its disadvantage is the cut rate to be slower than the circular saw; therefore, it needs a longer thinning time, the shorter cutting bit, by the way, may have to repeat the roughing process several times before the required depth is reached, it might not be suitable for copying large blades with a sharp twist angle. Even the manufacturing processes were efficient to reproduce the blades, their accuracy was not assessed.

Although the manufactured blades have not been tested in a turbine, the feasibility of making low-cost timber blades with reasonable replication cost has been demonstrated, using common tools, and simple templates. The simplified method allows it to be done in remote villages, significantly the total cost of a turbine.

The blade made using PUR as core coated with vinyl ester resin reinforced by jute fiber fabric [90] was manufactured in two molds, shown in Figure 16.12. This mold, in turn, was made from a model blade. The manufacture of the blades went through the stages of cutting the jute fabric in the size of the suction and pressure surfaces, as

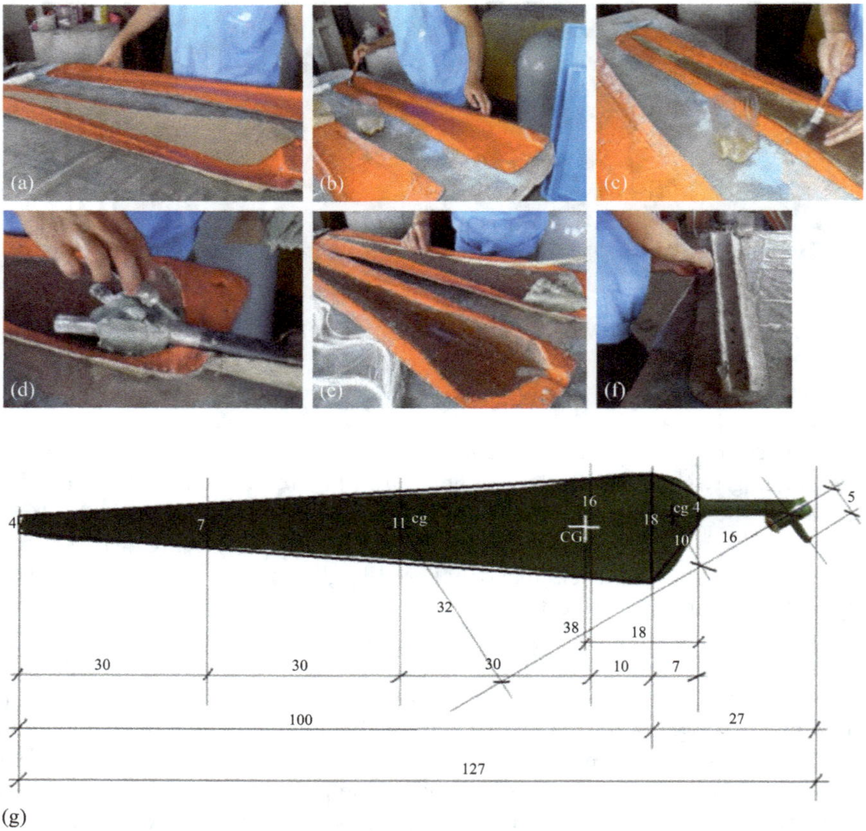

Figure 16.12 Steps of the manufacturing process of the blade shells in a bipartite mold: (a) cutting the fabrics, (b) applying a gel coat, (c) depositing the layers of jute fabric, (d) fixation of the metallic insert, (e) placement of E-glass fibers strands to join the regions of the blade and (f) closing the mold. In (g) the dimensions of the blade manufactured [90]

shown in Figure 16.12(a), the preparation of the mold surface with mold release wax and manual application, with brush, of gel coat, as presented in Figure 16.12(b).

Described by [90] as follows, this gel consisted of pre-accelerated vinyl ester resin with 0.3% cobalt octoate and 0.2% dimethyl aniline as a cocatalyst and addition of methyl ethyl ketone peroxide, in a proportion of 2% by weight, making it possible to work for 30 min before curing started. On the gel coat, on both sides of the mold, two layers of jute fiber fabric were deposited, laminated with catalyzed vinyl ester resin, as shown in Figure 16.12(c). This step was carefully performed in order to guarantee a satisfactory interweaving between the fabric and the resin and their good fit in the mold, in order to avoid bubbles, bad impregnation or the formation of folds in the fabric.

After curing the shell of the blade, an inspection was carried out on the trailing and leading edges. Subsequently, the blade was sanded to ensure that there were no protrusions that would prevent the fit of the mold sides in the next step. To attach the blade to the rotor shaft, a metallic insert was fixed inside of the blade, Figure 16.12(d), with terephthalic polyester resin mixed with talc and calcium carbonate in a ratio of 70:30.

Once the shells were manufactured and the metallic insert fixed, strands of fiberglass were prepared, in the form of wicks, Figure 16.12(e), impregnated with vinyl ester resin and arranged around the edges of the mold, to promote the union and sealing the two regions of the blade. The mold was closed and sealed by screws distributed along the trailing edge and leading edge, as seen in Figure 16.12(f). After curing and demolding the part, two holes were made, one for injecting PUR resin, and one outlet in order to bond the internal surfaces of the shell and guarantee bending stiffness.

The blade and dimension, with geometric center (cg) and center of mass (CG), are shown in Figure 16.12(g). It was manufactured with 1.07 m and 1.547 kg. The blade was statically loaded using weights equivalent to classes I–IV of wind loads, defined by IEC 61400-1 [194]. The blade was tested in a period of 15 min and sustained all simulated wind loads, with a safety margin of 10%, which is equivalent to a maximum total force of 395 N, with no sign of structural failure.

However, after loading, the blade showed a slight permanent strain, measured at the tip of the blade, of 1.5 cm, in the flatwise direction, because the excessively long time (15 min) of test allows an internal delamination of the composite. This strain is not considered a failure that disqualifies the blade for these classes of winds, considering they occurs briefly and in the 10 s required by guideline for the certification of WTs [52,194], the strain did not happen.

In making a blade of polyester reinforced by E-glass fiber, in [11] is used a CNC machined core, structural bubble caps in composite with unidirectional fibers and shell constant thickness, reinforced by multiaxial fiber. The core resists buckling, the caps promote bending stiffness and the outer shells provide resistance to torsional shear loads. The blades were manufactured using unsaturated polyester resin in a light RTM process. Its structural integrity was assessed through design load analysis and full-scale mechanical tests, according to BS-EN 61400-2:2006 [79] and BS-EN 61400-23:2002 [80]. Table 16.5 shows some properties of the blades and the materials used.

The lower density of vegetable fibers reduces the blade weight; the flax/polyester blade is 10% lighter than the E-glass/polyester blade, which represents a 45% fiber mass saving. The static test of the blades, according to the certification standards, confirmed that both blades met the design and structural integrity requirements for an 11-kW turbine under normal operating conditions and ultimate loading. In addition, methods to overcome the deflection of the tip of the flax/polyester blade were suggested. For example, to improve its stiffness, it is suggested that a spar using webs and caps for shear strength be incorporated.

Shah *et al.* [11] also proposed that linen is structurally efficient and suitable for replacing E-glass for applications in small WTs blades. It was concluded that the

Table 16.5 Some properties of the flax/polyester and E-glass/polyester blades manufactured and the materials for manufacturing a 3.5-m blade [11]

Material	Density (g/cm^3)	Mass (kg)	Volumetric fraction of fiber in the blade (%)
Flax/polyester blade	1.28	23.3	–
Flax	1.31	4.2	22.9
Polyester resin	–	10.7	–
E-glass/polyester blade	1.59	25.8	–
E-glass fiber	1.79	7.7	26.4
Polyester	–	9.7	–
Core	–	8.4	–

development of low-cost vegetable fibers is a limiting factor for the industrial use of these fibers, in structural composites. More studies were recommended to understand the behavior of these materials when employed in specific applications/structures, rather than limiting the analysis of materials to data extracted from static coupon tests.

These studies, [11,62,63,90], use methodologies for the manufacture of blades, that differ both in processes, equipment, tools and the materials used, which have a direct influence on the final product. The starting time depends on the blade's moment of inertia, which is a function of the blade geometry and its density [58]. The density is intrinsic to the materials and the way they are combined and arranged in the blade.

With this in mind, using a low-density raw material, the petiole of the miriti, a methodology was developed for the manual manufacture of turbine blades for HKTs and WTs [91], presented as follows. The blade shells were manufactured in separate molds, shown in Figure 16.13(a), through vacuum lamination. Special attention was paid to preventing air bubbles.

For vacuum lamination, it was necessary to prepare the surfaces of the molds, applying, respectively, layers of release wax, and release agent. The shape of the surface of the mold can be transferred to the surface of the shell, so it is important to avoid roughness, which is one of the crucial problems for turbine blade performance [200]. After drying the surface of the mold, polyester resin mixed with aerosol (hydrophilic pyrogenic silica) and industrial talc was applied, to provide rigidity, resistance to thinning and lightness to the outer surface of the shell.

A spiral tube was anchored on one side of each mold, to distribute the vacuum pressure over the mold surfaces. In the middle of this tube, a T-connection to a hose segment was attached to both sides of the mold. A new T-connection connected these segments to the hose from the vacuum pump, to allow manufacture of the two shells at the same time. Then, as shown in Figure 16.13(b), a layer of bidirectional fiberglass fabric was spread over these surfaces to apply and distribute, with a brush, polyester resin mixed with aerosol, as shown in Figure 16.13(c). This was to roughen the inner surface of the shell to increase adhesion to the core.

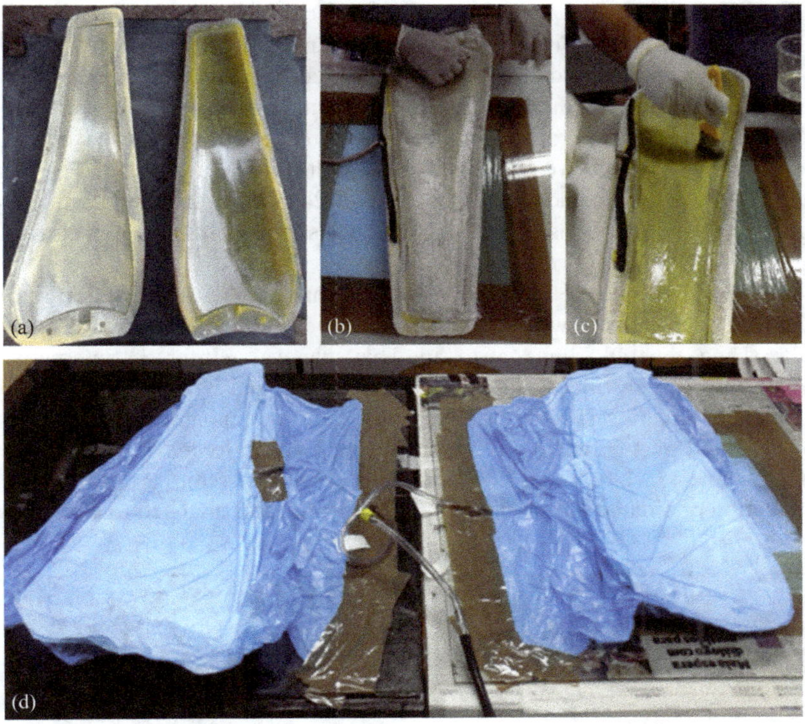

Figure 16.13 *(a) Bipartite mold, (b) bidirectional E-glass fibers fabric layer on the mold surface, (c) application of polyester resin and (d) supply of vacuum pressure to the system [91]*

The peel ply fabric was a release polyamide woven fabric that absorbs some of the matrix resin and leaves a fresh, clean, roughened inner surface of the shell, avoiding irregularities. In addition, a "resin runner" a perforated film was added to assist the passage of resin to and into the absorbent fabric. The runner also helps remove excess resin from the laminate and, thus, reduces its thickness. All connections and hoses were sealed with tape. The sides of the mold were then placed inside a vacuum bag and on a glass plate, as seen in Figure 16.13(d). The vacuum bags were also sealed with tape.

The vacuum pump, which is not shown in the figure, was then activated, providing the system with negative gauge pressure of the order of −0.8 bar, for 2 h. Subsequently, the shells, still in the mold, but with hoses disconnected from the vacuum pump and blocked with pressure pliers, were taken on the glass base to the oven for post-curing. After curing the material, the blade halves were demolded by lightly tapping with a rubber hammer on the outer surface. The detachment took place from tip to root. The excess material at the edges of the shells were cut and finished by sanding.

For the blade core, the petiole of miriti was sectioned and subsequently cut into blocks shown in detail in Figure 16.14(a), according to the dimensions of the blade sections to be manufactured. The profiles, obtained from the blade's design shape, were glued to the sides of these blocks, as shown in the example in Figure 16.14(b).

The centroids of the profiles were connected by a metal pin. Consequently, the contours of the profiles were cut and a through hole was drilled, as shown in Figure 16.14(c). This method was used with all 24 pairs of profiles for the manufacture of 24 blade sections. After manufacture, surface finishing was carried out using manual sanding.

The central spar of the blade, which passes through the centroid of all profiles to replace the metal pin, was also manufactured with petiole of miriti, having a circular shape, tapered from root to tip. This spar was reinforced, internally and externally, by carbon fibers, aligned unidirectionally and wrapped transversely in relation to its length, respectively. The fibers were impregnated in polyester resin then cured and post-cured in an oven. The internal reinforcement aimed to give bending stiffness, while the external one, resistance to torsion. After manufacturing, a section of circular aluminum cylindrical profile was attached to the spar at its

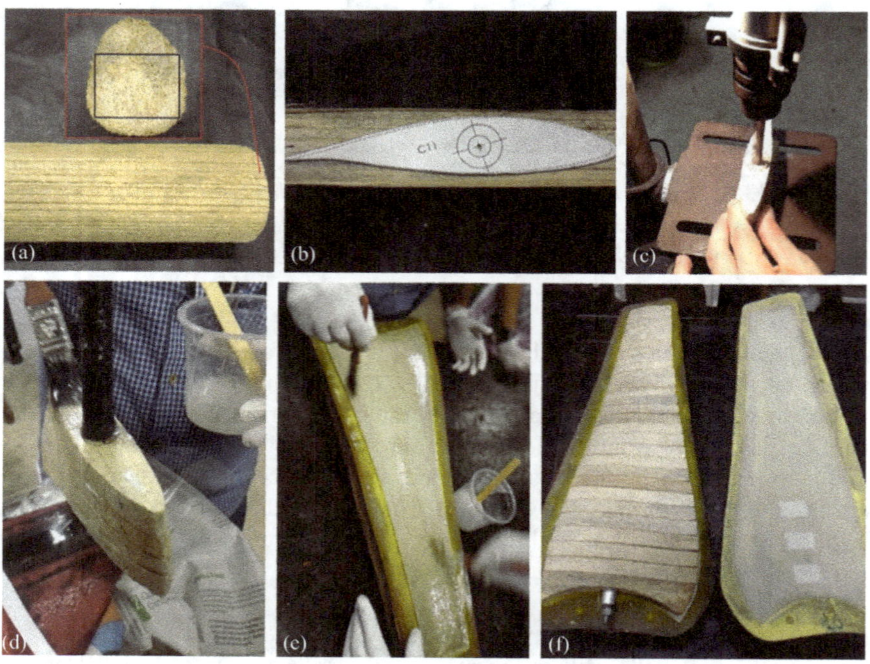

Figure 16.14 (a) Cutting the petiole of the miriti in a block, (b) placing an NACA 653-618 profile on the side of the block, (c) making a through hole in a bench drill, (d) joining the miriti sections between itself and the spar, (e) application of polyester resin on the inner surface of the shell to (f) anchor the core to the inside of the blade [91]

root, in which a metal screw was anchored, to allow the blade to be fitted to the turbine hub. This coupling was provided by polyester resin and anchoring by polyester resin mixed with milled glass fiber and industrial talc.

The sections of miriti were joined to each other and to the spar using polyester resin. For this, the stringer was attached to a vise and the sections were inserted one by one, through the hole in the centroid, as shown in Figure 16.14(d). The faces, the hole and the region where the section was placed were brushed with polyester resin during the bonding of each section to the spar. The shells of the blade were placed on the sides of the two molds, as in Figure 16.14(e), applying layers of polyester resin to the internal surface of it. Figure 16.14(f) shows the miriti core fitted on one side of the mold. Strips of bidirectional fiberglass fabric were placed on the trailing and leading edges, wetted with polyester resin, for joining the shells together. Then the mold was closed until the resin cured.

After demolding, to eliminate the imperfections and cracks in the joint and to waterproof the blade, hand finishing was done using polyester putty, followed by sanding on the trailing edges, the leading edges, the tip and the root of the blade. Table 16.6 shows some characteristics of the manufactured blade, and the rotational moment of inertia, *J*, calculated according to a method from [18] and developed from experimental data verified from computational modeling. This calculation is simplified as it assumes uniform density and the connection accessories with the turbine hub are not taken into account.

The blade using petiole of miriti as core has a low density and consequently a low rotational moment of inertia that is the key element for the starting performance of a turbine [19,201]. However, it is necessary to improve the methodology used, considering that, even the hand manufacturing can be used when there is a lack of technological devices, it can lead to errors. These include insufficient distribution and deposition of resin for bonding the parts or for producing a constant shell thickness, errors in the hole drilled in a bench drill, through the centroid, and irregularities in the profiles bonded to the central spar of the blade. Even though the petiole of miriti is easily worked and adapts to different geometries, all of this can contribute to the inaccuracy of the final blades.

Most of these errors would be avoided by good-quality control during mass production but very few small WTs and no HKTs that we know of have reached this stage. Further, the properties of the petiole of miriti have not been fully

Table 16.6 Some characteristics of the blade manufactured using petiole of miriti as core and polyester resin reinforced by E-glass fiber fabric as shells [91]

Material	**Blade length (m)**	**Mass (kg)**	**Blade density (kg/m^3)**	**J ($kg\ m^2$)**
Petiole of miriti/E-glass fiber/polyester	0.598	0.611	348.55	0.1189

determined, so that more information is needed on its composition, structure, elastic properties, damage behavior, fracture, debonding in sandwich structures, static and fatigue life behavior, all of which are better known for balsa [202–209].

16.4 Final considerations

This chapter aimed to draw a parallel between conventional materials with manufacturing methods and sustainable materials with alternative methodologies for manufacturing HKT and WT blades. The materials and methods of manufacturing WT blades can be used in underwater applications, as long as corrosion protection, lightness and impermeability to water, neutral buoyancy and antifouling coating are considered.

Whatever method used for manufacturing blades, considerable care is needed to reproduce the design foil profile especially near the tip, where significant errors in the blade shape could have large impact on the power output.

Synthetic fibers are currently used to make small HKT and WT blades due to their excellent mechanical properties and specific resistance. However, these fibers are far from being suitable for local manufacture, making them economically unfeasible in applications that generate energy on a small scale. For this reason, we discuss the potential use of vegetable fibers from Brazil, with their good mechanical properties, low specific mass, low processing cost. They are biodegradable and come from renewable sources, which makes them attractive for this type of application.

Nevertheless, the properties of sustainable materials by themselves do not make them useful in HKT and WT blades, considering that it is not enough that they have low density, good mechanical properties when associated with polymeric matrices and cause a low rotational moment of inertia to the blades. More detailed studies are needed, which directly evaluate the behavior of these materials in such applications, from tests in static conditions to verify if the blades are capable of resisting efforts, such as traction, flexion, impact and others, and tests in cyclical (dynamic) conditions, to verify the fatigue life and, also, what is the influence, in these properties, of the environment where the technology can be applied.

We describe the manufacture of small HKT and WT blades using a range of material from aluminum, through timber, conventional fiberglass and petroleum resin, through the use of natural fibers and the petiole of miriti for the foam core and spar. So far, the developments have been encouraging, with the resulting blades possessing the required strength but issues associated with small volume production need to be resolved. In addition, it was pointed out that more information on the properties of natural materials is needed. This requires additional experimental tests conducted on coupons, subcomponents and full-scale blades. As noted, coupon tests are just the initial stage to understand the behavior. In view of the variability in the properties of natural materials, anomalies and uncertainties can occur. For this reason, many coupon tests are needed, for different types of fibers, including their several possibility of configurations, environmental factors, as temperature, contact with various fluids, moisture content and others, to determine the stiffness and strength of the material in ultimate and fatigue limit state, and forward subcomponents and full-scale blades tests, to make possible the development of numerical models capable of predicting the performance of the blades in different operational conditions. These comments were directed at the basic material properties, but it was pointed out that good fatigue behavior is also important for small blades.

Despite the problems we have outlined, we believe that the future of natural materials is bright and the research described in this chapter will lead to further applications. Though it is not the primary focus of the chapter, we again emphasize that these developments are much easier to pursue on blades that are around 1-m long rather than the much longer blades for large turbines.

References

[1] Murray, R. E., Snowberg, D., Berry, D., Beach, R., and Rooney, S. (2017) Manufacturing a 9-meter thermoplastic composite wind turbine blade, National Renewable Energy Laboratory. *In: American Society for Composites, 32nd Technical Conference West Lafayette*, Indiana.

[2] Daniel, I. M., and Ishai, O. (2006) *Engineering Mechanics of Composite Materials.* Oxford University Press, Oxford.

[3] Okokpujie, I. P., Okonkwo, U. C., Bolu, C. A., Ohunakin, O. S. Agboola, M. G., and Atayero, A. A. (2020) Implementation of multi-criteria decision method for selection of suitable material for development of horizontal wind turbine blade for sustainable energy generation. Heliyon, v. 6, n. 1, pp. e031402.

[4] Jawaid, M., Khalil, H. P. S. A., Bakar, A. A., and Khanam, P. N. (2011) Chemical resistance, void content and tensile properties of oil palm/jute fibre reinforced polymer hybrid composites. *Materials and Design*, v. 32, pp. 1014.

[5] Gowda, T. M., Naidu, A. C. B., and Rajput, C. (1999) Some mechanical properties of untreated jute fabric-reinforced polyester composites. *Composites: Part A*, v. 30, pp. 277.

[6] Hun, T. Q. (2008) *Characterization of Lignocellulosic Materials*. Wiley-Blackwell, New Jersey.

[7] Huq, T., Khan, A., Akter, T., *et al.* (2011) Thermo-mechanical, degradation and interfacial properties of jute fiber reinforced PET-based composite. *Journal of Thermoplastic Composite Materials*, v. 24, pp. 889–898.

[8] Satyanarayana, K. G., Guimarães, J. L., and Wypych, F. (2007) Studies on lignocellulosic fibers of Brazil. Part I: Source, production, morphology, properties and applications. *Composites: Part A*, v. 38, pp. 1694–1709.

[9] Love, M., and Bohlmann, J. (1989) Aeroelastic tailoring and integrated wing design. *General Dynamics*, n. 89, pp. 25–167.

[10] Fernandez, G., Usabiaga, H., and Vandepitte, D. (2017) Subcomponent development for sandwich composite wind turbine blade bonded joints analysis. *Composite Structures*, v. 180, pp. 41–62.

[11] Shah, D. U., Schubel, P. J., and Clifford, M. J. (2013) Can flax replace E-glass in structural composites? A small wind turbine blade case study. *Composite Part B: Engineering*, v. 55, n. 1, pp. 117–181.

[12] Shokrieh, M. M., and Lessard, L. B. (2003) *Fatigue Under Multiaxial Stress Systems, in Harris B Fatigue in Composites*. England Woodhead Publishing Limited, Abington, Cambridge Chapter 3.

[13] Mandell, J. F., Samborsky, D. D., Agastra, P., Sears, A. T., and Wilson, T. J. (2010) *Analysis of SNL/MSU/DOE Fatigue Database Trends for Wind Turbine Blade Materials*, Sandia Contractor Report SAND2010–7052.

[14] Clausen, P. D., and Wood, D. H. (1999) Research and development issues for small wind turbines. *Renewable Energy*, v. 16, pp. 922–927.

[15] Amélie, B., Petera, D., and Frédéric, T. (2012) Sea water ageing of composites for ocean energy conversion systems: Influence of glass fibre type on static behaviour. *Applied Composite Materials*, pp. 459–473.

[16] Anyi, B, and Kirke, M. (2010) Evaluation of small axial flow hydrokinetic turbines for remote communities. *Energy for Sustainable Development*, v. 14, pp. 110–116.

[17] Davies, P., Germain, G., Gaurier, B., Boisseau, A., and Perreux, D. (2013) Evaluation of the durability of composite tidal turbine blades. *Proceedings of Royal Society of London A*, pp. 1–15.

[18] Wood, D. (2011) *Small Wind Turbines: Analysis, Design, and Application*. Springer, London 270p.

[19] Clausen, P. D., Reynal, F., and Wood, D. H. (2013) Design, manufacture, and testing of small wind turbine blades, Chapter 13 In: *Wind Turbine Blade Design and Materials*, Woodhead Publishing.

[20] Gómez, U. E. Y., López, Z. J. A., Jimenez, R. A., López, G. V., and Villalon, L. J. J. (2013) Design and manufacturing of wind turbine blades of low capacity using CAD/CAM techniques and composite materials. *Energy Procedia*, v. 57, pp. 682–690.

[21] Joseph, S., Sreekala, M. S., Koshy, P., and Thomas, S. (2008) Mechanical properties and water sorption behavior of phenol–formaldehyde hybrid composites reinforced with banana fiber and glass fiber. *Journal of Applied Polymer Science*, v. 109, n. 3, pp. 1439–1446.

[22] Biswas, S., and Satapathy. A. (2010) A comparative study on erosion characteristics of red mud filled bamboo–epoxy and glass–epoxy composites. *Materials and Design*, v. 31, pp. 1752–1767.

[23] Yadav, M., Kumar, D., Butola, R., and Singari, R. M. (2020) Effect of the impact strength of glass fibre reinforced plastic composite using wet layup process. *Materials Today: Proceedings*, v. 25, pp. 919–924.

[24] Shinomiya, L. D., Vaz, J. R. P., Mesquita, A. L. A., de Oliveira, T. F., Brasil Júnior, A. C. P., and Silva, P. A. S. F. (2015) An approach for the optimum hydrodynamic design of hydrokinetic turbine blades. *Engenharia Térmica (Thermal Engineering)*, v. 14, n. 2, pp. 43–46.

[25] Tejeda, K. R., Turcotte, D. A., and Pike, S. (2017) Unsustainable wind turbine blade disposal practices in the United States. *New Solutions*, v. 26, pp. 581–598.

[26] Cherrington, R., Goodship, V., Meredith, J., *et al.* (2012) Defining the incentive for recycling composite wind turbine blades in Europe. *Energy Policy*, v. 47, pp. 13–21.

[27] Abdi, F., Baid, H., and Eftekharian, A. (2020) Advanced composite wind turbine blade design and certification based on durability and damage tolerance, Chapter 5, In: *Durability of Composite Systems*. Woodhead Publishing Series in Composites Science and Engineering, pp. 203–270.

[28] Barros, A. S. (2010) *Estudo do Desalinhamento das Fibras nas Propriedades Mecânicas de Compósitos Estruturais de Pás Eólicas*. 180p. Thesis, Space Engineering and Technology/Materials and Sensors Science and Technology, INPE, São José dos Campos, SP, Brazil.

[29] Nijssen, R. P. L., van Delft, D. R. V., Joose, P. A., and van Wingerde, A. M. (2003) Optimat blades – Optimal and reliable use of composite materials for wind turbines. *In: International Conference on Composite Materials*, n. 14, San Diego. Proceedings.

[30] Sjølund, J. H., and Lund, E. (2018) Structural gradient based sizing optimization of wind turbine blades with fixed outer geometry. *Composite Structures*, v. 203, pp. 725–739.

[31] Nijssen, R., Westphal, T., Stammes, E., Lekou, D., and Brondsted, P. (2008) Rotor structures and materials—Strength and fatigue experiments and phenomenological modelling. *In: European Wind Energy Conference*.

[32] Barnes, R. H., and Morozov, E. V. (2016) Structural optimisation of composite wind turbine blade structures with variations of internal geometry configuration. *Composite Structures*, v. 152, pp. 158–167.

[33] Berggreen, C., Branner, K., Jensen, J. F., and Schultz, J. P. (2007) Application and analysis of sandwich elements in the primary structure of large wind turbine blades. *Journal of Sandwich and Structural Materials*, v. 9, pp. 525–52.

[34] Li, G., and Jones, N. (2007) Development of rubberized syntactic foam. *Composites: Part A*, v. 38, pp. 1483–92.

[35] Lahuerta, F., Koorn, N., and Smissaert, D. (2018) Wind turbine blade trailing edge failure assessment *with sub-component test on static and fatigue load conditions*. *Composite Structures*, v. 204, pp. 755–766.

[36] Zhu, J., Zhou, Z., and Cai, X. (2020) Multi-objective aerodynamic and structural integrated optimization design of wind turbines at the system level through a coupled blade tower model. *Renewable Energy*, v. 150, pp. 523–537.

[37] Chang, R. R., Chiang, T. H., Tseng, Y. C., and Su, C. Y. (2011) Normalization process technique of composite foam-filled sandwich wind turbine blades. *Procedia Engineering*, v. 14, pp. 1988–1995.

[38] Angst, P., and Gätzi, R. (2009) One core does not fit all. *Reinforced Plastics*, v. 53, pp. 36–38.

[39] Hoevel, B., and Sagnard, A. (2011) New materials tackle weight, strength and cure time issues for wind turbine blades. *Reinforced Plastics*, v. 55, pp. 38–41.

[40] Asseff, N. S. (2009) *Design and Finite Element Analysis of an Ocean Current Turbine Blade*. MSc. Thesis, Florida Atlantic University, Florida, US.

[41] Ghosh, A. K., and Gandhi, C. (1986) Strength of metals and alloys (ICSMA 7). In: *Proceedings of the 7th International Conference on the Strength of Metals and Alloys*. Editor: McQueen, H. J., Oxford Pergamon Press, pp. 2065.

[42] Costa, D. I. G., Levy Neto, F., Morais, M. V. G., Ferreira, G. V., Oberg, M. B. A. M., and Cunha, J. T. L. (2013) Stress analysis of a river current hydrokinetic turbine blade. *In: 22nd International Congress of Mechanical Engineering*, COBEM 2013, Ribeirão Preto, SP, Brazil.

[43] Payne, G. S., Stallard, T., and Martinez, R. (2017) Design and manufacture of a bed supported tidal turbine model for blade and shaft load measurement in turbulent flow and waves. *Renewable Energy*, v. 107, pp. 312–326.

[44] Pierson, S. H. (2009) *Composite Rotor Design for a Hydrokinetic Turbine*. Thesis Projects, University of Tennessee Honors.

[45] Murthy, V. S. R., Jena, A. K., and Gupta, K. P. (2003) *Structure and Properties of Engineering Materials*. Tata McGraw-Hill Publishing Company Limited, New Delhi.

[46] Campbell, F. C. (2008) *In: Elements of Metallurgy and Engineering Alloys*. ASM International, Metal Park, OH, USA.

[47] Zakaria, K. A., Abdullah, S., and Ghazali, M. J. (2013) Comparative study of fatigue life behaviour of AA6061 and AA7075 alloys under spectrum loadings. *Materials and Design*, v. 49, pp. 48–57.

[48] Schubel, P. J. (2012) Cost modelling in polymer composite applications: Case study – Analysis of existing and automated manufacturing processes for a large wind turbine blade. *Composites: Part B*, v. 43, pp. 953–960.

[49] Brondsted, P., Lilholt, H., and Lystrup, A. (2005) Composite materials and wind power turbine blades. *Annual Review of Materials Research*, v. 35, pp. 505–538.

[50] Dutton, A. G., Bonnet, P. A., Hogg, P., and Lleong, Y. L. (2010) Novel materials and modelling for large wind turbine blades. *Journal of Power Energy*, v. 224, pp. 203–210.

[51] Shah, D. U. (2013) Developing plant fibre composites for structural applications by optimising composite parameters: A critical review. *Journal of Materials Science*, v. 48, pp. 6083–6107.

[52] Germanischer Lloyd. (2010) *Guideline for the Certification of Wind Turbines Edition*. Germanischer Lloyd Industrial Services GmbH Renewables Certification, Hamburg.

[53] Nijssen, R. P., Van Wingerde, A. M., and Van Delft, D. R. V. (2007) Wind turbine rotor blade materials: Estimating service lives. *SAMPE Journal*, v. 43, n. 2, pp. 7–15.

[54] Peterson, P., and Clausen, P. D. (2004) Timber of high efficiency small wind turbine blades. *Wind Engineering*, v. 28, n. 1, pp. 87–96.

[55] Sinha, R., Acharya, P., Freere, P., Sharma, R., Ghimire, P., and Mishnaevsky Jr, L. (2010) Selection of Nepalese timber for small wind turbine blade construction. *Wind Engineering*, v. 34, n. 3, pp. 263–76.

[56] Astle, C., Burge, I., Chen, M., *et al.* (2013) Timber for small wind turbine blades. *Energy for Sustainable Development*, v. 17, n. 6, pp. 671–676.

[57] Mishnaevsky Jr, L., Freere, P., Sinha, R., Acharya, P., Shrestha, R., and Manandhar, P. (2011) Small wind turbines with timber blades for developing countries: Materials choice, Development, Install and Experiences. *Renewable Energy*, v. 36, n. 8, pp. 2128–38.

[58] Pourrajabiana, A., Maziar Dehghanb, M., Javedc, A., and Wood, D. H. (2019) Choosing an appropriate timber for a small wind turbine blade: A comparative study. *Renewable and Sustainable Energy Reviews*, v. 100, pp. 1–8.

[59] Casey, T. (2019) *Absolute Beast of a Wooden Wind Turbine Blade Rolls Off the Assembly Line*. Retrieved from Website: https://cleantechnica.com/2019/04/19/absolute-beast-of-a-wooden-wind-turbine-blade-rolls-off-the-assembly-line/.

[60] McKellep, D., and Coleman, V. (2010) *Research on Wooden Tidal Energy Turbine Blades*. Ocean Energy Council, Inc. Retrieved from Website http://oceanenergycouncil.com/.

[61] Liu, W. Y., Tang, B. P., Han, J. G., Lu, X. N., Hu, N. N., and He, Z. Z. (2015) The structure healthy condition monitoring and fault diagnosis methods in wind turbines: A review. *Renewable and Sustainable Energy Reviews*, v. 44, pp. 466–472.

[62] Anyi, M., and Kirke, B. (2011) Hydrokinetic turbine blades: Design and local construction techniques for remote communities. *Energy for Sustainable Development*, v. 15, n. 1, pp. 223–230.

[63] Muñoz, A. H., Chiang, L. E., and De la Jara, E. A. (2014) A design tool and fabrication guidelines for small low cost horizontal axis hydrokinetic turbines. *Energy for Sustainable Development*, v. 22, pp. 21–33.

[64] Song, Q. (2012) *Design, Fabrication, and Testing of a New Small Wind Turbine Blade*. 134f. Thesis, University of Guelph, Guelph, ON, Canada.

[65] Bond, I. P., and Ansell, M. P. (1998) Fatigue properties of jointed wood composites. Part I: Statistical analysis, fatigue master curves and constant life diagrams. *Journal of Material Science*, v. 33, pp. 2751–2762.

[66] Lam, F., and Varoglu, E. (1991) Variation of tensile strength along the length of lumber. Part 1: Experimental. *Wood Science Technology*, v. 25, pp. 351–356.

[67] Kohler, J., Brandner, R., Thiel A. B., and Schickhofer, G. (2013) Probabilistic characterisation of the length effect for parallel to the grain tensile strength of central European spruce. *Engineering Structures*, v. 56, pp. 691–697.

[68] Kepler, J. A. (2011) Simple stiffness tailoring of balsa sandwich core material. *Composite Science and Technology*, v. 71, pp. 46–51.

[69] Dimitrov, N., and Berggreen. C. (2015) Probabilistic fatigue life of balsa cored sandwich composites subjected to transverse shear. *Journal of Sandwich Structures and Materials*, v. 17, pp. 562–77.

[70] Zaharia, S. M., Morariu, C. O., Nedelcu, A., and Pop, M. A. (2017) Experimental study of static and fatigue behavior of CFRP-balsa sandwiches under three-point flexural loading. *Bioresources*, v. 12, pp. 2673–89.

[71] Ozdemir, O., Oztoprak, N., and Kandas, H. (2018) Single and repeated impact behaviors of bio-sandwich structures consisting of thermoplastic face sheets and different balsa core thicknesses. *Composites Part B*, v. 149, pp. 49–57.

[72] Chen, X. (2019) Experimental observation of fatigue degradation in a composite wind turbine blade. *Composite Structures*, v. 212, pp. 547–551.

[73] Christou, P. (2007) Advanced materials for turbine blade manufacture. *Reinforced Plastics*, v. 51, pp. 22–24.

[74] Clausen, P. D., and Wood, D. H. (2000) Recent advances in small wind turbine technology. *Wind Engineering*, v. 24, n. 3, pp. 189–201.

[75] Habali, S. M., and Saleh, I. A. (2000) Local design, testing and manufacturing of small mixed airfoil wind turbine blades of glass fiber reinforced plastics Part I: Design of the blade and root. *Energy Conversion & Management*, v. 41, pp. 249–280.

[76] Clifton-Smith, M. (2010) Aerodynamic noise reduction for small wind turbine rotors. *Wind Engineering*, v. 34, pp. 403–420.

[77] Huber, T., Biedermann, U., and Muessig, J. (2010) Enhancing the fibre matrix adhesion of natural fibre reinforced polypropylene by electron radiation analyzed with the single fibre fragmentation test. *Composite Interfaces*, v. 17, n. 4, pp. 371–81.

[78] Samuel, O. D., Agbo, S., and Adekanye, T. A. (2012) Assessing mechanical properties of natural fibre reinforced composites for engineering applications. *Journal of Minerals and Materials Characterization and Engineering*, v. 11, pp. 780–784.

[79] BS-EN 61400-23:2002. In wind turbine generator systems – Part 23: Full-scale structural testing of rotor blades. British Standards Institution, London.

[80] BS EN 61400-2:2006. In wind turbines – Part 2: Design requirements for small wind turbines. British Standards Institution, London.

[81] Rodrigues, J. S. (2014) *Estudo da técnica de infusão de resina aplicada à fabricação de compósitos de matriz poliéster reforçados por fibras naturais da Amazônia*. Thesis, Engineering of Natural Resources of the Amazon, University of Pará, Belém, PA, Brazil.

[82] Margem, J. I. (2013) *Estudo das características estruturais e propriedades de compósitos poliméricos reforçados com fibras de malva*. 139p. Thesis, Estadual University of Norte Fluminense, Campos dos Goytacazes, RJ, Brazil.

[83] D'almeida, J. R. M., Aquino, R. C. M. P., and Monteiro, S. N. (2006) Tensile mechanical properties morphological aspects and chemical characterization of piassava (*Attalea funifera*) fibers. *Composites: Part A*, v. 37, pp. 1473–1479.

[84] Tomczak, F., Satyanarayana, G., and Sydenstricker, T. H. D. (2007) Studies on lignocellulosic fibers of Brazil: Part III – Morphology and properties of Brazilian curauá fibers. *Composites: Part A*, v. 38, pp. 2227–2236.

[85] Santos, N. S., Dias, C. G. B. T., Sanches, E. M. S., Zavaglia, C. A., and Albuquerque, E. L. (2008) Caracterização microestrutural e mecânica da fibra de Miriti para utilização como reforço em compósitos poliméricos. *In: 18º CBECiMat*, pp. 3565–3573, Porto de Galinhas, PE, Brazil.

[86] Li, Y., Mai, Y., and Ye, L. (2000) Sisal fibre and its composites: A review of recent developments. *Compos Science and Technology*, v. 60, n. 11.

[87] Gehlen, L. R. (2014) *Efeito da utilização de fibras lignocelulósicas (açaí e curauá) em compósitos com matriz de resina poliéster insaturado*. MSc. Thesis, Materials Science and Engineering, Federal University of Paraná, PR, Brazil.

[88] Duarte, C. (2011) *Fabricação e caracterização de material compósito de resina poliéster reforçada por tecido de fibra de tururi extraído da palmeira de ubuçu (Manicaria saccifera Gaert)*. 76p. MSc. Thesis, Materials and Manufacturing Processes, Federal University of Pará, Belém, PA, Brazil.

[89] Santos, L. F. G. (2016) *Estudo sistemático do Miriti (Mauritia flexuosa) para o desenvolvimento de ECO-VANT*. 76p. MSc. Thesis, Federal University of Ouro Preto, Minas Gerais, MG, Brazil.

[90] Vogt, H. H. (2010) *Análise estrutural de pás de gerador eólico de pequeno porte feitas de fibra vegetal brasileira*. 133p. MSc. Thesis, Applied Physical Sciences, Federal University of Ceará, CE, Brazil.

[91] Gomes, I. S. (2019) *Metodologia de fabricação de pás de turbinas hidrocinéticas e eólicas empregando novos materiais*. 100p. MSc. Thesis, Mechanical Engineering, Federal University of Pará, Belém, PA, Brazil.

[92] Thomas, S., Paul, S. A., Pothan, L. A., and Deepa, B. (2011) Natural fibres: Structure, properties and applications. *In: Cellulose fibers – Bio and Nano Polymer Composites, Green Chemistry and Technology*. Edited by Kalia, S., Kaith, B. S., and Kaur, I., Springer-Verlag.

[93] John, M. J., and Anandjiwal, R. D. (2008) Recent developments in chemical modification and characterization of natural fiber-reinforced composites. *Polymer Composites*, pp. 187–207.

[94] Zimmermann, T., Pohler, E., and Geiger, T. (2004) Cellulose fibrils for polymer reinforcement. *Advanced Engineering Materials*, v. 6, pp. 754–761.

[95] Dittenber, D. B., and Gangarao, H. V. S. (2012) Critical review of recent publications on use of natural composites in infrastructure. *Composites: Part A*, v. 43, n. 8, pp. 1419–1429.

[96] Defoirdt, N., Biswas, S., Vriese, L., *et al.* (2010) Assessment of the tensile properties of coir, bamboo and jute fibre. *Composites: Part A*, v. 41, pp. 588–595.

[97] Sydenstricker, T. H. D., Mochnaz, S., and Amico, S. C. (2003) Pull-out and other evaluations in sisal-reinforced polyester biocomposites. *Polymer Testing*, v. 22, pp. 375–380.

[98] Wang, W., Cai, Z., and Yu, J. (2008) Study on the chemical modification process of jute fiber. *Journal of Engineered Fibers and Fabrics*, v. 3, pp. 1–11.

[99] Ochi, S. (2008) Mechanical properties of kenaf fibers and kenaf/PLA composites. *Mechanics of Materials*, v. 40, pp. 446–452.

[100] Costa, D. S. (2016) *Estudo da influência de resíduos gerados pela indústria de mineração nas propriedades de compósitos de matriz poliéster reforçados com fibras naturais*. Thesis, Engineering of Natural Resources of the Amazon, University of Pará, Belém, PA, Brazil.

[101] Nascimento, D. C. O., Ferreira, A. S., Monteiro, S. N., Aquino, R. C. M. O., and Kestur, S. G. (2012) Studies on the characterization of piassava fibers and their epoxy composites. *Composites: Part A*, v. 43, pp. 353–362.

[102] Fidelis, M. E. A., Pereira, T. V. C., Gomes, O. F. M., Silva, F. D. S., and Toledo Filho, R. D. (2013) The effect of fiber morphology on the tensile strength of natural fibers. *Journal of Materials Research and Technology*, v. 2, n. 2, pp. 149–157.

[103] Demosthenes, L. C. C., Nascimento, L. F. C., Monteiro, S. N., *et al.* (2020) Thermal and structural characterization of buriti fibers and their relevance in fabric reinforced composites. *Journal of Materials Research and Technology*, v. 9, pp. 115–123.

[104] Costa, D. S., El Banna, W. R., Almeida, D. F., Santos, E. J. S., Lopes, C. E. P., and Fujiyama, R. T. (2013) Características de compósitos de fibras de palha da costa (Raphia vinifera) com resina poliéster. *In: 68° ABM International Annual Congress*, Belo Horizonte, MG, Brazil.

[105] Costa, D. S., El Banna, W. R., Lima, L. S., Almeida, D. F., Santos, E. J. S., Lopes, C. E. P., and Fujiyama, R. T. (2013) Compósito de fibras de açaí e resina poliéster. *In: 68° ABM International Annual Congress*, Belo Horizonte, MG, Brazil.

[106] Martins, M. A., and Pessoa, J. D. C. (2008) Thermal and mechanical properties of the acai fiber/natural rubber composites. *Journal of Materials Science*, v. 43, pp. 6531–6538.

[107] Singh, B. K., Komal, U. K., Singh, Y., Banwait, S. S., and Singh, I. (2021) Development of banana fiber reinforced composites from plastic waste. *Materials Today: Proceedings*, Elsevier, v. 44, pp. 2194–2198.

[108] Merlini, C., Soldi, V., and Barra, G. M. O. (2011) Influence of fiber surface treatment and length on physico-chemical properties of short random banana fiber-reinforced castor oil polyurethane composites. *Polymer Testing*, v. 30, pp. 833–840.

[109] Porras, A., Maranon, A., and Ashcroft, I. A. (2015) Characterization of a novel natural cellulose fabric from *Manicaria saccifera* palm as possible reinforcement of composite materials. *Composites Part B: Engineering*, v. 74, n. 1, pp. 66–73.

[110] Pires, E. N. (2009) *Efeito do tratamento de superfície em fibras de juta no comportamento mecânico de compósitos de matriz epóxi*. MSc. Thesis, Federal University of Santa Catarina, Florianópolis, SC, Brazil.

[111] Ferreira, A. S., and Homma, A. K. O. (2019) A colonização britânica e o declínio da produção de juta indiana: singularidades e possibilidades para a Amazônia. *Fronteiras (Campo Grande)*, v. 21, n. 37, pp. 39–57.

[112] Fidelis, M. E. A. (2014) *Desenvolvimento e caracterização mecânica de compósitos cimentícios têxteis reforçados com fibras de juta*. Thesis, Federal University of Rio de Janeiro, RJ, Brazil.

[113] Savastano Jr., H. (2000) *Materiais à base de cimentos reforçados com fibras vegetais: Reciclagem de resíduos para a construção de baixo custo*. 144p. Thesis, University of São Paulo, SP, Brazil.

[114] Conab. (2018) Companhia Nacional de Abastecimento. *Análise mensal da produção de fibras de juta, malva e sisal*. Retrieved from Website: https://conab.gov.br/info-agro/analises-do-mercado-agropecuario-e-extrativista/analises-do-mercado/historico-mensal-de-fibras/item/download/22582_c625a2e446783a36627571 1d8d78f08d.

[115] Araújo. C. R., and Mothé, C. G. (2003) Uso de programa computacional aliado às técnicas de análise térmica para determinação de parâmetros cinéticos de compósitos de PU/fibra de curauá. *Revista Analytica*, n. 4.

[116] EcoDebate. (2008) *Projeto de cultivo do curauá (Ananas erectifolius) transforma realidade da agricultura familiar na Amazônia*. Retrieved from Website: https://ecodebate.com.br/2008/08/12/projeto-de-cultivo-do-curaua-ananas-erectifolius-transforma-realidade-da-agricultura-familiar-na-amazonia/.

[117] Mourão, L. (2004) Açaizeiro: Açaí e Palmito no estuário amazônico. *In: Açaí (Euterpe oleracea Mart.): possibilidades e limites para o desenvolvimento sustentável no estuário amazônico*. Editor: Jardim, M. A. G, pp. 181–204.

[118] Jardim, M. (2004) Pesquisas com palmeira açaí (*Euterpe oleracea*), no Museu Paraense Emilio Goeldi. *In: Açaí (Euterpe oleracea Mart.): possibilidades e limites para o desenvolvimento sustentável no estuário amazônico*. Editors: Jardim, M. A. G., Mourão, L., and Grossmann, M. Museu Paraense Emilio Goeldi, Coleção Adolpho Ducke, Belém, PA, Brazil. 274, pp. 181–204.

[119] Mesquita, A. L. (2013) *Estudos de processos de extração e caracterização de fibras do fruto do açaí (Euterpe oleracea Mart.) da amazônia para produção de ecopainel de partículas homogêneas de média densidade*. Thesis, Engineering of Natural Resources of the Amazon, University of Pará, Belém, PA, Brazil.

[120] Balick, M. (1986) As palmeiras economicamente importantes do Maranhão. In: PRANCE, *Ghillean (Org.). Manual de Botânica Econômica do Maranhã*. Federal University of Maranhão, São Luís, MA Brazil, pp. 199–226.

[121] Santos, R. S., and Ferreira, M. C. (2011) Artefatos de miriti (*Mauritia flexuosa* L. f.) em Abaetetuba, Pará: da produção à comercialização. *Boletim do Museu Paraense Emílio Goeldi: Ciências Humanas*, v. 6, n. 3, pp. 559–571.

[122] Dias, E. C., Da Silva, G. P. C., Ribeiro, R. P., and Fernandes, L. L. A. (2017) Utilização sustentável da palmeira de miriti como forma de geração de renda para artesãos no município de Abaetetuba Pará. *Revista SEAF*, v. 1, n. 1.

[123] Chukwudi, A. D. (2012) *Effect of Two Pretreated Raffia Palm (Raphia) Fibres on Polyester Composite*. 92p. MSc. Thesis, Science in Pure and

Industrial (Chemistry), Faculty of Physical Sciences, Nnamdi Azikiwe University, Awka.

[124] Correa, M. P. (1978) *Dicionário das plantas úteis do Brasil e das exóticas cultivadas*. Rio de Janeiro, Imprensa Nacional, v. 6.

[125] Gomes, I. S., and Fujiyama, R. T. (2019) Utilização de fibras naturais provenientes folha primária e secundária da palmeira do ubuçú em compósitos de matriz poliéster. *In: Meio Ambiente, Sustentabilidade e Agroecologia 7*, Ed. Atena, v. 7, pp. 299–315.

[126] Pereira, L. C. O. (2018) *A influência da metodologia de fabricação na resistência à tração de compósitos reforçados com fibras de piaçava*. 54p. Undergraduate Thesis, Mechanical Engineering, Federal University of Pará, Belém, PA, Brazil.

[127] Vinha, S. G., and Silva, L. A. M. (1998) *A piaçaveira da Bahia*. Editus, Ilhéus.

[128] IBGE. (2014) Instituto Brasileiro De Geografia E Estatística. *Produção da extração vegetal e da silvicultura*. ISSN 0103-8435 Produção Extrativista Vegetal e Silvicultura, v. 29, pp. 1–56.

[129] Silva, O. R. R. F., and Beltrão, N. E. M. (1999) *O agronegócio do sisal no Brasil*. Embrapa-SPI, Brasília.

[130] Levy Neto, F., and Pardini, L. C. (2006) *Compósitos estruturais: ciência e tecnologia*. 1ª Ed. Editora Edgard Blücher, São Paulo, Brazil.

[131] Silva, R. V. (2003) *Compósito de Resina Poliuretano Derivada de Óleo de Mamona e Fibras Vegetais*. 157p. Thesis, Materials Science and Engineering, University of São Paulo, São Paulo, SP, Brazil.

[132] Agrianual. (2012) *Anuário estatístico da agricultura brasileira*. Instituto FNP, São Paulo, SP, Brazil, pp. 177–188.

[133] Cordioli, S. (2008) *Oficina de Avaliação e Planejamento do Projeto Fibra da bananeira 2008/2009*. Santa Catarina, SC, Brazil.

[134] Souza, J. P. F. M. (2009) *Modelo analítico de resistência ao impacto em material composto de fibra de bananeira em matriz termoplástica*. Undergraduate Thesis, Federal University of Santa Catarina, Florianópolis, SC, Brazil.

[135] Summerscales, J., Dissanayake, N., Virk, A. S., and Hall, W. (2010) A review of bast fibres and their composites. Part 1 – Fibres as reinforcements. *Composites Part A: Applied Science and Manufacturing*, v. 41, n. 11, pp. 1329–1335.

[136] Faruk, O., Bledzki, A. K., Fink, H. P., and Sain, M. (2012) Biocomposites reinforced with natural fibers: 2000-2010. *Progress in Polymer Science*, v. 37, pp. 1552–1596.

[137] Roy, A., Chakraborty, S., Kundu, S. P., Basak, R. K., Majumder, S. B., and Adhikari, B. (2012) Improvement in mechanical properties of jute fibres through mild alkali treatment as demonstrated by utilization of the Weibull distribution model. *Bioresource Technology*, v. 107, pp. 222–228.

[138] Agopyan, V., and Savastano Jr., H. (1997) Uso de materiais alternativos à base e fibras vegetais na construção civil: experiência brasileira. *In:*

Seminário Iberoamericano de materiais Fibrorreforçados, 1., y Reunión Proyecto PIP VIII.5 Cyted, 3., Cali, Memoriais. Cali, Universidad Del Valle, pp. 23–40 (Artículo 03).

[139] Guimarães, B. M. G. (2014) *Estudo das características físico-químicas de fibras têxteis vegetais de espécies de Malvaceae*. 169p. MSc. Thesis, University of São Paulo, São Paulo, SP, Brazil.

[140] Monteiro, S. N., Perissé, F. D. L., Costa, L. L., Bevitori, A. B., Silva, I. A. S., and Barbosa, A. P. (2011) Natural Lignocellulosic Fibers as engineering materials – An overview. *The minerals, Metals & Materials Society and ASM International*, v. 42 A.

[141] Pereira, S. J., Muñiz, G. I. B., Kaminski, M., Klock, U., Nisgoski, S., and Fabrowski, F. J. (2003) Celulose de buriti (*Mauritia vinifera Martius*) Buriti (Mauritia vinifera Martius) pulp. *Scientia Forestalis*, n. 63, pp. 202–213.

[142] Rodrigues, J. S. (2008) *Comportamento mecânico de material compósito de matriz poliéster reforçado por sistema híbrido fibras naturais e resíduos da indústria Madeireira*. 91p. MSc. Thesis, Materials and Manufacturing Processes, Federal University of Pará, Belém, PA, Brazil.

[143] Costa, D. S., Peres, M. N. P. B., Barreira, R. M., *et al.* (2012) Compósitos reforçados com fibras de sisal e malva: análise da tensão x comprimento de fibras. *In: 20° Congresso Brasileiro de Engenharia e Ciência dos Materiais*, 2012c, Joinville, SC, Brazil.

[144] Siqueira, J. E. L. (2006) *Utilização de fibras naturais e sintéticas em argamassas de revestimento de alvenaria: estudo comparativo de desempenho*. MSc. Thesis, Federal University of Santa Catarina, Florianópolis, SC, Brazil.

[145] Azevedo, A. R. G., Marvila, M. T., Tayeh, B. A., Cecchin, D., Pereira, A. C., and Monteiro, S. N. (2021) Technological performance of açaí natural fibre reinforced cement-based mortars. *Journal of Building Engineering*, v. 33.

[146] Satyanarayana, K. G., Sukumaran, K., Mukherjee, P. S., Pavithran, C., and Pillai, S. G. K. (1990) Natural fiber-polymer composites. *Cement and Concrete Composites*, v. 12, n. 2, pp. 117– 136.

[147] Karthick, R., Adithya, K., Hariharaprasath, C., and Abhishek, V. (2018) Evaluation of mechanical behavior of banana fibre reinforced hybrid epoxy composites. *Materials Today: Proceedings*, v. 5, pp. 12814–12820.

[148] Barbosa, A. P. (2011) *Características estruturais e propriedades de compósitos poliméricos reforçados com fibras de buriti*. Thesis, Estadual University of Norte Fluminense, Campos dos Goytacazes, RJ, Brazil.

[149] Sousa, S. A. V., Silva, M C., Silva Júnior, O. G., Artur Caron Mottin, A. C., Oréfice, R. L., and Ayres, E. (2017) Design and characterization of biocomposites from poly(lactic acid) (PLA) and buriti petiole (*Mauritia flexuosa*). *Journal of Renewable Materials*, v. 5.

[150] Monteiro, S. N., Rodriguez, R. J. S., da Costa, L. L., Portela, T. G. R., and Santos, N. S. S. (2010) Thermal behavior of buriti biofoam. *Revista Matéria*, v. 15, pp. 104–109.

[151] Borrega, M., Ahvenainen, P., Serimaa, R., and Gibson, L. (2015) Composition and structure of balsa (*Ochroma pyramidale*) wood. *Wood Science and Technology*, v. 49, pp. 403–420.

[152] Borrega, M., and Gibson, L. J. (2015) Mechanics of balsa (*Ochroma pyramidale*) wood. *Mechanics of Materials*, v. 84, pp. 75–90.

[153] Bledzki, A. K., and Gassan, J. (1999) Composites reinforced with cellulose based fibres. *Progress in Polymer Science*, v. 4, pp. 221–274.

[154] Klason, C., Kubát, J., and Strömvall, H. E. (1984) The efficiency of cellulosic fillers in common thermoplastics. Part I. Filling without processing aids or coupling agents. *International Journal of Polymeric Materials*, v. 10, pp. 159–187.

[155] Collins, A. W., Cutler, D., Atherton, M., and Collins, M. (2007) The palm – A model for success? Chapter 10. *In: Design and Information in Biology*. Edited by Bryant, J. A., Atherton, M. A., and Collins, M. W., WIT Press.

[156] Zaman, H. U., Khan, A., Khan, R. A., *et al.* (2010) Preparation and characterization of jute fabrics reinforced urethane based thermoset composites: Effect of UV radiation. *Fibers and Polymers*, v. 11, n. 2, pp. 258.

[157] Misra, M., Pandey, J. K., and Mohanty, A. K. (2015) *Biocomposites: Design and Mechanical Performance*, 1 Ed., Elsevier Science, Oxford.

[158] Yusoff, R. B., Takagi, H., and Nakagaito, A. N. (2016) Tensile and flexural properties of polylactic acid-based hybrid green composites reinforced by kenaf, bamboo and coir fibers. *Industrial Crops and Products*, v. 94, pp. 562–573.

[159] Abilash, N., and Sivapragash, M. (2016) Optimizing the delamination failure in bamboo fiber reinforced polyester composite. *Journal King Saud University Science*, v. 28, n. 1, pp. 92–102.

[160] Ramachandran, M., Sahas, B., and Pramod, R. (2016) Experimental study of bamboo using banana and linen fiber reinforced polymeric composites. *Perspective on Science*, v. 8, pp. 313–316.

[161] Khalil, H. P. S. A., Alwani, M. S., Islam, M. N., *et al.* (2015) The use of bamboo fibers as reinforcements in composites. *In: Biofiber Reinforcements in Composite Materials*, Woodhead Publishing (Elsevier Store), USA, pp. 488–524. Chapter: 16.

[162] Yang, T., Wu, T., Hung, K., Chen, Y., and Wu, J. (2015) Mechanical properties and extended creep behavior of bamboo fiber reinforced recycled poly (lactic acid) composites using the time–temperature superposition principle. *Construction and Building Materials*, v. 93, pp. 558–563.

[163] Zakikhani, P., Zahari, R., Sultan, M. T. H., and Majid, D. L. (2014) Extraction and preparation of bamboo fiber-reinforced composites. *Materials and Design*, v. 63, pp. 820–828.

[164] Shah, D, Schubel, P. J., Licence, P., and Clifford, M. J. (2012) Hydroxyethyl cellulose surface treatment of natural fibres: The new 'twist' in yarn preparation and optimization for composites applicability. *Journal of Materials Science*, v. 47, pp. 2700–2711.

[165] Prasad, A. V. R., and Rao, K. M. (2011) Mechanical properties of natural fibre reinforced polyester composites: Jowar, sisal and bamboo. *Materials and Design*, n. 32, pp. 4658–4663.

[166] Stocchi, A., Lauke, B., Vázquez, A., and Bernal, C. (2007) A novel fiber treatment applied to woven jute fabric/vinylester laminates. *Composites: Part A*, v. 38, pp. 1337–1343.

[167] Almeida, L. M. (2018) *Compósitos híbridos com fibras naturais e sintéticas contínuas e alinhadas*. MSc. Thesis, Mechanical Sciences, Federal University of Pará, Belém, PA, Brazil.

[168] Ribeiro, M. M. (2015) *Comportamento em tração/flexão e aspecto fractográfico de matriz poliéster com pó de aerosil e de madeira com a inclusão de fibras curtas de sisal e juta*. 198p. MSc. Thesis, Mechanical Sciences, Federal University of Pará, Belém, PA, Brazil.

[169] Costa, D. S., El Banna, W. R., Santos, E. J. S., Lopes, C. E. P., Fernandes, E. A., and Fujiyama, R. T. (2013) Compósitos de fibras de juta e matriz poliéster. *In: 68° ABM International Annual Congress*, Belo Horizonte, MG, Brazil.

[170] Costa, D. S., El Banna, W. R., Lima, L. S., Santos, E. J. S., Lopes, C. E. P., and Fujiyama, R. T. (2013) Materiais compósitos reforçados com fibras de malva. *In: 68° ABM International Annual Congress*, Belo Horizonte, MG, Brazil.

[171] Moraes, Y. M., Ribeiro, C. G. D., Ferreira, C. L., *et al.* (2018) Mechanical behavior of mallow fabric reinforced polyester matrix composites. *Journal of Materials Research and Technology*, v. 7, n. 4, pp. 515–519.

[172] Nascimento, L. F. C., Monteiro, S. N., Louro, L. H. L., *et al.* (2018) Charpy impact test of epoxy composites reinforced with untreated and mercerized mallow fibers. *Journal of Materials Research and Technology*, v. 7, n. 4, pp. 520–527.

[173] Nascimento, L. F. C. (2017) *Caracterização do compósito epóxi-fibra de malva para emprego em blindagem balística multicamada*. Thesis, Military Engineering Institute, Rio de Janeiro, RJ, Brazil.

[174] Costa, D. S., El Banna, W. R., Santos, E. J. S., Lopes, C. E. P., Lira Junior, P. F., and Fujiyama, R. T. (2013) Compósito de poliéster com fibras de piaçava *(Attalea funifera)*. *In: 68° ABM International Annual Congress*, Belo Horizonte, MG, Brazil.

[175] Oliveira Filho, E. G. (2017) *Avaliação do comportamento mecânico de compósitos poliméricos reforçados com fibras naturais nas condições in natura e tratadas quimicamente*. MSc. Thesis, Mechanical Sciences, Federal University of Pará, Belém, PA, Brazil.

[176] Monteiro, S. N., Aquino, R. C. M. P, Lopes, F. P, D., and D'Almeida, J. R. M. (2006) Tenacidade ao Entalhe por Impacto Charpy de Compósitos de Poliéster Reforçados com Fibras de Piaçava. *Revista Matéria*, v. 11, n. 3, pp. 204–210.

[177] Oliveira Filho, G. C., Mota, R. C. S., Conceição, A. C. R., Leão, M. A., and Araújo Filho, O. O. (2019) Effects of hybridization on the mechanical

properties of composites reinforced by piassava fibers tissue. *Composites: Part B*, v. 162, pp. 73–7975.

[178] Monteiro, S. N., Aquino, R. C. M. P., Lopes, F. P. D., Carvalho, E. A., and d'Almeida, J. R. M. (2006) Comportamento Mecânico e Características Estruturais de Compósitos Poliméricos Reforçados com Fibras Contínuas e Alinhadas de Curauá. *Revista Matéria*, v. 11, n. 3, pp. 197–203.

[179] Rocha, I. G. (2015) *Influência do retardante de chamas nas propriedades mecânicas e flamabilidade em compósitos poliméricos de epóxi/fibra de curauá*. MSc. Thesis, Federal University of Rio Grande do Norte, RN, Brazil.

[180] Rodrigues, J. S., Souza, J. A., and Fujiyama, R. T. (2015) Compósitos poliméricos reforçados com fibras naturais da Amazônia fabricados por infusão. *Revista Matéria*, v. 20, n. 4, pp. 946–960.

[181] Santos, N. S. S. (2010) *Análise experimental e teórica do comportamento mecânico sob carregamentos quase-estáticos de compósitos reforçados com fibras vegetais*. 123p. Thesis, Estadual University of Campinas, SP, Brazil.

[182] Demosthenes, L. C. C. (2019) *Comportamento balístico de blindagem multicamada com cerâmica frontal e compósito de epóxi reforçado com tecido de buriti*. MSc. Thesis, Military Engineering Institute, RJ, Brazil.

[183] Honorato, P. A., Pereira, A. I., da Silva, I. S., *et al.* (2018) Caracterização mecânica em tração de lâminas produzidas a partir de pré-pregs de fibra de palha da costa e matriz poliéster. *In: 23° Congresso Brasileiro de Engenharia e Ciência dos Materiais*, Foz do Iguaçú, PR, Brazil.

[184] Kuruvilla, J., Medeiros, E. S., and Carvalho, L. H. (1999) Compósitos de Matriz Poliéster Reforçados por Fibras Curtas de Sisal. *Polímeros: Ciência e Tecnologia*, Out/Dez.

[185] Carvalho, M. J. C., and Fujiyama, R. T. (2012) Compósito de fibra de sisal alinhado 0° unidirecional: comparação da tensão máxima experimental e teórica. *In: 67° ABM International Annual Congress*, Rio de Janeiro, RJ, Brazil.

[186] Luz, L. M. C., Baganha, M. L. P. S., Rodrigues, M. R., Santos, N. S. S., and Silva, M. R. (2014) Comportamento mecânico sob flexão e impacto de compósitos de matriz polimérica e fibra de açaí. *In: 21° CBECIMAT – Congresso Brasileiro de Engenharia e Ciência dos Materiais*, Cuiabá, MT, Brazil.

[187] Costa, A. S. G., Rocha, K. B., Brito, S. C. S., and Andrade, S. M. C. (2014) Ensaios de tração e absorção de água em compósitos de resina poliéster e fibra de açaí. *In: 21° CBECIMAT – Congresso Brasileiro de Engenharia e Ciência dos Materiais*, Cuiabá, MT, Brazil.

[188] El Banna, W. R., Pereira, L. C. O., and Fujiyama, R. T. (2013) Aplicação de fibras da bananeira São-Tomé (*Musa sapientum*, Musaceae) em compósitos poliméricos. *In: 68° ABM International Annual Congress, Belo Horizonte*, MG, Brazil.

[189] Reddy, M. V. K., Lakshmi, R. V. S., Pragathi, Y., Devaraj, P., and Naresh, N. (2020) Evaluation of mechanical properties: Influence of glass fiber on

banana fiber reinforced with epoxy composite. *Materials Today: Proceedings*, v. 37, pp. 1305–1312.

[190] Murugan, T., and Kumar, B. S. (2020) Studies on mechanical and dynamic mechanical properties of banana fibre nonwoven composite. *Materials Today: Proceedings*, v. 39.

[191] Portela, T. G. R., Costa, L. L., Santos, N. S. S., Lopes, F. P. D., and Monteiro, S. N. (2010) Tensile behavior of lignocellulosic fiber reinforced polymer composites: Part II buriti petiole/polyester. *Revista Matéria*, v. 15, n. 2, pp. 195–201.

[192] Barbosa, A. P., Oliveira, M. P., Altoé, G. R., Margem, F. M., and Monteiro, S. N. (2014) Charpy impact test of epoxy matrix composites reinforced with buriti fibers. *Materials Science Forum*, v. 775–776, pp. 296–301.

[193] Seyam, A. F. M., Monteiro, A. S., Midani, M., and Ramos, J. B. (2017) Effect of structural parameters on the tensile properties of multilayer 3D composites from Tururi palm tree (*Manicaria saccifera Gaertn*) fibrous material. *Composites: Part B*, v. 111, pp. 17–2618.

[194] International Electrotechnical Commission (IEC). (2005) IEC 61400-1. Wind turbines – Part 1: Design requirements. International Standard.

[195] International Electrotechnical Commission (IEC). (2001) IEC 61400-23. Wind turbine generator systems – Part 23: Full-scale structural testing of rotor blades. International Standard.

[196] Vassilopoulos, A. P. (2015) *Fatigue and Fracture of Adhesively-Bonded Composite Joints. Behaviour, Simulation and Modelling*. Woodhead Publishing Series in Composites Science and Engineering, n. 52.

[197] da Silva, L. F. M., Öchsner, A., and Adams, R. D. (2011) *Handbook of Adhesion Technology*. Springer, 552p.

[198] Chiang, L. E. (2016) *Turbem 2.0* [computer software]. Santiago, Chile. Available from www.mecatronix.cl.

[199] Hernández, F., Chiang, L. E., and Corbalán, P. (2017) A general architecture for electric power management of small scale NCRE converters: Design methodology and validation. *Energy for Sustainable Development*, v. 41, pp. 128–138.

[200] Zidane, I. F., Saqr, K. M., Swadener, G., Ma, X., and Shehadeh, M. F. (2016) On the role of surface roughness in the aerodynamic performance and energy conversion of horizontal wind turbine blades: A review. *International Journal of Energy Research*, v. 40, pp. 2054–2077.

[201] Sessarego, M., and Wood, D. (2015) Multi-dimensional optimization of small wind turbine blades. *Renewables: Wind, Water, and Solar*, v. 2, n. 9.

[202] Jiangyan, Y., Qingmin, L., Zixin, G., Li, Z., and Weijia, Z. (2016) Damage mechanism of PVC and balsa wood used in wind turbine blade under thermal effect of lightning strikes. *In: 33rd International Conference on Lightning Protection*. Estoril, Portugal.

[203] Mohammadi, M. S., and Nairn, J. A. (2017) Balsa sandwich composite fracture study: Comparison of laminated to solid balsa core materials and debonding from thick balsa core materials. *Composites Part B*, v. 122, pp. 165–172.

[204] Malek, S., and Gibson, L. J. (2017) Multi-scale modelling of elastic properties of balsa. *International Journal of Solids and Structures*, v. 113–114, pp. 118–131.

[205] Antwi, M. O., Castro, J., Vassilopoulos, A. P., and Keller, T. (2014) Fracture in complex balsa cores of fiber-reinforced polymer sandwich structures. *Construction and Building Materials*, v. 71, pp. 194–201.

[206] Shi, H., Liu, W., and Fang, H. (2018) Damage characteristics analysis of GFRP-Balsa sandwich beams under Fourpoint fatigue bending. *Composites Part A*, v. 109, pp. 564–577.

[207] Yang, C., Li, X., Fang, H., Shi, H., Liu, W., and Zhu, D. (2020) An experimental investigation of static and fatigue behaviour of GFRP sandwich beams with a web reinforced wood core. *Construction and Building Materials*, v. 253, 119194.

[208] Shi, H., Fang, H., Liu, W., and Xia, Z. (2020) Flexural fatigue behavior and life prediction of web reinforced GFRP-balsa sandwich beams. *International Journal of Fatigue*, v. 136, 105592.

[209] Sayahlatifi, S., Rahimi, G. H., and Bokaei, A. (2020) The quasi-static behavior of hybrid corrugated composite/balsa core sandwich structures in four-point bending: Experimental study and numerical simulation. *Engineering Structures*, v. 210, 110361.

Index

www.ingramcontent.com/pod-product-compliance
Lightning Source LLC
LaVergne TN
LVHW011553200926
848484LV00001B/313